VOLUME NINETY SEVEN

ADVANCES IN BOTANICAL RESEARCH

Eutectic Solvents and Stress in Plants

ADVANCES IN BOTANICAL RESEARCH

Series Editor

Jean-Pierre Jacquot
Membre de l'Institut Universitaire de France, Unité Mixte de Recherche INRA, UHP 1136 "Interaction Arbres Microorganismes", Université de Lorraine, Faculté des Sciences, Vandoeuvre, France

VOLUME NINETY SEVEN

Advances in BOTANICAL RESEARCH

Eutectic Solvents and Stress in Plants

Volume Editors

ROBERT VERPOORTE

Natural Products Laboratory, Institute of Biology, Leiden University, 2300 RA Leiden, The Netherlands

GEERT-JAN WITKAMP

Biological and Environmental Sciences and Engineering Division, King Abdullah University of Science and Technology (KAUST), Thuwal, Saudi Arabia

YOUNG HAE CHOI

Natural Products Laboratory, Institute of Biology, Leiden University, 2300 RA Leiden, The Netherlands

ELSEVIER

ACADEMIC PRESS

An imprint of Elsevier

Academic Press is an imprint of Elsevier
125 London Wall, London, EC2Y 5AS, United Kingdom
The Boulevard, Langford Lane, Kidlington, Oxford OX5 1GB, United Kingdom
50 Hampshire Street, 5th Floor, Cambridge, MA 02139, United States
525 B Street, Suite 1650, San Diego, CA 92101, United States

First edition 2021

Notices

Knowledge and best practice in this field are constantly changing. As new research and experience broaden our understanding, changes in research methods, professional practices, or medical treatment may become necessary.

Practitioners and researchers must always rely on their own experience and knowledge in evaluating and using any information, methods, compounds, or experiments described herein. In using such information or methods they should be mindful of their own safety and the safety of others, including parties for whom they have a professional responsibility.

To the fullest extent of the law, neither the Publisher nor the authors, contributors, or editors, assume any liability for any injury and/or damage to persons or property as a matter of products liability, negligence or otherwise, or from any use or operation of any methods, products, instructions, or ideas contained in the material herein.

ISBN: 978-0-12-821691-0
ISSN: 0065-2296

For information on all Academic Press publications
visit our website at https://www.elsevier.com/books-and-journals

Publisher: Zoe Kruze
Acquisitions Editor: Sam Mahfoudh
Editorial Project Manager: Leticia M. Lima
Production Project Manager: James Selvam
Cover Designer: Victoria Pearson

Typeset by SPi Global, India

Contents

Contributors

Andrew P. Abbott
School of Chemistry, University of Leicester, Leicester, United Kingdom

Caprin Benoit
GATTEFOSSÉ Research & Development Centre, Saint-Priest Cedex, France

Joanne Bentley
Department of Molecular and Cell Biology, University of Cape Town, Rondebosch, South Africa

Vogelgesang Boris
GATTEFOSSÉ Research & Development Centre, Saint-Priest Cedex, France

Claire Bourlieu-lacanal
INRA, UMR IATE, Montpellier, France

Saffron J. Bryant
Department of Chemistry, University of Bath, Bath, United Kingdom; Centre for Molecular and Nanoscale Physics, College of Science, Engineering and Health, RMIT University, Melbourne, VIC, Australia

Frédéric Carrière
Aix Marseille Université, CNRS, UMR 7281 BIP, Marseille, France

Young Hae Choi
Natural Products Laboratory, Institute of Biology, Leiden University, Leiden, The Netherlands

Yuntao Dai
Natural Products Laboratory, Institute of Biology, Leiden University, Leiden, The Netherlands; Institute of Chinese Materia Medica, China Academy of Chinese Medical Sciences, Beijing, China

Ana Rita Cruz Duarte
LAQV@REQUIMTE, Departamento de Química, Faculdade de Ciências e Tecnologia, Universidade Nova de Lisboa, Caparica, Portugal

Erwann Durand
CIRAD, UMR IATE; IATE, Université Montpellier, CIRAD, INRAE, Institut Agro, Montpellier, France

S. Francois du Toit
Department of Molecular and Cell Biology, University of Cape Town, Rondebosch, South Africa

Karen J. Edler
Department of Chemistry, University of Bath, Bath, United Kingdom

Özlem Erol
Natural Products Laboratory, Institute of Biology, University of Leiden, The Netherlands

Jill M. Farrant
Department of Molecular and Cell Biology, University of Cape Town, Rondebosch, South Africa

Elena A. Golovina
Laboratory of Plant Physiology, Department of Plant Sciences, Wageningen University, Wageningen, The Netherlands

Katharina Häckl
University of Regensburg, Institute of Physical and Theoretical Chemistry, Regensburg, Germany

Riina Häkkinen
VTT Technical Research Centre of Finland Ltd, Tietotie, Finland

Oliver S. Hammond
Laboratoire de Chimie, École Normale Supérieure de Lyon, Lyon, France

Frank Hollmann
Department of Biotechnology, Delft University of Technology, The Netherlands

Verena Huber
University of Regensburg, Institute of Physical and Theoretical Chemistry, Regensburg, Germany

Andrew J. Jackson
European Spallation Source; Division of Physical Chemistry, Lund University, Lund, Sweden

Ana Rita Jesus
LAQV@REQUIMTE, Departamento de Química, Faculdade de Ciências e Tecnologia, Universidade Nova de Lisboa, Caparica, Portugal

Wim Jiskoot
Division of BioTherapeutics, Leiden Academic Centre for Drug Research, Leiden University, Leiden, The Netherlands

Peter G.L. Klinkhamer
Plant Ecology and Phytochemistry, Institute of Biology, Leiden University, Leiden, The Netherlands

Werner Kunz
University of Regensburg, Institute of Physical and Theoretical Chemistry, Regensburg, Germany

Tomas Laursen
Plant Biochemistry Laboratory, Department of Plant and Environmental Sciences, University of Copenhagen, Copenhagen, Denmark

Iva Manasi
Department of Chemistry, University of Bath, Bath, United Kingdom

Liane Meneses
LAQV@REQUIMTE, Departamento de Química, Faculdade de Ciências e Tecnologia, Universidade Nova de Lisboa, Caparica, Portugal

Aleksandra Mišan
Institute of Food Technology, University of Novi Sad, Novi Sad, Serbia

Birger Lindberg Møller
Plant Biochemistry Laboratory, Department of Plant and Environmental Sciences, University of Copenhagen, Copenhagen, Denmark

Natali Rianika Mustafa
Natural Products Laboratory, Institute of Biology, Leiden University, Leiden, The Netherlands

Filipe Silva Nunes de Oliveira
LAQV@REQUIMTE, Departamento de Química, Faculdade de Ciências e Tecnologia, Universidade Nova de Lisboa, Caparica, Portugal

Alexandre Paiva
LAQV@REQUIMTE, Departamento de Química, Faculdade de Ciências e Tecnologia, Universidade Nova de Lisboa, Caparica, Portugal

Milica Pojić
Institute of Food Technology, University of Novi Sad, Novi Sad, Serbia

Adrian Sanchez-Fernandez
Food Technology, Engineering and Nutrition, Lund, Sweden

Didier Touraud
University of Regensburg, Institute of Physical and Theoretical Chemistry, Regensburg, Germany

Henni Vanda
Natural Products Laboratory, Institute of Biology, Leiden University, Leiden, The Netherlands; Faculty of Veterinary Medicine, Universitas Syiah Kuala, Banda Aceh, Indonesia

Eleni Maria Varypataki
Division of BioTherapeutics, Leiden Academic Centre for Drug Research, Leiden University, Leiden, The Netherlands

Robert Verpoorte
Natural Products Laboratory, Institute of Biology, Leiden University, Leiden, The Netherlands

Pierre Villeneuve
CIRAD, UMR IATE; IATE, Université Montpellier, CIRAD, INRAE, Institut Agro, Montpellier, France

Charton Virginie
GATTEFOSSÉ Research & Development Centre, Saint-Priest Cedex, France

Geert-Jan Witkamp
Division of Biological and Environmental Science and Engineering, King Abdullah University of Science and Technology, Thuwal, Saudi Arabia

Preface: Natural deep eutectic solvents: A third liquid phase in living organisms? Discovery, theory, biology, and applications

Robert Verpoorte[a], Geert-Jan Witkamp[b], and Young Hae Choi[a]
[a]Natural Products Laboratory, Institute of Biology, Leiden University, Leiden, The Netherlands
[b]Biological and Environmental Sciences and Engineering Division, King Abdullah University of Science and Technology (KAUST), Thuwal, Saudi Arabia

1. Discovery

About 20 years ago the concept of metabolomics was introduced with the aim of qualitative and quantitative chemical analysis of all compounds present in an organism or any biological material. This approach should ideally give a total picture of all small molecules in an organism, and complement the data obtained by proteomics, transcriptomics, and genomics. The methods applied are mass spectroscopy (MS) and nuclear magnetic resonance-spectroscopy (NMR), either stand alone or coupled with a chromatographic system (LC-MS, GC-MS, and LC-NMR) (Schripsema & Verpoorte, 1991; Verpoorte, Choi, & Kim, 2007; Wolfender, Rudaz, Choi, & Kim, 2013). These methods are capable of analyzing complex mixtures, and in combination with various chemometric methods new insights can be obtained from the large data sets generated in various experiments. The omics were the beginning of a paradigm shift of doing research, moving from a hypothesis-based strategy to a systems biology approach. The latter is based on as many observations as possible and analyzing the huge data sets by chemometric methods, e.g., to identify which genes, proteins, and compounds correlate with the resistance of a plant against pests and diseases (Kim, Choi, & Verpoorte, 2010; Leiss, Choi, Verpoorte, & Klinkhamer, 2011; Mouden, Klinkhamer, Choi, & Leiss, 2017).

A major advantage of ^{1}HNMR-based metabolomics is that the integral of a proton signal is proportional to the molar concentration of the molecule it is part of and can be directly compared with internal standard. This advantage makes the quantitation of NMR metabolomics data much easier as all the compounds can be compared with each other. In our research using NMR-based metabolomics, we have built up a library with more than 25,000 plant extracts and including also some other organisms like marine organisms, microorganisms, fish, and mammalian cells. Also the spectra of some 2000 pure plant compounds are in our database. In all the spectra of biological materials we find that sugars, some amino acids, and organic bases and acids are present in relatively high levels compared to other common primary and secondary metabolites. Even, some of these abundant compounds keep their molar ratio constant to certain other compounds. This raised the question about the role of these compounds for cells and organisms. That was the point that we asked ourselves if the common acids like malic acid and citric acid might be forming ionic liquids (ILs) with the common bases like choline and betaine. The first experiment was made with malic acid and choline chloride, and indeed these two solid compounds did form an IL. A whole series of ILs was then made from compounds we identified by the NMR-based metabolomics in the biological materials. We then came across the work of Abbott, Boothby, Capper, Davies, and Rasheed (2004) who described a series of ILs as deep eutectic solvents. We continued making other combinations of common primary metabolites found in the biological samples, such as sugars, sugar alcohols, amino acids, organic bases, and organic acids. Many of these mixtures of solids indeed showed a lowering of the melting point to a temperature below room temperature (Choi et al., 2011; Dai, van Spronsen, Witkamp, Verpoorte, & Choi, 2013). Because these liquids turned out to be good solvents for various poorly water-soluble natural products, we named these mixtures "natural deep eutectic solvents" (NADES). The NMR-based metabolomics showed that in all organisms certain ingredients for NADES are present as major compounds, often keeping a constant molar ratio to some others. This brought us to the hypothesis that "Everywhere in living systems Natural Deep Eutectic Solvents (NADES) occur and form a third liquid phase of intermediate polarity" (Choi et al., 2011). This hypothesis can explain many biological phenomena, for example, biosynthesis, including transport and storage of nonwater-soluble compounds, like terpenoids and polysaccharides (Chapters 4, 5, 8); how seeds can survive long periods of drought and cold (Solberg, Brodal, et al., 2020; Solberg, Yndgaard, et al., 2020; Chapter 15); how resurrection plants may survive in the desert (Chen, Jung, Giarola, & Bartels, 2020; Chapter 9); and how cacti might be able to catch water from the atmosphere (Chapter 7).

2. Basics

The past decade showed an explosion in the number of studies on NADES, particularly for the various potential applications, such as a green extraction solvent and as solvent for formulations of poorly water-soluble compounds in cosmetics and medicines (Chapters 11–14). On the theoretical aspects new insights have been reported (Chapters 1–6). These applications concern hydrophilic NADES that are miscible with water. Hydrophobic eutectic mixtures are already known since long (Vanstone, 1909), e.g., camphor–borneol that occurred in old pharmacopeias. In the past years these were revisited (Chapter 2; Martins et al., 2019; Van Osch, Dietz, Warrag, & Kroon, 2020), which raised interest in further studies.

The hydrophilic NADES which are discussed in this book can be divided into five classes (Scheme 1). For finding the optimal solvent for a given compound, two approaches are used, one is modeling and the other is testing. Both are needed, but optimization for a single compound requires addressing both high selectivity and high yield, so sometimes a compromise is needed. The extraction of a plant material in which several compounds are important for the quality also might need a compromise as the most suited solvent. Measuring the solubility of a pure compound for finding the optimal extraction solvent does not necessarily result in the solvent that gives the highest yield. Apparently, liberating the desired molecules from the plant

	Sugars e.g. sucrose, fructose, glucose, trehalose Polyalcohols e.g. inositol, mannitol, glycerol	Organic acids e.g. malate, citrate, succinate, lactate, phytate, malonate	Organic bases, e.g. Choline, betaine	Amino acids, e.g. proline, alanine
Ionic liquids: acid and base		X	X	
Neutral compounds only	X			
Neutral compounds with base	X		X	
Neutral compounds with acid	X	X		
Neutral compounds with amino acids	X			X

Scheme 1 Classes of Natural Deep Eutectic Solvents.

matrix is also an important parameter to be considered (Gonzalez, Mustafa, Wilson, Verpoorte, & Choi, 2018). Concerning the modeling, the major problem is that there is an almost infinite number of combinations possible to make NADES, when considering the number of ingredients for making a NADES consisting of two or three components and having the water content as a further variable. In Chapter 1, some examples of a rational approach for optimization of extractions are discussed. Obviously, the field of molecular modeling for selecting the best solvent for a certain application should be developed further.

The NADES are more like a concept for a dynamic multiphase system in cells, than a well-defined static eutectic mixture. The NADES are miscible with water, so in nature they may occur with different percentages of water, which, e.g., makes a difference for the activity of enzymes and for the solubility of various precursors and products (Chapters 5, 6). The biological challenge is to prove the in situ presence of NADES in living cells and living organisms, which means that one should identify and quantify the NADES molecules through time and in space. In a static system, one can measure all kind of parameters, but in a dynamic system that is complicated. Understanding such dynamic systems is the other physicochemical challenge for the coming years.

3. Biology

Through the systems biology approach the various omics methods generate a lot of information on the DNA, RNA, proteins, metabolites, and various high-molecular-weight macromolecules such as starch, cellulose, and lignin. In fact, the omics allow the identification, characterization, and quantitation of all molecules in a cell. However, so far the efforts failed to make a synthetic cell out of these components. In fact, it is like a four-dimensional jigsaw puzzle (three dimensions of space and one dimension of time), we have all the pieces, but we have not been able yet to put all pieces into one model, though we know the cell morphology very well from direct observations. Cells are highly organized with a number of cell organelles with different functions. Compartments are shaped by cell membranes, which consist of small lipid molecules that may arrange themselves in liposomes in an aqueous environment. In searching the literature, one will find all kinds of transient cellular processes involving membranes. A large number of keywords is needed to find these, e.g., formation of vesicles, ER bodies, exosomes, ectosomes, extracellular vesicles, particles without membranes, metabolons, lipid droplets, vacuolar droplets, cytoplasmic droplets, fibrils, anthocyanic vascular inclusions, biocondensates, nanobubbles, lipid rafts, phase separation, exclusion zones, and heterogeneous water structure. Obviously, many dynamic processes occur in the life cycle of a cell. In the past years, there have been quite some discussions about these particles and various processes leading to the different structures mentioned. NADES may play a role in these processes. Some of the research on NADES include various types of liposomes, i.e., working with in vitro models, (e.g., see Chapters 2, 3, 6–8). In vivo research is done with different plant species and different types of plant cells (e.g., Ballesteros & Walters, 2019; Chapters 6, 8,10). But it is obvious that particularly in this field NADES might play an essential role. Because of the strong hydrogen bond donor and hydrogen bond acceptor character of the NADES components, they should be able to interact with the polar head groups of the membranes as well as with macromolecules such as proteins and the cell wall's cellulose and lignin. Considering that in cells several metabolites are present that are candidates for the formation of a NADES, different NADES may be formed as a third liquid phase in different cellular compartments under different conditions. In case of the vacuolar membrane one may think of a sugar–malic acid based NADES on the inside vacuolar membrane, whereas the outside carries a positively charged NADES, like choline chloride–sugar.

4. Applications

The real in situ presence of a NADES in cells has not been demonstrated yet. Most studies concern the interactions of proteins with membranes, not considering any effect of the various ions and molecules present in cells as major components. At least it is clear that proteins could be dissolved and stored in pure NADES for prolonged periods, and by dilution with water, enzyme activities are restored (Choi et al., 2011). The much better solubility of various medium polar compounds in partly diluted NADES compared to pure water, or to pure NADES, improves the substrate availability for enzyme reactions and thus the production (Chapters 4, 5).

Though there is still much work to do to understand the NADES and their biological roles, since 2012 there has been an exponential growth of the number of papers on applications of NADES. Most papers concern the use of NADES to extract various natural products from different materials. Particularly phenolic compounds have been subject of these studies. It was shown that NADES stabilizes such compounds (e.g., Dai, Verpoorte, & Choi, 2014). From the various studies it becomes clear that NADES are rather selective solvents; thus, for extracting highest amounts of certain compounds, one needs to develop a specific extraction protocol. The NADES have a number of advantages, a major one is their status as stable natural solvents. Moreover, they are nontoxic, which makes them suitable for various applications in agriculture, food, cosmetics, and medicines. However, their extremely low volatility is a disadvantage if one wants to extract a single compound, as a further step will be needed to isolate the desired compound from the NADES, e.g., by adding water as an antisolvent. But, on the other hand, this can be an advantage for the use of the extract itself, as the NADES extracts can be added to food, or be formulated in cosmetics or medicinal preparations (Chapters 11–14). The NADES can also be used to dissolve poorly water soluble medicines, including biologicals. Such NADES solutions, e.g., biologicals or small molecules, are more stable than solutions in alcohol, and can also be applied as such as they are nontoxic. The dissolved medicine may have a higher bioavailability (Chapters 11, 14). The NADES itself may be used as medicine, e.g., for wound healing (Chapters 11, 14). In line with the potential biological role of NADES in plant, cold resistance is the application of NADES in cryopreservation (Chapter 15).

5. Perspectives

The hypotheses on the potential roles of the NADES are based on a concept of a dynamic liquid system, rather than seeing NADES as a completely defined, static, in vitro deep eutectic mixture of two or three pure common primary metabolites. That two or more natural compounds can give a NADES shows that these compounds are eager to form hydrogen bonds. That characteristic seems to be used by Nature in many different ways. First of all, of course, the NADES itself, but also in interaction with other molecules, and in particular the small molecules that act as hydrogen bond acceptors (HBA) and donors (HBD) can be seen as important versatile building blocks in the organization of the cellular infrastructure. They could function as a reversible cement or glue that can easily be reallocated as part of a response to always changing conditions. The fact that NADES ingredients can be found in all organisms and that under various stress conditions the levels of some common NADES components increase shows that they play an important role in the cellular processes. Most evidence for the various hypothetical roles of NADES is still mostly indirect, i.e., the presence of the compounds was shown, but clear evidence for their direct interaction forming a NADES is still missing.

The quote "Learning from Nature, learning from our ancestors," applies very well to NADES. For example, the nectar–honey connection has a number of biological implications, and a number of applications were discovered by our ancestors (Chapter 14). Applications in cosmetics have already been developed. Pharmaceutical formulations for poorly water-soluble drugs and biologicals seem to be an interesting field for applications. NADES may also be useful for coatings of fruits to increase their shelf life. Seed and bulb coatings with NADES may improve their quality in terms of viability and protect them from microbial infections.

Understanding the NADES and their interaction with membranes, macromolecules such as proteins, lignin, polysaccharides, metabolites, and various ions will be instrumental for finding novel applications. The tools that will be developed for the basic physicochemical research can be applied in studying the possible biological roles of the NADES in, e.g., drought and cold resistance. The viability of seeds, spores, resurrection plants, and lichen would be interesting models for studying for possible biological roles of NADES. Obviously, much more research is needed to get insight in the various mentioned biological aspects of the NADES. Particularly, multidisciplinary collaborations are needed. Advanced microscopic methods are required to enable in situ chemical analysis to show the co-occurrence of the NADES ingredients in cells. Physical and chemical modeling of interactions between NADES, proteins, and lipids (membranes) requires both experimental and computational approaches. Chemistry and biochemistry are needed to identify and quantify potential NADES-forming molecules. In a systems biology approach, these disciplines may shape a model for the cellular fluid dynamics in which liposomes and an aqueous phase interact with the various hydrogen bond donors and acceptors. Such a model should explain the dynamic compartmentalized structures of living organisms and cells that are able to respond to various cues from the environment. For understanding the biological role of NADES, the major challenge will be studies on nanoscale of the dynamics of the metabolome, ionome, lipidome, proteome, and transcriptome in living cells.

References

Abbott, A. P., Boothby, D., Capper, G., Davies, D. L., & Rasheed, R. K. (2004). Deep eutectic solvents formed between choline chloride and carboxylic acids: Versatile alternatives to ionic liquids. *Journal of American Chemical Society*, *126*, 9142–9147.

Ballesteros, D., & Walters, C. (2019). Solid-state biology and seed Longevity: A mechanical analysis of glasses in pea and soybean embryonic axes. *Frontiers in Plant Science*. Article Number: 920. https://doi.org/10.3389/fpls.2019.00920.

Chen, P. L., Jung, N. U., Giarola, V., & Bartels, D. (2020). The dynamic responses of cell walls in resurrection plants during dehydration and rehydration. *Frontiers in Plant Science*, *10*. Article Number: 1698. https://doi.org/10.3389/fpls.2019.01698.

Choi, Y. H., van Spronsen, J., Dai, Y., Verberne, M., Hollmann, F., Arends, I. W., et al. (2011). Are natural deep eutectic solvents the missing link in understanding cellular metabolism and physiology? *Plant Physiology*, *156*, 1701–1705.

Dai, Y., van Spronsen, J., Witkamp, G. J., Verpoorte, R., & Choi, Y. H. (2013). Natural deep eutectic solvents as new potential media for green technology. *Analytica Chimica Acta*, *766*, 61–68.

Dai, Y., Verpoorte, R., & Choi, Y. H. (2014). Natural deep eutectic solvents providing enhanced stability of natural colorants from safflower (*Carthamus tinctorius*). *Food Chemistry*, *159*, 116–121.

Gonzalez, C. G., Mustafa, N. R., Wilson, E. G., Verpoorte, R., & Choi, Y. H. (2018). Application of natural deep eutectic solvents for the "green" extraction of vanillin from vanilla pods. *Flavor and Fragrance Journal*, *33*, 91–96.

Kim, H. K., Choi, Y. H., & Verpoorte, R. (2010). NMR-based metabolomic analysis of plants. *Nature Protocols*, *5*, 536–549.

Leiss, K. A., Choi, Y. H., Verpoorte, R., & Klinkhamer, P. G. L. (2011). An overview of NMR-based metabolomics to identify secondary plant compounds involved in host plant resistance. *Phytochemistry Reviews*, *10*, 205–216.

Martins, M. A. R., Silva, L. P., Schaeffer, N., Abranches, D. O., Maximo, G. J., Pinho, S. P., et al. (2019). Greener terpene–Terpene eutectic mixtures as hydrophobic solvents. *ACS Sustainable Chemical Engineering*, *7*, 17414–17423.

Mouden, S., Klinkhamer, P. G. L., Choi, Y. H., & Leiss, K. A. (2017). Towards eco-friendly crop protection: Natural deep eutectic solvents and defensive secondary metabolites. *Phytochemistry Reviews*, *16*, 935–951.

Schripsema, J., & Verpoorte, R. (1991). Investigation of extracts of plant cell cultures by ^{1}H NMR. *Phytochemistry Analysis*, *2*, 155–162.

Solberg, S. O., Brodal, G., von Bothmer, R., Meen, E., Yndgaard, F., Andreasen, C., et al. (2020). Seed germination after 30 years storage in permafrost. *Plants*, *9*(5), 579. https://doi.org/10.3390/plants9050579.

Solberg, S. O., Yndgaard, F., Andreasen, C., von Bothmer, R., Loskutov, G. G., & Asdal, A. (2020). Long-term storage and longevity of orthodox seeds: A systematic review. *Frontiers Plant Science*, *11*, 1007. https://doi.org/10.3389/fpls.2020.01007.

Van Osch, D. J. G. P., Dietz, C. H. J. T., Warrag, S. E. E., & Kroon, M. C. (2020). The curious case of hydrophobic deep eutectic solvents: A story on the discovery, design, and applications. *ACS Sustainable Chemistry & Engineering*, *8*, 10591–10612.

Vanstone, E. (1909). LXXVI—The miscibility of solids. *Journal Chemical Society, Transactions*, *95*, 590–604.

Verpoorte, R., Choi, Y. H., & Kim, H. K. (2007). NMR-based metabolomics at work in phytochemistry. *Phytochemistry Reviews*, *6*, 3–14.

Wolfender, J.-L., Rudaz, S., Choi, Y. H., & Kim, H. K. (2013). Plant metabolomics: From holistic data to relevant biomarkers. *Current Medicinal Chemistry*, *20*, 1056–1090.

Foreword

When Professor Verpoorte asked me to write the foreword to this book some months ago, I agreed, but not without some anxiety because, at the time, I had no clear idea what deep eutectic solvents (DESs) were, and the very words brought back memories of phase diagrams of multicomponent mixtures that gave me such headaches when I was an undergraduate chemistry student 60 years ago. But I figured I would learn a lot by studying just what DESs are, and I did.

Happily, a DES is a fascinating mixture of compounds with remarkable properties that gave me only a few headaches as I waded through some of the growing literature, now approaching 10,000 journal articles. Huber et al. (Chapter 2) said it well: "Isn't it fascinating to mix two powders in a mortar and to see that suddenly the mixture becomes liquid? All the more when the melting points of the two powders are far higher than room temperature." For example, the classic case that is starting to appear in textbooks is a mixture of choline chloride and urea in a 1:2 mole ratio. The resulting mixture has a melting point of 12 °C (far less than the melting point of choline chloride, 302 °C and urea, 133 °C). This mixture is liquid at room temperature, so the melting point is depressed by an astonishing 290 or 121 °C for the 2 components, respectively. This remarkable effect has been made prominent in the literature by the pioneering work of Abbott and his colleagues at the University of Leicester (Chapter 1).

Surprisingly, the existence of DES has been known for only a short time, mainly in the current century. The physical chemistry underlying the existence of DES is still under investigation, but briefly, DESs are systems formed from a eutectic mixture of Lewis or Brønsted acids and bases, such as the example above, which can contain a variety of anionic and/or cationic species. A listing of some of the known pairs of such molecules and what is understood of their physical chemistry appears in the excellent review from Abbott's laboratory (Smith, Abbott, & Ryder, 2014) and in the first three chapters of this book. In addition, Huber et al. (Chapter 2) and Edler et al. (Chapter 3) began to set the stage for the role of NADES in living cells through investigations of DES in more complex mixtures and with the addition of water.

Further investigations that set the stage for intact cells have shown that enzymes may function in DES, nonaqueously (Chapters 4 and 5). In a closely related work (Chapter 6), Durand et al. proposed that naturally occurring DES (designated as NADES) may occur near the surface of membranes. Thus, taken in conjunction with Chapters 4 and 5, it seems possible that catalysis in cells could be organized near surfaces by these means. Møller et al. (Chapter 8) presented some considerable evidence for organization of intracellular enzymes into "metabolons," which seems consistent with organization in NADES. This proposal has antecedents in the work of Drost-Hansen (2001), who suggested, with considerable evidence, that the organizing principles involved water near intracellular surfaces.

Succeeding chapters deal extensively with naturally occurring NADES. A concerted effort is underway to establish whether NADES exist in living cells. The evidence thus far has been circumstantial for the most part. Dai et al. (Chapter 7), for example, showed that the components to form NADES are present in intact plant cells at the required concentrations. While these findings are consistent with NADES in cells, they clearly do not establish unequivocally that they exist. Similarly, Francois du Toit et al. (Chapter 9) described concentrations of appropriate molecules for NADES formation in cells and suggested that during dehydration, NADES formation might be key to survival. This proposal is related to the long-established water replacement hypothesis involving interactions of sugars with dry membranes, particularly trehalose (e.g., Crowe, 2007). It does not seem mutually exclusive with the proposal that NADES are involved.

The first direct evidence that NADES exist as biological products is the finding that they are found in plant exudates (Chapter 10). While this is a promising finding, it still does not establish the existence of NADES in vivo. There is a good evidence that such exudates may be involved in a variety of biological processes such as wound healing and antibacterial activity (Chapter 11), so they are not without function, whether they are found in the intracellular environment or not.

Finally, applications of NADES are emerging and have been embraced by industry. Many applications that are coming out of the field are development of cosmetics (Chapter 12), preservation of food (Chapter 13), medicines (Chapter 14), and cryobiology (Chapter 15).

Clearly, NADES are maturing as a distinct discipline and will soon become a part of common knowledge. Over 50 years ago, when I spent two summers as a graduate student in Woods Hole, Massachusetts, I got to know Albert Szent-Gyorgyi, who was a regular summer resident of

Woods Hole in those days. I had some nascent ideas about stability of membranes, proteins, and nucleic acids in anhydrobiotic organisms, which I related to Szent-Gyorgyi. He listened to me with some interest and then spoke about his notions of the progress of scientific discovery: phase 1, "this is weird science, and it can't possibly be true" (which described my research at the time); phase 2, "maybe there's something to this after all"; and phase 3, "we knew it all the time." My own work on trehalose and anhydrobiosis was in phase 1 when I talked with Szent-Gyorgyi about it, but it has been decidedly in phase 3 for years now. It is gratifying that the work is being enhanced by further and much extended studies in laboratories and in industry all over the world.

Research on NADES is well into Szent-Gyorgyi's phase 2 and will be within the next decade advance to phase 3. These studies are bringing new perspectives to vast areas of scientific endeavor, and they will surely lead to revolutions in those areas.

John H. Crowe
Davis, CA, United States

References

Crowe, J. H. (2007). Trehalose as a "chemical chaperone": Fact and fantasy. *Advances in Experimental Medicine and Biology*, *594*, 143.

Drost-Hansen, W. (2001). Temperature effects on cell-functioning—A critical role for vicinal water. *Cellular and Molecular Biology*, *47*, 865.

Smith, E. L., Abbott, A. P., & Ryder, K. S. (2014). Deep eutectic solvents (DESs) and their applications. *Chemical Reviews*, *114*, 11060.

CHAPTER ONE

Deep eutectic solvents—Teaching nature lessons that it knew already

Riina Häkkinen[a] and Andrew P. Abbott[b,*]
[a]VTT Technical Research Centre of Finland Ltd, Tietotie, Finland
[b]School of Chemistry, University of Leicester, Leicester, United Kingdom
*Corresponding author: e-mail address: andrew.abbott@le.ac.uk

Contents

Abstract

When Brønsted and Lewis acids and bases are mixed strong hydrogen bonds develop and these deviate from ideal mixtures resulting in a significant depression of freezing point. Many of these mixtures are liquid at ambient temperatures. These so-called deep eutectic solvents have become an area of intense study since their discovery. Mixtures of Brønsted and Lewis acids and bases are commonly found in nature and the properties of these mixtures are similar to ionic liquids. This article explains the types of mixtures which form and how bioinspired formulations can be used to form materials which mimic natural structures and can be used for applications such as drug delivery. The ability of sugars to form concentrated complex liquid mixtures is discussed and compared to aqueous solutions. DESs have complex phase behavior when solutes of lower polarity are included and it is shown that microemulsions are formed when water is introduced into DESs. This may explain the behavior of proteins and biomacromolecules in DES-water mixtures. The thermodynamics of species partitioning into DESs is discussed together with methods to calculate enthalpies of interaction.

List of abbreviations

API	active pharmaceutical ingredient
ChCl	choline chloride ($HOC_2H_4N(CH_3)_3Cl$)
HBD	hydrogen bond donor
HOESY	heteronuclear overhauser spectroscopy
LTTM	low-transition-temperature Mixtures
NADES	natural deep eutectic solvents

Advances in Botanical Research, Volume 97
ISSN 0065-2296
https://doi.org/10.1016/bs.abr.2020.09.013

NOESY nuclear overhauser spectroscopy
QAS quaternary ammonium salt
UF urea-formaldehyde resin
ΔT_f depressions of freezing point

1. Overview

The term deep eutectic solvents was coined in 2004 to describe mixtures of quaternary ammonium salts with hydrogen bond donors (Abbott, Boothby, Capper, Davies, & Rasheed, 2004) although the systems were described the year before but not named (Abbott, Capper, Davies, Rasheed, & Tambyrajah, 2003). These systems were investigated as alternatives to ionic liquids and are of interest for their ease of manufacture, lower cost, lower toxicity and ease of registration. Eutectic mixtures as media for chemical processes naturally has a much longer history and the purpose of this book is to propose that nature uses eutectic mixtures to protect cell life. It is useful to reflect on synthetic eutectic mixtures and review how they were used and understand the behavior of solutes in them.

As with most scientific discoveries, deep eutectic solvents resulted from academic curiosity and good fortune. Our first breakthrough resulted from a piece of consultancy to investigate aluminum chloride eutectics. We were studying mixtures of quaternary ammonium salts, QAS, with $AlCl_3$ which were first reported in the 1950s (Hurley & WIer, 1951). The questions arose why are these mixtures liquid at room temperature and what else forms these liquid mixtures?

The answer was clearly that these mixtures form complex anions between the metal salt and the quaternary ammonium salt resulting in a eutectic as shown schematically in Fig. 1.

1.1 Classification of DESs

Aluminum chloride is particularly suitable because it has a relatively low melting point (192 °C). The depression of freezing point is related to the strength of the interaction between the Brønsted and Lewis acids and bases. The complex metal anions ($AlCl_4^-$ and $Al_2Cl_7^-$ in the case of $AlCl_3$) delocalize the charge and decrease the lattice energy and melting point of the mixture.

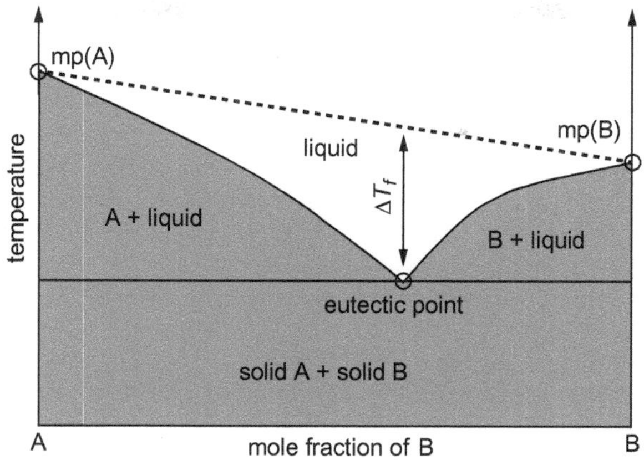

Fig. 1 Schematic diagram of the formation of a eutectic in a two component system.

This led to the obvious question "which other metal salts have low melting points"? It was found that quaternary ammonium (and by inference phosphonium) salts formed eutectics when mixed with metal chlorides of the form MCl_x, where suitable metal halides included zinc, tin, iron, aluminum, gallium and indium (see Table 1). These are not random salts but rather they are the metal halides with the lower melting points. To form mixtures which have a eutectic point at around ambient temperature the approximate melting point of the metal halide needs to be below about 320 °C.

Many of the interesting metals do not have halides with low melting points and so the idea can be extended by using metal salt hydrates. The waters of hydration interact with the metal ion to decrease the lattice energy but we have subsequently shown that some of these water molecules aggregate to form a separate, water-rich phase in the DES (Abbott, Alabdullah, Al-Murshedi, & Ryder, 2018; D'Agostino et al., 2015). Mixtures such as $Zn(NO_3)_2{\cdot}6H_2O$: 2 propylene glycol have higher conductivities ($\kappa = 20$ Scm^{-1} at 25 °C) than any other eutectic based ionic liquids probably due to increased mobility of ions in the water-rich phase. In these liquids the metal salt dissociates into a cationic and anionic species, e.g., in the $CrCl_3{\cdot}6H_2O$- urea system $[CrCl_2{\cdot}4H_2O]^+$ is the cationic species and chloride is the anion. These fluids have been used for metal electroplating and leather tanning but they are limited in their thermal stability by the loss of water from the mixtures (Abbott, Capper, Davies, & Rasheed, 2004; Abbott, Capper, Davies, Rasheed, Archer, et al., 2004).

Table 1 Classification of different types of deep eutectic solvent depending on their composition.

Eutectic	Acid	Base
1	Lewis MClx M=Zn, Sn, Fe (Abbott et al., 2001; Xu, Lü, Zhang, Gui, & Yang, 2006; Yang, Tian, He, & Xu, 2003), Al (Hurley & WIer, 1951), Ga (Xu et al., 2006), In (Yang et al., 2003)	Lewis R_4NCl
2	Lewis MClx·yH_2O, M=Cr (Abbott, Capper, Davies, & Rasheed, 2004), Co, Cu, Ni, Fe	Lewis R_4NCl
3	Brønsted RZ, Z=$CONH_2$ (Abbott et al., 2003), COOH (Abbott, Boothby, Capper, Davies, & Rasheed, 2004; Abbott, Capper, Davies, & Rasheed, 2004; Abbott, Capper, Davies, Rasheed, Archer, & John, 2004), OH (Abbott, Capper, Swain, & Wheeler, 2005)	Lewis R_4NCl
4	Lewis MClx or MClx·yH_2O M=Al (Abood, Abbott, Ballantyne, & Ryder, 2011), Zn (Abbott, Barron, Ryder, & Wilson, 2007), Cr (Abbott et al., 2014)	Lewis RZ, Z=$CONH_2$, COOH, OH
5	Brønsted ROH RZ, Z=COOH (Francisco, van den Bruinhorst, & Kroon, 2013)	Lewis RZ, Z=$CONH_2$, COOH, OH

The big breakthrough came with the discovery that eutectics form with Brønsted acids and halide salts (Abbott et al., 2003; Abbott, Boothby, et al., 2004). To differentiate these eutectics from metal based mixtures, the term deep eutectic solvents (DESs) was coined although this is unhelpful because some of the most useful eutectics do not have a particularly large depression of freezing point and the most useful systems do not use the DES as a solvent but rather as a liquid active ingredient. The potential hydrogen bond donors which form DESs include carboxylic acids, amides and alcohols. These are particularly pertinent to the idea of natural deep eutectic solvents (NaDESs) which form in nature as many naturally occurring compounds have hydrogen bonding moieties, e.g., fatty acids, aminoacids, polyphenols, tannins, sugars and ureas etc. Again larger depressions of freezing point, ΔT_f, arise from the stronger hydrogen bond donors such as carboxylic acids and amides rather than alcohols, e.g., the oxalic acid–choline chloride system, ΔT_f, is

212 °C and the choline chloride–urea system ΔT_f, is 178 °C (Abbott et al., 2003; Abbott, Boothby, et al., 2004). It is our opinion that the term NADES should only be applied to naturally occurring fluids found in living organisms rather than those made from compounds some of which can be found in nature.

In most cases the eutectic composition is related to the number of HBD moieties on the donor molecule. It appears that in most cases the eutectic composition is when 2 hydrogen bonding moieties complex with each chloride, e.g., for diprotic carboxylic acids such as malonic acid the eutectic composition is at 1 QAS: 1 HBD whereas monoprotic acids such as benzoic acid have a eutectic composition at 1 QAS: 2 HBD. Fig. 2 shows a scheme of how these components interact, however it should be noted that the HBD is labile and is constantly associating and dissociating with the anion in the same way that water does in aqueous electrolyte solutions.

These DESs tend to be easy to form by simply mixing the two components with gentle heating. It has however been shown that these liquids can also be formed using an extruder to give a fast, continuous flow process which can be advantageous for HBD compounds which are not particularly thermally stable, e.g., sugars (Crawford, Wright, James, & Abbott, 2016). These eutectics do not react with moisture, they tend to have low toxicity and many are biodegradable. It is possible to create eutectics from compounds which occur in nature.

Many DESs have been made using choline chloride ($HOC_2H_4N(CH_3)_3Cl$), ChCl, as the QAS because it is readily available, non-toxic, produced as a large scale as an animal feed additive and tends to produce more fluid liquids. ChCl-derived liquids tend to be hygroscopic but several studies have shown that hydrophobic DESs can be created using longer chain QASs or HBDs based on fatty acids (van Osch, Zubeir, van den Bruinhorst, Rocha, & Kroon, 2015).

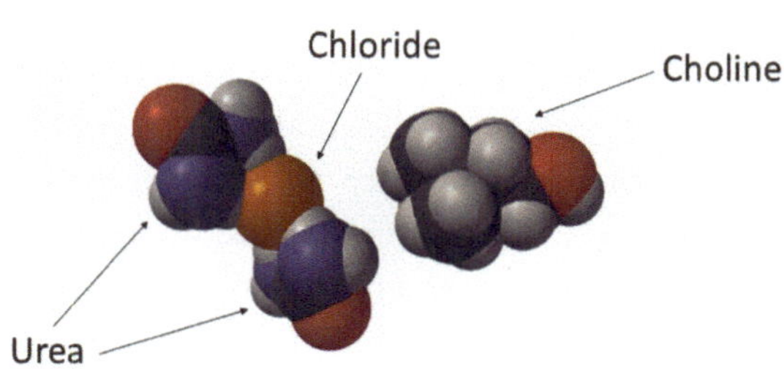

Fig. 2 Schematic diagram of how urea complexes the anion of the quaternary ammonium salt.

Many of the active pharmaceutical agents, APIs, which are commonly used are either Brønsted or Lewis acids or bases. Many APIs are amine salts which are quaternized and sold as HCl salts to increase their solubility in the stomach. Likewise many other APIs contain functional groups which are hydrogen bond donors such as carboxylic acids or polyphenols. It has been shown that pharmaceutic DESs can be synthesized using one or more API. This maintains the active ingredient in a liquid form can enhance uptake in the body and overcome issues associated with polymorphs of a drug having different activities (Abbott, Ahmed, Prasad, Qader, & Ryder, 2017).

The first eutectics without quaternary ammonium salts were found to form with Lewis acidic metal salts when mixed with Brønsted bases, e.g., $ZnCl_2$ with a range of amides (Abbott et al., 2007). The eutectics were formed between metal salts and simple organic alcohols or amides. These eutectics are unusual as they have both metal-containing anions and cations (Abood et al., 2011), e.g., for the aluminum chloride urea system the following equilibrium is established.

$$AlCl_3 + n(urea) \rightleftharpoons AlCl_2 + \ \cdot n\ (urea) + AlCl_4^-$$

The final type of eutectic have been termed Low-Transition-Temperature Mixtures (LTTMs) which are mixtures of proton donors and proton acceptors (Francisco et al., 2013). These mixtures have a depression of glass transition temperature rather than of freezing point. A subsection of these has been termed Type 5 DESs (Abranches et al., 2019). These are somewhat different from the previous DESs in that most have limited conductivity and are non-ionic, however, they are more akin to systems found in nature. It was shown that alcoholic mixtures of compounds such as phenol, menthol, thymol and hydroquinone, all of which are found in plants, could form eutectic mixtures where the depression of freezing point was approximately 100 °C lower than that for either of the pure components.

One final mixture which can be discussed is that some hydrogen bond donors form highly structured liquids when mixed with lithium and sodium salts (Abbott, D'Agostino, Davis, Gladden, & Mantle, 2016). These do not strictly behave in the same way as deep eutectic solvents in that the salts interact strongly with the HBD and cause the liquid to become more structured, rather than less structured which is the case with QASs. This is pertinent to the behavior of salts in plant and animal cells which are primarily sodium and potassium cations. The interactions between the polar organic molecules and the group 1 cations mean that at low temperatures the mixtures do not freeze to form crystalline solids, but rather they form glasses. This may explain why NaDESs are important in animal and plant physiology as they prevent the formation of crystals at low temperature of low water content. The volume and structural changes when crystalline solids melt are large whereas the changes occurring when glasses liquefy are relatively minor. Smith et al. has recently reviewed the properties and applications of DESs in more depth (Smith, Abbott, & Ryder, 2014).

2. DES modified biomaterials

To understand part of the importance of NaDESs in biomaterials it is useful to reflect on the types of interactions that can arise when mixtures of Brønsted and Lewis acids and bases interact with biopolymers. Most biomaterials are made from carbohydrates or proteins and these have polar functional groups which can strongly interact with the Brønsted and Lewis acids and bases found in DESs. In their pure dried form, most biomaterials are crystalline and rigid but they become plasticized by the incorporation of moisture and polar organic compounds. The composition of carbohydrate-based materials change with composition and modifiers such as sugars, polar molecules and salts will depend on the position of the polymer in the plant, i.e., stem, trunk or roots. Wood tends to have 40–55% cellulose with a wide variety of compounds able to modify the mechanical flexibility and strength. The fact that DESs are very hydroscopic may explain how they retain moisture in the cellular structure by decreasing the vapor pressure of water.

Polar organic molecules such as alcohols and amides (El-Sonbati, 2012; Janssen & Moscicki, 2009; Zhang, Rempel, & Liu, 2014) can be used to produce thermoplastic materials from biopolymers such as starch. Using thermomechanical processes such as extrusion can produce thermoplastic materials with significantly modified glass transition temperatures (Lourdin, Coignard, Bizot, & Colonna, 1997). In some cases, the polar modifiers can migrate in or evaporate from the material and the biomaterial can recrystallize changing its properties. When these polar modifiers are incorporated in combination with QASs, in the form of DESs, they result in materials with long-term stability. The DESs act as lubricants or plasticizers, breaking up the hydrogen bond structure of the biopolymers and decreasing the degree of crystallinity. This has been demonstrated for starch modified using DESs. It produced a thermoplastic which was comparable in strength to low density polyethylene and could retain its properties for more than 2 years under

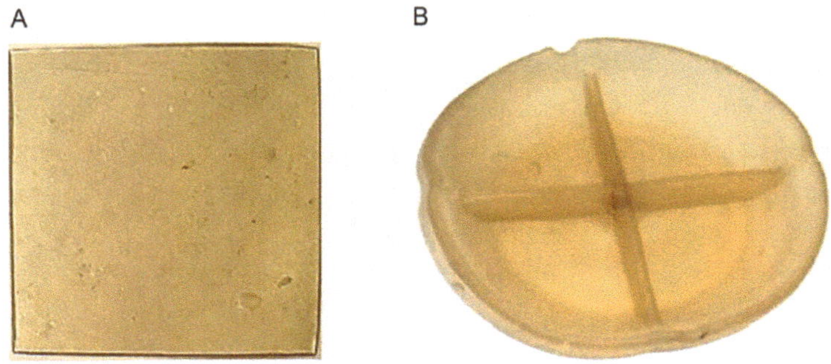

Fig. 3 Sample of starch plasticized with glycerol and choline chloride (A) presses into a 10 cm square and (B) injection molded (200 °C and 60 Bar).

anhydrous conditions. The fact that they are thermoplastics means that they can be reprocessed and recycled (Abbott et al., 2011; Abbott, Palazuela Conde, Davis, & Wise, 2012). Since components of the DESs can be tuned to be similar to those found in nature the plastic, thus derived, can be rapidly biodegraded or composted with the suitable conditions. Sheets of thermoplastic starch have been vacuum formed and injection molded as shown in Fig. 3.

In nature, starch is found in softer plant tissues and often acts as in intercellular modifier. It is also commonly used as an adhesive for products like cardboard and wallpaper paste. It has been shown that DES modified starch acts as a good resin in composite materials such as medium density fiberboard or particle board replacing the traditional urea-formaldehyde (UF) thermoset resins. This is to some extent a bioinspired solution to synthetic wood replacements. It has been shown that boards with comparable mechanical properties to conventional UF boards can be obtained (Abbott et al., 2012) but the DES based resins enable the boards to be recycled and composted. Resins based on group 1 inorganic salts have also been produced (Abbott et al., 2016) and used for composite board production (Davis, 2015).

In comparison to carbohydrates, comparatively few studies have modified proteins with DESs which is partly due to their different hydrophobicities. Starch is very hydrophilic and has a high solubility in some choline chloride (ChCl)-based DESs (Biswas, Shogren, Stevenson, Willett, & Bhowmik, 2006; Qin, Owyeung, Sonkusale, & Panzer, 2019) but globular proteins such as zein and soy are more hydrophobic and have poor solubilities in the DESs used to modify starch. DESs do not tend to improve the thermostability of proteins and can disrupt the tertiary structure of some proteins (Esquembre et al., 2013; Sanchez-Fernandez, Edler, Arnold, Alba Venero, & Jackson, 2017) but the secondary structure is generally preserved.

Qu studied the modification of zein and soy protein using a ChCl-glycerol eutectic and produced semi-transparent sheets which were weak and brittle (Qu, 2019; Qu, Häkkinen, Allen, D'Agostino, & Abbott, 2019). It was concluded that the hydrophilic DES did not wet the hydrophobic protein well. Studies with gelatin, however showed that a much more flexible and stronger material could be produced as the DES was able to penetrate into the coiled structure of the hydrophilic protein and fully plasticize the material. The properties of the material were dependent on the time that the protein and DES were mixed, showing that penetration was relatively slow. Qu also showed that pharmaceutical DESs could be blended with gelatin and these solid gels could be used for transdermal delivery systems which were 4 times more rapid at delivering the API than the corresponding solid API in the same matrix (Qu et al., 2019). Gelatin mixed with ethylene glycol-ChCl DES (Qin et al., 2019) produces a highly stretchable, conductive gel which they proposed could be suitable for ionic skin applications.

These limited studies of biomaterial modification may point the way to understanding how nature uses similar formulations to ensure how natural materials maintain flexibility in tissues during extremes in temperature and/or water content.

3. Solubility in DESs

With more than 10^6 binary combinations of QASs and HBDs it is possible to formulate the DES to dissolve a wide range of solutes. Very few fundamental studies have characterized solubility in DESs beyond the phenomenological "like dissolves like" approach. The factors affecting solubility are no different for DESs than any other solvent but it is useful to reflect on these when designing a DES for a given solute. Dissolution is a combination of enthalpic and entropic contributions. It should be remembered that DES formation is endothermic and the large positive entropy change of going from two ordered solids to a more disordered liquid drives the process.

The entropy of solute dissolution is usually positive, driven by a disordering of the solute in solution compared to its organized lattice structure, but it should also be acknowledged that DESs are generally ordered liquids which can become more disordered when a solute dissolves in them.

The enthalpy of dissolution should be negative if the enthalpy of solvation is larger than the lattice energy combined with the enthalpy of hole formation. It is often overlooked that DESs have a high surface tension so the enthalpy of hole formation is often quite large and as will be seen below, this tends to dominate the solubility of non-polar solutes.

DESs are being studied for the dissolution of a wide variety of solutes, both as an extraction tool and an application of an active ingredient. The applications of DESs have been reviewed recently (Smith et al., 2014). In general these can be split into inorganic solutes, mostly for metal processing, and organic for natural product extraction. In the former case, the interactions of the halide and HBD with the metal species dominate speciation and this affects the enthalpy of solvation of the solute. The solubility of a variety of metal oxides have been characterized (Abbott, Capper, Davies, McKenzie, & Obi, 2006; Abbott, Frisch, Hartley, & Ryder, 2011) and these are related to speciation (Hartley et al., 2014).

DESs have also been applied to the extraction of species from mineral oils and natural products (Dai, Van Spronsen, Witkamp, Verpoorte, & Choi, 2013; Duan, Dou, Guo, Li, & Liu, 2016; García, Rodríguez-Juan, Rodríguez-Gutiérrez, Rios, & Fernández-Bolaños, 2016; Tang, Zhang, & Row, 2015). This naturally encompasses a wide range of functional groups including phenolic compounds, anthraquinones, flavonoids, alkaloids, essential oils, carbohydrates and lignins. The components of the DES can be chosen to extract specific solutes. This is being carried out on a commercial scale by Naturex although there are no details in the open literature (Lavaud et al., 2016).

The thermodynamics of dissolution in DESs have been investigated in a limited number of studies. The simplified system which is the liquid-liquid extraction from an oil phase into a DES have investigated sulfur containing heterocycles, iodine, carboxylic acids, alcohols, esters and ketones into a variety of the most common DESs (Abbott et al., 2017). Using thiophene as an analogue of the compounds found in mineral diesel, extraction studied using 6 DESs found that the liquids with the lower surface tensions were more effective at extracting thiophene from decane. This shows that the enthalpy required to form a hole in the DES was the factor which limited the partition coefficient. It was also shown that larger solutes such as benzothiophene partition less into the DES as they require a larger hole. To show the complexity of the process, however it was shown that the same trend was not observed for the extraction of iodine from decane. This was argued from the perspective of the species formed when iodine dissolved in a DES, I_2Cl^-, which will partition preferentially into liquids which are better Lewis bases which tend to be those with higher surface tensions.

When studying the extraction of solutes with different polarities it was found that those with the larger enthalpy of solvation had a higher partition coefficient from decane into the DES, i.e., carboxylic acids > alcohols > ketones. This is clearly due to the ability for the solute to hydrogen bond to the chloride in the DES.

Of particular importance, from a botanical perspective, is the solubility of sugars and oligosaccharides. The multiple hydroxyl groups in those carbohydrates facilitate the hydrogen bond formation with protic solvents. As a result, low molecular weight carbohydrates dissolve well in water. However, the solubility of sugars in other protic solvents, e.g., alcohols, is not as good as in water. In our recent study, we studied the solubility of different sugars in five ChCl-based DESs, and compared these values to water and ethylene glycol. (Häkkinen & Abbott, 2019). It was found that the solubility of sugars was highest in water when compared in wt%, but actually in mol% a higher solubility was obtained in the DESs. By using molecular modeling, it was shown that this was because of the similarities in sizes of the DES and the glucose molecule, whereas in water, the surface of the glucose could not be fully covered by the small water molecules. When ChCl was added to ethylene glycol to form an ethylene glycol-ChCl DES, the solubility of the sugars significantly increased compared to the solubility in pure ethylene glycol. Therefore, the chloride anion was found to be the driving force for the solubility of sugars, which was confirmed using Nuclear Overhauser spectroscopy ($\{^1H-^1H\}$-NOESY) and $\{^1H-^{19}F\}$-Heteronuclear NOESY (HOESY). The results implied that the chloride anion could create strong hydrogen bonding with both the sugar molecule and the DES components.

A calorimetric method was developed to calculate the enthalpies of interactions between the QAS and HBD. Basically, DESs are formed via hydrogen bonds, which can store energy leading to a higher heat capacity of the liquid. With the help of Hess cycles and through heat capacity measurements of the eutectic mixture and the pure components, the thermodynamics of hydrogen bond formation in DES could be measured. The hydrogen bond strength of the DES was also found to control the dissolution of sugars. Due to the strong hydrogen bonding between the DES and the sugars, the DES-sugar mixtures formed highly viscous, syrup-like liquids, which did not crystallize even at high sugar concentrations and low temperatures.

The presence of water, as would be the case in plant and animal cells, decreases the viscosity of the cellular fluids and enables the transport system for sugars and other cellular metabolites. Sugars with reducing ends cannot stand too acidic or basic conditions and so the high electrolyte concentrations act as buffers in the cellular fluids but they also allow the cell to act as a glass as low temperatures and water contents.

A surprising aspect of DESs is their behavior when water is added to them. It may be expected that water will be totally miscible when added to a DES as both are polar and hydrophilic. It has, however been determined using dynamic light scattering (DLS) that water forms a separate phase when mixed with DESs (Abbott et al., 2018). The same is, however observed for other hydrogen bonding liquids. When alcoholic solutes including glucose, phenol, and 1-pentanol were added to ethylene glycol-ChCl DES, their phase behavior differed significantly (Häkkinen, Alshammari, Timmermann, D'Agostino, & Abbott, 2019). It was shown by DLS and NMR spectroscopy that only glucose formed a homogeneous mixture with ethylene glycol-ChCl DES while phenol and 1-pentanol formed heterogeneous mixtures. Moreover, larger clustering was observed above the melting point of phenol (at 40 °C). It was concluded that solid solutes distribute homogenously in DESs while liquid solutes form heterogeneous phases. It has been shown that mass transport is different in the heterogeneous phases, i.e., species dissolved in the water dominant phase diffuse faster than in the ion dominant phase. This could be extremely important in plant physiology and the phase behavior and composition of cellular fluids many require further investigation.

Rheology and NMR spectroscopy can be used to study the solvent-solute, solute-solute, and solvent-solvent interactions (Häkkinen & Abbott, 2019). Generally, the higher the viscosity, the higher the intermolecular forces. In their pure form, many DESs have viscosities in the range of 30–800 cP. The addition of a solute into a DES changes those intermolecular interactions. The enthalpies of interactions include not only those between the solute and the DES, but the DES should be considered as a binary mixture formed via hydrogen bonds. It was recently shown that depending on the system, the solute might interact equally with both the QAS and the HBD that form the DES, or just with one of these components (Häkkinen & Abbott, 2019). However, the interactions that hold the DES structure together can be maintained up to 40 wt% addition of solute (Hammond, Bowron, & Edler, 2017). Increasing temperature weakens all these interactions, but the DES-DES interactions might actually reduce quicker than the solute-solute interactions (Häkkinen et al., 2019).

4. Conclusions

This article has highlighted that DESs can be formed from a wide range of Brønsted and Lewis acids and bases. These fluids have unusual phase behavior that may be important to cell physiology. When they freeze they tend to form glasses which do retain some fluidity. They can be made to dissolve a range of solutes encompassing both hydrophilic and hydrophobic moieties. The phase behavior with liquid solutes such as water leads to separate phases on the 10–100 s of nm scale. DESs have been shown to plasticize both carbohydrates and proteins making materials which are less brittle. All of these factors may be important in explaining why cells can survive in extremes of temperature and areas of low humidity. It may also explain why crystalline ice does not form in plants at low temperatures.

References

Abbott, A. P., Capper, G., Davies, D. L., Munro, H., Rasheed, R. K., & Tambyrajah, V. (2001). Preparation of novel, moisture-stable, Lewis-acidic ionic liquids containing quaternary ammonium salts with functional side chains. *Chemical Communications*, *19*(19), 2010–2011. https://doi.org/10.1039/B106357J.

Abbott, A. P., Capper, G., Davies, D. L., Rasheed, R. K., & Tambyrajah, V. (2003). Novel solvent properties of choline chloride/urea mixtures. *Chemical Communications*, *1*, 70–71. https://doi.org/10.1039/b210714g.

Abbott, A. P., Capper, G., Davies, D. L., & Rasheed, R. K. (2004). Ionic liquid analogues formed from hydrated metal salts. *Chemistry - A European Journal*, *10*(15), 3769–3774. https://doi.org/10.1002/chem.200400127.

Abbott, A. P., Boothby, D., Capper, G., Davies, D. L., & Rasheed, R. K. (2004). Deep eutectic solvents formed between choline chloride and carboxylic acids: Versatile alternatives to ionic liquids. *Journal of the American Chemical Society*, *126*(29), 9142–9147. https://doi.org/10.1021/ja048266j.

Abbott, A. P., Capper, G., Davies, D. L., Rasheed, R. K., Archer, J., & John, C. (2004). Electrodeposition of chromium black from ionic liquids. *Transactions of the Institute of Metal Finishing*, *82*, 14.

Abbott, A. P., Capper, G., Swain, B. G., & Wheeler, D. A. (2005). Electropolishing of stainless steel in an ionic liquid. *Transactions of the IMF*, *83*(1), 51–53. https://doi.org/10.1179/002029605X17657.

Abbott, A. P., Capper, G., Davies, D. L., McKenzie, K. J., & Obi, S. U. (2006). Solubility of metal oxides in deep eutectic solvents based on choline chloride. *Journal of Chemical and Engineering Data*, *51*(4), 1280–1282. https://doi.org/10.1021/je060038c.

Abbott, A. P., Barron, J. C., Ryder, K. S., & Wilson, D. (2007). Eutectic-based ionic liquids with metal-containing anions and cations. *Chemistry - A European Journal*, *13*(22), 6495–6501. https://doi.org/10.1002/chem.200601738.

Abbott, A. P., Frisch, G., Hartley, J., & Ryder, K. S. (2011). Processing of metals and metal oxides using ionic liquids. *Green Chemistry*, *13*(3), 471. https://doi.org/10.1039/c0gc00716a.

Abbott, A. P., Harris, R. C., Ryder, K. S., D'Agostino, C., Gladden, L. F., & Mantle, M. D. (2011). Glycerol eutectics as sustainable solvent systems. *Green Chemistry*, *13*(1), 82–90. https://doi.org/10.1039/C0GC00395F.

Abbott, A. P., Palazuela Conde, J., Davis, S. J., & Wise, W. R. (2012). Starch as a replacement for urea-formaldehyde in medium density fibreboard. *Green Chemistry*, *14*(11), 3067. https://doi.org/10.1039/c2gc36194a.

Abbott, A. P., Al-Barzinjy, A. A., Abbott, P. D., Frisch, G., Harris, R. C., Hartley, J., et al. (2014). Speciation, physical and electrolytic properties of eutectic mixtures based on CrCl3·6H2O and urea. *Physical Chemistry Chemical Physics*, *16*, 9047–9055.

Abbott, A. P., D'Agostino, C., Davis, S. J., Gladden, L. F., & Mantle, M. D. (2016). Do group 1 metal salts form deep eutectic solvents? *Physical Chemistry Chemical Physics*, *18*(36), 25528–25537. https://doi.org/10.1039/C6CP05880A.

Abbott, A. P., Ahmed, E. I., Prasad, K., Qader, I. B., & Ryder, K. S. (2017). Liquid pharmaceuticals formulation by eutectic formation. *Fluid Phase Equilibria*, *448*, 2–8. https://doi.org/10.1016/j.fluid.2017.05.009.

Abbott, A. P., Al-Murshedi, A. Y. M., Alshammari, O. A. O., Harris, R. C., Kareem, J. H., Qader, I. B., et al. (2017). Thermodynamics of phase transfer for polar molecules from alkanes to deep eutectic solvents. *Fluid Phase Equilibria*, *448*, 99–104. https://doi.org/10.1016/j.fluid.2017.05.008.

Abbott, A. P., Alabdullah, S. S. M., Al-Murshedi, A. Y. M., & Ryder, K. S. (2018). Brønsted acidity in deep eutectic solvents and ionic liquids. *Faraday Discussions*, *206*(September), 365–377. https://doi.org/10.1039/C7FD00153C.

Abood, H. M. A., Abbott, A. P., Ballantyne, A. D., & Ryder, K. S. (2011). Do all ionic liquids need organic cations? Characterisation of [$AlCl_2$·nAmide]+ $AlCl_4^-$ and comparison with imidazolium based systems. *Chemical Communications*, *47*(12), 3523. https://doi.org/10.1039/c0cc04989a.

Abranches, D. O., Martins, M. A. R., Silva, L. P., Schaeffer, N., Pinho, S. P., & Coutinho, J. A. P. (2019). Phenolic hydrogen bond donors in the formation of non-ionic deep eutectic solvents: The quest for type V DES. *Chemical Communications*, *55*, 10253–10256. https://doi.org/10.1039/C9CC04846D.

Biswas, A., Shogren, R. L., Stevenson, D. G., Willett, J. L., & Bhowmik, P. K. (2006). Ionic liquids as solvents for biopolymers: Acylation of starch and zein protein. *Carbohydrate Polymers*, *66*(4), 546–550. https://doi.org/10.1016/j.carbpol.2006.04.005.

Crawford, D. E., Wright, L. A., James, S. L., & Abbott, A. P. (2016). Efficient continuous synthesis of high purity deep eutectic solvents by twin screw extrusion. *Chemical Communications*, *52*(22), 4215–4218. https://doi.org/10.1039/C5CC09685E.

D'Agostino, C., Gladden, L. F., Mantle, M. D., Abbott, A. P., Ahmed, E. I., Al-Murshedi, A. Y. M., et al. (2015). Molecular and ionic diffusion in aqueous-deep eutectic solvent mixtures: Probing inter-molecular interactions using PFG NMR. *Physical Chemistry Chemical Physics*, *17*(23), 15297–15304. https://doi.org/10.1039/c5cp01493j.

Dai, Y., Van Spronsen, J., Witkamp, G., Verpoorte, R., & Choi, Y. H. (2013). Ionic liquids and deep eutectic solvents in natural products research: Mixtures of solids as extraction solvents. *Journal of Natural Products*, *76*, 2162–2173.

Davis, S. J. (2015). *Deep eutectic solvents derived from inorganic salts*. PhD Thesis University of Leicester.

Duan, L., Dou, L.-L., Guo, L., Li, P., & Liu, E.-H. (2016). Comprehensive evaluation of deep eutectic solvents in extraction of bioactive natural products. *ACS Sustainable Chemistry & Engineering*, *4*(4), 2405–2411. https://doi.org/10.1021/acssuschemeng.6b00091.

El-Sonbati, A. (Ed.). (2012). *Thermoplastic elastomers*. London: InTech Open. https://doi.org/10.5772/2038.

Esquembre, R., Sanz, J. M., Wall, J. G., del Monte, F., Mateo, C. R., & Ferrer, M. L. (2013). Thermal unfolding and refolding of lysozyme in deep eutectic solvents and their aqueous dilutions. *Physical Chemistry Chemical Physics*, *15*(27), 11248. https://doi.org/10.1039/c3cp44299c.

Francisco, M., van den Bruinhorst, A., & Kroon, M. C. (2013). Low-transition-temperature mixtures (LTTMs): A new generation of designer solvents. *Angewandte Chemie International Edition*, *52*(11), 3074–3085. https://doi.org/10.1002/anie.201207548.

García, A., Rodríguez-Juan, E., Rodríguez-Gutiérrez, G., Rios, J. J., & Fernández-Bolaños, J. (2016). Extraction of phenolic compounds from virgin olive oil by deep eutectic solvents (DESs). *Food Chemistry*, *197*(Pt A), 554–561. https://doi.org/10.1016/j.foodchem.2015.10.131.

Häkkinen, R., & Abbott, A. (2019). Solvation of carbohydrates in five choline chloride-based deep eutectic solvents and the implication for cellulose solubility. *Green Chemistry*, *21*(17), 4673–4682. https://doi.org/10.1039/C9GC00559E.

Häkkinen, R., Alshammari, O., Timmermann, V., D'Agostino, C., & Abbott, A. (2019). Nanoscale clustering of alcoholic solutes in deep eutectic solvents studied by nuclear magnetic resonance and dynamic light scattering. *ACS Sustainable Chemistry & Engineering*, 7(17), 15086–15092. https://doi.org/10.1021/acssuschemeng.9b03771.

Hammond, O. S., Bowron, D. T., & Edler, K. J. (2017). The effect of water upon deep eutectic solvent nanostructure: An unusual transition from ionic mixture to aqueous solution. *Angewandte Chemie International Edition*, *56*(33), 9782–9785. https://doi.org/10.1002/anie.201702486.

Hartley, J. M., Ip, C.-M., Forrest, G. C. H., Singh, K., Gurman, S. J., Ryder, K. S., et al. (2014). EXAFS study into the speciation of metal salts dissolved in ionic liquids and deep eutectic solvents. *Inorganic Chemistry*, *53*(12), 6280–6288. https://doi.org/10.1021/ic500824r.

Hurley, F. H., & WIer, T. P. (1951). The electrodeposition of aluminum from nonaqueous solutions at room temperature. *Journal of the Electrochemical Society*, *98*(5), 207. https://doi.org/10.1149/1.2778133.

Janssen, L., & Moscicki, L. (2009). *Thermoplastic starch: A green material for various industries. (L. Janssen & L. Moscicki, Eds.).* Weinheim, Germany: WILEY-VCH Verlag.

Lavaud, A., Laguerre, M., Birtic, S., Fabiano Tixier, A. S., Roller, M., Chemat, F., et al. (2016). *WO 2016/162703 Al. France.*

Lourdin, D., Coignard, L., Bizot, H., & Colonna, P. (1997). Influence of equilibrium relative humidity and plasticizer concentration on the water content and glass transition of starch materials. *Polymer*, *38*(21), 5401–5406. https://doi.org/10.1016/S0032-3861(97)00082-7.

Qin, H., Owyeung, R. E., Sonkusale, S. R., & Panzer, M. J. (2019). Highly stretchable and nonvolatile gelatin-supported deep eutectic solvent gel electrolyte-based ionic skins for strain and pressure sensing. *Journal of Materials Chemistry C*, 7(3), 601–608. https://doi.org/10.1039/C8TC05918G.

Qu, W. (2019). *DES modified protein-based materials and their use in pharmaceutical applications.* University of Leicester.

Qu, W., Häkkinen, R., Allen, J., D'Agostino, C., & Abbott, A. P. (2019). Globular and fibrous proteins modified with deep eutectic solvents: Materials for drug delivery. *Molecules*, *24*(19), 3583. https://doi.org/10.3390/molecules24193583.

Sanchez-Fernandez, A., Edler, K. J., Arnold, T., Alba Venero, D., & Jackson, A. J. (2017). Protein conformation in pure and hydrated deep eutectic solvents. *Physical Chemistry Chemical Physics*, *19*(13), 8667–8670. https://doi.org/10.1039/C7CP00459A.

Smith, E. L., Abbott, A. P., & Ryder, K. S. (2014). Deep eutectic solvents (DESs) and their applications. *Chemical Reviews*, *114*(21), 11060–11082. https://doi.org/10.1021/cr300162p.

Tang, B., Zhang, H., & Row, K. H. (2015). Application of deep eutectic solvents in the extraction and separation of target compounds from various samples. *Journal of Separation Science*, *38*(6), 1053–1064. https://doi.org/10.1002/jssc.201401347.

van Osch, D. J. G. P., Zubeir, L. F., van den Bruinhorst, A., Rocha, M. A. A., & Kroon, M. C. (2015). Hydrophobic deep eutectic solvents as water-immiscible extractants. *Green Chemistry*, *17*(9), 4518–4521. https://doi.org/10.1039/C5GC01451D.

Xu, W.-G., Lü, X.-M., Zhang, Q.-G., Gui, J.-S., & Yang, J.-Z. (2006). Studies on the thermodynamic properties of the ionic liquid BMIGaCl4. *Chinese Journal of Chemistry*, *24*(3), 331–335. https://doi.org/10.1002/cjoc.200690063.

Yang, J.-Z., Tian, P., He, L.-L., & Xu, W.-G. (2003). Studies on room temperature ionic liquid InCl3–EMIC. *Fluid Phase Equilibria*, *204*(2), 295–302. https://doi.org/10.1016/S0378-3812(02)00265-0.

Zhang, Y., Rempel, C., & Liu, Q. (2014). Thermoplastic starch processing and characteristics—A review. *Critical Reviews in Food Science and Nutrition*, *54*(10), 1353–1370. https://doi.org/10.1080/10408398.2011.636156.

CHAPTER TWO

Natural deep eutectic solvents: From simple systems to complex colloidal mixtures

Verena Huber, Katharina Häckl, Didier Touraud, and Werner Kunz*
University of Regensburg, Institute of Physical and Theoretical Chemistry, Regensburg, Germany
*Corresponding author: e-mail address: werner.kunz@ur.de

Contents

Abstract

In this chapter we try to embed deep eutectic solvents (DES) of synthetic and natural origin in the large group of solvents, solubilisers and surfactants. We show the relationship of DES to ionic liquids and classical hydrotropes as well as their possibility to create ephemeral solvents during a reaction process. A particular consideration to hydrophobic DES is also given. Finally, we address the compatibility of both hydrophilic and hydrophobic DES with hydrophilic and hydrophobic liquids and their miscibility induced by hydrotropes and surfactants. Possible colloidal structures occurring in these mixtures are discussed. Within all subsections, we critically consider the sustainability of the liquid systems and possible practical applications, e.g., for extraction processes, chemical reactions or drug delivery.

Advances in Botanical Research, Volume 97
ISSN 0065-2296
https://doi.org/10.1016/bs.abr.2020.09.001

1. Introduction

Isn't it fascinating to mix two powders in a mortar and to see that suddenly the mixture becomes liquid? All the more when the melting points of the two powders are far higher than room temperature. Ok, it is not always as easy as that. Often, it is necessary to heat the mixture for a while to get it liquid, but then it remains liquid even after cooling down to room temperature, where the liquid state is the thermodynamically stable one.

In any case, this is probably one of the simplest ways to "create" a liquid or solvent. Although, there might be some drawbacks such as very high viscosity or usually a very high polarity, which may prevent the solubilization of hydrophobic species, this simplicity is not only remarkable but also very useful, especially in the research for "greener" liquids and solvents. This is all the more valuable, since the ingredients for more or less deep eutectics can— and often— are natural, low toxic, even edible substances.

Note that sometimes three solids or more are mixed to form liquids at room temperature. In this case, it is better to speak of low melting mixtures (LMM) rather than deep eutectics, since no one will make the effort to detect the absolute eutectic point of these ternary mixtures. We will restrict our discussion to those deep eutectic solvents (DES) and LMM, whose ingredients are solid at room temperature. We are aware that also concentrated (often electrolyte) solutions, be it in water or glycerol or any other liquid, are termed DES today, but then, the definition is somewhat deviated from its strict thermodynamic meaning, because rarely the true eutectic composition is determined. Also, the main purpose of producing a liquid out of solids is no longer visible. It just became fashionable to use the word natural deep eutectic solvent (NADES) for nearly any liquid mixture or solution of natural substances, but we do not want to follow this fashion here.

Be it as it is, in the following, we will consider a DES just as another kind of solvent. As such, it potentially can replace other solvents. Therefore, the question is not, whether a DES is a good solvent, but rather if it is a better solvent than a classical one and in which respect.

2. Ephemeral DES-based solvents

Doubtlessly, in general, the best solvent is no solvent, or, if this is not possible, an ephemeral solvent. It can save money and be the most environmentally friendly use of a solvent. How can this be achieved with a DES?

Originally, many deep eutectic solvents were based on choline chloride (ChCl) in combination with a hydrogen bond donor. Chloride, as the hydrogen bond acceptor, undergoes a strong interaction with the donor (e.g., urea, sugar, acids, …), thus leaving the choline cation "alone." As a consequence, the crystallization of the mixture is disfavoured and the melting point goes down. We think that as far as chemical reactions are concerned, the main advantage of choline chloride is the fact that it forms DES with many different classes of molecules. If such an appropriate molecule takes part as a reactant in a chemical reaction, then this reactant can be transformed into a liquid with ChCl and ideally into the "ephemeral" solvent for the whole reaction.

If one of the reagents of an organic reaction is, for example, an acid, it may be liquefied by adding a salt such as choline chloride. The second reactant might then be readily soluble in this liquid mixture and the reaction can start. In the end, by simply adding water, the salt can be separated from the hydrophobic water-insoluble product. This approach has been applied to synthesize caffeic acid phenethyl ester (CAPE), which is a molecule of pharmaceutical relevance (Fischer, Touraud, & Kunz, 2016). Fig. 1 illustrates the

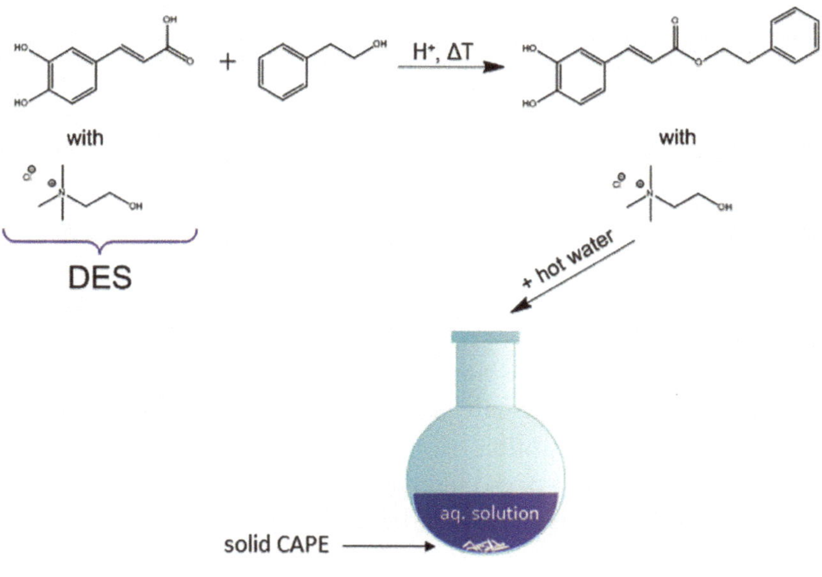

Fig. 1 Acid-catalyzed esterification of caffeic acid with phenethyl alcohol and subsequent collection of CAPE by the addition of water: a simple, DES-based method of a transient in-situ creation of a solvent allows an environmentally friendly synthesis of a pharmacologically interesting molecule. *Fig. reproduced from Häckl, K., & Kunz, W. (2018). Some aspects of green solvents.* Comptes Rendus Chimie, 21*(6), 572–580. https://doi.org/10.1016/j.crci.2018.03.010 with permission.*

approach: as caffeic acid is almost insoluble in the second reactant phenethyl alcohol, a DES with choline chloride and caffeic acid (molar ratio 2:1) is built first, in which phenethyl alcohol is completely soluble. The reaction is started by the addition of an acid catalyst. When the reaction is complete, the reaction mixture is poured into hot water, where choline chloride and residual phenethyl alcohol solubilize. The CAPE product precipitates upon cooling and can be collected by filtration.

3. DES as green solvents

Today, the literature is full of papers about "green solvents" praising the particular "greenness" of a certain liquid. However, there is no absolute greenness. First, the criteria for this expression must be clearly defined and second, it must be checked, if a new solvent such as a NADES fulfills these criteria better than an already existing solvent. Concerning the first point, Gu and Jérôme proposed 12 criteria that green solvents should fulfill, related to availability, price, recyclability, grade, synthesis, toxicity, biodegradability, performance, stability, flammability, storage, and renewability (Gu & Jérôme, 2013).

Doubtlessly, this is a reasonable collection of conditions. However, it also demonstrates the dilemma of classifying a solvent as green, as in most cases, not all requested parameters can be fulfilled by one solvent. For example, choline-based NADES may start to decompose at higher temperatures and may develop an unpleasant fishy smell. Further, in terms of toxicity regulations in Europe, it should be kept in mind that choline is forbidden in cosmetics, as it is a quaternary ammonium ion. Although, it is often argued that choline is natural (part of the B vitamins), industrially, it is produced from ammonia and petrol-based sources, so it is not really biogenic. As a consequence, the question is not, if a particular NADES is very green or not, the question is, if it is "greener" than other solvents for a particular process or product formulation.

One of the 12 criteria of green solvents is their performance. Here, DES and also NADES may have advantages and disadvantages depending on their application. Viscosity may be a disadvantage that for some applications can be offset by adding a certain amount of water. The significant number of chemical functionalities in the molecules constituting DES can also be an issue, since the DES may not be inert enough and rather easily react with other chemicals. On the other hand, there may be some intrinsic advantages over other solvents that have nothing to do with greenness or sustainability.

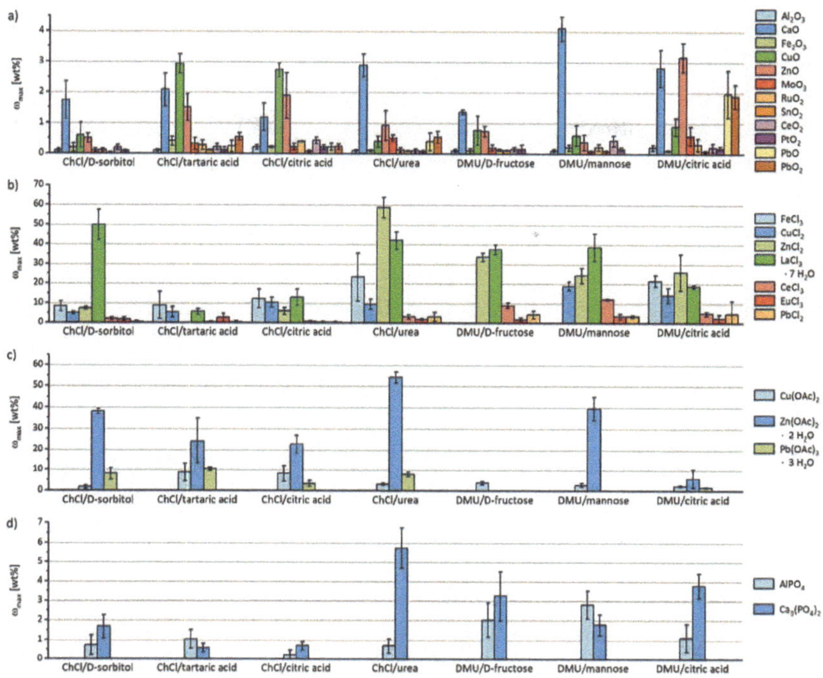

Fig. 2 Maximum solubility as the mass fraction ω_{max} (wt%) of metal compounds in seven different DESs: (A) oxides, (B) chlorides, (C) acetates, and (D) phosphates. The mass fractions represent an average of three independent measurements. *Picture reproduced from Söldner, A., Zach, J., & König, B. (2019). Deep eutectic solvents as extraction media for metal salts and oxides exemplarily shown for phosphates from incinerated sewage sludge ash.* Green Chemistry, 21*(2), 321–328. https://doi.org/10.1039/c8gc02702a with copyright permission from the Royal Society of Chemistry.*

For example, Söldner et al. found an astonishingly high solubility of metals and metal salts in DES, see Fig. 2 (Söldner et al., 2019). Among other potential applications, they showed the feasibility to extract phosphorous from incinerated sewage sludge ash with low energy cost and with readily biodegradable NADES as solvents.

4. Hydrophobic DES

Commonly, most of the currently known and in this monograph considered DES and NADES are of hydrophilic nature and water-miscible. However, in this section, we will give an overview of the still scarce literature in the field of hydrophobic DES and NADES.

The molecules capable of forming DES and NADES are usually of small size, often possess charges, and do not exhibit an extended organic nonpolar part. They are capable of hydrogen bonding, which is the basis of forming DES of any kind. Since lipophilic molecules, in general, form less hydrogen bonds, they remain rather passive in the formation of NADES. However, they may bear electronegative groups, such as alcohol or carboxy groups, which may undergo interactions with their neighbouring molecules, leading to the formation of hydrophobic NADES. Since the intermolecular interactions between the constituents of hydrophobic and hydrophilic DES are similar, their way of synthesis is similar, too. It consists of stirring and heating the primary materials before letting the mixtures cool down to room temperature (Liu et al., 2018).

The research on hydrophobic DES only started few years ago, with the examination of quaternary ammonium salts (QAS) with varying alkyl chains as HBA (hydrogen bond acceptor), mixed with fatty acids as HBD (hydrogen bond donor), yielding DES with a melting temperature lower than room temperature and a manageable viscosity, comparable to sour cream. It was found that with ammonium salts with short alkyl chains (smaller than C_7), no stable DES upon contact with water could be generated, due to the problem of leaching, originating from the water solubility of the ammonium salts. However, this property, as well as the hydrophobicity, viscosity, and polarity of the mixtures, can be customized by changing the lengths of the hydrophobic moieties. As expected, the leaching and water solubility decrease with increasing carbon chain length. With this knowledge, a hydrophobic N_{8888}-Cl/decanoic acid DES, characterized by its low water content and leaching (both <2 wt%), could be obtained (Van Osch, Zubeir, Van Den Bruinhorst, Rocha, & Kroon, 2015). Similar results were found, when mixing QAS with fatty alcohols. With an increasing chain length of the fatty alcohol, even more hydrophobic DES could be generated, as compared to the fatty acid-based ones. For example, a DES consisting of N_{8881}-Cl/dodecanol is only miscible with less than 1.5 wt% water. Ultimately, the QAS-based DES were employed in extraction experiments. The results showed that the N_{8881}-Cl/alcohol DES were the most efficient in retrieving antibiotics from aqueous solutions. The extraction efficiency, of course, was dependent on the carbon chain lengths of the chosen constituents, where an optimum was found at C_8. The extraction efficiency of the DES was even better than the efficiency of commonly used, halogenated solvents like chloroform. Thus, these tailor-made solvents proved that novel, greener extraction methods are capable of competing with the common ones (Tang, Dai, & Row, 2018).

In the last years, the research turned towards the development of naturally derived, hydrophobic DESs resulting in the search for alternative HBAs, since quaternary ammonium salts also comprise one significant drawback: they are not naturally derived and under prolonged exposure toxic to humans and the environment (Melin et al., 2014; SIGMA-ALDRICH, 2012). Soon, substitute HBAs were found in naturally occurring alcohols like the terpenoids menthol and thymol and pharmaceuticals like ibuprofen and lidocaine. Fig. 3 gives an overview of typical HBDs and HBAs of hydrophobic (NA)DES.

While keeping it simple, Florindo et al. found that exclusively combining fatty acids, namely octanoic acid (C_8), nonanoic acid (C_9), decanoic acid (C_{10}), and dodecanoic acid (C_{12}), which can simultaneously act as HBA and HBD, can yield chemically stable, hydrophobic DES. A particularity of these fatty acid-based NADES is their high hydrophobicity and extremely low viscosity, which is even lower than the viscosity of water. The fatty acid NADES are the most hydrophobic deep eutectic mixtures reported so far. They can only solubilize 1 wt% of water, before phase separation occurs. It is possible to tune the hydrophobicity of the mixture by altering the alkyl chain length of the acids, yielding more hydrophobic mixtures with longer hydrophobic moieties. This finding correlates with the melting points of the NADES. Fatty acid mixtures with long alkyl chains (C_{12} and above) yield eutectic mixtures with reduced melting temperatures, which exceed room temperature. The melting point depression of the fatty acid-based NADES with respect to the pure fatty acids is not as pronounced as the depression of some other hydrophilic DESs. Taking the combination of lauric acid ($T_m = 43.2\,°C$) with decanoic acid ($T_m = 31.6\,°C$) as an example: The melting temperature of the generated NADES is 18.0 °C, which is less than 30 °C lower than the melting temperature of the pure lauric acid. Consequently, they are rather eutectic solvents than deep ones, but there is a lack of proper definition in this field. Therefore, they are also referred to as DES in literature. In any way, fatty acid combinations of C_{8-10} acids with lauric acid are liquid at room temperature and of low viscosity, giving them desirable characteristics of commonly known solvents (Florindo, Romero, Rintoul, Branco, & Marrucho, 2018).

Another possibility to create low viscous, hydrophobic NADES is the use of terpenoids as HBAs. The most extensively studied ones are menthol and thymol (cf. structures in Fig. 3). Martins et al. discovered that both terpenoids form eutectic solvents upon mixing with long-chain fatty acids, which can be characterized by the typical eutectic solid-lipid phase diagrams (cf. Fig. 4) (Martins et al., 2018). Just as mentioned before, the melting point

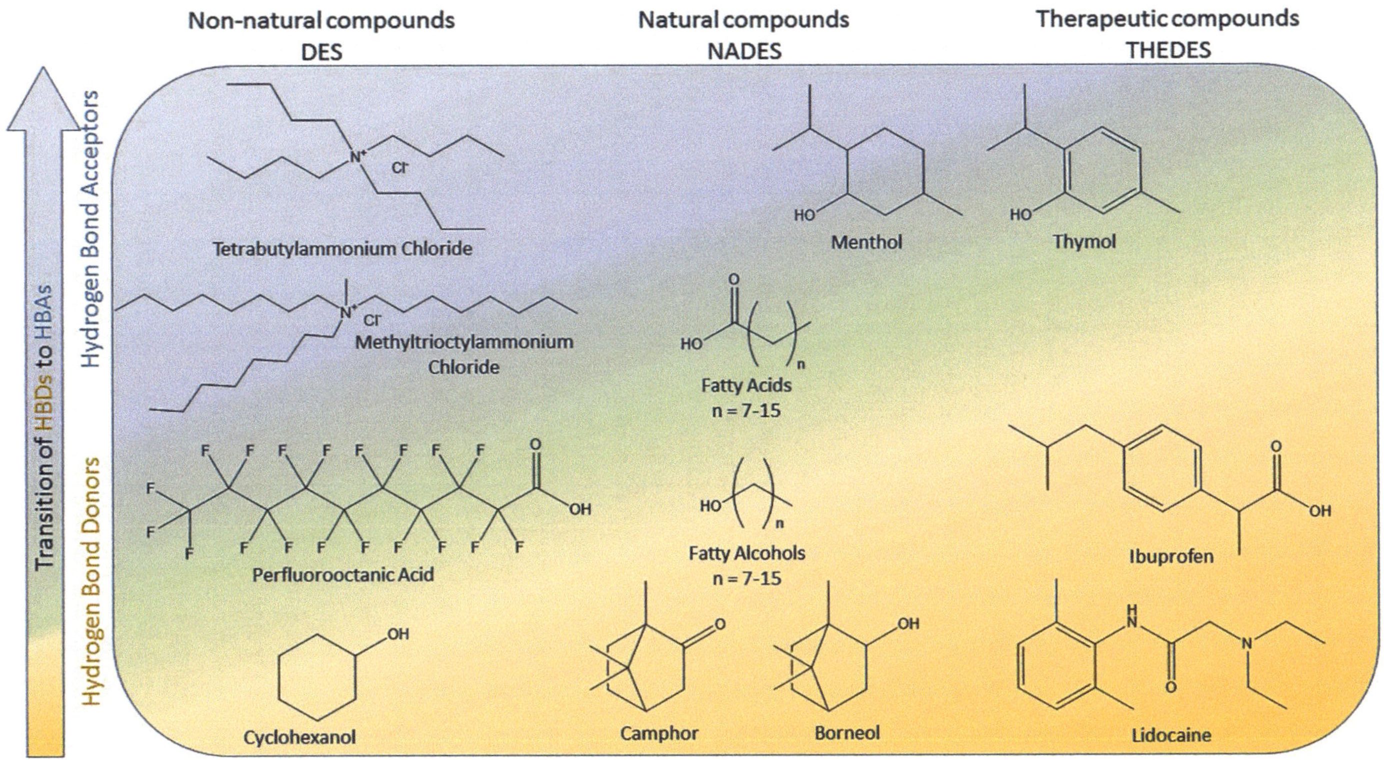

Fig. 3 Overview of some currently investigated HBDs and HBAs for (NA)DES formation. The color profile marks the transition of hydrogen bond donors (yellow) to hydrogen bond acceptors (blue).

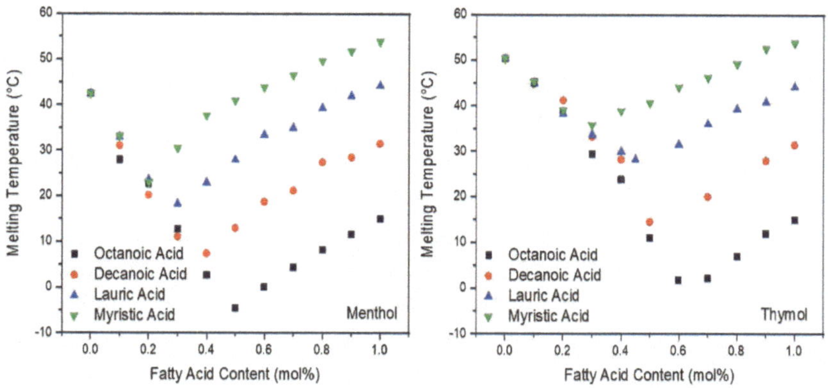

Fig. 4 Solid-liquid phase diagrams of monocarboxylic acids with L(−)-menthol (left) and thymol (right); the compositions are given in mole percent. *Figure reproduced from Martins, M. A. R., Crespo, E. A., Pontes, P. V. A., Silva, L. P., Bülow, M., Maximo, G. J., et al. (2018). Tunable hydrophobic eutectic solvents based on terpenes and monocarboxylic acids.* ACS Sustainable Chemistry & Engineering, *6(7), 8836–8846. https://doi.org/10.1021/acssuschemeng.8b01203 with permission of the American Chemical Society.*

depression of fatty acid-based NADES is not very pronounced; the decrease in the melting temperature of all terpenoid/fatty acid mixtures with respect to the fatty acids is less than 30 °C. According to Florindo et al., it seems as if salts like QAS are necessary to obtain real hydrophobic deep eutectic mixtures. Nevertheless, for fatty acids with a chain length between C_8 and C_{12}, it was possible to obtain liquid eutectic mixtures with a melting temperature below or around room temperature.

The obtained liquid NADES could be characterized as fairly low viscous in the range of 8–34 mPa s, for C_8-C_{14} at 25 °C, while longer-chained fatty acids are more viscous than the short-chained ones. At elevated temperatures, the differences in viscosities become negligible. In general, the thymol-based DES are of lower viscosity than the menthol-based ones. The rule of ideal mixture for viscosity (Eq. 1) is applicable for these NADES mixtures:

$$\ln \eta_{mix} = x_1 \cdot \ln \eta_1 + x_2 \cdot \ln \eta_2 \quad (1)$$

with η_{mix} being the viscosity of the mixture and η_i the viscosity and x_i the mole fraction of component i, respectively.

The hydrophobicity of the terpenoid NADES was also examined. The terpenoid-based NADES were not as hydrophobic as the exclusively fatty acid-based ones, as mentioned before. A maximum water content of

1.3–1.8 wt% for menthol-based NADES and 2.0–2.4 wt% for thymol-based ones could be reached. Also, the hydrophobicity, as mentioned before, was increased by longer alkyl moieties of the fatty acids, which means that the hydrophobicity of the NADES is dependent on the hydrophobicity of the HBD. Based on this knowledge, the polarity of the NADESs was examined. Through experiments concerning the solvatochromic parameters of the terpenoid-based DESs, it was found that the polarity was highly influenced by the terpenoids and less by the chain length of the fatty acids. It was found that the thymol-based NADES showed higher non-specific interactions and could be characterized as hydrogen bond acidic as opposed to the menthol-based ones, which preferably accept hydrogen bonds. With this polarity dependence on the terpenoids, the NADES can be tuned to various specific applications (Martins et al., 2018).

Beyond the fatty acid-based NADES, menthol can be found to be able to form eutectic mixtures with several different compounds, including pharmaceutically active ingredients (API). This new class of therapeutically active, natural and non-natural eutectic solvents is called THEDES, short for therapeutic deep eutectic solvents. Common API are, among others, acetylsalicylic acid, benzoic acid, ibuprofen, and lidocaine. THEDES consisting of menthol, combined with ibuprofen or lidocaine, even yield liquid eutectic solvents with a melting temperature close to room temperature, 19 and 26 °C, respectively (Ribeiro, Florindo, Iff, Coelho, & Marrucho, 2015). Some THEDES can as well be classified as NADES, if all constituents are of natural origin. Thus, the DES consisting of menthol as HBA and borneol or camphor (natural compounds, cf. Fig. 3) as HBD, which are used as transdermal drug delivery vehicles, can be considered as natural THEDES. The use of DES in the pharmaceutical industry is of rising importance, as the solubility of API in DES is higher, in comparison to aqueous media. This and the combination with new, sustainable drug delivery media is a promising step towards a more efficient administration of drugs (Dwamena, 2019).

The research on hydrophobic DES is starting to outgrow its infancy, as significant progress has been made in the past five years. Some more effort has to be put into the complete understanding of the fundamentals of hydrophobic DES formation. However, in the field of applications, the development is thriving. Among numerous potential applications, the extraction of organic acids, natural compounds and pollutants like metal ions or antibiotics of aqueous solutions is one big field of interest. The extraction of all these compounds has already been successfully carried out by employing the variety of the above mentioned DES, NADES, and THEDES (Ribeiro et al., 2015; Van Osch et al., 2016, 2015). As illustrated in this short review, all types of hydrophobic DES offer their own set of advantages. And since it is possible to customize their properties for a given application, they can be a competitive alternative to the already established solvents, due to their sustainability, greenness and easy production.

5. The interplay of DES and ionic liquids (IL)

5.1 A continuous pathway from DES to IL

In the case of choline chloride, the chloride anion is the HBA and the choline cation is sufficiently inert (in the sense of Hofmeister specific ion effects) and bulky to stay away from the chloride-HBD complex. This explains the significant melting point depression.

The bulkiness of the choline cation also explains, why choline can easily form ionic liquids. Of course, the two liquids, shown in Fig. 5 could be mixed in different ratios and as such, one could go from a pure NADES to a pure "NA-IL" (an ionic liquid composed of only natural substances). But can we find systems that intrinsically behave as a Janus-like mixture of DES and IL? To conceive such systems, we focused on the following question:

What happens, if the cation bears its own counterion? This is the case with zwitterions. The corresponding species closest to choline are betaine and carnitine, see Fig. 6, both natural components that can be found even in the human body.

Fig. 5 Left: the NADES lactic acid/choline chloride (2,1), right: the IL choline lactate at room temperature. *Picture reproduced from Francisco, M., van den Bruinhorst, A., & Kroon, M. C. (2013). Low-transition-temperature mixtures (LTTMs): A new generation of designer solvents.* Angewandte Chemie International Edition, 52*(11), 3074–3085. https://doi.org/10.1002/anie.201207548 with permission.*

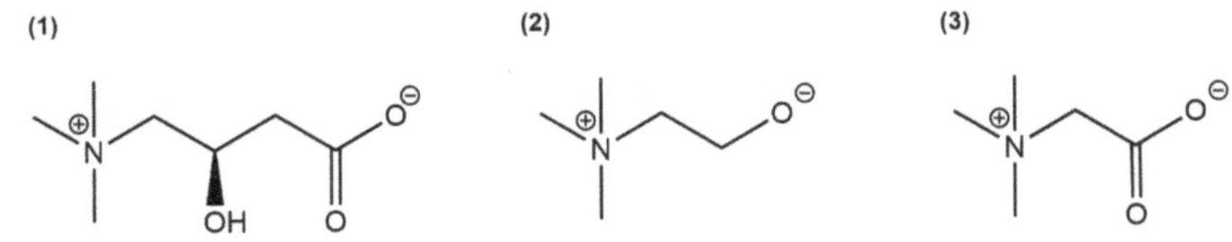

Fig. 6 Molecular structures of L-carnitine (1), choline (2) and betaine (3).

Here the carboxylic group is now the HBA. Fig. 7 contains some NADES formed by combining carnitine and betaine with various acids, acting as HBD, when combined in a certain mole ratio (HBA:HBD = 1:1, 1:2 or 2:1). The interesting feature here is that the proton of the HBD acid may exchange with the HBA and bind on the zwitterion, see Fig. 8.

As will be described in detail in Häckl, Nardello-Rataj, and Kunz (2020), the equilibrium can be influenced by partly replacing carnitine by a mixture of carnitine and carnitine hydrochloride. In this way, a continuous crossover from a NADES to a NA-IL can be achieved and, of course, a mixture of both. For the moment, this is just a proof of concept, but it could serve as a starting point to differentiate the behavior of both types of mixtures and to see the importance of the strong hydrogen bond interactions in the case of DES for its mascroscopic properties, in contrast to the even stronger ionicity in the case of IL.

5.2 Combining the effects of DES and IL

Dicarboxylic acids such as succinic, glutaric, or adipic acid are interesting natural chemicals. Their diesters are promising, green solvents. The acids form NADES together with choline chloride. Except with oxalic and malonic acid, their melting temperature is above room temperature. On the other hand, Ionic Liquids can be formed by replacing one of the protons of the diacids by choline. However, they still remain solid at room temperature. In order to conceive a liquid at room temperature out of these chemicals, the NADES can be mixed with the IL. Indeed, it turned out that, e.g., choline glutarate-choline chloride mixtures are liquids at room temperature at compositions containing 95–98 wt% of choline glutarate. Finally, by adding another HBD, such as urea, the melting point can further be reduced. Fig. 9 illustrates the strategy (Rengstl et al., 2014).

Admittedly, the obtained low melting mixture of a NADES and a NA-IL shows a high viscosity at room temperature. On the other hand, it might be a

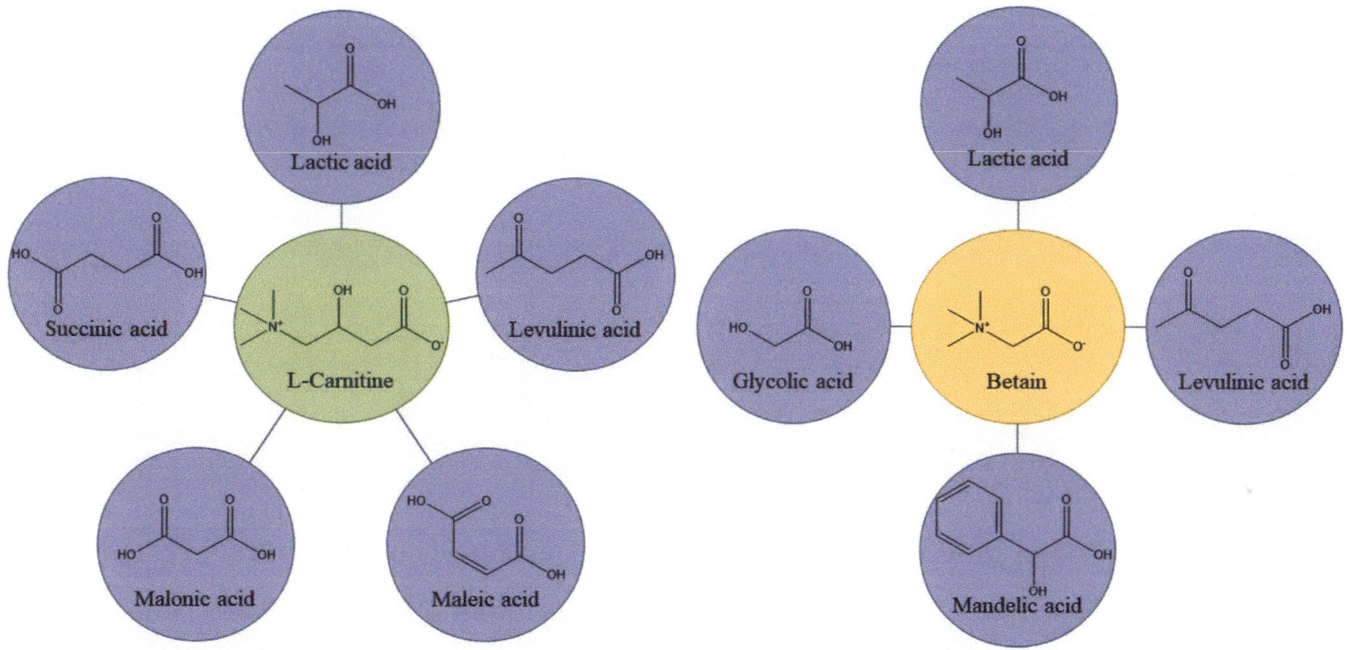

Fig. 7 Combinations of L-carnitine (left) and betaine (right) with carboxylic acids yielding NADES.

Fig. 8 Equilibrium between a eutectic solvent (two overall neutral species), where the HBA is part of the zwitterion (left side) and an ionic liquid, composed of two ions, in the case where the proton is exchanged from the initial HBD to the HBA.

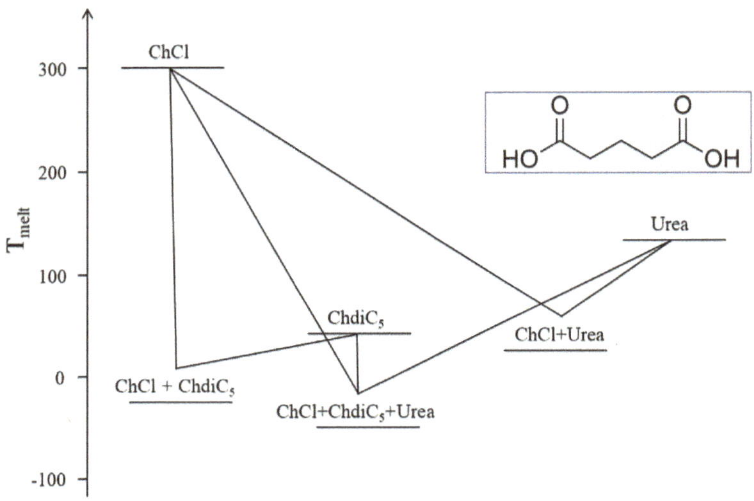

Fig. 9 Melting points of the different combinations of choline chloride with choline glutarate and urea. The inserted molecule is glutaric acid (abbreviated here with diC_5). *Figure reproduced from Rengstl, D., Fischer, V., & Kunz, W. (2014). Low-melting mixtures based on choline ionic liquids.* Physical Chemistry Chemical Physics, 16*(41), 22815–22822. https://doi.org/10.1039/c4cp02860k.*

promising medium to dissolve pharmaceutically relevant molecules. Especially, because both of the used DES and IL are composed of entirely natural constituents.

6. DES and hydrotropes

Both hydrophilic and hydrophobic DES can be suitable solvents for various substances. For practical applications, it is often desirable to dissolve a hydrophobic component in a polar solvent or, vice versa, a hydrophilic component in a nonpolar solvent. This can be achieved by adding either a surfactant to the mixture of both components or, if a surfactant is to be avoided (e.g., because foaming might be an issue), a so-called hydrotrope. A hydrotrope (or more general a solvotrope) typically is a co-solubilizer that is highly soluble in the solvent, but has also a strong affinity to the scarcely soluble solute so that the latter is surrounded (or aggregated) by the hydrotrope and thus being better solubilized in the solvent.

Four cases are conceivable: (i) the DES is hydrophilic and should be mixed with an organic substance (or the hydrophobic substance should be dissolved in the hydrophilic DES) or (ii) vice versa, i.e., the DES is hydrophobic and should be mixed with water or another polar solvent or (iii) mixing a hydrophobic with a hydrophilic DES with the help of a convenient hydrotrope or (iv) the DES could even be the hydrotrope making a hydrophilic and a hydrophobic liquid miscible.

6.1 Case (i) Hydrophilic DES mixed with a hydrophobic solvent

An example of this type of mixtures is given in Fig. 10. The phase diagram shows that the classical NADES urea-choline chloride can be mixed with the green solvent diethyl adipate, when a sufficient (and here unfortunately relatively high amount) of the hydrotrope THFA is added

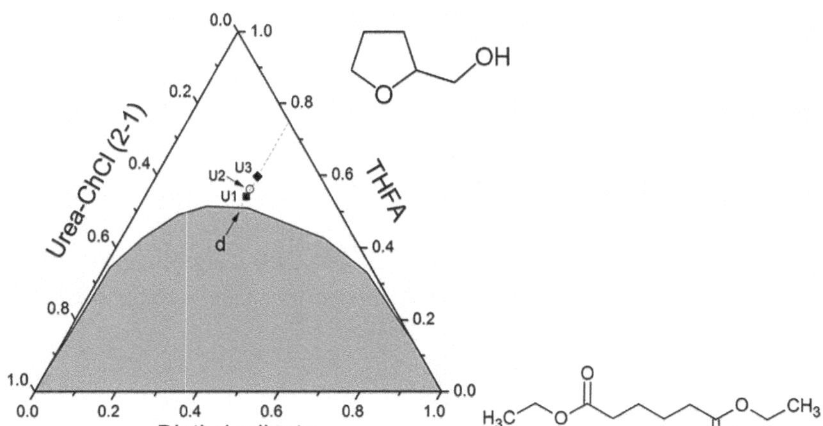

Fig. 10 Making the NADES urea-choline chloride miscible with the green solvent diethyl adipate with the help of the hydrotrope tetrahydrofurfuryl alcohol (THFA). The upper white part is the monophasic region, the lower gray part the diphasic one. The points U1-U3 are compositions, at which DLS and SAXS measurements have been performed. The phase diagram has been recorded at room temperature. The compositions are given in weight percent. *Picture taken from Fischer, V., Marcus, J., Touraud, D., Diat, O., & Kunz, W. (2015). Toward surfactant-free and water-free microemulsions.* Journal of Colloid and Interface Science, 453, *186–193. https://doi.org/10.1016/j.jcis.2015.04.069 with permission.*

(Fischer, Marcus, Touraud, Diat, & Kunz, 2015). The molecular organization is the same as in classical water/hydrotrope/co-solvent systems (Fischer et al., 2015). A monophasic mixture, as represented in white, in Fig. 10, and a gray lower area of biphasic mixtures are reported. Interestingly, both dynamic light scattering (DLS) and small-angle X-ray scattering (SAXS) confirm that there are distinct structures at the compositions U1-U3, which are similar to classical microemulsions. So, this is one of the first examples of both water- and surfactant-free microemulsions. As such, these mixtures may have a promising future for a variety of applications.

Fig. 11A shows another example of such a mixture of a hydrophilic NADES (proline + oxalic acid) with an edible oil (triacetin) and ethanol as the hydrotrope. This system has the remarkable advantage that it is entirely food agreed and can also be advantageously used for plant extraction. The phase diagram shows the classic overall shape of diagrams with two non-miscible solvents, where the addition of a significant amount of hydrotrope closes the miscibility gap (the gray lower region). However, the extension of the biphasic region changes significantly, when water is replaced by proline +oxalic acid as the polar phase (c.f. Fig. 11B). The miscibility gap is much larger when the NADES is used. The bottom line of the phase triangle shows that the oil phase is not miscible with the NADES at all, whereas triacetin can be dissolved in water up to 10 wt%. With an increasing water content of the proline + oxalic acid mixture, the area of the miscibility gap decreases. It can be assumed that with a further increasing amount of water, the biphasic domain of the NADES system will approach the shape of the ternary water mixture. Pronounced correlation functions can be detected using DLS experiments both in the two monophasic binary mixtures and in the ternary mixtures. As the hydrophilic NADES is significantly different from water, new partitions of extracted compounds can be expected in both phases after liquid/liquid phase separation.

6.2 Case (ii) Mixing hydrophobic DES with water

When comparing the diagram of a hydrophilic NADES/ethanol/triacetin (Fig. 11A) and the diagram water/ethanol/hydrophobic NADES (Fig. 12A) to the diagram water/ethanol/triacetin, it can be observed that both the hydrophilic and hydrophobic NADES reduce the area of monophasic mixtures.

It is not surprising that the two-phasic region of the water and ethanol mixed with the hydrophobic NADES is larger than for the mixture with

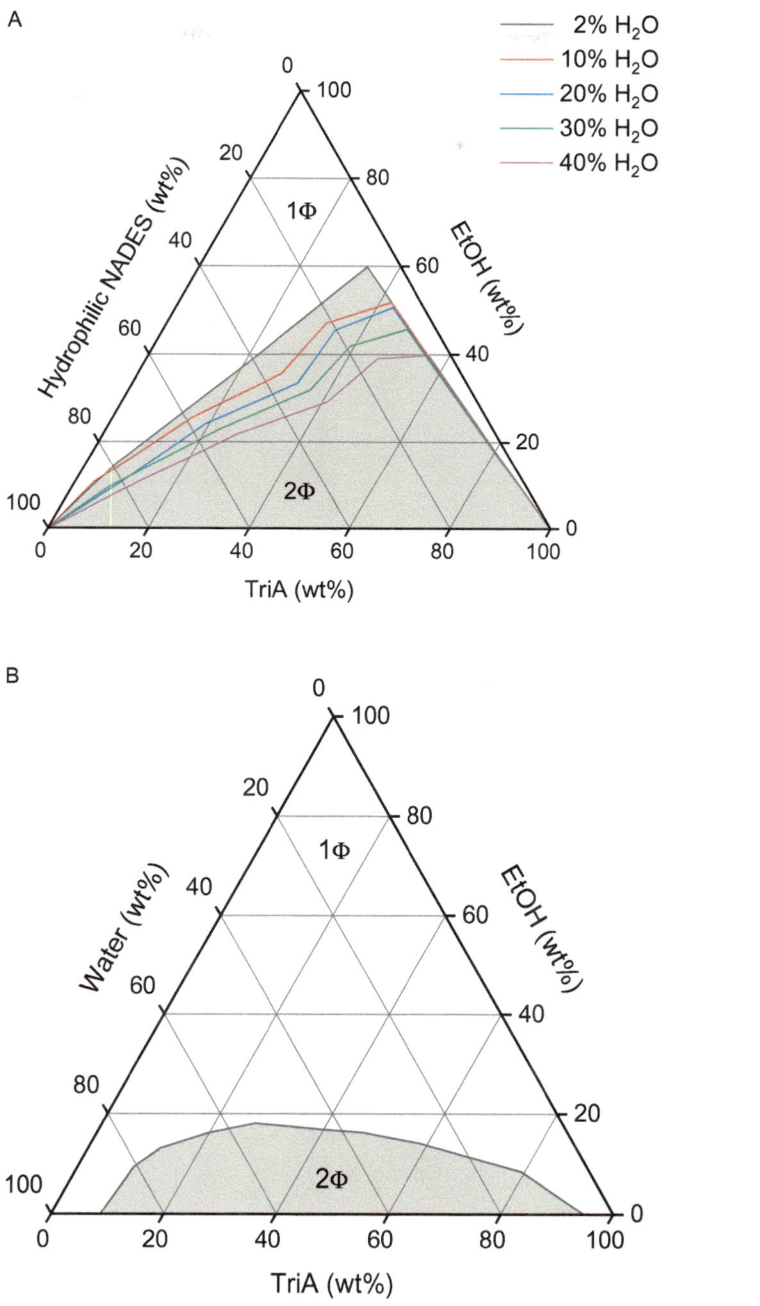

Fig. 11 (A) Ternary phase diagram of entirely edible components at room temperature. The NADES proline-oxalic acid is made miscible with the oil triacetin with ethanol as the hydrotrope, with different amounts of added water. (B) Ternary phase diagram of the reference system at room temperature consisting of water, triacetin as the oil, and ethanol as the hydrotrope. In the white area above the grey lines, both systems are transparent, stable and monophasic. The compositions are given in weight percent.

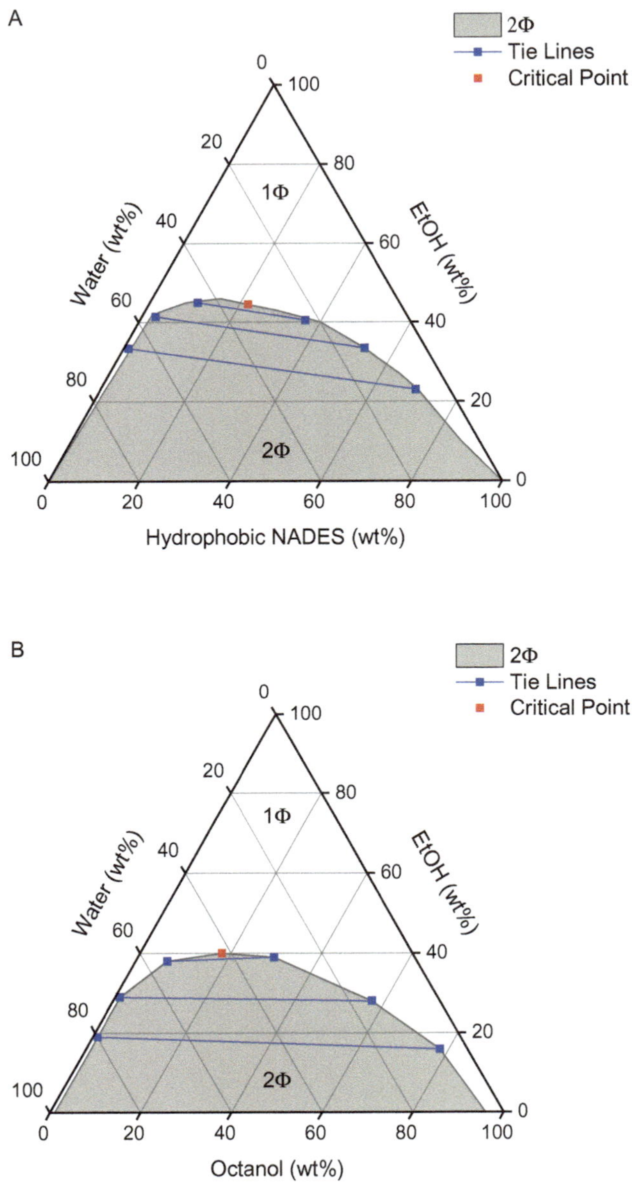

Fig. 12 (A) Ternary phase diagram of entirely edible components at room temperature. Here the hydrophobic NADES menthol-lauric acid is made miscible with water with ethanol as the hydrotrope. The blue lines represent the tie lines with the endpoints giving the compositions of the two phases in equilibrium in the lower two-phasic system. (B) Ternary phase diagram of the reference system consisting of water, 1-octanol as the oil phase and ethanol as the hydrotrope. The compositions are given in weight percent.

triacetin. The hydrophobic part of the menthol + lauric acid NADES used in Fig. 12A is more extended than the one of triacetin. Thus, the ternary system consisting of water, ethanol and 1-octanol (cf. Fig. 12B) serves as a better reference. These two systems are highly similar in their appearance. The biphasic region of both systems is of similar size and also shape, indicating a similar capability of ethanol to act as a co-solvent for 1-octanol and for the hydrophobic NADES. The angle of the tie lines differs slightly, yet they are tilted in the same direction and nearly parallel in both systems. Also, the critical point was determined to be in the mid to left part of both ternary diagrams. In dynamic light scattering experiments above the miscibility gap, correlation functions could be found, which are most pronounced around the critical point. In further studies of the hydrophobic NADES ternary system, the similarities with the 1-octanol mixture will be examined in more detail.

6.3 Case (iii) Mixing a hydrophilic with a hydrophobic DES with an appropriate hydrotrope

In the example presented in Fig. 13, a hydrophilic NADES consisting of proline and oxalic acid is mixed with a hydrophobic NADES made of menthol and lauric acid, by use of a hydrotrope. As can be seen, the biphasic region is enormous as compared to the monophasic one, when ethanol is

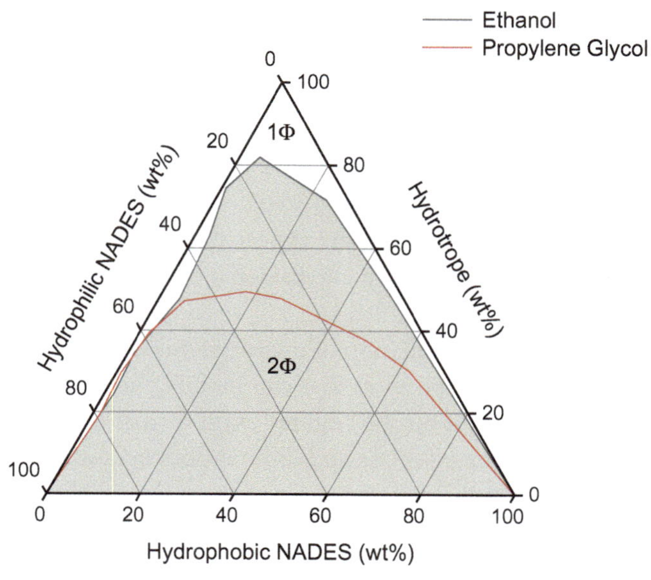

Fig. 13 Ternary phase diagram of immiscible hydrophilic (proline + oxalic acid) and a hydrophobic (menthol + lauric acid) DES, made miscible by the addition of a hydrotrope, at room temperature. The black line represents ethanol as the hydrotrope and the red one propylene glycol.

used. Remarkably, ethanol is a bad hydrotrope for this mixture of two NADES, despite the fact that it is miscible with both NADES. Of course, this is a very complex multi-component mixture, and for the moment, it is not clear how all these substances interact or mix with each other. As will be discussed in more details in following publications, other hydrotropes are more convenient to enlarge the monophasic region in pseudo-ternary mixtures of these two NADES.

6.4 Case (iv) Using a DES to mix a polar phase with a nonpolar one

To our knowledge, this case is not yet described in literature. However, it would be interesting to find an appropriate NADES to mix for example water and an edible oil.

7. DES and surfactants

Surfactants can form micelles and various types of lyotropic liquid crystalline phases in aqueous solutions as well as in various other solvents. Also in IL, these structures have often been detected. Further, various mixtures of hydrophobic IL with water or hydrophilic IL with an oil show the typical signs of microemulsions. Even surfactant-IL can be used to form structures in another IL used as a solvent. Today, there is a considerable amount of literature on this topic.

In light of the structural similarity, it is not astonishing that surfactants induce similar structures in DES as well. The first papers appeared only in 2014 and 2015. Rengstl et al. dissolved 2 wt% of choline dodecylsulfate in a mixture consisting of 96% choline adipate and 4% of choline chloride. SAXS experiments clearly showed the appearance of micelles above 50 °C. However, these were preliminary experiments and not pursued for more detailed analysis (Rengstl et al., 2014).

Pal et al. presented the formation of microemulsions, consisting of probably the most classical composition that can be imagined: a 1:2 mixture of ChCl and urea as the hydrotrope, diluted with a small amount of added water, cyclohexane as the oil phase and sodium dodecyl sulfate as surfactant (Fig. 14) (Pal et al., 2014). Unsurprisingly, they could produce transparent and stable oil-in-DES/water microemulsions and determine the structure by various experimental techniques.

Of course, also W/O microemulsions are possible, as shown by Sakuragi, Tsutsumi, and Kusakabe, 2018 and Sakuragi, Yoshimura, and Kusakabe (2020). They "blew up" water-in-isopropyl myristate micelles by adding

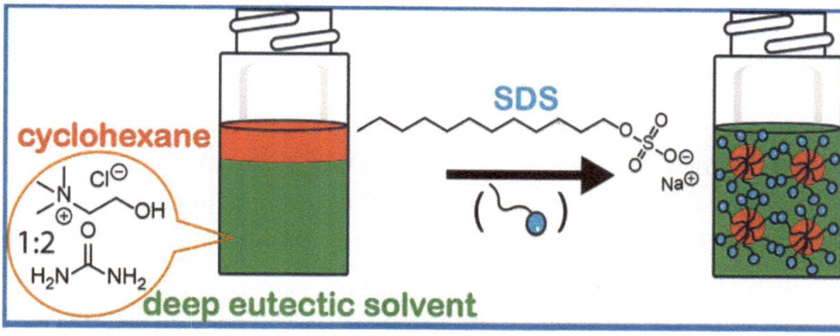

Fig. 14 Cyclohexane in DES/water microemulsions made with sodium dodecyl sulfate. *Picture taken from Pal, M., Rai, R., Yadav, A., Khanna, R., Baker, G. A., & Pandey, S. (2014). Self-aggregation of sodium dodecyl sulfate within (choline chloride + urea) deep eutectic solvent.* Langmuir, 30*(44), 13191–13198. https://doi.org/10.1021/la5035678 with permission.*

a hydrophilic DES (ChCl-glycerol), which was solubilized in the micelles. The rationale was the finding that various pharmaceutical compounds are better soluble in DES than they are in pure water. The thus obtained microemulsion was supposed to be a suitable candidate for a transdermal drug delivery system.

Similarly, Wang et al. prepared DES microemulsion-based gels for enhanced transdermal delivery of paeonol (Wang et al., 2017). Paeonol is an antioxidant and anti-inflammatory agent with a melting point T_m of 51 °C and menthol a commonly used skin penetration enhancer with T_m around 37 °C. To our knowledge, the T_m of the mixture is not determined and it might not be very low, but at least the melting point depression of various of these binary mixtures is sufficiently pronounced so that the mixtures are liquid at room temperature. Consequently, it is not astonishing that therefore the transdermal drug delivery is enhanced without the addition of a further solvent, which would be necessary, because paeonol is hydrophobic (as is its mixture with menthol) and almost insoluble in water. So this is another example of a hydrophobic solvent. The strategy of liquifying a chemical substance is similar to the approach of ephemeral DES-solvents, presented in Section 2. With the exception that there, the liquefied substance is a reactant and here a drug to be delivered through the skin.

As Wang et al. showed, drug permeation across the skin was further facilitated by the microemulsion formulation. So, the authors combined two effects to increase transdermal transport of paeonol: first, taking advantage of liquefying it by transforming it into a eutectic mixture and second, by micro-emulsifying this mixture then with water and a surfactant system (Wang et al., 2017).

8. Conclusion and outlook

Numerous applications of DES and NADES can be thought up. In any application, one takes advantage of liquefying a mixture of powders. The thus gained liquid may serve (i) as a new and in general easy to prepare solvent. And (ii), it can also be used to avoid an additional classical solvent, see, e.g., Section 2.

Concerning the first point: as described in almost all chapters in this monograph, DES may have useful solvent properties. However, these properties should be always compared to the solubilization power of classical "green" solvents. The question is not, if a NADES is a good solvent, but if it has some superior properties compared to classical ones.

In this chapter, we gave some examples of making more or less colloidal liquid systems based on NADES. This is in line with the classical use of solvents to form microemulsions or other mixtures between hydrophilic and hydrophobic liquids made compatible with the help of hydrotropes or surfactants. The results are promising, but not really astounding, in light of the abundant literature about colloidal mixtures with ionic liquids. Promising applications could be in the field of extraction (plant extraction, etc.) and decontamination processes, further to galenic and cosmetics formulations. Even household cleaners could be envisaged. Concerning extraction followed by separation, probably new systems could be fine-tuned to increase the specificity of liquid/liquid separation.

Strange enough, it seems that up to now nobody looked at the appearance of lyotropic liquid crystalline phases in surfactant systems with NADES. There could be some interest, because these would be naturally gelified systems, probably useful, e.g., in pharmacy and cosmetics.

Abbreviations

°C degrees Celsius
API pharmaceutically active ingredient
CAPE caffeic acid phenethyl ester
ChCl choline chloride
C_n alkyl moiety of chainlength n
DES deep eutectic solvent
DLS dynamic light scattering
HBA hydrogen bond acceptor
HBD hydrogen bond donor
IL ionic liquid

LMM	low melting mixture
NADES	natural deep eutectic solvent
NA-IL	ionic liquid composed of only natural substances
N_{nnnn}-Cl	quaternary ammonium chloride with alkyl moieties of chain length n
QAS	quaternary ammonium salt
SAXS	small-angle X-ray scattering
THEDES	therapeutic deep eutectic solvent
T_m	melting temperature
W/O	water in oil
wt%	weight fraction
x_i	mole fraction of compound i
η_i	viscosity of compound i

References

Dwamena, A. K. (2019). Recent advances in hydrophobic deep eutectic solvents for extraction. *Separations*, *6*(1), 3. https://doi.org/10.3390/separations6010009.

Fischer, V., Marcus, J., Touraud, D., Diat, O., & Kunz, W. (2015). Toward surfactant-free and water-free microemulsions. *Journal of Colloid and Interface Science*, *453*, 186–193. https://doi.org/10.1016/j.jcis.2015.04.069.

Fischer, V., Touraud, D., & Kunz, W. (2016). Eco-friendly one pot synthesis of caffeic acid phenethyl ester (CAPE) via an in-situ formed deep eutectic solvent. *Sustainable Chemistry and Pharmacy*, *4*, 40–45. https://doi.org/10.1016/j.scp.2016.08.002.

Florindo, C., Romero, L., Rintoul, I., Branco, L. C., & Marrucho, I. M. (2018). From phase change materials to green solvents: Hydrophobic low viscous fatty acid-based deep eutectic solvents. *ACS Sustainable Chemistry & Engineering*, *6*(3), 3888–3895. https://doi.org/10.1021/acssuschemeng.7b04235.

Gu, Y., & Jérôme, F. (2013). Bio-based solvents: An emerging generation of fluids for the design of eco-efficient processes in catalysis and organic chemistry. *Chemical society reviews*, *Vol. 42*, 9550–9570. The Royal Society of Chemistry, issue 24. https://doi.org/10.1039/c3cs60241a.

Häckl, K., Nardello-Rataj, V., & Kunz, W. (2020). *L-Carnitine—A natural substance offering various opportunities for the formation of renewable liquid media and functional molecules.* Manuscript in preparation Germany: University of Regensburg.

Liu, Y., Friesen, J. B., McAlpine, J. B., Lankin, D. C., Chen, S. N., & Pauli, G. F. (2018). Natural deep eutectic solvents: Properties, applications, and perspectives. *Journal of Natural Products*, *81*(3), 679–690. https://doi.org/10.1021/acs.jnatprod.7b00945.

Martins, M. A. R., Crespo, E. A., Pontes, P. V. A., Silva, L. P., Bülow, M., Maximo, G. J., et al. (2018). Tunable hydrophobic eutectic solvents based on terpenes and monocarboxylic acids. *ACS Sustainable Chemistry & Engineering*, *6*(7), 8836–8846. https://doi.org/10.1021/acssuschemeng.8b01203.

Melin, V. E., Potineni, H., Hunt, P., Grisworld, J., Siems, B., Werre, S., et al. (2014). Exposure to common quaternary ammonium disinfectants decreases fertility in mice. *Reproductive Toxicology*, *50*, 163–170. https://doi.org/10.1016/j.reprotox.2014.07.071.

Pal, M., Rai, R., Yadav, A., Khanna, R., Baker, G. A., & Pandey, S. (2014). Self-aggregation of sodium dodecyl sulfate within (choline chloride + urea) deep eutectic solvent. *Langmuir*, *30*(44), 13191–13198. https://doi.org/10.1021/la5035678.

Rengstl, D., Fischer, V., & Kunz, W. (2014). Low-melting mixtures based on choline ionic liquids. *Physical Chemistry Chemical Physics*, *16*(41), 22815–22822. https://doi.org/10.1039/c4cp02860k.

Ribeiro, B. D., Florindo, C., Iff, L. C., Coelho, M. A. Z., & Marrucho, I. M. (2015). Menthol-based eutectic mixtures: Hydrophobic low viscosity solvents. *ACS Sustainable Chemistry & Engineering*, *3*(10), 2469–2477. https://doi.org/10.1021/acssuschemeng.5b00532.

Sakuragi, M., Tsutsumi, S., & Kusakabe, K. (2018). Deep eutectic solvent-induced structural transition of microemulsions explored with small-angle X-ray scattering. *Langmuir*, *34*(42), 12635–12641. https://doi.org/10.1021/acs.langmuir.8b02565.

Sakuragi, M., Yoshimura, H., & Kusakabe, K. (2020). Effects of structures of microemulsions containing a deep eutectic solvent on the entrapment amount and the skin permeation of resveratrol. *Japanese Journal of Applied Physics*, *59*(3), 035002. https://doi.org/10.35848/1347-4065/AB6B74.

SIGMA-ALDRICH. (2012). *Material safety data sheet*. https://www.nwmissouri.edu/naturalsciences/sds/t/Tetraoctylammoniumbromide.pdf.

Söldner, A., Zach, J., & König, B. (2019). Deep eutectic solvents as extraction media for metal salts and oxides exemplarily shown for phosphates from incinerated sewage sludge ash. *Green Chemistry*, *21*(2), 321–328. https://doi.org/10.1039/c8gc02702a.

Tang, W., Dai, Y., & Row, K. H. (2018). Evaluation of fatty acid/alcohol-based hydrophobic deep eutectic solvents as media for extracting antibiotics from environmental water. *Analytical and Bioanalytical Chemistry*, *410*(28), 7325–7336. https://doi.org/10.1007/s00216-018-1346-6.

Van Osch, D. J. G. P., Parmentier, D., Dietz, C. H. J. T., Van Den Bruinhorst, A., Tuinier, R., & Kroon, M. C. (2016). Removal of alkali and transition metal ions from water with hydrophobic deep eutectic solvents. *Chemical Communications*, *52*(80), 11987–11990. https://doi.org/10.1039/c6cc06105b.

Van Osch, D. J. G. P., Zubeir, L. F., Van Den Bruinhorst, A., Rocha, M. A. A., & Kroon, M. C. (2015). Hydrophobic deep eutectic solvents as water-immiscible extractants. *Green Chemistry*, *17*(9), 4518–4521. https://doi.org/10.1039/c5gc01451d.

Wang, W., Cai, Y., Liu, Y., Zhao, Y., Feng, J., & Liu, C. (2017). Microemulsions based on paeonol-menthol eutectic mixture for enhanced transdermal delivery: Formulation development and in vitro evaluation. *Artificial Cells, Nanomedicine and Biotechnology*, *45*(6), 1241–1246. https://doi.org/10.1080/21691401.2016.1226178.

CHAPTER THREE

Interactions of water and amphiphiles with deep eutectic solvent nanostructures

Iva Manasi[a], Saffron J. Bryant[a,b], Oliver S. Hammond[c], and Karen J. Edler[a,*]

[a]Department of Chemistry, University of Bath, Bath, United Kingdom
[b]Centre for Molecular and Nanoscale Physics, College of Science, Engineering and Health, RMIT University, Melbourne, VIC, Australia
[c]Laboratoire de Chimie, École Normale Supérieure de Lyon, Lyon, France
[*]Corresponding author: e-mail address: k.edler@bath.ac.uk

Contents

Abstract

Deep eutectic solvents (DES) are extensively hydrogen-bonded eutectic mixtures with low melting points. The presence of H-bonding makes them highly susceptible to interactions with water and also allows them to support amphiphile self-assembly, both for surfactants and biological lipids. Water disrupts the DES nework through intermolecular H-bonding, leading to changes in physicochemical properties: melting point, density and viscosity decrease; whereas electrical conductivity increases as water is added. Trends in these properties and structural studies suggest an upper hydration limit, where a transition is observed, around 60–80 mol% water, from water-in-DES to DES-in-water behavior. Amphiphiles form aggregates in DES depending on solvophobicity, water content and Gordon parameter with structures ranging from globular micelles to highly ordered hexagonal/cubic phases. Lipid vesicle formation is also observed in DES with and without water. These interactions find important applications in fields such as biology, nanoparticle synthesis and electrodeposition.

Advances in Botanical Research, Volume 97
ISSN 0065-2296
https://doi.org/10.1016/bs.abr.2020.09.002

Abbrevations

ATPPB allyltriphenyl phosphonium bromide
C_{12}TAB dodecyltrimethylammonium bromide
C_{14}TAB tetradecyltrimethylammonium bromide
C_{16}TAB hexadecyltrimethylammonium bromide
ChCl choline chloride
DEAC N,N-diethylethanol ammonium chloride
DEG diethylene glycol
EG ethylene glycol
LA lactic acid
MA malonic acid
GA glutaric acid
Gluc glucose
Gly glycerol
PD 1,2-propanediol
SDS sodium dodecylsulfate
Suc sucrose
TG triethylene glycol
U urea

1. Introduction

Deep eutectic solvents (DES) are prepared by mixing a hydrogen bond acceptor (HBA) and a hydrogen bond donor (HBD), such as choline chloride and urea, respectively. At the eutectic molar ratio these mixtures have a melting point much lower than those of the individual components. DES are often rich in charge and H-bonding functionality, which makes most DES extremely hygroscopic. Latent absorbed water is unavoidable and impacts the physicochemical properties of DES, such as the density, melting point, viscosity and conductivity (Bobrova, Danilov, & Protsenko, 2016; Dai, Witkamp, Verpoorte, & Choi, 2015; Ghaedi et al., 2017; Kuddushi, Nangala, Rajput, Ijardar, & Malek, 2019; Ma, Laaksonen, Liu, Lu, & Ji, 2018; Siongco, Leron, & Li, 2013; Yadav, Kar, Verma, Naqvi, & Pandey, 2015; Yadav, Trivedi, Rai, & Pandey, 2014). Physicochemical changes give clues as to the effect of added water on the DES structures, which can be studied using techniques such as simulations, NMR, FTIR and neutron scattering (Fetisov et al., 2018; Hammond, Bowron, & Edler, 2017; Pandey & Pandey, 2014; Posada et al., 2019; Roldán-Ruiz, Jiménez-Riobóo, Gutiérrez,

Ferrer, & del Monte, 2019; Shah & Mjalli, 2014; Zhekenov, Toksanbayev, Kazakbayeva, Shah, & Mjalli, 2017).

Amphiphile behavior in DES is still a relatively unexplored field. Preliminary investigations suggest many factors contribute toward surfactant and lipid self-assembly in DES, including solvophobicity, Gordon parameter, freedom of movement of the DES components and water content (Arnold et al., 2015; Bryant, Atkin, & Warr, 2016, 2017; McCluskey et al., 2019; Pal et al., 2014; Pal, Singh, & Pandey, 2015; Sanchez-Fernandez, Arnold, et al., 2016; Sanchez-Fernandez, Edler, et al., 2016; Sanchez-Fernandez, Hammond, et al., 2017; Sanchez-Fernandez, Hammond, et al., 2018; Sanchez-Fernandez, Moody, et al., 2018).

Here we discuss the effects of water on the physicochemical properties of DES, and how it can be linked to DES nanostructure. We also detail the interactions of surfactants and biological lipids with DES and how water affects self-assembly in these solvents. The most widely reported DES in literature are 1ChCl:2Urea (ChCl:U), 1ChCl:2Glycerol (ChCl:Gly) and 1ChCl:2Ethylene Glycol (ChCl:EG). The main discussion revolves around these three DES, but where data are available, we also include others such as ChCl with acids, sugars and metal salts and other HBAs with glycerol and ethylene glycol. Due to the hygroscopic nature of polar DES such as these, significantly different experimental values for various properties are reported for the same DES from different groups: e.g., melting points of ChCl:U reported by Abbott, Capper, Davies, Rasheed, and Tambyrajah (2003) and Meng, Ballerat-Busserolles, Husson, and Andanson (2016); or viscosities of ChCl:U by Xie, Dong, Zhang, Lu, and Ji (2014) and Yadav et al. (2014). Therefore, careful attention should be paid to experimental methods, preparation routes and water content of DES.

2. Effect on physical properties

Studies of the effects of water on DES focus on changes in melting point, density, viscosity and electrical conductivity, to understand the interactions of water with DES constituents. We summarize literature on these four important physicochemical properties, along with other properties such as enthalpy of mixing.

a. *Melting Point*: Melting point is a defining characteristic of DES. The significant lowering of melting points in DES result from non-ideal mixing, typically from H-bonding between the constituent molecules. Therefore, it is greatly affected by H-bonding additives like water.

Prior to Meng et. al's (Meng et al., 2016) work, the eutectic temperature for ChCl:U was widely accepted as 12 °C (285 K) (D'Agostino et al., 2015; Shah & Mjalli, 2014) with some studies finding a melting temperature of 17 °C (290 K) (Morrison, Sun, & Neervannan, 2009). By carefully monitoring the water content, Meng et al. found the melting point of pure ChCl:U to be 25 °C (298 K) and showed that addition of 5 wt% water decreased the melting point by 10 °C. This drastic change in the melting point with water content can explain the differences between previously reported data, as water content was not specified or controlled in those studies. Meng et al. also showed that the melting point decreases linearly with concentration of water up to 10 wt% where a melting point of ~0 °C (273 K) is observed. Studies of the effect of water on ChCl:U over a wider concentration range by Smith, Arroyo, Lopez Hernandez, and Goeltz (2019) show a similar effect, albeit their starting point is 12 °C (285 K), consistent with Abbott et al. (2003), possibly due to moisture present in their "pure" DES. They found a minimum melting point at 67 mol% water (−48 °C (225 K)) before it gradually increased to the melting point of ice (0 °C (273 K)) at 100 wt% water. This is indicative of a transition from DES-with-water to a DES-components-in-water behavior. Systematic studies on other DES-water mixtures are limited, however, since the dip in melting point is attributed to the H-bonding capability of water, similar effects are expected. Studies on ChCl + acid-based DES (Itaconic, L-(+)-Tartaric, 4-Hydrobenzoic, Caffeic, p-Coumaric & *trans*-Cinnamic acid) show a significant lowering of melting point (5–45 °C) upon the addition of 14–33 mol% of another H-bonding solvent, glycerol (Maugeri & Domínguez De María, 2012). In some cases, e.g., suberic acid/gallic acid with ChCl DES (Maugeri & Domínguez De María, 2012), addition of glycerol is essential for formation of a liquid DES at ambient temperatures. Similarly, water is sometimes added to aid preparation of certain DES (Hammond, Bowron, & Edler, 2019). Thus addition of water or glycerol decreases the melting point of DES, probably due to modified H-bond interactions.

b. *Density*: Density plays a key role in understanding DES interactions with water and also in structural and chemical processes, which is understood better by using the excess molar volume (V_E). V_E for a two-component system is calculated from the density of the mixture (ρ_{mix}), the densities (ρ_1 & ρ_2), the mole fractions (x_1 & x_2) and the molecular weights (M_1 & M_2) of individual components using the equation:

$$V_E = \frac{x_1 M_1 + x_2 M_2}{\rho_{mix}} - \frac{x_1 M_1}{\rho_1} - \frac{x_2 M_2}{\rho_2}$$

The density of most DES ranges from 1.1 to 1.2 g/cm^3 barring some DES composed of metal salts where densities of 1.4–1.8 g/cm^3 are observed (Abbott, Capper, Davies, & Rasheed, 2004; Bobrova et al., 2016; Hammond et al., 2019; Marcus, 2017; Qin et al., 2019). Addition of water to DES results in a gradually decreasing density which asymptotically plateaus to the density of water ($\rho_w = 1.0$ g/cm^3). However, V_E over the complete compositional range ($x_{DES} = 0$–1) gives more insight into the mixing of DES with water, showing curves with one or three turning points, "u"-shaped or "w"-shaped, respectively. Table 1 summarizes the density variation with water for various DES; the dry DES density (ρ_{DES}), shape of the V_E vs x_{DES} curve, position and value of the V_E minimum are reported.

A negative V_E could arise from the disruption of DES-DES and H_2O-H_2O interaction in the initial state by addition of water. Strong H-bonding interactions between water and DES components, the interstitial accommodation of H_2O molecules within the H-bonded DES network, or a combination of the two can occur (Kuddushi et al., 2019). These interactions are dependent on the DES composition, molecular size and water concentration. Table 1 shows that the minimum in V_E for all DES occurs at $x_{DES} = 0.25$–0.4, i.e., in the water-rich region. This suggests there is a point where the H-bonding between water-DES components is optimized, lying in the same DES concentration range as the minima in melting point. This could arise when the HBA (e.g., choline ion) is maximally hydrated, to enhance intermolecular H-bonding, over bonding between DES components (in the DES-rich region) or H_2O-H_2O interactions (in the water-rich region).

c. *Viscosity*: Viscosity is an important transport property and DES often have extremely high viscosities which poses difficulty in their practical application, leading to time-consuming solvent transfer operations and slow mass transfer. This can be overcome by either increasing the temperature or, where temperature increase is not feasible (e.g., due to stability issues), by adding water. Addition of water can change viscosity drastically, sometimes by orders of magnitude, generally showing a power-law type dependence (Kuddushi et al., 2019; Shah & Mjalli, 2014). For ChCl:U, the viscosity is reduced by up to an order of magnitude upon addition of 5 wt% water and up to two orders of magnitude

Table 1 Density and excess volumes of DES-H_2O binary systems.

DES	ρ_{DES} at 303 K (g/cm^3)	Temp range (K)	V_E shape	V_E min [low T, high T] (cm^3/mol)	x_{DES} at min	Reference
ChCl:U (1:2)	1.1952, 1.1945, 1.216, 1.1946	293–363	u	[−0.2, −0.1]	0.3	Shah and Mjalli (2014), Shekaari, Zafarani-Moattar, and Mohammadi (2017), Xie et al. (2014), Yadav et al. (2014)
ChCl:Gly (1:2)	1.1885, 1.1885, 1.1884	283–363	u	[−0.4, −0.3]	0.35–0.4	Kim and Park (2018), Leron, Soriano, and Li (2012), Yadav et al. (2014)
ChCl:EG (1:2)	1.1142, 1.1114	283–363	u	[−0.33, −0.25]	0.4	Leron et al. (2012) and Yadav et al. (2015)
ChCl:TG (1:2)	1.1254	303–353	u	[−0.9, −0.7]	0.3	Mjalli and Ahmad (2017)
ChCl:Gluc (2:1)	1.2397	303–353	w	[−0.75, −0.7], [−0.6, −0.58]	0.7, 0.3	Mjalli and Ahmad (2017)
ChCl:MA (1:1)	1.1899, 1.2283	293–323	u	[−0.45, −0.35]	0.3	Kuddushi et al. (2019) and Lin, Leron, Caparanga, and Li (2014)
ChCl:MA (1:2)	1.1786	283–363	u	[−0.27, −0.18]	0.3	Yadav et al. (2015)

ChCl:GA (1:1)	1.1804	293–323	u	[−0.38, −0.22]	0.3	Kuddushi et al. (2019)
DEAC:Gly (1:2)	1.1735	298–343	u	[−0.45, −0.4]	0.4	Siongco, Leron, and Li (2013)
DEAC:EG (1:2)	1.0968	298–343	u	[−0.54, −0.45]	0.4	Siongco, Leron, and Li (2013)
ATPPB: DEG (1:4)	1.1938	293–343	w	[−0.72, −0.44]	0.25, 0.85	Ghaedi et al. (2017)
ATPPB: DEG (1:10)	1.1526	293–343	u	[−0.66, −0.44]	0.3	Ghaedi et al. (2017)
ATPPB: DEG (1:16)	1.1391	293–343	u	[−0.74, −0.53]	0.3	Ghaedi et al. (2017)

upon the addition of 20 wt% water, depending on temperature (Yadav et al., 2014). The effect on ChCl:Gly is smaller and ~15 wt% water is required to reduce the viscosity by an order of magnitude and ~40 wt % for two orders of magnitude (Yadav et al., 2014). Similar effects are observed for other DES, e.g., ChCl:EG but the magnitude of the reduction is dependent on the viscosity of the dry DES (Mjalli & Mousa, 2017). As with density, to further illustrate the effect of water on DES viscosity, excess viscosity (η_E) or excess logarithmic viscosity ($\ln \eta_E$) are calculated. These are defined as:

$$\eta_E = \Delta\eta = \eta_{mix} - x_1\eta_1 - x_2\eta_2$$

$$\ln \eta_E = \ln \eta_{mix} - x_1 \ln \eta_1 - x_2 \ln \eta_2$$

where η_{mix} is the viscosity of the mixture and x_n and η_n are the mole fraction and viscosity of component "n," respectively. η_E can be either positive or negative and the shape of η_E vs x_{DES} curve (one or two turning points) gives important information about DES-water interactions, with the position of the extremum detailing the water content which has the largest effect on transport properties. Fig. 1A shows the dependence of dynamic viscosity on x_{DES} (x_1 in the figure) for ChCl:U/water mixtures at temperatures ranging from 293 to 363 K with Fig. 1B depicting $\ln \eta_E$ variation with x_{DES}; exhibiting u-shaped curves for $T < 343$ K, s-shaped for $T = 343$ K and n-shaped for $T > 343$ K.

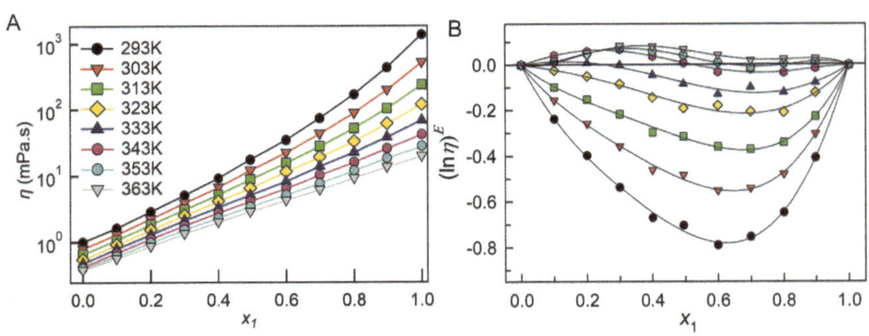

Fig. 1 Dynamic viscosity (A) and $\ln \eta_E$ (B) as a function of ChCl:U mole fraction (x_1) in ChCl:U/water mixtures at different temperatures. Legend in (A) also applies to (B). *Reproduced with permission from Yadav, A., Trivedi, S., Rai, R., & Pandey, S. (2014). Densities and dynamic viscosities of (choline chloride+glycerol) deep eutectic solvent and its aqueous mixtures in the temperature range (283.15-363.15) K.* Fluid Phase Equilibria, 367, *135–142. doi:10.1016/j.fluid.2014.01.028. Copyright 2014 American Chemical Society.*

Table 2 summarizes the viscosity and excess (logarithmic) viscosity for the most widely reported DES along with shape of the η_E or ln η_E curve and the position of the extremum.

Negative deviations in viscosity (η_E) may be due to unequal DES and water molecular sizes. In addition, large negative η_E values could signify weaker H-bonding between the DES constituents due to hydration by water molecules, with the resulting more labile H-bonds with water increasing fluidity of the solution (Mjalli & Mousa, 2017). This behavior is very similar to that found for addition of water to many ionic liquid mixtures (Carvalho, Regueira, Santos, Fernandez, & Coutinho, 2010; Gómez-Díaz, Navaza, & Sanjurjo, 2007). In most studies the extremum in excess viscosity occurs at $x_{DES} = 0.3–0.4$, which is approximately one water molecule per choline ion (HBA) and may be significant (D'Agostino et al., 2015). The positive values of ln η_E observed for ChCl:Gly and ChCl:U for $T > 343$ K suggest stronger interspecies H-bonding between water and ChCl:Gly/ChCl:U as compared to H-bonding among water or DES components, whereas negative values of ln η_E, observed for ChCl:U at $T < 343$ K, indicate that the intraspecies interactions are disrupted, and the interspecies interactions are not significant (Yadav et al., 2014).

d. *Electrical Conductivity*: Electrical conductivity provides information about the mobility of ions and charges, so is an important transport property for applications such as electroplating. The conductivity of aqueous solutions of DES are lower than those of strong electrolyte solutions, such as aqueous solutions of ionic salts (NaCl, Na_2SO_4), but higher than those of weak electrolytes, such as solutions of organic acids (CH_3COOH) and similar to those of ionic liquids (Ma et al., 2018). Unlike traditional electrolytes where electrical conductivity increases with increasing electrolyte concentration, the electrical conductivity of aqueous DES solutions initially increases and thereafter decreases as the DES concentration in the solution increases, showing a maximum in the water-rich region, at $x_{DES} = 0.1$ or weight fraction; $w_{DES} = 0.2–0.4$ for different DES. Their behavior is akin to ionic liquids which also show a maximum electrical conductivity at molar concentrations of 0.02–0.1 (Ma et al., 2018). Table 3 summarizes properties related to electrical conductivity for DES-water systems. The conductivity of dilute aqueous DES solutions at their maximum can be 10–100 times that of the dry DES (Shah & Mjalli, 2014; Siongco, Leron, Caparanga, & Li, 2013), attributed to a combined effect of viscosity reduction and increased mobility of ions.

Table 2 Viscosities and excess viscosities for DES-H_2O binary systems.

DES	Viscosity at 303 K (mPa s)	Viscosity at $x_{DES} = 0.5$ at 303 K (mPa s)	Temp range (K)	η_E (ln η_E) shape	Extremum x_{DES}	Reference
ChCl:U (1:2)	530–552	60–66	293–363	u, $T < 343$ K s, $T = 343$ K n, $T > 343$ K	0.60.3, 0.7 0.4	D'Agostino et al. (2015), Mjalli and Mousa (2017), Shah and Mjalli (2014), Xie et al. (2014), Yadav et al (2014)
ChCl:Gly (1:2)	188–250	50–70	283–363	n	0.35–0.4	D'Agostino et al. (2015), Mjalli and Mousa (2017), Yadav et al. (2014)
ChCl:EG (1:2)	32	14	283–363	u	0.3	D'Agostino et al. (2015), Mjalli and Mousa 2017), Yadav et al. (2015)
ChCl:LA (1:2)	372	91 ($x_{DES} = 0.65$)	298–358	–	–	Francisco, van den Bruinhorst, Zubeir, Peters, and Kroon (2013)
ChCl: $CrCl_3 \cdot 6H_2O$ (2.5:1)	1960 (313 K)	~212 (313 K)	298–353	–	–	Protsenko, Bobrova, and Danilov (2017)
ChCl: $CrCl_3 \cdot 6H_2O$ (1:2)	307.6 (333 K)	~26 (333 K)	398–353	–	–	Bobrova et al. (2016)
DEAC:Gly (1:2)	351	59	298–343	u	0.5	Siongco, Leron, and Li (2013)
DEAC:EG (1:2)	41	16	298–343	u	0.3	Siongco, Leron, and Li (2013)
ChCl:Gluc (5:2)[a]	479[a] (313 K)	7.5[a] ($w_{DES} = 0.75$)	313	–	–	Dai et al. (2015)
ChCl:PD (1:1)[a]	36[a] (313 K)	6.2[a] ($w_{DES} = 0.75$)	313	–	–	Dai et al. (2015) and Dai, van Spronsen, Witkamp, Verpoorte, and Choi (2013)

Table 3 Electrical conductivity for DES-H_2O binary systems.

DES	Conductivity at 303 K (mS cm^{-1})	Temp range (K)	x_{DES}/w_{DES} range	Max x_{DES}/w_{DES}	Max conductivity	Reference
ChCl:Urea (1:2)	2.31	293–363	0–1	$x_{DES}=0.1$	59.70	Shah and Mjalli (2014)
DEAC:Gly (1:2)	0.98	298–343	0–1	$x_{DES}=0.1$	36.3	Siongco, Leron, Caparanga, and Li (2013)
DEAC:EG (1:2)	5.65	298–343	0–1	$x_{DES}=0.1$	44.1	Siongco, Leron, Caparanga, and Li (2013)
ChCl:$CrCl_3 \cdot 6H_2O$ (2.5:1)	0.28	298–353	0.34–1	–	–	Protsenko et al. (2017)
ChCl:$CrCl_3 \cdot 6H_2O$ (1:2)	0.36	298–353	0.39–1	–	–	Bobrova et al. (2016)
ChCl:Gluc (5:2)	6.81 (300 K)	300	0–1	$w_{DES}=0.4$	~115[a]	Dai et al. (2015)
ChCl:PD (1:1)	12.09 (300 K)	300	0–1	$w_{DES}=0.4$	~120[a]	Dai et al. (2015)
Suc:ChCl (4:1)	3.39 (300 K)	300	0–1	$w_{DES}=0.4$	~106[a]	Dai et al. (2015)
Proline:Malic Acid (1:1)	1.06 (300 K)	300	0–1	$w_{DES}=0.4$	~13.5[a]	Dai et al. (2015)
LA:Gluc (3:1)	0.11 (300 K)	300	0–1	$w_{DES}=0.2$	~8.5[a]	Dai et al. (2015)

[a]Read from the graph in Dai et al. (2015).

e. *Other Properties*: Other important properties of DES include refractive index, speed of sound in the DES, mixing enthalpy, isentropic compressibility and interfacial tension. While these properties are extremely important for both fundamental understanding and application contexts of DES, the studies on the impact of water on these properties are less extensive than on those discussed above. Refractive index and speed of sound can be inferred from other properties, namely density and viscosity. Studies show that the refractive index (ChCl:U (Shah & Mjalli, 2014) & ChCl:Gly, ChCl:EG and 1ChCl:1MA (Ma et al., 2017)) and speed of sound (ChCl:U (Shah & Mjalli, 2014), 1ChCl:1MA & 1ChCl:1GA (Kuddushi et al., 2019)) decrease rapidly with increasing water concentration in DES + H_2O mixtures, akin to density. Studies on mixing enthalpy (H_m) for DES + H_2O mixtures show more interesting trends; upon increasing the DES content H_m can both increase and decrease, with ChCl:Gly and ChCl:EG showing a "u" type curve, with a minimum at $x_{DES} = 0.4$ (Ma et al., 2017), while ChCl:U and 1ChCl:1MA show "s" type curve with minima at $x_{DES} = 0.7$ & $x_{DES} = 0.4$ and maxima at $x_{DES} = 0.0.14$ & $x_{DES} = 0.05$, respectively (Ma et al., 2016, 2017). This suggests that mixing of water with DES is generally endothermic, but in some cases can be exothermic at high water contents. Studies on the effect of water on interfacial tension for DES are scarce, however investigations of ionic liquid + H_2O mixtures show significant variation of surface tension with water content (Tariq et al., 2012; Wang, Zhang, Liu, & Hu, 2011). Studies of ChCl:$CrCl_3 \cdot 6H_2O$ show a decrease in interfacial tension of the DES as the number of moles of water (n_w) is increased: a 2.5:1 DES showed a decrease from 74 to 64 mN/m (Protsenko et al., 2017) while a 1:2 DES showed a decrease from 82 to 50 mN/m (Bobrova et al., 2016) as n_w was increased from 6 to 18. It is therefore also expected that water will alter the surface tension of other DES, which should be investigated. It must be noted that these studies can be challenging as some water absorption from the atmosphere can be expected during the course of the experiment.

3. Nanostructure

For pure DES, a 3D intermolecular H-bonding network has been experimentally confirmed by many studies (Hammond, Bowron, & Edler, 2017; Hammond, Bowron, Jackson, et al., 2017; Turner &

Holbrey, 2019) and is now the most widely accepted form of DES structure. This structure can be used to explain the dip in melting point at the eutectic ratio, the high viscosity and other physicochemical properties of these solvents. Addition of water is expected to disrupt this 3D H-bonding structure. Molecular dynamics (MD) simulations for ChCl:U, ChCl:Gly and ChCl:EG by Zhekenov et al. (2017) suggest that water molecules in small amounts (<5 wt%) are absorbed in the molecular matrix of the DES, forming H-bonds with the ions and HBDs without affecting diffusion of ions. At higher concentrations water dampens intermolecular and intramolecular interactions, which result in significant changes in the physicochemical properties. In ChCl:U Fetisov et al. (2018) suggest, based on first principles MD simulations, that addition of water breaks strong H-bonds between urea and chloride ions by preferentially hydrating chloride ions and forming urea-water H-bonds. MD simulations by Shah et al. on ChCl:U (Shah & Mjalli, 2014) also show that the chloride anion is preferentially hydrated by water compared to urea or choline ions. They classified the effect of water into three regimes. In the low water regime (<5 wt%) urea-urea and urea-choline H-bonding show a non-monotonic behavior with a maximum at $\sim$2.5 wt%. Thereafter, a gradual decrease in the number of H-bonds is observed. Similar behavior was seen by Zhekenov et al. for ChCl:U, ChCl:Gly and ChCl:EG, who also observed an increase in the number of H-bonds between choline/HBD/chloride and water as water content in the mixture is increased. At intermediate water fraction (5–25 wt%) the components of ChCl:U are hydrated with water but the anion and urea both have low diffusivity which results in a gradual change in physical properties, whereas at high water fractions (>25 wt%) both urea and chloride ions have high diffusivity resulting in drastic changes in transport properties. These are consistent with other simulations performed on ChCl:U (Gao et al., 2018; Kumari, Shobhna, Kaur, & Kashyap, 2018), which suggest that at lower water content the ion pairing of cation and anion is important whereas at high water contents it is the hydration of ions that plays the more important role. In the case of ChCl:Gly and ChCl:EG the MD simulations suggest that ChCl prefers to form H-bonds with water instead of the HBD (Ahmadi et al., 2018; Kaur, Gupta, & Kashyap, 2020), with some hydration of the polar groups around glycerol/EG at high water contents leading to self-association of the alkyl groups (Celebi, Vlugt, & Moultos, 2019; Zhekenov et al., 2017).

In addition to the simulations, experimental studies have also been conducted on ChCl:U, ChCl:Gly and ChCl:EG to understand the effect of water on DES structure. Pandey et al. studied the effect of water on solvatochromatic probe behavior in ChCl:U, ChCl:Gly and ChCl:EG using FTIR and Raman spectroscopy (Pandey & Pandey, 2014). In ChCl:U the addition of water helps to form interspecies H-bonding but does not greatly alter the overall H-bonding landscape. They suggest that in ChCl: U, interstitial accommodation of water appears more important whereas in ChCl:Gly and ChCl:EG interspecies H-bonding dominates. Neutron scattering studies done by Edler's group (Hammond, Bowron, & Edler, 2017) on the effect of water (25–91 mol%; 6.5–67.5 wt%) show that ChCl:U can be somewhat hydrated without loss of structural properties, and that there is a water-in-DES to DES-in-water transition point. They did not find any nanostructured water domains as suggested by NMR studies (D'Agostino et al., 2015; Posada et al., 2017), and, as with the pure DES structural data (Hammond, Bowron, & Edler, 2016), the picture is one of local ordering rather than any regular, segregated long-range ordering, consistent with the MD simulations discussed above. Snapshots of the molecular model for ChCl:U at various water contents are shown in Fig. 2. At low water contents (25 mol%; 6.5 wt%) the DES structure undergoes only slight disruption, and certain intermolecular interactions increase in strength, such as choline-urea H-bonding, which can explain the sigmoidal viscosity vs DES mole fraction curve for ChCl:U discussed above. This suggested that the innate water content of DES is not a huge issue for the field, as long as authors make efforts to measure and report it for their system.

In the hydrated DES regime, water plays the role of an additional HBD/HBA which results in a lowering of melting point as compared to

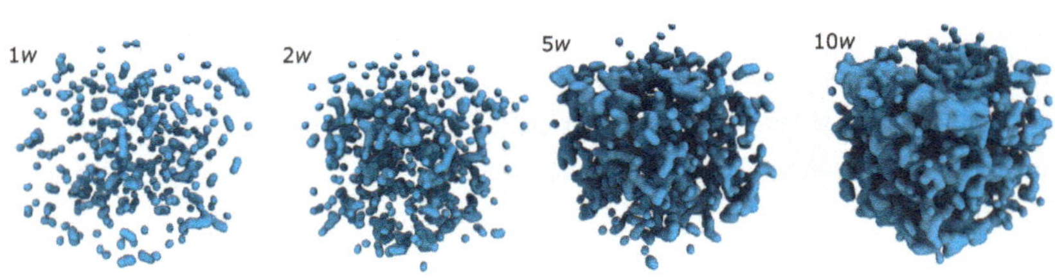

Fig. 2 Model snapshots from Empirical Potential Structure Refinement (EPSR) structures derived from work in Hammond et al. (2016), showing transient clustering of water molecules for four different molar water concentrations in ChCl:U: 1w (6.5 wt%), 2w (12.2 wt %), 5w (25.7 wt%) and 10w (41 wt%). Water molecules only are shown, depicted as blue isosurfaces rather than molecular structures. Individual dots represent single molecules, whereas large blots and chains depict H-bonded clusters.

the "pure" DES or water; an effect observed in ChCl:U (Hammond, Bowron, & Edler, 2017), ChCl:MA (Hammond, Bowron, Jackson, et al., 2017) and 3 Resorcinol:2 Urea:1 ChCl (López-Salas et al., 2019). At higher water concentrations a discontinuity was observed at 41–51 wt% water (77–83 mol%) in ChCl:U; below which the DES interactions systematically weakened but remained intact (Hammond, Bowron, & Edler, 2017). Above this, water-water interactions plateau and the length scale of these is the same as in aqueous solution. This was interpreted as a "transition point" where a hydrated DES becomes an aqueous solution of DES components, at a water concentration (41–51 wt%; ~50 vol%) where water structure would naturally begin to percolate throughout the system. Brillouin and NMR spectroscopic studies by del Monte's group suggest a similar transition is observed for ChCl:MA (Roldán-Ruiz et al., 2019) and 1.25 Resorcinol:1 Hexylresorcinol:1 Tetraammonium Bromide (Posada et al., 2019) at DES concentration of ~70 wt%.

These structural studies are consistent with physicochemical investigations which also indicate a point at which the H-bonding between DES and water components is maximally optimized, which lies in the region of 60–80 mol% water, depending on the DES.

4. Amphiphiles and self-assembly

Deep eutectic solvent nanostructure and the presence of water can also influence amphiphile behavior and micelle formation (Sanchez-Fernandez, Edler, et al., 2016). Table 4 shows the criticial aggregation concentrations (CAC) for a range of surfactants in various DES with different water contents.

Pal and co-workers found that non-ionic surfactants (Triton X-100, Brij-35, and Tween-80) were insoluble in ChCl:U, while cationic surfactants (alkyltrimethylammonium bromides (C_nTAB)) were only sparingly soluble and anionic surfactants (sodium dodecyl sulfate (SDS) and sodium dodecylbenzenesulfonate) were soluble up to 170 and 80 mM, respectively (Pal et al., 2014). Further characterization in ChCl:U/water mixtures (5, 15, and 50 wt%) showed that the CAC, measured by pyrene fluorescence, decreased with increasing water content to well below that of SDS in water. This is probably due to free DES ions shielding electrostatic repulsion between SDS molecules, facilitating aggregation at lower concentrations, as expected in the presence of salt. At low water concentrations, DES ions are not sufficiently free to interact with the surfactant molecules.

Table 4 Critical aggregation concentration (CAC) of amphiphiles in DES and DES/water mixtures.

Deep eutectic solvent	Surfactant	CAC (mM)
ChCl:U + Water (5 wt%)	Sodium dodecylsulfate (SDS)	13.9 (±1.0)[a]
ChCl:U + Water (15 wt%)	SDS	9.2 (±0.8)[a]
ChCl:U + Water (50 wt%)	SDS	3.8 (±0.7)[a]
Water	SDS	10.0 (±0.5)[a] 8.2 (±0.2) (25 °C (298 K))[b]
ChCl:U	SDS	2 (±1) (30–50 °C (303 − 323 K))[c]
ChCl:U	Lithium dodecylsulfate	2.4 (±0.2)[d]
ChCl:U	Emim dodecylsulfate	1.8 (±0.2)[d]
ChCl:U	Choline dodecylsulfate	1.8 (±0.1)[d]
ChCl:U	$Mg_{1/2}$ dodecylsulfate	1.2 (±0.1)[d]
ChCl:U	Cesium dodecylsulfate	1.9 (±0.1)[d]
ChCl:U	Bmim dodecylsulfate	1.3 (±0.2)[d]
ChCl:Gly	Lithium dodecylsulfate	4.2 (±0.3)[d]
ChCl:Gly	Emim dodecylsulfate	3.7 (±0.1)[d]
ChCl:Gly	Choline dodecylsulfate	2.3 (±0.2)[d]
ChCl:Gly	$Mg_{1/2}$ dodecylsulfate	2.1 (±0.2)[d]
ChCl:Gly	Cesium dodecylsulfate	3.1 (±0.2)[d]
ChCl:Gly	Bmim dodecylsulfate	2.1 (±0.1)[d]
ChCl:Gly	Dodecyltrimethylammonium bromide ($C_{12}TAB$)	>100 (25 °C (298 K))[e], 22 (±2) (40 °C (313 K))[f]
ChCl:Gly	Tetradecyltrimethylammonium bromide ($C_{14}TAB$)	20.9 (25 °C (298 K))[e], 3.9 (±0.1) (40 °C (313 K))[f]
ChCl:Gly	Hexadecyltrimethylammonium bromide ($C_{16}TAB$)	15.2 (25 °C (298 K))[e], 16.1 (40 °C (313 K))[e], 0.9 (±0.1) (40 °C (313 K))[f]

Table 4 Critical aggregation concentration (CAC) of amphiphiles in DES and DES/water mixtures.—cont'd

Deep eutectic solvent	Surfactant	CAC (mM)
ChCl:Gly	Hexadecyltrimethylammonium chloride (C_{16}TAC)	15.4 (25 °C (298 K))[e]
ChCl:Gly	1-Hexadecyl-3-methylimidazolium chloride (C16mimCl)	25.7[g]
Water	C_{12}TAB	15.0 (30 °C)[e], 15.0 (40 °C (313 K))[h]
Water	C_{14}TAB	3.83 (30 °C)[e], 3.51 (30 °C (303 K))[i]
Water	C_{16}TAB	1.00 (30 °C)[e], 9.2 (30 °C (303 K))[j]
Water	C_{16}TAC	1.30 (30 °C (303 K))[e]
ChCl:MA (1:1)	C_{12}TAB	54 (±6)[k]
ChCl:MA (1:1)	C_{14}TAB	4.6 (±0.5)[k]
ChCl:MA (1:1)	C_{16}TAB	1.5 (±0.3)[k]
ChCl:MA + Water (1:1:2)	C_{16}TAB	1.0 (±0.1)[k]
ChCl:MA + Water (1:1:5)	C_{16}TAB	0.6 (±0.1)[k]
ChCl:MA + Water (1:1:10)	C_{16}TAB	0.26 (±0.07)[k]
ChCl:MA + Water (1:1:20)	C_{16}TAB	0.18 (±0.04)[k]

[a]Pal et al. (2014).
[b]Elworthy and Mysels (1966).
[c]Arnold et al. (2015).
[d]Sanchez-Fernandez, Hammond, et al. (2018).
[e]Pal et al. (2015).
[f]Sanchez-Fernandez, Arnold, et al. (2016).
[g]Tan et al. (2016).
[h]Schick (1963).
[i]Venable and Nauman (1964).
[j]Lu, Hromadova, Simister, Thomas, and Penfold (1994).
[k]Sanchez-Fernandez, Hammond, et al. (2017).
Measurements made at room temperature unless otherwise specified.

Micelle formation of SDS in ChCl:U was further explored by Arnold and co-workers, who characterized micelles in the dry DES (Arnold et al., 2015) and found a critical micelle concentration (CMC) of 2 mM, which is much lower than that of SDS in water. They propose that the CMC is primarily determined by the cohesive energy density, quantified by the Gordon Parameter and the π^* solvent polarizability parameter. For ChCl:U, the Gordon parameter is lower than that of water, but higher than many other solvents which facilitate micellisation, and the polarizability parameter is higher than that of water, thus driving micellisation and leading to a lower CMC. Characterization studies showed that while SDS micelles in water are spherical, in ChCl:U they are cylindrical, with a similar radius (14.7 ± 1.9 Å) but a length that increased with increasing SDS concentration up to 268 ± 12 Å at 17 mM. A follow-on study found that further increases to the SDS concentration resulted in a reversal of this trend. This was attributed to there being insufficient choline counterions to shield the sulfate headgroups, thus increasing headgroup area, decreasing packing parameter, and leading to more spherical assemblies (Sanchez-Fernandez, Edler, et al., 2016). This lack of counterions may be due to competition between interaction of choline with the surfactant and its place in the H-bond network of the solvent itself. Addition of water to this system resulted in more spherical aggregates.

The Edler group explored the effect of changing the counterion of dodecylsulfate surfactants on assembly in ChCl:U and ChCl:Gly. They found that counterions with stronger binding to the surfactant headgroup, or greater lyophobicity, result in lower CMCs (Sanchez-Fernandez, Hammond, et al., 2018). In almost all cases, CMCs were lowest in ChCl:U, followed by ChCl:Gly, with solutions in water having the highest CMCs. In ChCl:U, the micelles tended to be elongated in contrast to ChCl:Gly where they were globular. This was attributed to the higher solubility of surfactant counterions in ChCl:Gly compared to ChCl:U, as a result of the extensive glycerol network.

In contrast to the above anionic surfactants, cationic surfactants appear to have higher CMCs in DES compared to water. Alkyltrimethylammonium halides are more soluble in ChCl:Gly than ChCl:U, with CACs 10 times higher in ChCl:Gly than in water (Pal et al., 2015). The authors attributed this to weaker structuring of the DES around the surfactant monomers compared to water. As in water, increasing the tail length of the surfactant caused a decrease in CAC, although not to the same extent, while larger counterions led to lower CACs. However, subsequent work by Edler and co-workers found CMCs for these surfactants that were an order of magnitude smaller, and in fact quite similar to the CMCs in water (Sanchez-Fernandez, Arnold, et al., 2016). These micelles were prolate ellipsoids that were larger and more elongated than those formed in water. Studies by the same group on cationic surfactants in carboxylic acid-based DES found that the CMCs of C_nTABs were slightly higher in ChCl:MA compared to ChCl:U or water, suggesting greater solubility (Sanchez-Fernandez, Hammond, et al., 2017). Adding water to ChCl:MA caused a decrease in the CMC, even below the CMC in pure water. In these cases the DES may act more like an electrolyte, shielding electrostatic repulsion between surfactant headgroups, thus promoting aggregation and decreasing the CMC. Longer alkyl chains resulted in formation of more highly elongated micelles in the DES compared to those for the same surfactant in water. Addition of water resulted in a transition from elongated to more globular micelles. As with SDS in ChCl:U, this was attributed to a reduction in headgroup screening, and thus a smaller packing parameter after water addition.

More complex phase behavior has been observed for hexadecylpyridinium bromide in ChCl:Gly and ChCl:EG as shown in Fig. 3 (Li et al., 2018). They observed not only a micellar phase (L_1), but also hexagonal (H_1), lamellar (L_α)

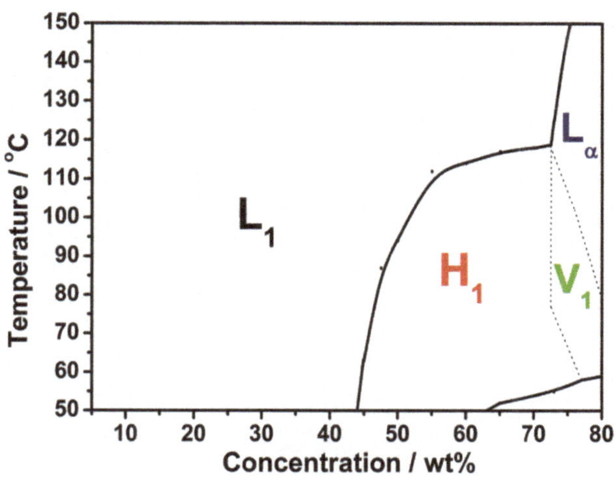

Fig. 3 Phase diagram of hexadecylpyridinium bromide in ChCl:Gly. *Reproduced from Li, Q., Wang, J., Lei, N., Yan, M., Chen, X., & Yue, X. (2018). Phase behaviours of a cationic surfactant in deep eutectic solvents: From micelles to lyotropic liquid crystals.* Physical Chemistry Chemical Physics, 20*(17), 12175–12181. doi:10.1039/C8CP00001H with permission from the PCCP Owner Societies.*

and cubic phases (V_1), depending on surfactant concentration and temperature in both DES. This is the same phase sequence observed in water. As discussed for SDS in ChCl:U, the Gordon parameter estimates the cohesive energy density, and in general a higher Gordon parameter is related to a greater phase diversity. The authors used this to explain why so many phases are observed for ChCl:Gly, ChCl:EG, and water, but not for the ionic liquid tested (ethylammonium nitrate). The critical concentration for formation of the hexagonal phase in these solvents followed the same order as the Gordon parameters, although they did not report CMC values.

Investigations into other types of surfactants in DES are more limited. Zwitterionic surfactants had higher CMCs in ChCl:Gly than in water and the CMC depended on both the headgroup and tail length (Sanchez-Fernandez, Moody, et al., 2018). Unlike cationic surfactants, zwitterionic surfactants formed globular, rather than elongated micelles, indicating that they were less affected by solvent screening and solvent-headgroup interactions. Rengstl et al. reported aggregation of the unusual surfactant choline dodecylsulfate in choline glutarate:choline chloride:urea DES, with SAXS at low temperatures showing a crystal peak at 3.07 nm^{-1} which dissolved, forming large aggregates above 50 °C (323 K) (Rengstl, Fischer, & Kunz, 2014). CMCs were also measured by Tan et al. for $C_n$mimCl surfactant ionic liquids (n = 8–16) in ChCl:Gly and found to be higher than in water. Teubner and Strey analysis of SAXS data of concentrated micellar solutions gave periodicities of 10.6, 8.2 and 7.5 nm for C_{16}mimCl solutions of 5, 20, and 30 wt%, respectively (Tan et al., 2016).

The first vesicles observed in a DES were made by freeze drying an aqueous solution of dilute ChCl:U or ChCl:thiourea containing unilamellar vesicles made from 1,2-dimyristoyl-sn-glycero-3-phosphocholine (DMPC) (Gutiérrez, Ferrer, Mateo, & del Monte, 2009). This removed water, leaving the vesicles suspended in DES. The authors stated that water is required for vesicle formation, however this was later disproved by Bryant et al. who demonstrated spontaneous vesicle formation by a number of lipids in ChCl:U (Bryant et al., 2016). The transition temperature of the lipids was higher in ChCl:U compared to water and this effect was further explored in subsequent work (Bryant et al., 2017). The follow-up research systematically looked at the effect of DES nanostructure on lipid behavior. Alkylammonium chloride/bromide DES with either glycerol or EG were tested against a range of lipids. As the alkyl chain length of the DES increased, the lipid transition temperature also increased, due to increased nanostructure in the DES which makes it harder for the lipids to swell. However, butyl- and pentylammonium

DES resulted in a decrease in transition temperature because the DES chain is sufficiently long that it acts as a cosurfactant (Bryant et al., 2017). Unlike previous work using ionic surfactants, this research did not find trends consistent with changes in the Gordon parameter of the DES. Therefore, other factors such as nanostructure and solvophobic effects dominate amphiphile behavior in these systems. A further study using AFM also found that these DES supported liposome formation, finding that vesicle size increased as the DES became more hydrophobic (Eid, El Achkar, Fourmentin, Greige-Gerges, & Jraij, 2020).

As well as bulk lipid behavior, one study has examined lipid monolayers at the air-ChCl:Gly interface (McCluskey et al., 2019). They found that while the tail layer was much the same as at an air-water interface, the headgroup size of the phosphatidylglycerol (PG) lipid was bigger in ChCl:Gly than in water, an effect not seen for phosphatidylcholine lipids. This was attributed to the anionic nature of the PG headgroup, which interacts with the choline groups of the DES, as observed for anionic surfactants, leading to apparent swelling of the PG headgroup.

5. Conclusions

Deep eutectic solvents interact with both water and amphiphiles due to their H-bonding capabilities and polar nature. The presence of water leads to a change in the DES nanostructure, thus altering their physicochemical properties. Melting point, density and viscosity decrease with increasing water content whereas electrical conductivity increases. Excess molar volume and excess (logarithmic) viscosity show non-monotonic behavior, typically with a minimum in the water-rich region ($x_{DES} = 0.25–0.4$) suggesting an optimum water concentration to tune these interactions.

Structural studies, both experimental and simulations, confirm that water disrupts the H-bonding network in DES, replacing DES-DES interactions with DES-water interactions. Broadly two regimes occur: low water concentrations where water can be interstitially accommodated and contributes to the H-bonding; and high water concentrations where DES components are hydrated, molecular interactions are damped and a solution-like system is observed.

Based on the structural studies and trends in physicochemical properties, a schematic for the structure of a typical DES and how it changes at low (~20 vol%) and high (~50 vol%) water fractions is shown in Fig. 4 depicting the transition from water-in-DES behavior to DES-in-water behavior.

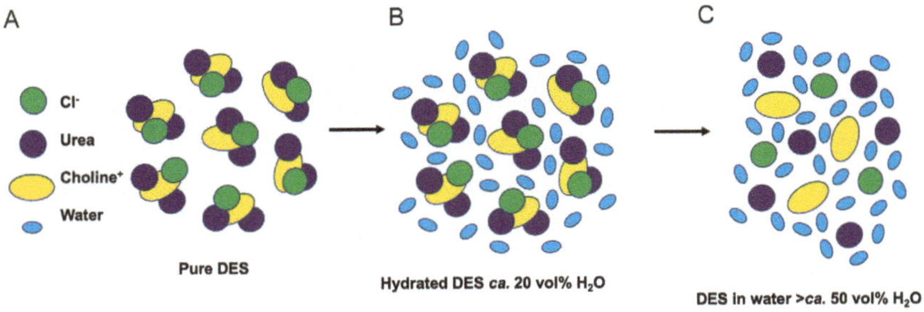

Fig. 4 Cartoon depicting the structure for ChCl:U (A) and the changes upon addition of water: (B) low water concentration regime (20vol%) and (C) high water concentration regime (50vol%).

It must be noted here that the DES concentration at transition may vary for different DES and different co-solvents; this will be a fruitful area for future study.

DES nanostructure and water content can also influence amphiphile behavior. A variety of amphiphiles; anionic, cationic, non-ionic, and biological lipids aggregate in DES with and without water, forming micelles (globular, spherical and elliptical) as well as more ordered hexagonal/cubic phases and vesicles or liposomes in case of lipids.

Addition of water not only affects the properties and nanostructure of DES but also is important in many applications: concentration of water can effect shape and size of metal oxide nanoparticles synthesized in DES (Hammond, Edler, Bowron, & Torrente-Murciano, 2017; Hammond, Eslava, Smith, Zhang, & Edler, 2017) and water content affects nickel nanostructures electrodeposited from ChCl:U (Cherigui et al., 2018). Amphiphile self-organization in DES is important for emulsification and in templating interactions, and is also critical to understand DES interactions with cells, and organs such as skin or mucus membranes to highlight mechanisms of toxicity (Cardellini et al., 2015; Hayyan et al., 2016), or the potential for using DES in transdermal delivery systems (Banerjee, Ibsen, Iwao, Zakrewsky, & Mitragotri, 2017). In addition, the burgeoning field of enzymatic reactions in DES would benefit from an improved understanding of solvophobic effects and specific ion binding in such solvents, since these will also control protein folding in these novel media (Sanchez-Fernandez, Edler, Arnold, Alba Venero, & Jackson, 2017). Therefore to understand the role of DES in applications from biology to materials science, one must first understand their interactions with water and amphiphiles.

References

Abbott, A. P., Capper, G., Davies, D. L., & Rasheed, R. K. (2004). Ionic liquid analogues formed from hydrated metal salts. *Chemistry—A European Journal*, *10*(15), 3769–3774. https://doi.org/10.1002/chem.200400127.

Abbott, A. P., Capper, G., Davies, D. L., Rasheed, R. K., & Tambyrajah, V. (2003). Novel solvent properties of choline chloride/urea mixtures. *Chemical Communications*, *9*(1), 70–71. https://doi.org/10.1039/b210714g.

Ahmadi, R., Hemmateenejad, B., Safavi, A., Shojaeifard, Z., Shahsavar, A., Mohajeri, A., et al. (2018). Deep eutectic-water binary solvent associations investigated by vibrational spectroscopy and chemometrics. *Physical Chemistry Chemical Physics*, *20*(27), 18463–18473. https://doi.org/10.1039/c8cp00409a.

Arnold, T., Jackson, A. J., Sanchez-Fernandez, A., Magnone, D., Terry, A. E., & Edler, K. J. (2015). Surfactant behavior of sodium dodecylsulfate in deep eutectic solvent choline chloride/urea. *Langmuir*, *31*(47), 12894–12902. https://doi.org/10.1021/acs.langmuir.5b02596.

Banerjee, A., Ibsen, K., Iwao, Y., Zakrewsky, M., & Mitragotri, S. (2017). Transdermal protein delivery using choline and geranate (CAGE) deep eutectic solvent. *Advanced Healthcare Materials*, *6*(15), 1601411. https://doi.org/10.1002/adhm.201601411.

Bobrova, L. S., Danilov, F. I., & Protsenko, V. S. (2016). Effects of temperature and water content on physicochemical properties of ionic liquids containing CrCl3·xH2O and choline chloride. *Journal of Molecular Liquids*, *223*, 48–53. https://doi.org/10.1016/j.molliq.2016.08.027.

Bryant, S. J., Atkin, R., & Warr, G. G. (2016). Spontaneous vesicle formation in a deep eutectic solvent. *Soft Matter*, *12*(6), 1645–1648. https://doi.org/10.1039/C5SM02660A.

Bryant, S. J., Atkin, R., & Warr, G. G. (2017). Effect of deep eutectic solvent nanostructure on phospholipid bilayer phases. *Langmuir*, *33*(27), 6878–6884. https://doi.org/10.1021/acs.langmuir.7b01561.

Cardellini, F., Germani, R., Cardinali, G., Corte, L., Roscini, L., Spreti, N., et al. (2015). Room temperature deep eutectic solvents of (1S)-(+)-10-camphorsulfonic acid and sulfobetaines: Hydrogen bond-based mixtures with low ionicity and structure-dependent toxicity. *RSC Advances*, *5*(40), 31772–31786. https://doi.org/10.1039/C5RA03932K.

Carvalho, P. J., Regueira, T., Santos, L. M. N. B. F., Fernandez, J., & Coutinho, J. A. P. (2010). Effect of water on the viscosities and densities of 1-butyl-3-methylimidazolium dicyanamide and 1-butyl-3-methylimidazolium tricyanomethane at atmospheric pressure. *Journal of Chemical and Engineering Data*, *55*(2), 645–652. https://doi.org/10.1021/je900632q.

Celebi, A. T., Vlugt, T. J. H., & Moultos, O. A. (2019). Structural, thermodynamic, and transport properties of aqueous reline and ethaline solutions from molecular dynamics simulations. *Journal of Physical Chemistry B*, *123*(51), 11014–11025. https://doi.org/10.1021/acs.jpcb.9b09729.

Cherigui, E. A. M., Sentosun, K., Mamme, M. H., Lukaczynska, M., Terryn, H., Bals, S., et al. (2018). On the control and effect of water content during the electrodeposition of Ni nanostructures from deep eutectic solvents. *Journal of Physical Chemistry C*, *122*(40), 23129–23142. https://doi.org/10.1021/acs.jpcc.8b05344.

D'Agostino, C., Gladden, L. F., Mantle, M. D., Abbott, A. P., Ahmed, E. I., Al-Murshedi, A. Y. M., et al. (2015). Molecular and ionic diffusion in aqueous-deep eutectic solvent mixtures: Probing inter-molecular interactions using PFG NMR. *Physical Chemistry Chemical Physics*, *17*(23), 15297–15304. https://doi.org/10.1039/c5cp01493j.

Dai, Y., van Spronsen, J., Witkamp, G. J., Verpoorte, R., & Choi, Y. H. (2013). Natural deep eutectic solvents as new potential media for green technology. *Analytica Chimica Acta*, *766*, 61–68. https://doi.org/10.1016/j.aca.2012.12.019.

Dai, Y., Witkamp, G. J., Verpoorte, R., & Choi, Y. H. (2015). Tailoring properties of natural deep eutectic solvents with water to facilitate their applications. *Food Chemistry*, *187*, 14–19. https://doi.org/10.1016/j.foodchem.2015.03.123.

Eid, J., El Achkar, T., Fourmentin, S., Greige-Gerges, H., & Jraij, A. (2020). First investigation of liposomes behavior and phospholipids organization in choline chloride-based deep eutectic solvents by atomic force microscopy. *Journal of Molecular Liquids*, *306*, 112851. https://doi.org/10.1016/j.molliq.2020.112851.

Elworthy, P. H., & Mysels, K. J. (1966). The surface tension of sodium dodecylsulfate solutions and the phase separation model of micelle formation. *Journal of Colloid and Interface Science*, *21*(3), 331–347. https://doi.org/10.1016/0095-8522(66)90017-1.

Fetisov, E. O., Harwood, D. B., Kuo, I. F. W., Warrag, S. E. E., Kroon, M. C., Peters, C. J., et al. (2018). First-principles molecular dynamics study of a deep eutectic solvent: Choline chloride/urea and its mixture with water. *Journal of Physical Chemistry B*, *122*(3), 1245–1254. https://doi.org/10.1021/acs.jpcb.7b10422.

Francisco, M., van den Bruinhorst, A., Zubeir, L. F., Peters, C. J., & Kroon, M. C. (2013). A new low transition temperature mixture (LTTM) formed by choline chloride + lactic acid: Characterization as solvent for CO2 capture. *Fluid Phase Equilibria*, *340*, 77–84. https://doi.org/10.1016/j.fluid.2012.12.001.

Gao, Q., Zhu, Y., Ji, X., Zhu, W., Lu, L., & Lu, X. (2018). Effect of water concentration on the microstructures of choline chloride/urea (1:2)/water mixture. *Fluid Phase Equilibria*, *470*, 134–139. https://doi.org/10.1016/j.fluid.2018.01.031.

Ghaedi, H., Ayoub, M., Sufian, S., Shariff, A. M., Murshid, G., Hailegiorgis, S. M., et al. (2017). Density, excess and limiting properties of (water and deep eutectic solvent) systems at temperatures from 293.15 K to 343.15 K. *Journal of Molecular Liquids*, *248*, 378–390. https://doi.org/10.1016/j.molliq.2017.10.074.

Gómez-Díaz, D., Navaza, J. M., & Sanjurjo, B. (2007). Density, kinematic viscosity, speed of sound, and surface tension of hexyl, octyl, and decyl trimethyl ammonium bromide aqueous solutions. *Journal of Chemical and Engineering Data*, *52*(3), 889–891. https://doi.org/10.1021/je060486k.

Gutiérrez, M. C., Ferrer, M. L., Mateo, C. R., & del Monte, F. (2009). Freeze-drying of aqueous solutions of deep eutectic solvents: A suitable approach to deep eutectic suspensions of self-assembled structures. *Langmuir*, *25*(10), 5509–5515. https://doi.org/10.1021/la900552b.

Hammond, O. S., Bowron, D. T., & Edler, K. J. (2016). Liquid structure of the choline chloride-urea deep eutectic solvent (reline) from neutron diffraction and atomistic modelling. *Green Chemistry*, *18*(9), 2736–2744. https://doi.org/10.1039/c5gc02914g.

Hammond, O. S., Bowron, D. T., & Edler, K. J. (2017). The effect of water upon deep eutectic solvent nanostructure: An unusual transition from ionic mixture to aqueous solution. *Angewandte Chemie*, *129*(33), 9914–9917. https://doi.org/10.1002/ange.201702486.

Hammond, O. S., Bowron, D. T., & Edler, K. J. (2019). Structure and properties of "type IV" lanthanide nitrate hydrate:urea deep eutectic solvents. *ACS Sustainable Chemistry and Engineering*, 7(5), 4932–4940. https://doi.org/10.1021/acssuschemeng.8b05548.

Hammond, O. S., Bowron, D. T., Jackson, A. J., Arnold, T., Sanchez-Fernandez, A., Tsapatsaris, N., et al. (2017). Resilience of malic acid natural deep eutectic solvent nanostructure to solidification and hydration. *Journal of Physical Chemistry B*, *121*(31), 7473–7483. https://doi.org/10.1021/acs.jpcb.7b05454.

Hammond, O. S., Edler, K. J., Bowron, D. T., & Torrente-Murciano, L. (2017). Deep eutectic-solvothermal synthesis of nanostructured ceria. *Nature Communications*, *8*(May 2016), 1–7. https://doi.org/10.1038/ncomms14150.

Hammond, O. S., Eslava, S., Smith, A. J., Zhang, J., & Edler, K. J. (2017). Microwave-assisted deep eutectic-solvothermal preparation of iron oxide nanoparticles for photoelectrochemical solar water splitting. *Journal of Materials Chemistry A*, *5*(31), 16189–16199. https://doi.org/10.1039/c7ta02078c.

Hayyan, M., Mbous, Y. P., Looi, C. Y., Wong, W. F., Hayyan, A., Salleh, Z., et al. (2016). Natural deep eutectic solvents: Cytotoxic profile. *Springerplus, 5*(1), 913. https://doi.org/10.1186/s40064-016-2575-9.

Kaur, S., Gupta, A., & Kashyap, H. K. (2020). How hydration affects the microscopic structural morphology in a deep eutectic solvent. *The Journal of Physical Chemistry B, 124*(11), 2230–2237. https://doi.org/10.1021/acs.jpcb.9b11753.

Kim, K. S., & Park, B. H. (2018). Volumetric properties of solutions of choline chloride + glycerol deep eutectic solvent with water, methanol, ethanol, or iso-propanol. *Journal of Molecular Liquids, 254*, 272–279. https://doi.org/10.1016/j.molliq.2018.01.087.

Kuddushi, M., Nangala, G. S., Rajput, S., Ijardar, S. P., & Malek, N. I. (2019). Understanding the peculiar effect of water on the physicochemical properties of choline chloride based deep eutectic solvents theoretically and experimentally. *Journal of Molecular Liquids, 278*, 607–615. https://doi.org/10.1016/j.molliq.2019.01.053.

Kumari, P., Shobhna, K., Kaur, S., & Kashyap, H. K. (2018). Influence of hydration on the structure of reline deep eutectic solvent: A molecular dynamics study. *ACS Omega, 3*(11), 15246–15255. https://doi.org/10.1021/acsomega.8b02447.

Leron, R. B., Soriano, A. N., & Li, M. H. (2012). Densities and refractive indices of the deep eutectic solvents (choline chloride + ethylene glycol or glycerol) and their aqueous mixtures at the temperature ranging from 298.15 to 333.15K. *Journal of the Taiwan Institute of Chemical Engineers, 43*(4), 551–557. https://doi.org/10.1016/j.jtice.2012.01.007.

Li, Q., Wang, J., Lei, N., Yan, M., Chen, X., & Yue, X. (2018). Phase behaviours of a cationic surfactant in deep eutectic solvents: From micelles to lyotropic liquid crystals. *Physical Chemistry Chemical Physics, 20*(17), 12175–12181. https://doi.org/10.1039/C8CP00001H.

Lin, C. M., Leron, R. B., Caparanga, A. R., & Li, M. H. (2014). Henry's constant of carbon dioxide-aqueous deep eutectic solvent (choline chloride/ethylene glycol, choline chloride/glycerol, choline chloride/malonic acid) systems. *Journal of Chemical Thermodynamics, 68*, 216–220. https://doi.org/10.1016/j.jct.2013.08.029.

López-Salas, N., Vicent-Luna, J. M., Imberti, S., Posada, E., Roldán, M. J., Anta, J. A., et al. (2019). Looking at the "water-in-deep-eutectic-solvent" system: A dilution range for high performance eutectics. *ACS Sustainable Chemistry and Engineering*, 7(21), 17565–17573. https://doi.org/10.1021/acssuschemeng.9b05096.

Lu, J. R., Hromadova, M., Simister, E. A., Thomas, R. K., & Penfold, J. (1994). Neutron reflection from hexadecyltrimethylammonium bromide adsorbed at the air/liquid interface: The variation of the hydrocarbon chain distribution with surface concentration. *The Journal of Physical Chemistry, 98*(44), 11519–11526. https://doi.org/10.1021/j100095a037.

Ma, C., Guo, Y., Li, D., Zong, J., Ji, X., & Liu, C. (2017). Molar enthalpy of mixing and refractive indices of choline chloride-based deep eutectic solvents with water. *Journal of Chemical Thermodynamics, 105*, 30–36. https://doi.org/10.1016/j.jct.2016.10.002.

Ma, C., Guo, Y., Li, D., Zong, J., Ji, X., Liu, C., et al. (2016). Molar enthalpy of mixing for choline chloride/urea deep eutectic solvent + water system. *Journal of Chemical and Engineering Data, 61*(12), 4172–4177. https://doi.org/10.1021/acs.jced.6b00569.

Ma, C., Laaksonen, A., Liu, C., Lu, X., & Ji, X. (2018). The peculiar effect of water on ionic liquids and deep eutectic solvents. *Chemical Society Reviews, 47*(23), 8685–8720. https://doi.org/10.1039/c8cs00325d.

Marcus, Y. (2017). Unconventional deep eutectic solvents: Aqueous salt hydrates. *ACS Sustainable Chemistry & Engineering, 5*(12), 11780–11787. https://doi.org/10.1021/acssuschemeng.7b03528.

Maugeri, Z., & Domínguez De María, P. (2012). Novel choline-chloride-based deep-eutectic-solvents with renewable hydrogen bond donors: Levulinic acid and sugar-based polyols. *RSC Advances, 2*(2), 421–425. https://doi.org/10.1039/c1ra00630d.

McCluskey, A. R., Sanchez-Fernandez, A., Edler, K. J., Parker, S. C., Jackson, A. J., Campbell, R. A., et al. (2019). Bayesian determination of the effect of a deep eutectic solvent on the structure of lipid monolayers. *Physical Chemistry Chemical Physics*, *21*(11), 6133–6141. https://doi.org/10.1039/C9CP00203K.

Meng, X., Ballerat-Busserolles, K., Husson, P., & Andanson, J. M. (2016). Impact of water on the melting temperature of urea + choline chloride deep eutectic solvent. *New Journal of Chemistry*, *40*(5), 4492–4499. https://doi.org/10.1039/c5nj02677f.

Mjalli, F. S., & Ahmad, O. (2017). Density of aqueous choline chloride-based ionic liquids analogues. *Thermochimica Acta*, *647*, 8–14. https://doi.org/10.1016/j.tca.2016.11.008.

Mjalli, F. S., & Mousa, H. (2017). Viscosity of aqueous ionic liquids analogues as a function of water content and temperature. *Chinese Journal of Chemical Engineering*, *25*(12), 1877–1883. https://doi.org/10.1016/j.cjche.2017.09.008.

Morrison, H. G., Sun, C. C., & Neervannan, S. (2009). Characterization of thermal behavior of deep eutectic solvents and their potential as drug solubilization vehicles. *International Journal of Pharmaceutics*, *378*(1–2), 136–139. https://doi.org/10.1016/j.ijpharm.2009.05.039.

Pal, M., Rai, R., Yadav, A., Khanna, R., Baker, G. A., & Pandey, S. (2014). Self-aggregation of sodium dodecyl Sulfate within (choline chloride + urea) deep eutectic solvent. *Langmuir*, *30*(44), 13191–13198. https://doi.org/10.1021/la5035678.

Pal, M., Singh, R. K., & Pandey, S. (2015). Evidence of self-aggregation of cationic surfactants in a choline chloride+glycerol deep eutectic solvent. *ChemPhysChem*, *16*(12), 2538–2542. https://doi.org/10.1002/cphc.201500357.

Pandey, A., & Pandey, S. (2014). Solvatochromic probe behavior within choline chloride-based deep eutectic solvents: Effect of temperature and water. *Journal of Physical Chemistry B*, *118*(50), 14652–14661. https://doi.org/10.1021/jp510420h.

Posada, E., López-Salas, N., Jiménez Riobóo, R. J., Ferrer, M. L., Gutiérrez, M. C., & Del Monte, F. (2017). Reline aqueous solutions behaving as liquid mixtures of H-bonded co-solvents: Microphase segregation and formation of co-continuous structures as indicated by Brillouin and 1H NMR spectroscopies. *Physical Chemistry Chemical Physics*, *19*(26), 17103–17110. https://doi.org/10.1039/c7cp02180a.

Posada, E., Roldán-Ruiz, M. J., Jiménez Riobóo, R. J., Gutiérrez, M. C., Ferrer, M. L., & del Monte, F. (2019). Nanophase separation in aqueous dilutions of a ternary DES as revealed by Brillouin and NMR spectroscopy. *Journal of Molecular Liquids*, *276*, 196–203. https://doi.org/10.1016/j.molliq.2018.11.139.

Protsenko, V. S., Bobrova, L. S., & Danilov, F. I. (2017). Physicochemical properties of ionic liquid mixtures containing choline chloride, chromium (III) chloride and water: Effects of temperature and water content. *Ionics*, *23*(3), 637–643. https://doi.org/10.1007/s11581-016-1826-7.

Qin, H., Hu, X., Wang, J., Cheng, H., Chen, L., & Qi, Z. (2019). Overview of acidic deep eutectic solvents on synthesis, properties and applications. *Green Energy and Environment*, *5*(1), 8–21. https://doi.org/10.1016/j.gee.2019.03.002.

Rengstl, D., Fischer, V., & Kunz, W. (2014). Low-melting mixtures based on choline ionic liquids. *Physical Chemistry Chemical Physics*, *16*(41), 22815–22822. https://doi.org/10.1039/C4CP02860K.

Roldán-Ruiz, M. J., Jiménez-Riobóo, R. J., Gutiérrez, M. C., Ferrer, M. L., & del Monte, F. (2019). Brillouin and NMR spectroscopic studies of aqueous dilutions of malicine: Determining the dilution range for transition from a "water-in-DES" system to a "DES-in-water" one. *Journal of Molecular Liquids*, *284*, 175–181. https://doi.org/10.1016/j.molliq.2019.03.133.

Sanchez-Fernandez, A., Arnold, T., Jackson, A. J., Fussell, S. L., Heenan, R. K., Campbell, R. A., et al. (2016). Micellization of alkyltrimethylammonium bromide surfactants in choline chloride:glycerol deep eutectic solvent. *Physical Chemistry Chemical Physics*, *18*(48), 33240–33249. https://doi.org/10.1039/C6CP06053F.

Sanchez-Fernandez, A., Edler, K. J., Arnold, T., Alba Venero, D., & Jackson, A. J. (2017). Protein conformation in pure and hydrated deep eutectic solvents. *Physical Chemistry Chemical Physics*, *19*(13), 8667–8670. https://doi.org/10.1039/C7CP00459A.

Sanchez-Fernandez, A., Edler, K. J., Arnold, T., Heenan, R. K., Porcar, L., Terrill, N. J., et al. (2016). Micelle structure in a deep eutectic solvent: a small-angle scattering study. *Physical Chemistry Chemical Physics*, *18*(20), 14063–14073. https://doi.org/10.1039/C6CP01757F.

Sanchez-Fernandez, A., Hammond, O. S., Edler, K. J., Arnold, T., Doutch, J., Dalgliesh, R. M., et al. (2018). Counterion binding alters surfactant self-assembly in deep eutectic solvents. *Physical Chemistry Chemical Physics*, *20*(20), 13952–13961. https://doi.org/10.1039/C8CP01008K.

Sanchez-Fernandez, A., Hammond, O. S., Jackson, A. J., Arnold, T., Doutch, J., & Edler, K. J. (2017). Surfactant–solvent interaction effects on the Micellization of cationic surfactants in a carboxylic acid-based deep eutectic solvent. *Langmuir*, *33*(50), 14304–14314. https://doi.org/10.1021/acs.langmuir.7b03254.

Sanchez-Fernandez, A., Moody, G. L., Murfin, L. C., Arnold, T., Jackson, A. J., King, S. M., et al. (2018). Self-assembly and surface behaviour of pure and mixed zwitterionic amphiphiles in a deep eutectic solvent. *Soft Matter*, *14*(26), 5525–5536. https://doi.org/10.1039/C8SM00755A.

Schick, M. J. (1963). Effect of temperature on the critical micelle concentration of nonionic detergents. Thermodynamics of micelle formation1. *The Journal of Physical Chemistry*, *67*(9), 1796–1799. https://doi.org/10.1021/j100803a013.

Shah, D., & Mjalli, F. S. (2014). Effect of water on the thermo-physical properties of reline: An experimental and molecular simulation based approach. *Physical Chemistry Chemical Physics*, *16*(43), 23900–23907. https://doi.org/10.1039/c4cp02600d.

Shekaari, H., Zafarani-Moattar, M. T., & Mohammadi, B. (2017). Thermophysical characterization of aqueous deep eutectic solvent (choline chloride/urea) solutions in full ranges of concentration at T = (293.15–323.15) K. *Journal of Molecular Liquids*, *243*, 451–461. https://doi.org/10.1016/j.molliq.2017.08.051.

Siongco, K. R., Leron, R. B., Caparanga, A. R., & Li, M. H. (2013). Molar heat capacities and electrical conductivities of two ammonium-based deep eutectic solvents and their aqueous solutions. *Thermochimica Acta*, *566*, 50–56. https://doi.org/10.1016/j.tca.2013.05.023.

Siongco, K. R., Leron, R. B., & Li, M. H. (2013). Densities, refractive indices, and viscosities of N,N-diethylethanol ammonium chloride-glycerol or -ethylene glycol deep eutectic solvents and their aqueous solutions. *Journal of Chemical Thermodynamics*, *65*, 65–72. https://doi.org/10.1016/j.jct.2013.05.041.

Smith, P. J., Arroyo, C. B., Lopez Hernandez, F., & Goeltz, J. C. (2019). Ternary deep eutectic solvent behavior of water and urea choline chloride mixtures. *Journal of Physical Chemistry B*, *123*(25), 5302–5306. https://doi.org/10.1021/acs.jpcb.8b12322.

Tan, X., Zhang, J., Luo, T., Sang, X., Liu, C., Zhang, B., et al. (2016). Micellization of long-chain ionic liquid in deep eutectic solvent. *Soft Matter*, *12*(24), 5297–5303. https://doi.org/10.1039/C6SM00924G.

Tariq, M., Freire, M. G., Saramago, B., Coutinho, J. A. P., Lopes, J. N. C., & Rebelo, L. P. N. (2012). Surface tension of ionic liquids and ionic liquid solutions. *Chemical Society Reviews*, *41*(2), 829–868. https://doi.org/10.1039/c1cs15146k.

Turner, A. H., & Holbrey, J. D. (2019). Investigation of glycerol hydrogen-bonding networks in choline chloride/glycerol eutectic-forming liquids using neutron diffraction. *Physical Chemistry Chemical Physics*, *21*(39), 21782–21789. https://doi.org/10.1039/C9CP04343H.

Venable, R. L., & Nauman, R. V. (1964). Micellar weights of and solubilization of benzene by a series of tetradecylammonium bromides. The effect of the size of the charged Head1a. *The Journal of Physical Chemistry*, *68*(12), 3498–3503. https://doi.org/10.1021/j100794a010.

Wang, J. Y., Zhang, X. J., Liu, Y. M., & Hu, Y. Q. (2011). Interfacial tensions of imidazolium-based ionic liquids with N -alkanes and cyclohexane. *Journal of Chemical and Engineering Data, 56*(10), 3734–3737. https://doi.org/10.1021/je200349t.

Xie, Y., Dong, H., Zhang, S., Lu, X., & Ji, X. (2014). Effect of water on the density, viscosity, and CO2 solubility in choline chloride/urea. *Journal of Chemical and Engineering Data, 59*(11), 3344–3352. https://doi.org/10.1021/je500320c.

Yadav, A., Kar, J. R., Verma, M., Naqvi, S., & Pandey, S. (2015). Densities of aqueous mixtures of (choline chloride + ethylene glycol) and (choline chloride + malonic acid) deep eutectic solvents in temperature range 283.15–363.15 K. *Thermochimica Acta, 600*, 95–101. https://doi.org/10.1016/j.tca.2014.11.028.

Yadav, A., Trivedi, S., Rai, R., & Pandey, S. (2014). Densities and dynamic viscosities of (choline chloride+glycerol) deep eutectic solvent and its aqueous mixtures in the temperature range (283.15-363.15)K. *Fluid Phase Equilibria, 367*, 135–142. https://doi.org/10.1016/j.fluid.2014.01.028.

Zhekenov, T., Toksanbayev, N., Kazakbayeva, Z., Shah, D., & Mjalli, F. S. (2017). Formation of type III deep eutectic solvents and effect of water on their intermolecular interactions. *Fluid Phase Equilibria, 441*, 43–48. https://doi.org/10.1016/j.fluid.2017.01.022.

CHAPTER FOUR

Proteins in deep eutectic solvents: Structure, dynamics and interactions with the solvent

Adrian Sanchez-Fernandez[a,*] and Andrew J. Jackson[b,c]
[a]Food Technology, Engineering and Nutrition, Lund, Sweden
[b]European Spallation Source, Lund, Sweden
[c]Division of Physical Chemistry, Lund University, Lund, Sweden
*Corresponding author: e-mail address: adrian.sanchez_fernandez@food.lth.se

Contents

Abstract

The structure and dynamics of proteins in solution are closely related to their activity and, ultimately, to their performance in products and technologies, for example, drug formulations and enzymatic catalysis. With this in mind, the study of the fundamental behavior of proteins in deep eutectic solvents not only provides insight into a novel alternative to aqueous and molecular organic solvents, but also contributes to understanding the potential behavior of these macromolecules in non-aqueous environments. In particular, it has relevance for systems with high levels of dehydration, liquid phases for dispersions at cryogenic temperatures, or systems with high ionic strength.

In this chapter, we will explore current knowledge of the physicochemical behavior of proteins in deep eutectic solvents. The local and global structure of proteins in these

Advances in Botanical Research, Volume 97
ISSN 0065-2296
https://doi.org/10.1016/bs.abr.2020.09.003

solvents will be presented and compared to those in aqueous solutions, organic solvents and ionic liquids. Furthermore, the characteristic behavior of proteins in deep eutectic solvents will be linked to solvation effects and the interaction of the macromolecules with the solvent, where specific electrostatic and hydrogen bond interactions can affect the structure and dynamics of proteins. Finally, challenges and opportunities in the field will be discussed, with the aim of guiding a common effort to develop a better understanding of the fundamental behavior of proteins in these interesting environments.

Abbreviations

BLC bovine liver catalase
BSA bovine serum albumin
CALB *Candida antartica* lipase B
CD circular dichroism
DES deep eutectic solvent
HRP horseradish peroxidase
IL ionic liquid
Lys lysozyme
SANS small-angle neutron scattering

1. Introduction

Deep eutectic solvent (DES) have been established as potential candidates for protein incorporation, leading to a substantial amount of work being performed in the area. In addition to their widely recognized sustainable character, DES offer a mild, hydrogen-bonded environment where proteins may remain stable and functional (Dai, van Spronsen, Witkamp, Verpoorte, & Choi, 2013). Interestingly, the different combinations of precursors used to prepare DES also open the possibility of tuning solvent characteristics, with the goal of rationally designing the solvent for particular applications. This brings myriad possibilities in terms of solvent properties (network charge, proton chemical activity, solvent polarity, and so on) and, as DES are complex solvents formed through the combination of salts and neutral compounds, the landscape of interactions between the solvent and dispersed macromolecules can greatly vary between different DES (Pandey, Rai, Pal, & Pandey, 2014; Zahn, Kirchner, & Mollenhauer, 2016). Thus, the protein microenvironment can be modified and optimized through changes in solvent composition. These changes may result in a different balance of intermolecular interactions between the solvent and the protein amino acids (electrostatic, hydrogen bond, Van der Waals) to that in water, resulting in a variety of protein structure, conformation and dynamic landscapes.

Another interesting aspect of DES is their compatibility with water, where the addition of water to DES enables tuning the physicochemical properties of the solvent (Dai, Witkamp, Verpoorte, & Choi, 2015). Despite what could be initially thought, that a DES-water mixture just simply forms an aqueous solution of DES, or vice versa, the interactions between water and DES are inherently more complex (Hammond, Bowron, & Edler, 2017). As such, the addition of water not only provides the possibility of tuning the macroscopic properties of the system (e.g. viscosity), but also the molecular interactions that occur between solvent and biomolecules. Hence, the use of hydrated DES increases the possibilities in terms of protein solvation, protein-solvent interactions and, thus, protein behavior (Dai et al., 2015).

Due to these attractive characteristics, there is a recent rising interest in the behavior of proteins, amongst other biomolecules, in these solvents. Such systems of protein in DES find many possible applications including preservation, cryoprotection, solubilization of biomacromolecules, drug delivery, protein-based functional materials, and enzymatic catalysis. Therefore, increasing the fundamental understanding of protein behavior in DES has the potential to drive the development of novel technologies.

2. On protein structure, conformation and dynamics

Proteins are the essential building blocks of life. They play an important structural role, as well as serving as catalytic sites in chemical reactions occurring in the cell. The collection of proteins in a cell environment includes a huge diversity of structures and functionalities. Therefore, disentangling the characteristics of proteins helps to better understand biological processes and disease. Furthermore, the functional properties of proteins can be harnessed in applications such as enzymatic catalysis and biosensing.

In order to discuss the effect of DES on protein structure, conformation and dynamics, we must first consider those characteristics in water. Here we will look in general terms at proteins under biologically relevant conditions i.e. in aqueous solutions buffered to biological pH. This *native conformation and dynamics* are the baseline for comparison when proteins are placed in non-aqueous environments. It also serves to direct our attention to where, based on whether the structure and dynamics are similar to those in water, we might expect protein function in DES to be similar to that in water.

2.1 Protein structure

The underlying structure of proteins is that of a linear polypeptide chain, with a series of amino acids linked through peptide bonds. The sequence of amino acids that make up a protein is called the *primary structure*. The variety of possible amino acids presents a variety of functional groups that can be incorporated in the chain. Thus, each protein has a combination of polar—both anionic and cationic—and non-polar domains. This results in proteins having the possibility to interact through electrostatic, dispersion, and hydrogen-bond forces. Proteins are intrinsically amphiphilic molecules and hydrophobic interactions are a key driver in protein structure formation.

The interactions between amino acid functional groups within a given protein molecule can result in the chain forming locally organized structures. These structures are referred to as the *secondary structure* of the protein. Common motifs are alpha helices and beta sheets, with both the result of intra-molecular hydrogen bonds between carbonyl (C=O) and amino (N–H) moieties in the protein backbone. Alpha-helices form where there is a pattern of amino acids that repeatedly allows for a hydrogen bond to occur between one amino acid and another that is four positions along the chain. Beta sheets form where the pattern of amino acids allows for two or more longer sections of the protein to align with one another and form a series of hydrogen bonds.

Most proteins contain multiple helices and sheets, in addition to other less common patterns. The ensemble of those, driven by interactions between active sites within the secondary structure, and interactions with the solvent environment, causes the protein to fold into a more compact form. This lower energy state is known as the *tertiary structure* of the protein.

Finally, the *quaternary structure* refers to the conformation of several polypeptide chains in a supramolecular architecture. The *conformation* of a protein usually refers to the combination of secondary, tertiary and quaternary structure.

As an example, we consider the protein hemoglobin which is responsible for oxygen transport in blood. This protein has two forms, alpha hemoglobin with a primary structure consisting of 141 amino acids and beta hemoglobin with a primary structure consisting of 146 amino acids. Under biological conditions, these each form a secondary structure consisting of a series of alpha-helices. Each of the two types of subunits has a specific tertiary structure that positions the hem functionality such that the molecule can perform its role as a carrier of oxygen. The quaternary structure is essential for functioning and consists of a tetramer made up of two each

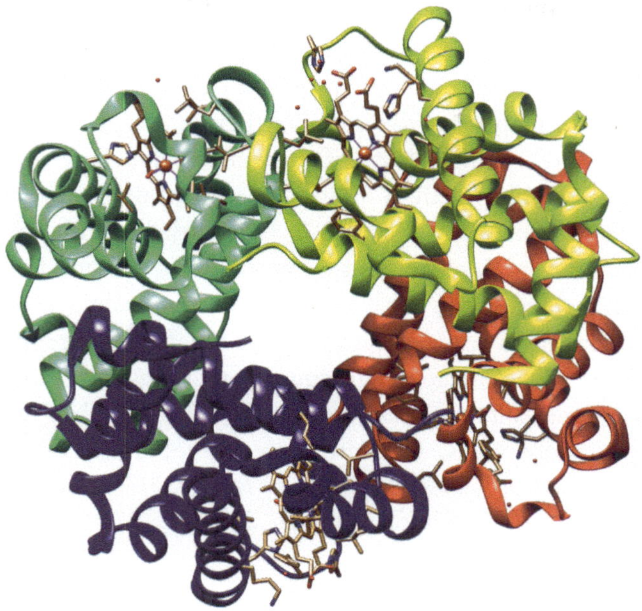

Fig. 1 Ribbon representation of the quaternary structure of hemoglobin (PDB: 1GZX). Each subunit is represented using different colors and the molecular structure of the heme groups are shown as molecular silhouettes.

of alpha-hemoglobin and beta-hemoglobin. Conformational changes—restructuring of the protein tertiary and quaternary structure—form part of the mechanism of oxygen uptake and release (Fig. 1).(Paoli, Liddington, Tame, Wilkinson, & Dodson, 1996; Perutz, 1972; Perutz et al., 1960).

2.2 Protein dynamics

The three-dimensional structure of a protein is not static, and it can change based on the environment the molecule is in and the balance of forces that it is subject to (McCammon, 1984). Most importantly, the structure of a protein can change based upon the presence or absence of a ligand bound to an active site in the molecule.

These conformational fluctuations have kinetic and thermodynamic parameters associated with them that determine the ease of the transitions and thus drive the function of many proteins, in particular those with enzymatic functions. The flexibility or rigidity of the protein has a major role in determining how readily structural changes can happen.

Returning to the example of hemoglobin we see that the structure and dynamics of the protein are key to the function. The ability of the hem pocket to reshape on ligand binding enables uptake of oxygen through co-operative restructuring of the tetramer (Perutz, 1972).

The static and dynamic study of protein structure is crucial to their biological and chemical activity (Kubitzki, de Groot, & Seeliger, 2017; Richards, 1977; Rupley & Careri, 1991). Due to the high number of functional sites and the amphiphilicity of the proteins, the surrounding environment strongly influences the behavior of the macromolecule. However, the solvation of proteins is very complex, and the mechanisms are still not fully understood for a wide range of proteins. Protein dynamics, and solvation in aqueous buffered solutions, has been widely investigated both experimentally, using primarily NMR and neutron spectroscopy, and computationally (Berendsen & Hayward, 2000; Ebbinghaus et al., 2007; Frauenfelder et al., 2009; Gabel et al., 2002; Khodadadi et al., 2008).

3. Proteins in non-aqueous solvents

It was generally held to be true that that water (even at low concentrations) was required to allow enzyme activity, and indeed the addition of catalytic amounts of water was shown to enable enzyme activity in hydrophobic organic solvents (Antonini, Carrea, & Cremonesi, 1981). The idea of developing a system for catalysis where the continuous medium was a neat organic solvent became reality with the studies performed on the activity of porcine pancreatic lipase in an alcohol-polyol solution (Zaks & Klibanov, 1984). These studies not only showed that the catalytic activity was retained at low water contents but also demonstrated that water-free organic solvents enhanced the activity, selectivity and thermal stability. Although this cannot be taken as a general rule by which any enzyme will better perform in an organic solvent, a completely new world of possibilities was born. The advances in genetic and enzyme engineering, coupled to the versatility that organic solvents offer in terms of properties (polarity, cohesivity and viscosity, among others), allow the design of systems that perform efficiently under specific, chosen, conditions (Stepankova et al., 2013).

There have been many efforts to study proteins in organic solvents in order to correlate enzymatic activity with the fundamental behavior of proteins. Experimental and computational methods have been combined to study the effects of organic solvents on protein structure, conformation and dynamics, as these directly relate to the stability and activity of proteins

(Tanford, 1968). The general belief is that, even in neat organic solvents, a few water molecules remain bound to the protein and allow the protein to remain folded, and thus active, and the total absence of water would lead to inactivity (Klibanov, 2001). As such, different observations were reported for organic solvents. Hydrophobic solvents tend to preserve the structure of the protein as these protein-bound water molecules remain associated (Antonini et al., 1981). Unlike hydrophobic organic solvents, hydrophilic organic solvents are capable of removing these water molecules from the protein at certain concentrations, which result in a disruption of the local native environment of the protein. Furthermore, these are capable of interacting with specific protein amino acids and induce structural and conformational changes that affect protein function and stability by affecting the intraprotein and solvent-protein interactions (Serdakowski & Dordick, 2008).

The use of ionic liquids (ILs) opened up the possibility of designing solvents for particular applications, as their properties can be finely tuned by preparing them through the combination of different ions. When compared to molecular organic solvents, they are considered green solvents due to their low vapor pressure and recyclable character (Rogers & Seddon, 2003). Also, the compatibility of some of them with water leads to an extra degree of freedom in terms of solvent design. Due to these interesting characteristics, ILs have been proved suitable for performing biocatalyzed reactions, with a dominating role in biotechnological research for the production of biofuels and in reactions where recalcitrant compounds are present (Schindl, Hagen, Muzammal, Gunasekera, & Croft, 2019). Furthermore, there has been a recent heightened interest in the use of these solvents for protein preservation (Benedetto & Ballone, 2015).

One interesting aspect when it comes to protein behavior in ILs is the inherent complexity of the cybotactic region. ILs are commonly composed of bulky organic ions with relatively weak ion-pair correlations, which often present very different molecular structure. Therefore, ILs offer a vast combination of possibilities for molecular interactions with biomolecules: electrostatic, Van der Waals, hydrogen- and π-bonding (Schindl et al., 2019). This results in a very complex interaction network that potentially drives the behavior of proteins in these solvents, which in turn may be very different to that in organic molecular solvents. Some relevant aspects of protein behavior in ILs are presented here. For more detailed information on the topic, refer to recent reviews (Benedetto & Ballone, 2015; Schindl et al., 2019; Zhao, 2016).

The combination of experimental and computational approaches has been used to gain an understanding of the behavior of proteins and enzymes in neat and hydrated ILs. Generally, it is thought that solvation in neat hydrophilic ILs alters the native conformation of the protein, resulting in different secondary and tertiary structure (Benedetto & Ballone, 2015; Debeljuh, Barrow, Henderson, & Byrne, 2011). Alterations in the local electrostatic and hydrogen bond environment of the protein compared to that in aqueous solutions are thought to be the driving force of the non-native folding. The presence of charged moieties in the solvent can, in principle, result in specific ion-ion interactions between the ions and charged residues, with preferential solvation environments that affect protein structure and dynamics (Ghosh, Parui, Jana, & Bhattacharyya, 2015). Furthermore, the type of ions that constitute the solvent affects the resulting conformation, where ions with higher charge density are more prone to interact with the protein, and thus affect protein structure, than lower charge density ions. These interactions have been hypothesized to follow an analogous trend to that described by the Hofmeister series (Kumar & Venkatesu, 2014; Zhao, 2016). Another important aspect is the solvophobic effect, analogous to the hydrophobic effect that promotes amphiphile self-assembly in water. As such, it was found that solvent cohesivity plays an important role in protein folding. More cohesive ILs, where the solvophobic effect is more dominant, promote a more compact conformation, similar to the native folding.(Wijaya, Separovic, Drummond, & Greaves, 2016). Similarly, hydrogen bonding between the protein and protic ILs have been found to play an important role in the behavior of proteins in ILs, inducing increased protein stability.

When moving from neat ILs to hydrated ones, the addition of water leads to various structural transitions in the protein. At low water contents, even catalytic amounts, the scenario is similar to that in organic solvents, where water forms a solvation layer around the protein active sites and enables it to partially or totally retain its native conformation (Benedetto & Ballone, 2015; Debeljuh et al., 2011). Upon further increase the hydration level, the protein generally retrieves its native conformation, even at relatively high IL concentration (e.g. 75 wt% IL) (Debeljuh et al., 2011; Mann, McCluskey, & Atkin, 2009). These effects are commonly attributed to the ion correlation that exists between the ionic liquid constituents, which hinders the interaction of free ions with the protein even in the presence of water. On the other hand, when sufficiently diluted, the decorrelation of the ions that form the IL leads to a water-like behavior, where the interaction of those with the protein is driven by the Hofmeister series. Thus, the solvent behaves as a dilute electrolyte (Zhao, 2016).

4. Proteins in DES

DES present an evolution in the green character of ILs. By combining the designable properties of ILs with the possibility of synthesizing them using mild, biodegradable compounds, DES open new possibilities in terms of protein-based technologies compared to ILs. While ILs have been more oriented to biocatalysis (e.g. industrial production of biofuels), (Schindl et al., 2019) the biocompatible character of (some) DES opens up new possibilities in terms of technological applications, such as drug delivery, preservation of therapeutic proteins and formulation of cosmetic products. However, many challenges must be overcome to exploit the beneficial characteristics of DES in these technologies.

4.1 Protein incorporation

One of the main disadvantages of DES in technological applications is often their high viscosity, which hinders the dissolution of molecules such as proteins. Some approaches for dissolving a variety of non-biological compounds have previously been used (e.g. heating the mixture for dissolving surfactants) (Sanchez-Fernandez et al., 2016). However, such approaches cannot always be used when biomacromolecules are present due to their thermal sensitivity. Therefore, alternative approaches for incorporating such molecules into DES are required.

Different approaches have been utilized for the incorporation of proteins to neat and hydrated DES. In the case of hydrated systems, the inherently lower viscosity of the solvent facilitates the process of dissolution (Dai et al., 2015). As such, proteins in DES-water systems have been prepared through direct mixing of the protein powder in the solvent or by mixing aqueous stock solutions of proteins with DES to the desired concentration (Harifi-Mood, Ghobadi, & Divsalar, 2017; Wu, Wen, Xu, & Yang, 2014). This approach holds the advantage of having a minimum impact on the protein as it does not involve any mechanical or thermal stress. The obvious limitation of this approach is the presence of water in the system and thus it cannot be applied the protein systems in neat DES. As such, the presence of water must be taken into account even at very low hydration levels, as this affects the physicochemical characteristics of the solvent and potentially protein behavior (Hammond et al., 2017).

An alternative approach is the dispersion of the solid powder in the DES. For some DES, the dissolution of the powder may be achieved with some mild heating and stirring (Parnica & Antalik, 2014). Nonetheless, this is limited to low viscosity systems. This approach may also be used in enzymatic catalysis, where some substrates may act as thinning agents and assist the solvation of the protein (Gorke, Srienc, & Kazlauskas, 2008).

An approach that has been (often indirectly) used is the two-phase extraction approach. This method consists in preparing a protein solution in a liquid matrix that is insoluble in DES but for which the protein has a partition coefficient that facilitates the transfer of it to the DES (Li et al., 2016). This approach has been mostly used for the selective recovery of proteins from other solvents (e.g. BSA extraction using a betaine-urea DES), but it can be established as a method for protein incorporation to DES. However, this may be limited to a range of suitable proteins and the role of interfacial stress between the two phases must be taken into consideration for sensitive proteins.

The freeze-drying approach has become one of the most promising methods for protein incorporation. Upon the preparation of a hydrated DES that easily solubilizes the protein, this method takes advantage of the low vapor pressure of the DES to remove the water through sublimation under vacuum at low temperatures (Gutierrez, Ferrer, Mateo, & del Monte, 2009). The approach is relatively mild and allows the preparation of protein solutions in neat DES (Esquembre et al., 2013; Sanchez-Fernandez, Edler, Arnold, Alba Venero, & Jackson, 2017). Similarly, it has also been used for the incorporation of lipid vesicles and bacteria, showing the suitability for sensitive dispersants (Gutierrez et al., 2009; Gutierrez, Ferrer, Yuste, Rojo, & del Monte, 2010). The limitations presented for this approach relates to the presence of residual amounts of water due to insufficient drying or the stress induced by the creation of interfacial tension if the sample melts during drying.

4.2 Local and global structuring in DES

Although a diverse set of biomolecules have been recently probed to retain activity and function in neat or hydrated DES, only a few investigations were focused on studying the structural behavior of proteins in these solvents. As it is expected, the conformation of the protein in DES must be intrinsically related to its stability and chemical activity. Therefore, understanding the influence of the solvent environment in protein folding is of great importance for predicting the behavior of a given protein in terms of solvent properties: polarity, network charge, hydrophobicity, hydrogen ion activity and so on (Pandey et al., 2014; Zahn et al., 2016).

The remarkable complexity of biological macromolecules is built through a hierarchical set of atomic and molecular interactions, as discussed previously. As such, integrative approaches are required to probe the full range of structure of the protein, from the amino acid environment and local structure to 3D conformation and protein self-association. Esquembre et al. investigated the local and global structure of lysozyme (Lys) in 1:2 choline chloride:glycerol and 1:2 choline chloride:urea, to connect the enzymatic activity in neat and hydrated DES to the protein structure (Esquembre et al., 2013). Intrinsic tryptophan fluorescence spectroscopy is a common tool to probe conformational changes in proteins in solution as the fluorescent response of the chromophore depends on the cybotactic region of the specific amino acid. Their fluorescence results showed conformational changes in Lys in neat DES compared to aqueous solutions. Interestingly, the change in the emission spectrum was connected to the formation of a more compact, isotropic conformation. Nonetheless, the fluorescence method presents some limitations as it does not provide direct structural information. The emission of tryptophan in DES is of great complexity even by itself and strongly depend on the molecular architecture of the cybotactic region, unlike in more simple solvation environments such as water or other molecular organic solvents (Kadyan, Juneja, & Pandey, 2019).

Further structural investigations were performed using circular dichroism (CD) spectroscopy. Far-UV CD results showed the formation of secondary structure similar to those observed for Lys in water, i.e. a combination of α-helices, β-sheets and random coils (White, 1976). The differences observed in the data probably relate to different populations of each of the local structures, but overall the local folding into secondary structure still occurs in DES (Fig. 2). The tertiary structure of the protein, as investigated by near-UV CD, showed that 3D folding of the lysozyme happened in neat DES in a very similar way to the aqueous behavior of the protein (Esquembre et al., 2013).

The structural studies of protein in DES have become the target of subsequent investigations by experimental and computational methods. Investigations that aimed to link protein structure (secondary and tertiary structure) to conformation were performed combining CD and small-angle neutron scattering (SANS). The behavior of two model proteins, bovine

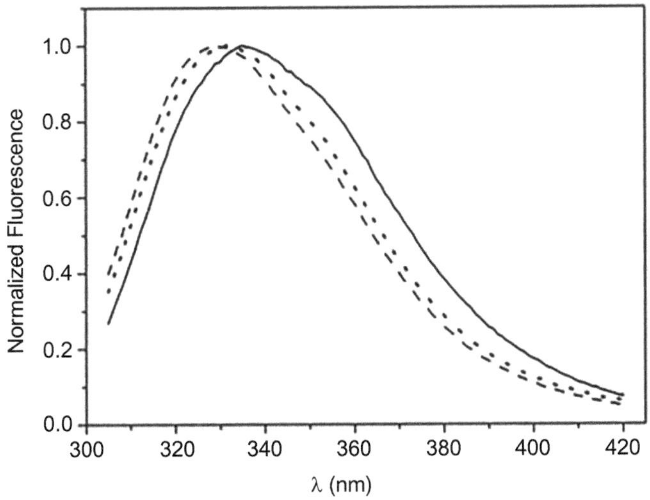

Fig. 2 Far-UV CD spectra of lysozyme at 20 °C dissolved in 1:2 choline chloride:glycerol (dashed line), 1:2 choline chloride:urea (dotted line) and MES buffered solution (solid line). *Reproduced from Esquembre, R., Sanz, J. M., Wall, J. G., del Monte, F., Mateo, C. R., & Ferrer, M. L. (2013). Thermal unfolding and refolding of lysozyme in deep eutectic solvents and their aqueous dilutions.* Physical Chemistry Chemical Physics, 15*(27), 11248–11256. doi:10.1039/c3cp44299c with permission from the PCCP Owner Societies.*

serum albumin (BSA) and hen egg-white Lys, were investigated in 1:2 choline chloride:glycerol, and 1:2 choline chloride:urea (Sanchez-Fernandez, Edler, et al., 2017). The secondary structure of the proteins in neat DES were found to be slightly different in DES to those in water, but the formation of local structures through intra-protein interactions still occurs for both proteins. Near-UV CD showed similar behavior for the protein tertiary structure, where conformational changes are observed in neat DES compared to the aqueous solution and hydrated DES. Nonetheless, the CD signal is different to that of the denatured protein, thus confirming that the protein folding still occurs but in a different fashion from the native structure of the protein.

CD is sensitive to changes in the environment of aromatic chromophores and does not provide quantitative structure determination. Therefore, further investigations were performed using SANS. The results proved that protein conformation in neat DES is different from that in aqueous buffer. However, partial folding is reported for BSA and a certain degree of globularity is retained compared to the denatured protein (Fig. 3).

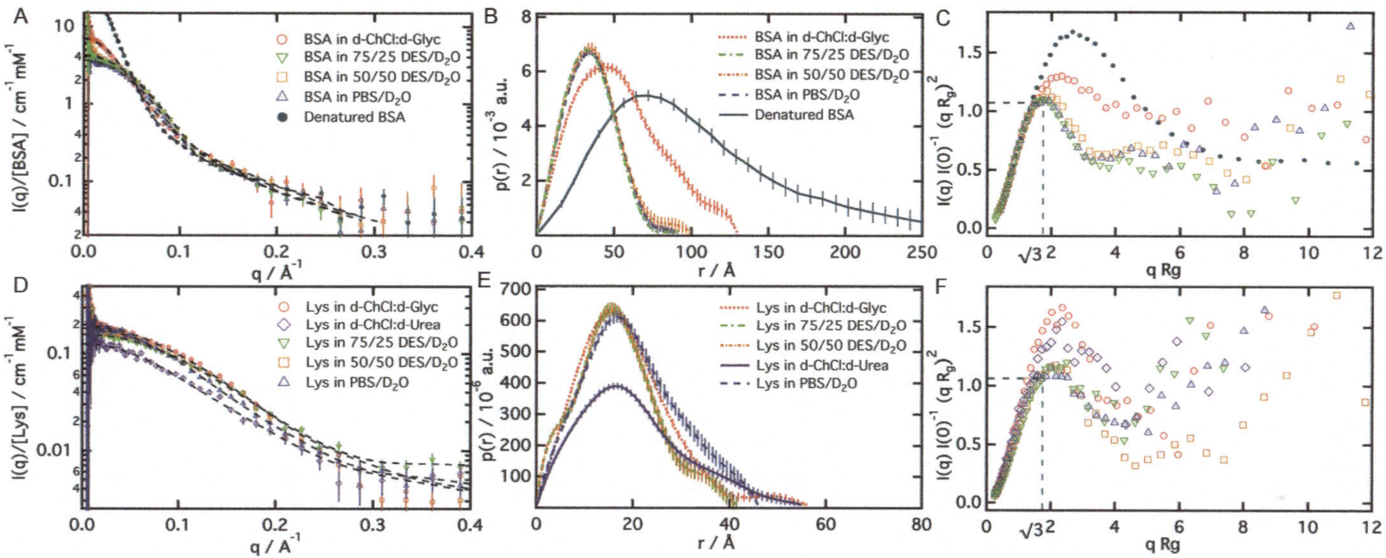

Fig. 3 SANS data (A and D), pair-distance distribution function (B and E) and Kratky plots (C and F) of BSA and Lys, respectively, in neat and hydrated choline chloride:glycerol DES, and phosphate buffer. *Reproduced from Sanchez-Fernandez, A., Edler, K. J., Arnold, T., Alba Venero, D., & Jackson, A. J. (2017). Protein conformation in pure and hydrated deep eutectic solvents.* Physical Chemistry Chemical Physics, 19*(13), 8667–8670. doi:10.1039/c7cp00459a with permission from the PCCP Owner Societies.*

For Lys, the formation of a different structure to that in aqueous solvent was reported in glycerol- and urea-based DES, although the structural differences were significantly less pronounced than the conformational loss of BSA. Although the local structure is minimally affected in neat DES, the global folding of the protein was found to be significantly affected. As the variations are observed in both levels of protein structure (i.e. small changes in secondary structure and different folding to those in aqueous buffer), it can be concluded that the connection between the local structure to global conformation in DES still applies and that changes in the secondary structure of the protein also prompt variations in the global folding (Sanchez-Fernandez, Edler, et al., 2017).

Computational studies of protein structure and folding were performed using MD simulations. The secondary structure of *Candida antartica* lipase B (CALB) in 1:2 choline chloride:urea was reported to be similar to that in aqueous buffer (Monhemi, Housaindokht, Moosavi-Movahedi, & Bozorgmehr, 2014). Interestingly, when comparing the behavior of the protein in neat DES to an aqueous solution of urea with equivalent concentration to the DES, it was found that the secondary structure was lost in the presence of the chemical denaturant when in aqueous solution. These simulations also showed that the protein globally folds into a compact structure in DES, with similar characteristics (but not identical) to the conformation in aqueous buffer.

Although the structure of these proteins has been shown to remain relatively unaffected in DES, it should not be taken as a general rule as they are strongly system dependent. Parnica et al. developed guanidinium salt-based DES as media for protein storage at room temperature (Parnica & Antalik, 2014). The CD studies showed that the secondary and tertiary structure of cytochrome *c* were greatly affected upon incorporation into the DES, displaying similar spectra to those of the denatured protein. However, subsequent dialysis with aqueous buffer of the protein-DES system showed that the structure of the protein was retrieved to the native form. Similarly, the structure of Lys was shown to be recovered to the native using extensive dialysis upon incorporation into neat choline chloride-based DES (Esquembre et al., 2013).

Apart from the global and local structure of proteins in neat DES, the behavior of those in hydrated DES have been probed. The investigations performed for Lys in 1:2 choline chloride:urea and 1:2 choline chloride: glycerol showed that the addition of water restored the native structure of the protein, as seen by CD (Esquembre et al., 2013). Similarly, the structure and conformation of BSA in hydrated 1:2 choline chloride:glycerol (25 and 50 wt% water) was practically the same as in aqueous buffer (Sanchez-Fernandez, Edler, et al., 2017). Therefore, the use of DES with moderate levels of hydration prompted protein folding as in aqueous buffer. In some cases, this was achieved even though urea/aqueous mixtures are well-known to act as a protein denaturant. Nonetheless, at higher hydration levels it was shown that the presence of DES altered the secondary structure of bovine liver catalase (BLC) and horseradish peroxidase (HRP) (Harifi-Mood et al., 2017; Wu et al., 2014). This characteristic difference at high hydration levels (>50 wt% water) may relate to a different interaction mechanism between the protein and the DES components compared to systems with lower hydration levels.

One of the most interesting features that DES offer is the possibility of developing solvents with similar compounds to those used for thermal protection and cryoprotection of proteins (e.g. sugars). As such, the structural resilience of proteins in DES under thermal stress has been also investigated. Xin et al. reported the preparation of neat and hydrated trehalose-based DES to develop a medium suitable for protein storage (Xin et al., 2017), since the stabilizing effect of trehalose in aqueous solutions of water is relatively well-known (Patist & Zoerb, 2005). The structural investigations performed using CD showed that, in hydrated DES, the tertiary structure of Lys was indeed preserved upon heating and cooling. In addition, it was shown that for this trehalose-based DES lower levels of hydration increased the stabilizing effect of the system. However, other investigations have reported opposite results for other systems. The structural resilience of Lys in 1:2 choline chloride:urea and 1:2 choline chloride:glycerol was found to be different. While the tertiary structure of Lys was recovered in the aqueous buffer after heating to 80 °C and subsequent cooling to 20 °C, the use of urea-based DES as the solvent resulted in the non-reversible unfolding of the protein and of glycerol-based DES in partial refolding (Esquembre et al., 2013). The global conformation of proteins has been also investigated upon applying thermal stress to the system. Investigations using SANS demonstrated that the conformation of the BSA was preserved at cryogenic temperatures in 1:2 choline chloride:glycerol. However, the protein was irreversibly unfolded at high temperatures and the folding was not retrieved upon cooling at room temperature (Sanchez-Fernandez, Edler, et al., 2017).

4.3 The role of protein-solvent and solvent-solvent interactions

One of the relevant questions to explore in protein-DES systems is the interaction of the biomolecules with the solvent. As it has been shown for some ionic liquids, the presence of charged ions in the solvent influences protein folding, as well as solvent structural changes in the solvent are induced by the presence of the protein (Ghosh et al., 2015; Schindl et al., 2019; Wijaya et al., 2016). Besides, DES contain neutral molecules capable of establishing hydrogen bonds, which may also interact with proteins.

4.3.1 Interactions in neat DES

Experimental methods are often limited in resolution to observe molecular solvation mechanisms in detail. As such, the utilization of computational approaches, such as MD simulations, aid in resolving questions regarding protein-solvent interactions, solvent-solvent interactions and, ultimately, how these relate to protein function. For instance, the hydrolase CALB has been shown to remain catalytically active in ethylammonium chloride- and choline chloride-based DES (Gorke et al., 2008). Interestingly, the presence of extremely high salt and urea concentrations in 1:2 choline chloride:urea DES did not induce loss of enzymatic activity. Also, the reaction shows excellent substrate selectivity toward the reactant, 1-butanol, despite the presence of polyalcohols in the solvent, ethylene glycol or glycerol. When these reactions were compared to analogous processes in *tert*-butanol solutions of ethylene glycol or glycerol (absence of halide salts), these were found to compete with 1-butanol for the reaction. Thus, the formation of a DES was hindering protein deactivation and side reactions with solvent components. This was worth further investigation, as the formation of DES was reducing side reactions by a 600-fold factor (Gorke et al., 2008).

With this goal, Monhemi et al. performed a pioneering MD investigation on protein-solvent interactions in 1:2 choline chloride:urea (Monhemi et al., 2014). This study showed that high concentrations of urea in aqueous solution (8 M) induced protein denaturation through the loss of the secondary and tertiary structure. The interaction of freely diffusing urea with the amino acid residues disrupts the intraprotein hydrogen bond network that promotes protein folding. MD simulations of CALB in DES showed that secondary structure was preserved as urea was strongly associated within the solvent structure (Hammond, Bowron, & Edler, 2016), thus inhibiting interactions with amino acid residues and hindering protein denaturation.

Therefore, the stabilizing effect observed in DES is subjected to solvent composition. The diffusion of discrete components in the bulk solvent, and thus the interaction of those with the protein, is subjected to the supramolecular solvent network. Sakar et al. demonstrated through MD simulations that, whilst a small helix-rich protein undergoes denaturation in the presence of urea in aqueous solution, the addition of ammonium salts (triethylammonium acetate or choline chloride) protects the protein against chemical denaturation (Sarkar, Ghosh, & Chakrabarti, 2017). However, this study showed that protein unfolding was more prone to happen at salt:urea ratios outside the eutectic condition for the system. The molecular mechanism of such an effect was attributed to the preferential binding of urea to the salts in the bulk solvent and the subsequent depletion of urea from the protein binding sites.

These interactions between the solvent and the protein surface have been shown to not only affect the behavior of the protein but also to alter the structure of the solvent around the protein. As such, the creation of a solvation layer with a different solvent structure to that of the bulk solvent has been reported using MD simulations (Monhemi et al., 2014). This phenomenon results from the change in the hydrogen bond and electrostatic landscape induced by the presence of the protein from the original solvent network. This solvation layer has been also reported in the solvation mechanism of surfactant micelles and nanoparticles (Hammons, Zhang, & Ilavsky, 2018; Sanchez-Fernandez et al., 2016, 2017). The preferable arrangement of ions or hydrogen bond donors is therefore driven by electrostatic and hydrogen bond interactions with the protein surface, and induce a reduction of solvent and residue mobility (Fig. 4).

Interestingly, it is believed that the solvation network confers the protein the ability to fold in DES, sometimes with structural similarities to proteins in aqueous solutions (Monhemi et al., 2014; Sarkar et al., 2017). Structural investigations also showed that those interactions may be responsible for the distinct folding of the protein compared to that of water, where proteins remain partially folded (e.g. BSA in 1:2 choline chloride:glycerol) or folded into a more compact state (e.g. lysozyme in 1:2 choline chloride:glycerol).(Esquembre et al., 2013; Sanchez-Fernandez, Edler, et al., 2017). Furthermore, it has been proposed that such interactions, in some cases, are responsible for an increased protein thermal and physical stability (Esquembre et al., 2013; Monhemi et al., 2014).

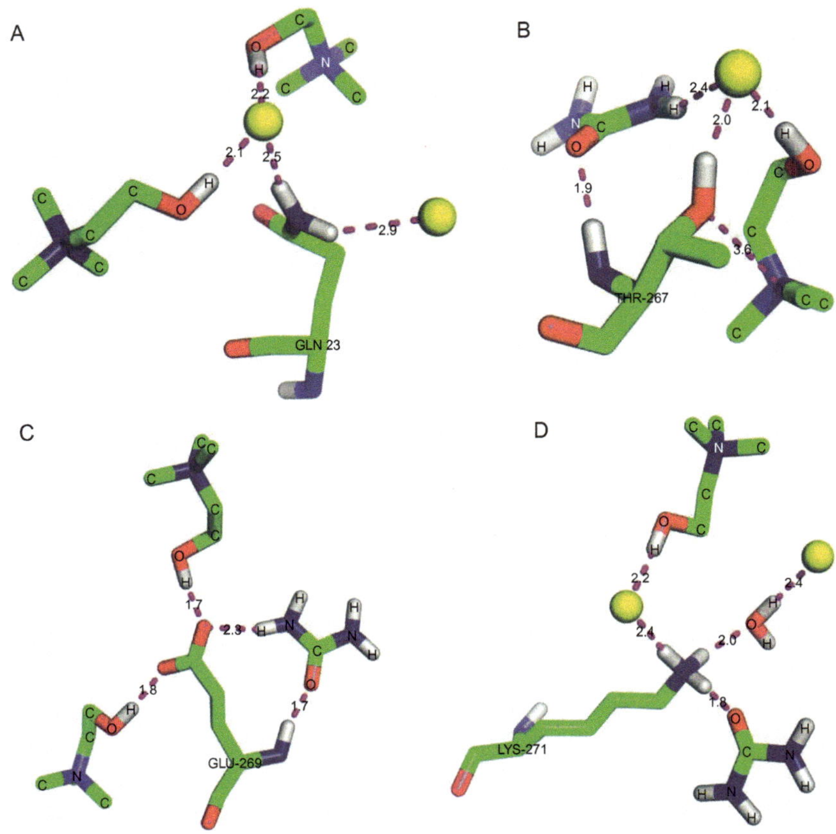

Fig. 4 Local interactions between the DES components (urea, choline and chloride) and different amino acids of CALB: (A) Gln23, (B) Thr267, (C) Glu269, and (D) Lys271; at 300 K. *Reproduced from Monhemi, H., Housaindokht, M. R., Moosavi-Movahedi, A. A., & Bozorgmehr, M. R. (2014). How a protein can remain stable in a solvent with high content of urea: Insights from molecular dynamics simulation of Candida antarctica lipase B in urea: Choline chloride deep eutectic solvent.* Physical Chemistry Chemical Physics, *16(28), 14882–14893. doi:10.1039/c4cp00503a with permission from the PCCP Owner Societies.*

4.3.2 Interactions in hydrated DES

Similarly, it has been hypothesized that DES-water systems may behave in a similar way, where DES-rich or water-rich domains are formed around specific sites of the protein at certain hydration levels and drive specific protein behavior. When mixed with water, DES go through complex structural transitions (Hammond et al., 2017). At low levels of hydration ($<\sim$5 wt%

water), the supramolecular structure of the DES is affected, and water takes part in the formation of the solvent network. At increased hydration levels (~5–50 wt% water), the DES-water structure remains relatively unchanged and water starts to form channels in the interstitial space between DES-water clusters. If the water content is further increased, the DES structure is lost, and the system behaves as discrete compounds. Due to this characteristic behavior, the interaction landscape between the protein and the solvent can differ at different hydration levels.

The structural characteristics of BSA in hydrated 1:2 choline chloride: glycerol (25 and 50 wt% water) were investigated (Sanchez-Fernandez, Edler, et al., 2017). The results showed that the addition of water, even at the lowest hydration level, prompted the protein to fold into its native structure. As such, BSA secondary and tertiary structure were retrieved, and the solution conformation was the same to that in aqueous buffer despite the relatively high concentration of DES in the system. Similarly, the structure of Lys was recovered in hydrated DES compared to that in neat DES (Esquembre et al., 2013). This characteristic behavior was hypothesized to arise from the formation of heterogeneous solvent domains around the protein (Hammond et al., 2017). The presence of strongly coordinated DES components may hinder the interaction of those with the protein and water probably populates the protein interface. As such, an aqueous microenvironment is form and the protein folds into its native (aqueous-like) conformation.

Also, the interaction between the DES components at low water contents may also drive phenomena such as protein refolding upon thermal denaturation. As thermal stress drives the unfolding of the protein, DES components have been hypothesized to adsorb to different protein sites (e.g. hydrophobic residues from the core) and hinder protein refolding in 1:2 choline chloride:urea (Esquembre et al., 2013). However, it was shown that hydrated trehalose-based DES (25 wt% water) induced refolding upon cooling through solvent-solvent interactions. The mechanism of refolding in this case was attributed to the strong correlation between trehalose and choline chloride in the bulk solvent, as it promotes the desorption of discrete DES components from the protein (Xin et al., 2017).

Another possible scenario is that where the DES components preferentially interact with the protein instead of water. The interaction through hydrogen bonds or electrostatic forces may cause a preferential interaction between protein residues and specific DES components. This is commonly observed in diluted DES (>50 wt% water), where ion-ion and ion-hydrogen bond donor correlations vanish, and the solvent becomes a solution of the discrete DES constituents. In that regard, dilute DES have been shown to affect the local structure of BLC and HRP through protein-solvent interactions (Harifi-Mood et al., 2017; Wu et al., 2014). Furthermore, specific ion effects were observed when using either choline chloride or choline acetate as the halide salt in diluted DES. The presence of choline acetate resulted in different protein secondary structure for HRP, also affecting protein function (Wu et al., 2014). This demonstrates that the discrete DES components interact with the protein potentially depending on their hydrophilicity/hydrophobicity, as observed in ILs which roughly comply with the traditional Hofmeister series (Zhao, 2016).

5. Future outlook

Although an effort has recently been paid to developing a better understanding of the physicochemical behavior of proteins in DES, a long road lies ahead. Here, from the authors' perspective, some of the future goals of the field are discussed.

5.1 Development of methods for protein incorporation

The incorporation of proteins to neat DES is probably the initial challenge when it comes to investigating systems where the dissolution kinetics are extremely slow. The development of approaches that avoid or minimize any physical, chemical or thermal stress becomes of great importance to preserve the integrity of biomolecules upon incorporation. As such, the existing methodologies should be further improved, and novel approaches developed.

When incorporating biomolecules through heating and mixing (or incubating), an increase in temperature results in a reduction of viscosity, enhancing the solvation rates. However, thermal stress may affect the integrity of the biomolecule. The unfolding temperature of the protein must be considered. As a general rule, the system must not be heated beyond this temperature to avoid the thermal degradation of the protein if this happens irreversibly, which is very often system dependent. Similarly, other routes for protein degradation (e.g. chemical degradation) must be taken into account as these may be significant at high temperatures.

Similarly, if co-solvents, salts or polymers are present in the incorporation of biomolecules (e.g. two-phase extraction), the content of residual traces of solvent in the final product must be considered. The influence of those in protein structure and stability must be evaluated, as well as the impact on the characteristics of the solvents. Ultimately and for the sake of the green character of DES, the utilization of green alternatives is also preferred for these processes.

For example, when freeze-drying of DES-water systems is performed on DES with low transition temperatures (e.g. 1:2 choline chloride:glycerol), it is observed that the system melts during primary drying. This situation could result in unwanted bubbling, which causes interfacial stress in the system that may affect sensitive proteins. Thus, the development of optimized freeze-drying methods, where temperature, pressure and water activity are finely tuned, is required to ensure that the quality of the biomolecules is preserved upon drying.

5.2 Conformational and dynamic landscape

One of the key aspects to understand protein behavior and function is to determine the structural and dynamic characteristics that confer the active behavior to the protein. As such, the investigations on protein structure, folding, self-association and internal dynamics in DES must continue. In-house spectroscopy methods could be combined with more advanced structural characterization using scattering methods to shed some light on the structure and conformation of proteins in DES.

The unique interplay between protein structure and dynamics, and the stability and function of those must be further investigated. To date, structural investigations of proteins in DES have began to appear, but the internal and external dynamics of the macromolecules in solution are relatively unexplored. Therefore, those investigations should be expanded to connect the biophysical behavior of proteins to the performance of technologies using this type of systems.

5.3 The role of protein-solvent and solvent-solvent interactions

In order to develop a more profound knowledge that aims to predict protein behavior, understanding the interaction between proteins and the surrounding environment becomes of great importance. Most experimental techniques lack the resolution to study molecular interactions in such level of detail and for systems of such a great complexity. On the other hand, theoretical and computational approaches would provide enough resolution to explore the molecular landscape of the system. Nonetheless, they present other limitations, such as the necessity of accurate force fields for MD simulations to avoid mechanistic aberrations.

One of the options to refine the results from the computational methods is to use experimental results as restraints for the simulations (e.g. Monte Carlo or MD), as it has been performed for other (arguably simpler) systems. The combination of total neutron diffraction and Monte Carlo simulations was used in the investigation of the structure of neat and hydrated DES (Hammond et al., 2016, 2017). Similarly, the same approach has been used to investigate the structure of micelles in aqueous solutions at the atomistic level, showing the possibilities of combining these methods for linking mesoscopic structure and molecular interactions (Hargreaves, Bowron, & Edler, 2011).

As observed in IL-protein systems, electrostatics and solvent cohesivity play an important role in protein structure, stability and function (Schindl et al., 2019). Potentially, certain parallelism can be drawn to the behavior of proteins in DES, where the charged nature of the environment has an influence on the electrostatics of the protein. Also, the ability of the solvents to promote amphiphile self-assembly will definitely help to understand certain aspects of protein behavior, such as protein folding. Therefore, investigating specific ion effects and protein-solvent interactions in a systematic manner will become highly relevant for elucidating the behavior of proteins in DES.

6. Concluding remarks

A variety of research has been performed to probe the ability of DES to serve as solvent for performing biocatalyzed reactions. However, investigations that aim to study the fundamental behavior of proteins (structure, dynamics and interactions) are still rather scarce. The field is starting to gather some basic knowledge on protein behavior in DES, but a long road lies ahead, and further efforts must be paid to systematically study these systems. Probably, the most complicated scenario that the future of these investigations will face is to elucidate the stabilization/solvation mechanism of proteins and enzymes by these solvents. The remarkable complexity of biological systems in neoteric solvents undoubtedly requires integrative approaches to study the characteristics of these systems. As such, the combination of methods that aim to understand the interactions between the protein and the solvent, and to link those to the protein behavior will be required. Furthermore, the knowledge from protein behavior in ILs, more thoroughly investigated to date, may serve as the initial benchmark to provide a framework for the complex interactions between the solvent and the proteins. As the final (and ideal) goal in the field, a level of understanding that allows prediction of the behavior of a protein in a given DES will allow tailoring of the systems for applications the wide range of areas that have been envisaged for biomolecules and DES.

References

Antonini, E., Carrea, G., & Cremonesi, P. (1981). Enzyme catalysed reactions in water—Organic solvent two-phase systems. *Enzyme and Microbial Technology*, *3*(4), 291–296. https://doi.org/10.1016/0141-0229(81)90002-8.

Benedetto, A., & Ballone, P. (2015). Room temperature ionic liquids meet biomolecules: A microscopic view of structure and dynamics. *ACS Sustainable Chemistry & Engineering*, *4*(2), 392–412. https://doi.org/10.1021/acssuschemeng.5b01385.

Berendsen, H. J., & Hayward, S. (2000). Collective protein dynamics in relation to function. *Current Opinion in Structural Biology*, *10*(2), 165–169. https://doi.org/10.1016/s0959-440x(00)00061-0.

Dai, Y., van Spronsen, J., Witkamp, G. J., Verpoorte, R., & Choi, Y. H. (2013). Natural deep eutectic solvents as new potential media for green technology. *Analytica Chimica Acta*, *766*, 61–68. https://doi.org/10.1016/j.aca.2012.12.019.

Dai, Y., Witkamp, G. J., Verpoorte, R., & Choi, Y. H. (2015). Tailoring properties of natural deep eutectic solvents with water to facilitate their applications. *Food Chemistry*, *187*, 14–19. https://doi.org/10.1016/j.foodchem.2015.03.123.

Debeljuh, N., Barrow, C. J., Henderson, L., & Byrne, N. (2011). Structure inducing ionic liquids-enhancement of alpha helicity in the Abeta(1-40) peptide from Alzheimer's disease. *Chemical Communications (Cambridge, England)*, *47*(22), 6371–6373. https://doi.org/10.1039/c1cc10377f.

Ebbinghaus, S., Kim, S. J., Heyden, M., Yu, X., Heugen, U., Gruebele, M., et al. (2007). An extended dynamical hydration shell around proteins. *Proceedings of the National Academy of Sciences of the United States of America*, *104*(52), 20749–20752. https://doi.org/10.1073/pnas.0709207104.

Esquembre, R., Sanz, J. M., Wall, J. G., del Monte, F., Mateo, C. R., & Ferrer, M. L. (2013). Thermal unfolding and refolding of lysozyme in deep eutectic solvents and their aqueous dilutions. *Physical Chemistry Chemical Physics*, *15*(27), 11248–11256. https://doi.org/10.1039/c3cp44299c.

Frauenfelder, H., Chen, G., Berendzen, J., Fenimore, P. W., Jansson, H., McMahon, B. H., et al. (2009). A unified model of protein dynamics. *Proceedings of the National Academy of Sciences of the United States of America*, *106*(13), 5129–5134. https://doi.org/10.1073/pnas.0900336106.

Gabel, F., Bicout, D., Lehnert, U., Tehei, M., Weik, M., & Zaccai, G. (2002). Protein dynamics studied by neutron scattering. *Quarterly Reviews of Biophysics*, *35*(4), 327–367. https://doi.org/10.1017/s0033583502003840.

Ghosh, S., Parui, S., Jana, B., & Bhattacharyya, K. (2015). Ionic liquid induced dehydration and domain closure in lysozyme: FCS and MD simulation. *The Journal of Chemical Physics*, *143*(12). https://doi.org/10.1063/1.4931974.

Gorke, J. T., Srienc, F., & Kazlauskas, R. J. (2008). Hydrolase-catalyzed biotransformations in deep eutectic solvents. *Chemical communications (Cambridge, England)*, (10), 1235–1237. https://doi.org/10.1039/b716317g.

Gutierrez, M. C., Ferrer, M. L., Mateo, C. R., & del Monte, F. (2009). Freeze-drying of aqueous solutions of deep eutectic solvents: A suitable approach to deep eutectic suspensions of self-assembled structures. *Langmuir, 25*(10), 5509–5515. https://doi.org/10.1021/la900552b.

Gutierrez, M. C., Ferrer, M. L., Yuste, L., Rojo, F., & del Monte, F. (2010). Bacteria incorporation in deep-eutectic solvents through freeze-drying. *Angewandte Chemie, International Edition in English, 49*(12), 2158–2162. https://doi.org/10.1002/anie.200905212.

Hammond, O. S., Bowron, D. T., & Edler, K. J. (2016). Liquid structure of the choline chloride-urea deep eutectic solvent (reline) from neutron diffraction and atomistic modelling. *Green Chemistry, 18*(9), 2736–2744. https://doi.org/10.1039/c5gc02914g.

Hammond, O. S., Bowron, D. T., & Edler, K. J. (2017). The effect of water upon deep eutectic solvent nanostructure: An unusual transition from ionic mixture to aqueous solution. *Angewandte Chemie, International Edition in English, 56*(33), 9782–9785. https://doi.org/10.1002/anie.201702486.

Hammons, J. A., Zhang, F., & Ilavsky, J. (2018). Extended hierarchical solvent perturbations from curved surfaces of mesoporous silica particles in a deep eutectic solvent. *Journal of Colloid and Interface Science, 520*, 81–90. https://doi.org/10.1016/j.jcis.2018.02.078.

Hargreaves, R., Bowron, D. T., & Edler, K. (2011). Atomistic structure of a micelle in solution determined by wide Q-range neutron diffraction. *Journal of the American Chemical Society, 133*(41), 16524–16536. https://doi.org/10.1021/ja205804k.

Harifi-Mood, A. R., Ghobadi, R., & Divsalar, A. (2017). The effect of deep eutectic solvents on catalytic function and structure of bovine liver catalase. *International Journal of Biological Macromolecules, 95*, 115–120. https://doi.org/10.1016/j.ijbiomac.2016.11.043.

Kadyan, A., Juneja, S., & Pandey, S. (2019). Photophysical behavior and fluorescence quenching of l-tryptophan in choline chloride-based deep eutectic solvents. *Journal of Physical Chemistry B, 123*(35), 7578–7587. https://doi.org/10.1021/acs.jpcb.9b04659.

Khodadadi, S., Pawlus, S., Roh, J. H., Garcia Sakai, V., Mamontov, E., & Sokolov, A. P. (2008). The origin of the dynamic transition in proteins. *The Journal of Chemical Physics, 128*(19). https://doi.org/10.1063/1.2927871.

Klibanov, A. M. (2001). Improving enzymes by using them in organic solvents. *Nature, 409*(6817), 241–246. https://doi.org/10.1038/35051719.

Kubitzki, M. B., de Groot, B. L., & Seeliger, D. (2017). *Protein dynamics: From structure to function (D. Rigden Ed.)*. Springer Science.

Kumar, A., & Venkatesu, P. (2014). Does the stability of proteins in ionic liquids obey the Hofmeister series? *International Journal of Biological Macromolecules, 63*, 244–253. https://doi.org/10.1016/j.ijbiomac.2013.10.031.

Li, N., Wang, Y., Xu, K., Huang, Y., Wen, Q., & Ding, X. (2016). Development of green betaine-based deep eutectic solvent aqueous two-phase system for the extraction of protein. *Talanta, 152*, 23–32. https://doi.org/10.1016/j.talanta.2016.01.042.

Mann, J. P., McCluskey, A., & Atkin, R. (2009). Activity and thermal stability of lysozyme in alkylammonium formate ionic liquids—Influence of cation modification. *Green Chemistry, 11*(6), 785–792. https://doi.org/10.1039/b900021f.

McCammon, J. A. (1984). Protein dynamics. *Reports on Progress in Physics, 47*(1), 1–46. https://doi.org/10.1088/0034-4885/47/1/001.

Monhemi, H., Housaindokht, M. R., Moosavi-Movahedi, A. A., & Bozorgmehr, M. R. (2014). How a protein can remain stable in a solvent with high content of urea: Insights from molecular dynamics simulation of Candida antarctica lipase B in urea: Choline chloride deep eutectic solvent. *Physical Chemistry Chemical Physics, 16*(28), 14882–14893. https://doi.org/10.1039/c4cp00503a.

Pandey, A., Rai, R., Pal, M., & Pandey, S. (2014). How polar are choline chloride-based deep eutectic solvents? *Physical Chemistry Chemical Physics, 16*(4), 1559–1568. https://doi.org/10.1039/c3cp53456a.

Paoli, M., Liddington, R., Tame, J., Wilkinson, A., & Dodson, G. (1996). Crystal structure of T state haemoglobin with oxygen bound at all four haems. *Journal of Molecular Biology, 256*(4), 775–792. https://doi.org/10.1006/jmbi.1996.0124.

Parnica, J., & Antalik, M. (2014). Urea and guanidine salts as novel components for deep eutectic solvents. *Journal of Molecular Liquids, 197*, 23–26. https://doi.org/10.1016/j.molliq.2014.04.016.

Patist, A., & Zoerb, H. (2005). Preservation mechanisms of trehalose in food and biosystems. *Colloids and Surfaces B: Biointerfaces, 40*(2), 107–113. https://doi.org/10.1016/j.colsurfb.2004.05.003.

Perutz, M. F. (1972). Nature of haem-haem interaction. *Nature, 237*(5357), 495–499. https://doi.org/10.1038/237495a0.

Perutz, M. F., Rossmann, M. G., Cullis, A. F., Muirhead, H., Will, G., & North, A. C. (1960). Structure of haemoglobin: A three-dimensional Fourier synthesis at 5.5-A. resolution, obtained by X-ray analysis. *Nature, 185*(4711), 416–422. https://doi.org/10.1038/185416a0.

Richards, F. M. (1977). Areas, volumes, packing and protein structure. *Annual Review of Biophysics and Bioengineering, 6*, 151–176. https://doi.org/10.1146/annurev.bb.06.060177.001055.

Rogers, R. D., & Seddon, K. R. (2003). Chemistry. Ionic liquids—Solvents of the future? *Science, 302*(5646), 792–793. https://doi.org/10.1126/science.1090313.

Rupley, J. A., & Careri, G. (1991). Protein hydration and function. In C. B. Anfinsen, F. M. Richards, J. T. Edsall, & D. S. Eisenberg (Eds.), *Vol. 41. Advances in protein chemistry* (pp. 37–172). Academic Press.

Sanchez-Fernandez, A., Arnold, T., Jackson, A. J., Fussell, S. L., Heenan, R. K., Campbell, R. A., et al. (2016). Micellization of alkyltrimethylammonium bromide surfactants in choline chloride:glycerol deep eutectic solvent. *Physical Chemistry Chemical Physics, 18*(48), 33240–33249. https://doi.org/10.1039/c6cp06053f.

Sanchez-Fernandez, A., Edler, K. J., Arnold, T., Alba Venero, D., & Jackson, A. J. (2017). Protein conformation in pure and hydrated deep eutectic solvents. *Physical Chemistry Chemical Physics, 19*(13), 8667–8670. https://doi.org/10.1039/c7cp00459a.

Sanchez-Fernandez, A., Edler, K. J., Arnold, T., Heenan, R. K., Porcar, L., Terrill, N. J., et al. (2016). Micelle structure in a deep eutectic solvent: A small-angle scattering study. *Physical Chemistry Chemical Physics, 18*(20), 14063–14073. https://doi.org/10.1039/c6cp01757f.

Sanchez-Fernandez, A., Hammond, O. S., Jackson, A. J., Arnold, T., Doutch, J., & Edler, K. J. (2017). Surfactant-solvent interaction effects on the micellization of cationic surfactants in a carboxylic acid-based deep eutectic solvent. *Langmuir, 33*(50), 14304–14314. https://doi.org/10.1021/acs.langmuir.7b03254.

Sarkar, S., Ghosh, S., & Chakrabarti, R. (2017). Ammonium based stabilizers effectively counteract urea-induced denaturation in a small protein: Insights from molecular dynamics simulations. *RSC Advances*, 7(83), 52888–52906. https://doi.org/10.1039/c7ra10712a.

Schindl, A., Hagen, M. L., Muzammal, S., Gunasekera, H. A. D., & Croft, A. K. (2019). Proteins in ionic liquids: Reactions, applications, and futures. *Frontiers in Chemistry*, 7, 347. https://doi.org/10.3389/fchem.2019.00347.

Serdakowski, A. L., & Dordick, J. S. (2008). Enzyme activation for organic solvents made easy. *Trends in Biotechnology, 26*(1), 48–54. https://doi.org/10.1016/j.tibtech.2007.10.007.

Stepankova, V., Bidmanova, S., Koudelakova, T., Prokop, Z., Chaloupkova, R., & Damborsky, J. (2013). Strategies for stabilization of enzymes in organic solvents. *ACS Catalysis, 3*(12), 2823–2836. https://doi.org/10.1021/cs400684x.

Tanford, C. (1968). Protein denaturation. In C. B. Anfinsen, M. L. Anson, J. T. Edsall, & F. M. Richards (Eds.), *Vol. 23. Advances in protein chemistry* (pp. 121–282). Academic Press.

White, F. H., Jr. (1976). Studies on secondary structure in chicken egg-white lysozyme after reductive cleavage of disulfide bonds. *Biochemistry*, *15*(13), 2906–2912. https://doi.org/10.1021/bi00658a032.

Wijaya, E. C., Separovic, F., Drummond, C. J., & Greaves, T. L. (2016). Activity and conformation of lysozyme in molecular solvents, protic ionic liquids (PILs) and salt-water systems. *Physical Chemistry Chemical Physics*, *18*(37), 25926–25936. https://doi.org/10.1039/c6cp03334b.

Wu, B.-P., Wen, Q., Xu, H., & Yang, Z. (2014). Insights into the impact of deep eutectic solvents on horseradish peroxidase: Activity, stability and structure. *Journal of Molecular Catalysis B: Enzymatic*, *101*, 101–107. https://doi.org/10.1016/j.molcatb.2014.01.001.

Xin, R., Qi, S., Zeng, C., Khan, F. I., Yang, B., & Wang, Y. (2017). A functional natural deep eutectic solvent based on trehalose: Structural and physicochemical properties. *Food Chemistry*, *217*, 560–567. https://doi.org/10.1016/j.foodchem.2016.09.012.

Zahn, S., Kirchner, B., & Mollenhauer, D. (2016). Charge spreading in deep eutectic solvents. *ChemPhysChem*, *17*(21), 3354–3358. https://doi.org/10.1002/cphc.201600348.

Zaks, A., & Klibanov, A. M. (1984). Enzymatic catalysis in organic media at 100 degrees C. *Science*, *224*(4654), 1249–1251. https://doi.org/10.1126/science.6729453.

Zhao, H. (2016). Protein stabilization and enzyme activation in ionic liquids: Specific ion effects. *Journal of Chemical Technology and Biotechnology*, *91*(1), 25–50. https://doi.org/10.1002/jctb.4837.

CHAPTER FIVE

Natural deep eutectic solvents as performance additives for biocatalysis

Özlem Erol[a] and Frank Hollmann[b,*]
[a]Natural Products Laboratory, Institute of Biology, University of Leiden, The Netherlands
[b]Department of Biotechnology, Delft University of Technology, The Netherlands
[*]Corresponding author: e-mail address: f.hollmann@tudelft.nl

Contents

Abstract

Following ionic liquids, (natural) deep eutectic solvents ((NA)DES) are receiving significant attention as performance additives for biocatalytic reactions.

(NA)DES are increasingly evaluated as solvents to replace water in hydrolase-catalyzed esterification reactions thereby shifting the reaction equilibrium. They also frequently outperform water in terms of solubility properties of hydrophobic reagents and thereby enable higher space-time yields. Furthermore, (NA)DES frequently exceed stabilizing effects on enzymes and thereby enable more robust (and therefore economically more attractive) biocatalytic syntheses.

In this contribution, we will summarize and critically evaluate the recent literature on (NA)DES-supported biocatalysis.

1. Introduction

Cells, microbial and those of higher organisms, are roughly composed of 70% water while the remaining 30% share out to proteins (15%), DNA (1%), RNA (6%), (Phospho)lipids (2%) polysaccharides (2%) and small molecules (4%) (Anonymous, 2014). Therefore, microbial cells are broadly seen as aqueous solutions of these components. Even though this model may be too simplistic and the interior of a cell should be more seen as a gel rather than a dilute aqueous solution, biocatalysis is traditionally performed in

Advances in Botanical Research, Volume 97
ISSN 0065-2296
https://doi.org/10.1016/bs.abr.2020.09.004

aqueous media. This approach, however, severely limits the broad applicability of biocatalysis for the synthesis of useful chemicals as many of them are rather hydrophobic and therefore poorly soluble in aqueous media. Dilute product mixtures of a few grams per liter reaction broth, however, are not attractive neither from an economic, not environmental point-of-view. Therefore, limitation to aqueous reaction media poses a severe limitation *en route* to a broad applicability of biocatalysis for chemical synthesis.

Pioneering works by Klibanov and coworkers (Dordick, Marletta, & Klibanov, 1986; Zaks & Klibanov, 1984, 1985) that demonstrated that enzymes can be active under non-aqueous conditions first received mostly academic interest. In recent years, however, the interest in neoteric solvents for biocatalysis has been increasing steadily.

Following the, now fallen from grace, ionic liquids, deep eutectic solvents are enjoying a rapidly increasing popularity in the biocatalysis community as biobased and non-toxic alternatives (Durand, Lecomte, & Villeneuve, 2013; Gotor-Fernández & Paul, 2019; Ibn Majdoub Hassani, Amzazi, & Lavandera, 2019; Kourist & González-Sabín, 2020; María, Guajardo, & Kara, 2020; Mbous et al., 2017; Pätzold et al., 2019; Pätzold, Weimer, Liese, & Holtmann, 2019; Perna, Vitale, & Capriati, 2020; Tan & Dou, 2020; Xu, Zheng, Zong, Li, & Lou, 2017). The aim of this chapter is to critically summarize the current efforts on establishing, understanding and applying DES as neoteric solvents for biocatalysis.

Kazlauskas and coworkers probably were the first ones to use DES as solvents for biocatalytic reactions (Gorke, Srienc, & Kazlauskas, 2008), demonstrating that several lipases catalyse the transesterification of ethyl valerate to butyl valerate. These authors, however, also pointed out one possible (undesired) side reaction in this reaction, i.e. the participation of the solvent (e.g. ethylene glycol- or glycerol-based DES) in the lipase-catalyzed reaction. Interestingly, probably to thermodynamic stabilization in the DES-typical H-bond network, DES components such as ethylene glycol or glycerol were significantly less reactive than expected.

Ever since these pioneering works, the number of reports on biocatalysis in DES has been increasing steadily (Fig. 1).

2. Natural deep eutectic solvents in biocatalysis

Table 1 gives a representative, yet incomplete overview over the manifold applications of DES as solvents for biocatalytic reactions.

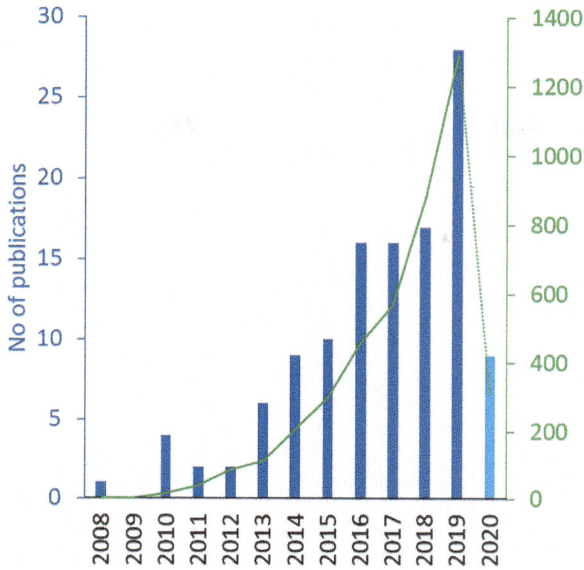

Fig. 1 Publications (blue bars) and their citations (green line) found within web of knowledge using the search terms "Deep Eutectic Solvent" and "Biocatalysis," accessed on 29.03.2020.

DES are rarely used as "neat solvents" and often water (buffer)/DES mixtures give best results as compared to "anhydrous" conditions. On the one hand, this may be ascribed to enzymes needing a certain amount of (non-bulk) water to maintain activity (water as lubricant) (María et al., 2020). On the other hand, the high viscosity of many DES necessitates dilution (with water) to attain acceptable viscosities for practical application. Above approximately 50% (v/v) of water as the reaction medium DES are more characterized as "performance additives" rather than as co-solvent.

In the following, we therefore focus on aspects of DESs with a clear advantage over existing reaction systems.

Clearly, one exciting aspect of DESs is their tuneable solvent properties (hydrophilicity, hydrophobicity). Carbohydrates, for example, are best soluble in aqueous media. If, however, the esterification of carbohydrates with carboxylic acids is the desired reaction, water is a very unfavorable (co-) solvent due to the unfavorable equilibrium of esterifications in aqueous media. In this respect, carbohydrate-based DES (in which the carbohydrates are liquefied in the absence of water) are very promising alternative solvents for the synthesis of glycolipids e.g. as surfactants (Scheme 1) (Siebenhaller et al., 2018, 2016).

Table 1 Biocatalytic reactions performed in DES.

Product	Enzyme	DES used	Remarks	Reference
Esterifications 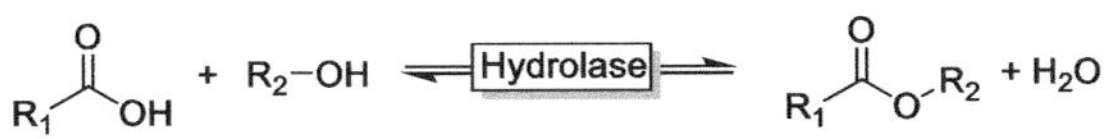				
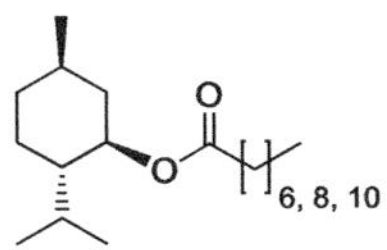	CRL	Men:OA (55:45)	DES as solvent and reagent, addition of water improves the enzymatic reaction significant, BCL, PCL and PFL showed no activity in DES,best results obtained with CRL with 10% water	Hümmer et al. (2018)
	N435	Men:DA (63:35)		
	CALB	Men:DDA (75:25)		
	BCL	H_2O-content:		
	PCL	0, 1, 5 or 10 wt%		
	PFL			
	CRL	Men:La (9:1 to 1:1)	As above, kinetic resolution of rac-menthol	Craveiro et al. (2019)
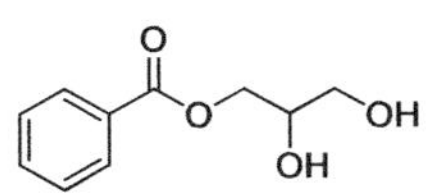	CalB immobilized	ChCl:Gly: Phosphate buffer	DES as solvent and reagent, positive influence of water on the reaction	Guajardo, Ahumada, and de Maria (2020)

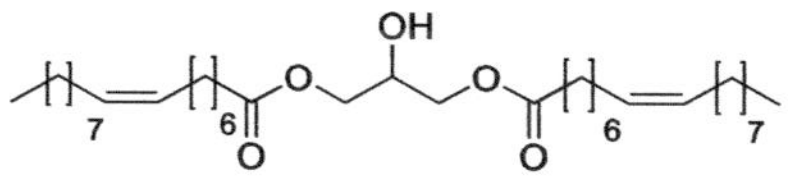	N435	B:Gly	Selectivity depends on DES	Zeng, Qi, Xin, Yang, and Wang (2015); Xu et al. (2017); Xu, Zheng, et al. (2017)
	RMIM	ChCl:Gly		
	TLIM	H_2O-content: 1–4%		
	Lipase G50			
	F-AP			
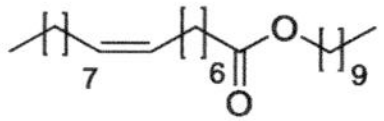	CalB immobilized	CHCL:G (1:1)	in ChCl:Gly DES	Kleiner and Schorken (2015)
		CHCl:U (1:2)	high enzyme stability (1200 h), -ChCl:U DES advantage is ester production free of by product	
		H_2O-content: 0–10%		
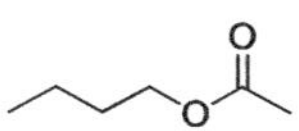	CalB	CHCL:G (1:2)	Higher esterification yield 80% (higher than reference solvent n-heptan (50%))	Bubalo et al. (2015)
		CHCL:EG (1:2)		
		CHCl:U (1:2)		
		H_2O-content: 0.5–2.5 mol%		

Continued

Table 1 Biocatalytic reactions performed in DES.—cont'd

Product	Enzyme	DES used	Remarks	Reference
Ester hydrolysis 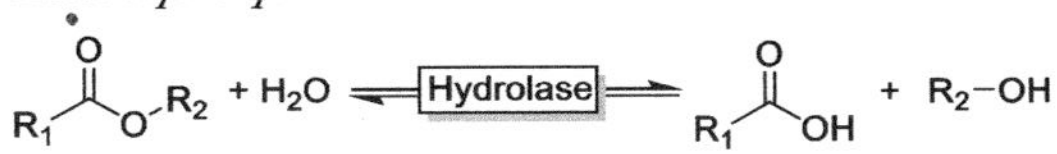				
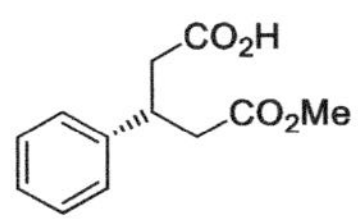	CalB	CHCL:U (1:2)	Desymmetrization reaction, increased enantioselectivity in DES	Fredes, Chamorro, and Cabrera (2019)
		H_2O-content: 50%		
Free fatty acids from hydrolysis of natural oils	Amino Lipase PS	ChCl:U (1:2)	Best reaction conditions reached in ChCl:U with 38% of water	Yang, Tong, Yang, Liu, and Wang (2019)
Pine nut oil		ChCl:Gly (1:2)		
		ChCl:EG (1:2)		
		ChCl:1,2-PG (1:2 – 1:3)		
		ChCl:CA (1:3)		
		ChCl:LA (1:1)		
		H_2O-content: 38%		

Palmitic acid hydrolysis of para-nitrophenyl palmitate	BCL	ChCl:U (1:2)	enzyme activity enhanced by up to 230%	Juneidi, Hayyan, Hashim, and Hayyan (2017)
		ChCl:Gly (1:2)		
		ChCl:EG (1:2)		
		ChCl:DEG (1:2)		
		EAC:Gly (1:2)		
		EAC:EG (1:2)		
		EAC:TEG (1:2)		
		H_2O-content: variable		
Transesterifications				
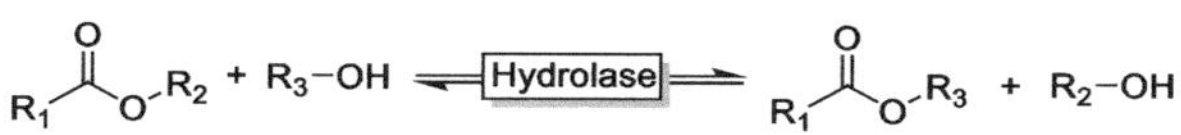				
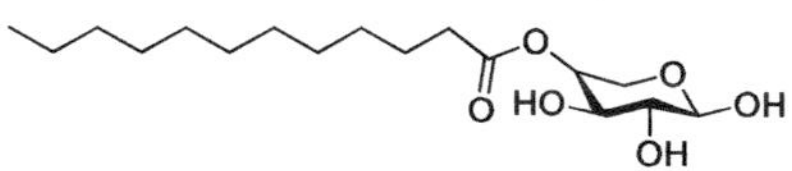	CalB	ChCl: carbohydrate (71.6% Glu and 16.6% Xyl)	DES as solvent and reagent, carbohydrate obtained from beech wood	Siebenhaller et al. (2018)

Continued

Table 1 Biocatalytic reactions performed in DES.—cont'd

Product	Enzyme	DES used	Remarks	Reference
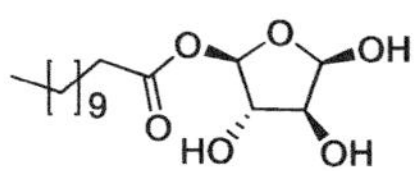	CalB	ChCl:Ara (1:1) ChCl:Glu (1:1) ChCl:Xyl (1:1) ChCl:Man (1:1) ChCl:Rha (1:1) ChCl:Lev (1:1)		Siebenhaller et al. (2016)
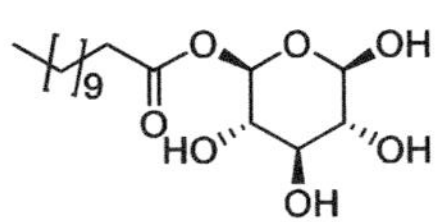	N435	ChCl:U	Increased stability of the biocatalyst	Andler, Wang, Rotello, and Goddard (2017)
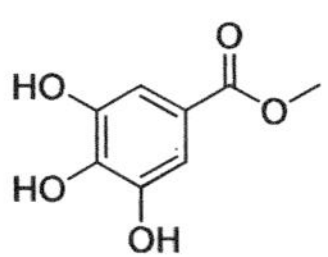	CalB	ChCl:Gly H_2O-content: variable	Optimized water content for maximized activity and minimized hydrolysis	Ulger and Takac (2017)

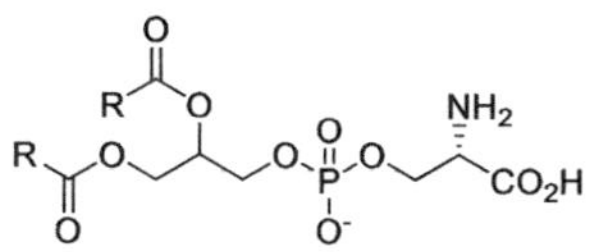	PLD	ChCl:U (1:2)	ChCl:EG DES best t for phosphatidylserine synthesis	Yang and Duan (2016)
		ChCl:A (1:2)		
		ChCL:EG (1:2)		
		ChCL:Gly (1:2)		
		ChCl:1,4-Bu (1:4)		
		ChCl:TEG (1:4)		
		ChCL:X (1:1)		
		ChCL:OA (1:1)		
		ChCl:LA (1:2)		
		ChCL:MAA (1:1)		
		ChCL:MA (1:1)		
		ChCl:CA (1:1)		
Biodiesel	Various lipases	ChCl:Gly (1:2)	best conditions were ChCl: Gly (1:1) in the presence of 30% MeOH	Zhao, Zhang, and Crittle (2013)
		CHAc:Gly (1:1.5)		
		MeOH: 20–50% (v/v)		
biodiesel	Lipozym TL	ChCl:U (1:2)		Kleiner, Fleischer, and Schorken (2016)
	CALB L	ChCl:Gly (1:2)		

Continued

Table 1 Biocatalytic reactions performed in DES.—cont'd

Product	Enzyme	DES used	Remarks	Reference
biodiesel	PEL	ChCl:U (1:1 – 2:1)		Huang, Wu, Wen, Yang, and Yang (2014)
		ChCl:A (1:1 – 2:1)		
	N435	ChCl:Gly (1:1 – 2:1)		
		ChCl:EG (1:1 – 2:1)		
		ChAc:U (1:1 – 2:1)		
		ChAc:A (1:1 – 2:1)		
		ChAc:Gly (1:1 – 2:1)		
		ChCl:EG (1:1 – 2:1)		
Amide synthesis				

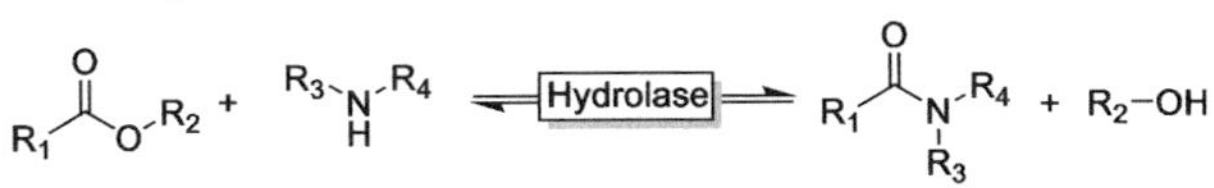

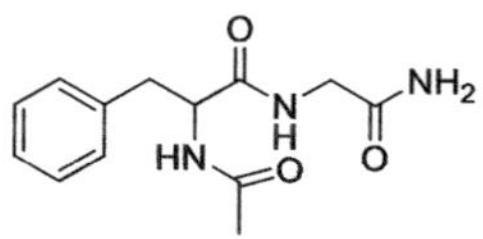	Chymotrypsin	ChCL:U (1:2)	productivities of approx. 20 gL^{-1} h^{-1}, presence of water absolutely crudial	Maugeri, Leitner, and Domínguez de María (2013)
		ChCL:Gly (1:2)		
		ChCL:X (1:1)		
		ChCL:Is (1:2)		
		H_2O-content: 4-50% (v/v)		
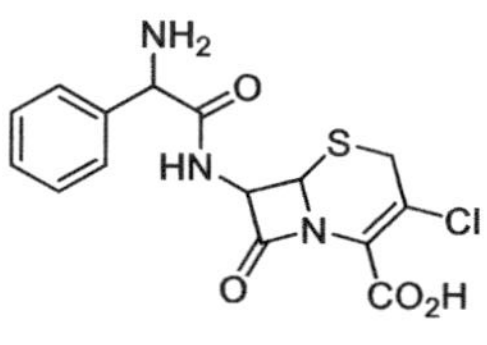	Penicillin acylase	ChCl: CA (1:1)	Higher solubility of 7-ACCA in DESthan in purely aqueous buffer	Wu et al. (2019)
		ChCl:OA (1:1)		
		ChCl:TA (1:1)		
		ChCl:MA (1:1)		
		ChCl:p-toluene		
		ChCL:X (1:1)		
		ChCL:Gly (1:2)		
		ChCl:PG (1:1)		
		ChCL:BG (1:4)		
		ChCL:Gl (1:2)		
		ChCL:I (1:2)		
		ChCl:U (1:2)		
		Buffer		

Continued

Table 1 Biocatalytic reactions performed in DES.—cont'd

Product	Enzyme	DES used	Remarks	Reference
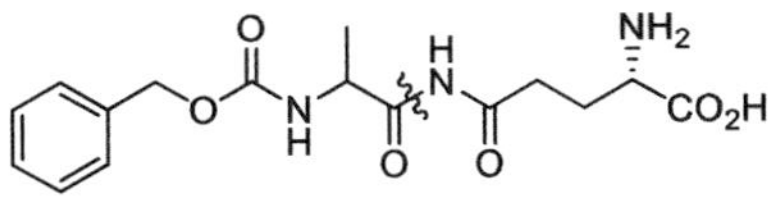	Papain	ChCl:U (1:2) H_2O-content: variable		Cao, Xu, Li, Lou, and Zong (2015)
Reduction reactions 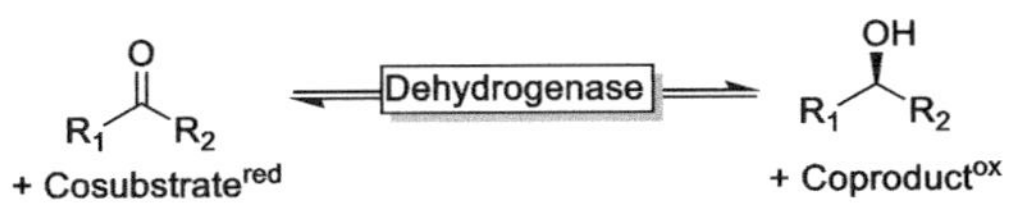				
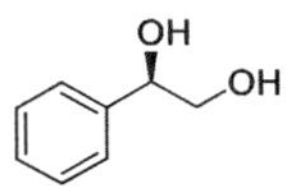	*Kurthia gibsonii* SC031	ChCl:1,4-Bu (1:4) ChCl:U (1:2) ChCl:Gly (1:2) ChCl:E (1:2) ChCL:TEG (1:4)	Increased activity due to cell permeabilization, DESs on whole-cell catalytic properties	Peng et al. (2020)

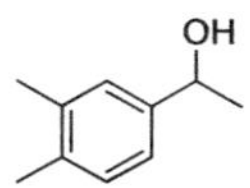	Carrot root	ChCl:G (1:1) ChCl:Xyl (2:1) ChCl:X (5:2) ChCl:G (1:2) ChCl:EG (1:2) H_2O-content 30–80%	Stereoselectivity depends on the DES used and the HBD used in DES	Panić, Elenkov, Roje, Bubalo, and Redovniković (2018)
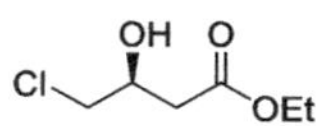	*E. coli* CCZU-T15	ChCL:U (1:2) ChCl:Gly (1:2) ChCL:EG (1:2) H_2O-content: variable	DES better than toluene-water solvent	Dai, Huan, Zhang, and He (2017)
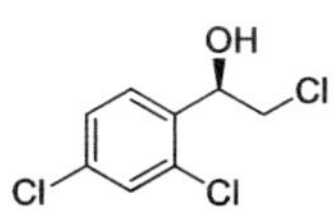 Several examples	*Lb*ADH, ADH-A, TeSADH	ChCl:Gly (1:2) H_2O-content: 50–80%	increased substrate concentration (up to 400 mM) in 20% v/v of DES, preparative scale	Ibn Majdoub Hassani, Amzazi, Kreit, and Lavandera (2020)

Continued

Table 1 Biocatalytic reactions performed in DES.—cont'd

Product	Enzyme	DES used	Remarks	Reference
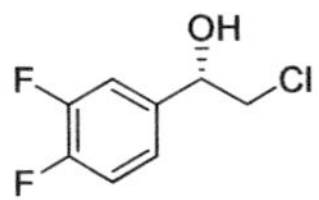	Recombinant *E. coli*	ChCl:GLy (1:1)	ChAc/Lys improves cofactor regeneration,	He, Huang, and Wang (2020)
		ChCl:Lys (1:1)	Increased cell membrane permeability	
		ChCl:GSH (1:1)		
		ChCl:Glu (1:1)		
		ChCl:Trp (1:1)		
		ChCl:Ala (1:1)		
		ChAc:Gly (1:1)		
		ChAc:Lys (1:1)		
		ChAc:GSH (1:1)		
		ChaC:Glu (1:1)		
		ChAc:Trp (1:1)		
		ChAc:Ala (1:1)		
		H_2O-content: <1%		

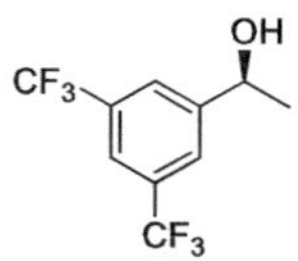	*Rhodococcus erythropolis* XS1012	ChCl:Ala (1:1)	Increased activity due to cell membrane permeabilization	Chen, Qian, Lin, Chen, and Wang (2020)
		ChCl:Cys (1:1)		
		ChCl:EG (1:1)		
		ChCl:Glu (1:1)		
		ChCl:GlY (1:1)		
		ChCl:GSH (1:1)		
		ChCl:IPA (1:1)		
		ChCl:Lys (1:1)		
		ChCl:Trp (1:1)		
		ChCL:Tyr (1:1)		
		ChCl:U (1:1)		
		ChCl:U (1:2)		
		ChCl:U (2:1)		
		H_2O-content: 99%		

Continued

Table 1 Biocatalytic reactions performed in DES.—cont'd

Product	Enzyme	DES used	Remarks	Reference
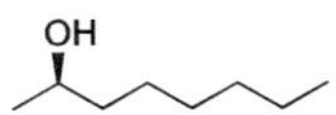	*Acetobacter pasteurianus* GIM1.158	ChCl:U (1:2)	DES increased initial rate	Xu, Du, Zong, Li, and Lou (2016)
		ChCl:Gly (1:2)		
		ChCl:EG (1:2)		
		CHCL:OA (1:2)		
		ChCl:MA (1:2)		
		ChCL:I (1:2)		
		H_2O-content: 90%		
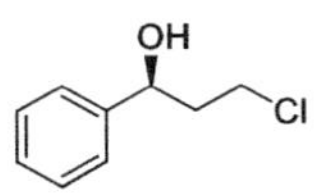	*Acetobacter* sp. CCTCC	ChCl:U (1:2)	Best results obtained with ChCl:U which also increased cell permeability, combining DES with ILs improved the reduction of CPE (85.2 v 93.3)	Xu et al. (2015)
	M209061	ChCl:Gly (1:2)		
		ChCl:EG (1:2)		
		CHCL:OA (1:2)		
		ChCl:MA (1:2)		
		ChCL:I (1:2)		
		H_2O-content: 95%		

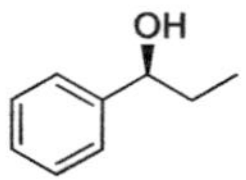	*Te*SADH	ChCl:Gly (1:2)	Significant influence of DES on enantioselectivity	Müller, Lavandera, Gotor-Fernández, and Domínguez de María (2015)
	HLADH			
	*Ras*ADH	H_2O-content: 20%		
	RasADH			
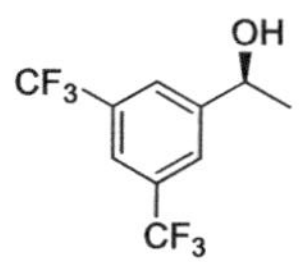	*T. asperellum* ZJPH0810	ChCl:GSH (1:1 – 1:2)		Li, Wang, He, Zhu, and Huang (2019)
		ChCL:Glu (1:1 – 1:2)		
	Candida tropicalis 104	CHCl:Cys (1:1 – 1:2)		
		ChCL:G (1:1 – 1:2)		
	Candida parapsilosis ZJPH1305	H_2O-content: <1%		

Oxidative kinetic resolution

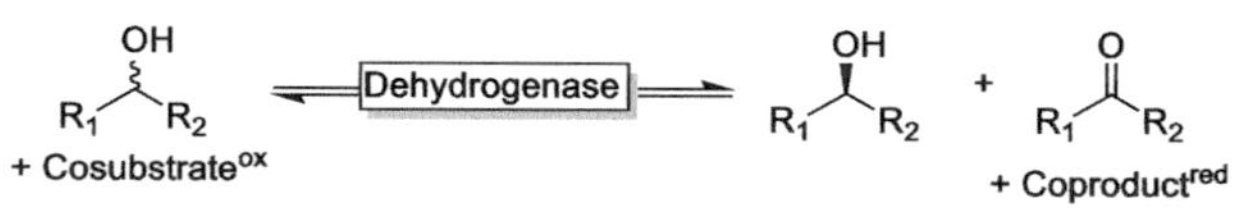

Continued

Table 1 Biocatalytic reactions performed in DES.—cont'd

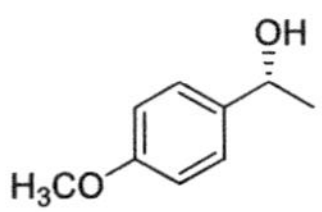	*Acetobacter* sp. CCTCC M209061	ChCL:Gly: [C_4MIM]PF_6	Oxidative kinetic resolution, a combination of ChCl:Gly with [C_4MIM] PF_6 gave best results in terms of substrate solubility and rate	Wei, Liang, Cheng, Zong, and Lou (2016); Xu, Cheng, Lou, and Zong (2015)
		Biphasic system with buffer		
(De)glycosylation reactions 				
Isoquercitin to rutin	*E. coli* BL21-pET21a-rhaB1	ChCl:U (1:2)		Zhang et al. (2020)
		ChCl:Gly (1:2)		
		ChCl:MA (1:1)		
		ChCl:EG (1:2)		
		ChCl:A (1:2)		
		H_2O content: 99%		
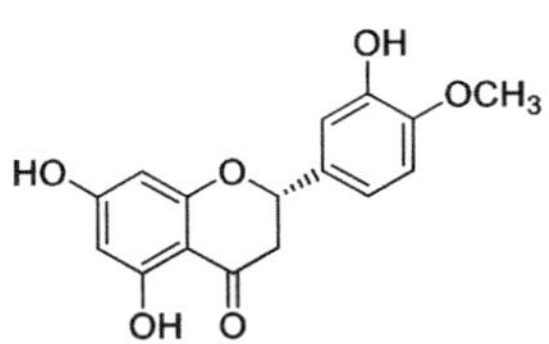	*Acremonium* sp. DSM24697	ChCl:U (1:2)		Weiz, Braun, Lopez, de María, and Breccia (2016)
		ChCL:EG (1:2)		
		ChCl:Gly (1:2)		

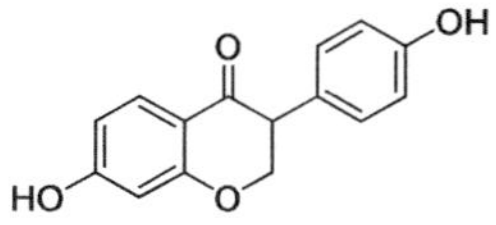	D-glucosidase	ChCl:PEG (1:1–1:3)		Cheng and Zhang (2017)
		ChCl:Glu (1:1–1:3)		
		ChCl:Gly (1:1–1:3)		
		ChCl:EG (1:1–1:3)		
		ChCl:U (1:1–1:3)		
		H_2O content: 80%		
Epoxide hydrolysis				
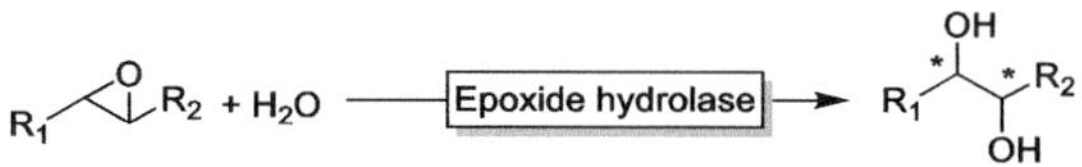				
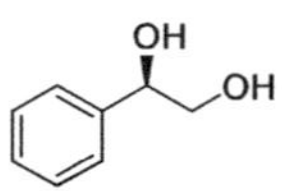	mEH	ChCl:U (1:2)	Improvement of enantiopurity achieved with 10–20% of DES ChCl:TEG in phosphate buffer	Peng, Zhao, Li, Zong, and Lou (2018)
		ChCl:EG (1:2)		
		ChCl:Gly (1:2)		
		ChCl:1,4-Bu (1:4)		
		ChCl:TEG (1:4)		
		ChCl:OA (1:1)		
		ChCl: LA (1:2)		
		ChCl:MA (1:1)		
		ChCl:MA (1:1)		
		ChCl:CA (1:1)		

Continued

Table 1 Biocatalytic reactions performed in DES.—cont'd

Product	Enzyme	DES used	Remarks	Reference
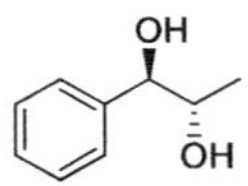	StEH1	ChCl:E (1:2)	DES enabled higher substrate concentrations	Lindberg, Revenga, and Widersten (2010)
		ChCl:Gly (1:2)		
		ChCL:U (1:2)		
		H_2O content: 40–80%		
Decarboxylation reactions 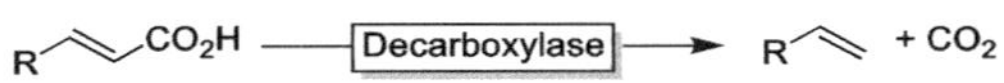				
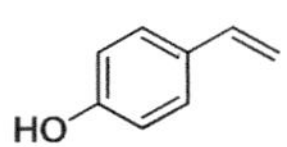	BsPAD	ChCl:Gly (1:2)	Significantly increased substrate solubility	Schweiger et al. (2019)
		ChCl:S (1:1)		
		ChCl:U (1:2)		
		DES:Water		
		H_2O content: 50%		

Desaturation reactions				

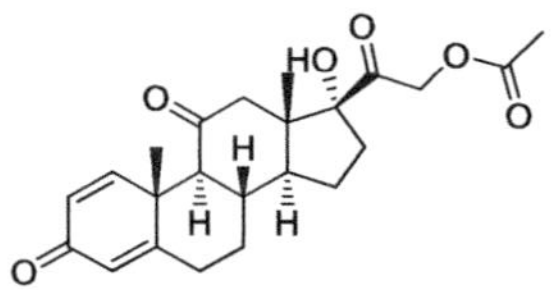	*Arthrobacter simplex*	ChCl:U (1:2) ChCl:EG (1:2) ChCl:Gly (1:2) H_2O content: 94%	Higher substrate solubility and increased rate due to permeabilization	Mao et al. (2018); Mao, Yu, Ji, Liu, and Lu (2016)
C–C-bond formation				
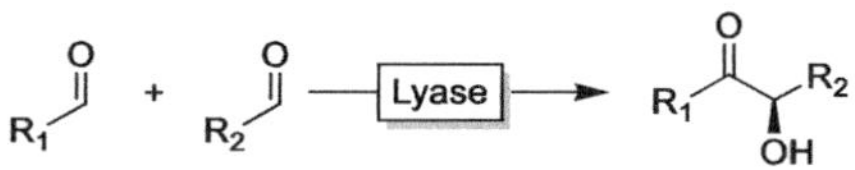				
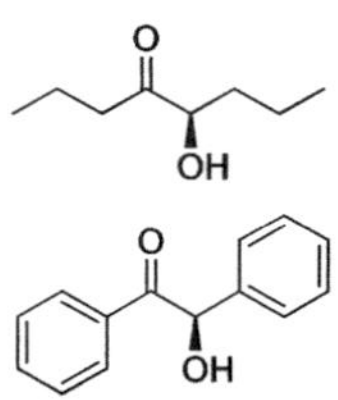	BAL	ChCl:Gly (1:2) ChCl:U (1:1) ChCl:Xyl (1:1) H_2O content: 40%		Maugeri and de Maria (2014)

Continued

Table 1 Biocatalytic reactions performed in DES.—cont'd

Product	Enzyme	DES used	Remarks	Reference
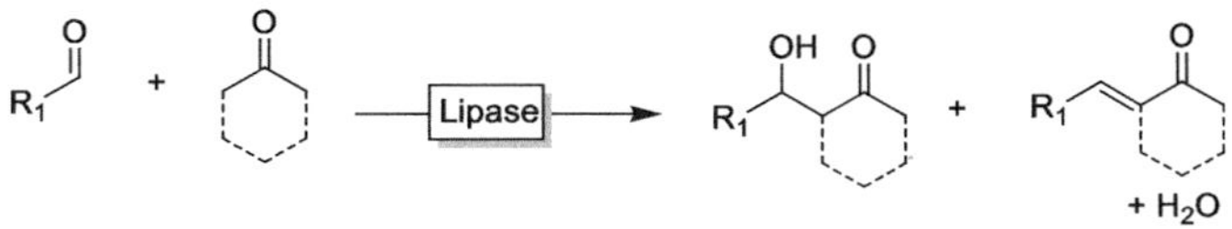				
	PPL	ChCl:Gly (1:1.5,1:2)	ChCl-Gly-based DES as solvent for benzaldehyde and other hydrophobic ketones to perform aldol reactions with molar concentrations	González-Martínez, Gotor, and Gotor-Fernández (2016)
	Alcalase-CLEA	H_2O content: 0–20%		
	CalB			
	PPL	ChCl:Gly (1:1.5)	Best results obtained in the co-solvent acetone	Milker, Pätzold, Bloh, and Holtmann (2019)
		TOABr:EG (1:3)		
		TOABr:1,5PD (1:3)		
		TOABr:4-NBA (2.2:1.5)		
		H_2O		
		Acetone content (up to 20% (v/v))		

Epoxidation reactions

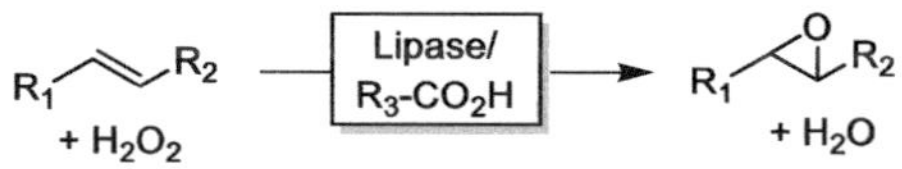

	CalB	ChCl:VA (1:2)	DES system ChCl:U withH_2O_2 achieved the fastest total conversion of reactants,	Ranganathan, Zeitlhofer, and Sieber (2017)
		ChCl:L (1:2)	DES have a stabilizing effect on the chemoenzymatic epoxidation of C=C-double bonds	
		ChCl:4-HPA (1:2)		
		ChCl:MAA (1:1)		
		ChCl:TA (2:1)		
		ChCl:GA (1:2)		
		ChCl:Gly (1:2)		
		ChCl:EG (1:2)		
		ChCl:U (1:2)		
		ChCl:F (1:2)		
		ChCl:Glu (1:2)		
		ChCl:X (1:1)		
		ChCl:S (1:1)		

Continued

Table 1 Biocatalytic reactions performed in DES.—cont'd

Product	Enzyme	DES used	Remarks	Reference
	CalB	ChCl:U (1:2)		Zhou, Wang, Yang, Hollmann, and Wang (2017)
		ChCl:EG (1:2)		
		ChCl:A (1:2)		
		ChCl:Gly (1:1)		
		ChCl:X (1:1)		
		ChCl:S (1:1)		
		ChCl:Xyl:H_2O (5:2:5)		
		ChCl:Glu:H2O (5:2:5)		
		ChCl:Su:H_2O (5:2:5)		
	Lipase G	ChCl:U (1:2)	DES increased the enzyme stability against H_2O_2	Zhou et al. (2017)
		ChCl:Gly (1:2)		
		ChCl:X (1:1)		
		ChCL:EG (1:2)		
		B:Gly (1:2)		

	PCL	ChCl:U (1:2)	Two liquid phase (2LP) approach, DES lower the surface tension of hydrophobic organic phases in aqueous reaction media and thereby enable more efficient biphasic biocatalytic reactions	Lan, Wang, Zhou, Hollmann, and Wang (2017) https://doi.org/10.1039/C7RA06755K
		ChCl:Gly (1:1)		
		ChCl:X (1:1)		
		ChCl:S (1:1)		

CRL, *Candida rugosa* lipase type VII; BCL, amano lipase PS from *Burkholderia cepacia*; PCL, lipase from *Pseudomonas cepacia*; PFL, amano lipase *Pseudomonas fluorescence*; CALB, Candida antarctica lipase B; TLIM, immobilized *Thermomyces lanuginosus* lipase; RMIM, immobilized *Rhizomucor miehei* lipase; G50, *Penicillium camemberti* lipase; F-AP, *Rhizopus oryzae* lipase; BCL, Amano lipase PS (free enzyme)from *Burkholderia cepacia*; CV-CALBY, *Candida antarctica* lipase B; PLD, Phospholipase D from *Streptomyces chromofuscus*; lipozyme TL, lipase from *T. lanuginosus*; PEL, lipase from *Penicillium expansum*; LBADH, alcohol dehydrogenase from *Lactobacillus brevis*; TeSADH, alcohol dehydrogenase from *Thermoanaerobacter ethanolicus*; ADH-A, alcohol dehydrogenase from *Thermoanaerobacter* sp.; RasADH, alcohol dehydrogenase from *Ralstonia* sp.; HLADH, horse liver ADH; StEH1, potato epoxide hydrolase; BsPAD, phenolic acid decarboxylase from *Bacillus subtilis*; mEH, mung bean epoxide hydrolases; BAL, benzaldehyde lyase from *P. fluorescens*; PPL, porcine pancreas lipase; Alcalase-CLEA, protease from *Bacillus licheniformis*; ChCl, choline chloride; ChAc, choline acetate; E, ethanediol; GI, glycol; I, imidazole; PEG, propylene ethylene glycol; BG, butyl glycol; PG, propylene glycol; 1,2-PG, 1,2-propylene glycol; TA, tartaric acid; Is, isosorbide; CA, citric acid; MA, malic acid; MAA, malonic acid; LA, lactic acid; CA, carbolic acid; LA, levulinic acid; GA, glutamic acid; VA, valeric acid; 4-HPA, 4-hydroxy phenyl acetic acid; Cys, cystein; Gln, glutamine; Lys, lysin; Ala, alanine; Trp, tryptophan; OCA, octanoic acid; DA, decanoic acid; DDA, dodecenoic acid; Su, sucrose; OA, oxalic acid; LA, lauric acid; X, xylitol; 1,4-Bu, 1,4-butanediol; 1,5-PD, 1,5-pentadiol; A, acetamide; Lev, levoglucosan; S, sorbitol; Rha, rhamnose; man, mannose; Xyl, xylose; Ara, arabinose; Glu, glucose; F, frucose; EAC, *N,N*-diethyl ethanol ammonium chloride; TEG, triethylene glycol; DEG, diethylene glycol; EG, ethylene glycol; U, urea; B, betaine; Gly, glycerol; 4-NBA, 4-nitrobenzaldehyde; TOABr, tetraoctylammonium bromide; GSH, glutathione, Men, menthol.

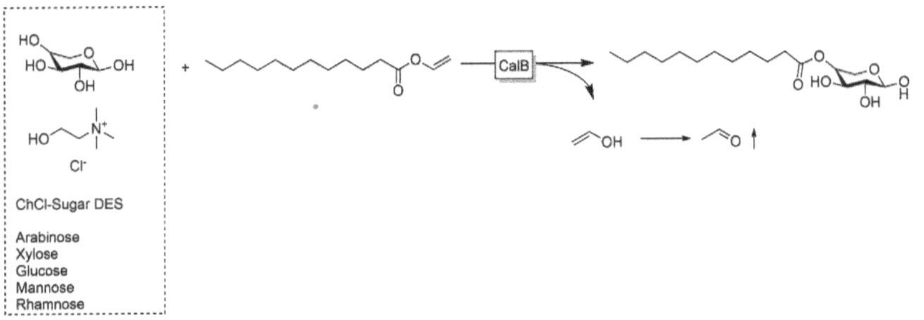

Scheme 1 Using carbohydrate-based DES as solvent for the synthesis of glycolipids.

Another class of reactions that is severely hampered by the presence of water is the amidation of carboxylic acid esters as catalyzed by amidases. In this reaction, the amine group of the starting amine nucleophilically attacks the carboxylate group of the acid ester, eventually substituting the ester alcohol and forming the desired amide. Water, however, competes as nucleophile yielding the free carboxylic acid, yielding thermodynamically stable and kinetically inert carboxylate salts of the starting amine (Scheme 2).

Performing enzymatic amidation reactions in DES offers the possibility of significantly reduced water contents, leading to increased yields of the desired amide products. This has been successfully exploited in the Chymotrypsin-(Maugeri et al., 2013) or Papain-(Cao et al., 2015) catalyzed synthesis of dipeptides or in the penicillin acylase catalyzed synthesis of the antibiotic Cefaclor (Wu et al., 2019).

The aforementioned tuneable solvent properties of DES can be exploited to increase the substrate loadings (and eventually the product titres). Rutin, for example, is practically insoluble in aqueous media while its solubility in aconitic acid-choline chloride is $>80\,\mathrm{g\,kg^{-1}}$ solvent (Choi et al., 2011).

The Gotor-Fernández group, for example, established ChCl-Gly-based DES as solvent for benzaldehyde and other hydrophobic ketones to perform aldol reactions with molar concentrations of these reagents (Scheme 3) (González-Martínez et al., 2016).

These results were confirmed later by Holtmann and coworkers (Milker et al., 2019). These authors, however also found that the highest productivities were found in the absence of DES and using acetone (one of the reagents) as solvent.

Further examples wherein DES enable higher substrate loadings are shown in Table 1.

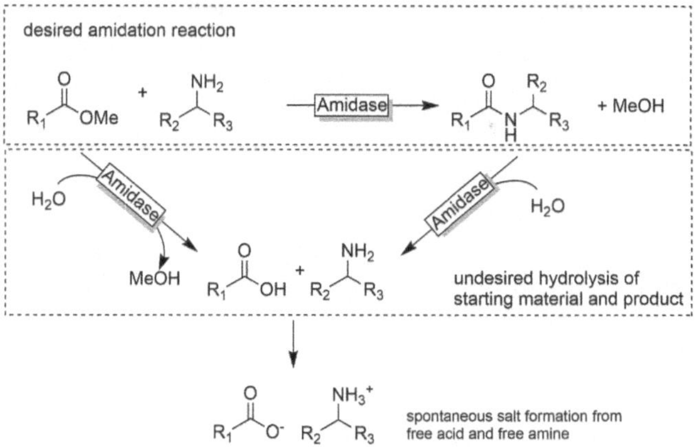

Scheme 2 Amidase-catalyzed acylation of amines using carboxylic acids and amines. In aqueous media also hydrolysis of the starting ester and the desired amide occur yielding the free acid and amine, which spontaneously form thermodynamically and kinetically inert salts.

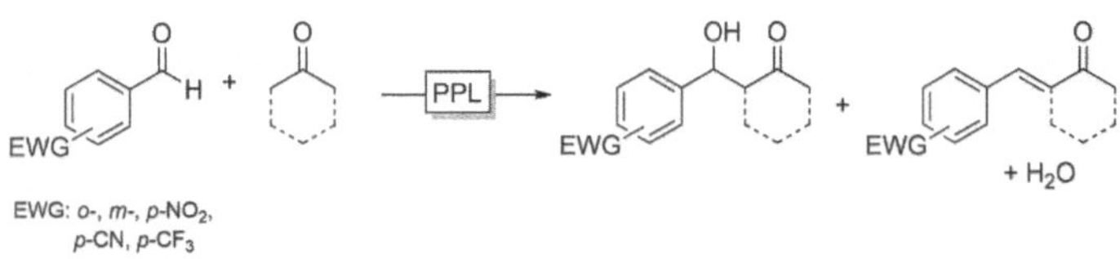

Scheme 3 Aldol reaction (condensation) catalyzed by porcine pancreatic lipase (PPL) in ChCl-Gly-DES.

The broad variability of components of which DES can be formed from also opens up a remarkable extension of DES beyond the mere solvent application: DES in a dual function as solvent and starting material for the (biocatalytic) reaction.

Holtmann and coworkers, for example, reported an esterification reaction of menthol and various fatty acids, forming a DES and hence providing a suitable medium for the solvent-free synthesis of menthol fatty acid esters (Scheme 4) (Hümmer et al., 2018). Addition of water increased the catalytic

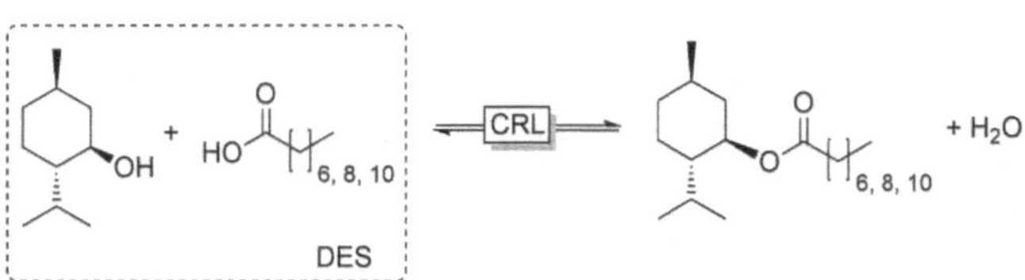

Scheme 4 Using a DES formed by menthol and fatty acids for the lipase (from *Candida rugosa*, CRL)-catalyzed synthesis of menthol esters in a solvent-free approach.

activity of the biocatalyst, which was attributed to the formation of a two liquid phase system and the resulting activity increase of the lipase due to interfacial activation. Under optimized conditions, full conversion of e.g. lauric acid and molar concentrations of the desired product have been achieved (Pätzold, Weimer, et al., 2019).

The original work used enantiomerically pure (-) menthol, which was later extended in work by Paiva and coworkers to the kinetic resolution of *rac*-menthol (Craveiro et al., 2019).

This two-in-one approach has also been used with sugar-based DES for the synthesis of fatty acid esters of carbohydrates (Pätzold, Siebenhaller, et al., 2019; Pätzold, Weimer, et al., 2019; Pöhnlein et al., 2015; Siebenhaller et al., 2018, 2016), structured lipids (Zeng et al., 2015), or benzoate glycerides (Guajardo et al., 2017).

In addition to forming part of the desired product, DESs can also be used to promote cofactor- or cosubstrate-dependent biocatalytic reactions. A very interesting application of glucose-based DESs to promote alcohol dehydrogenase (ADH)-catalyzed stereospecific reduction reactions of prochiral ketones was reported by Lavandera, Gotor and coworkers (Scheme 5) (Mourelle-Insua et al., 2019).

These dual-function "designer" NADES enabled stereoselective reductions of a range of ketones with various ADHs. On the one hand, the NADES served as cosubstrate enabling *in situ* regeneration of the reduced nicotinamide cofactors (NAD(P)H). On the other hand, the NADES also enabled significantly higher substrate concentrations than in aqueous reaction media.

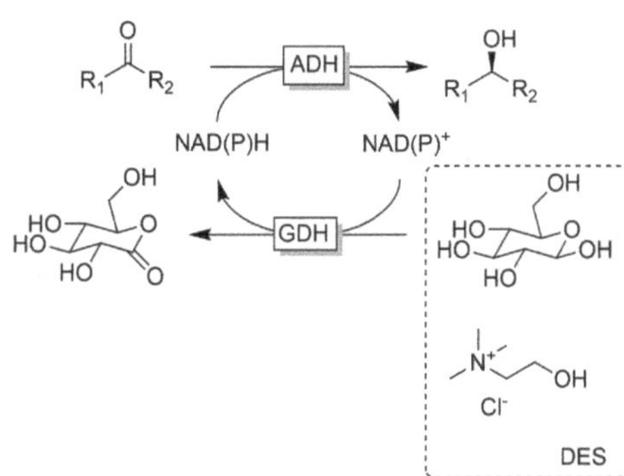

Scheme 5 Double use of a Glu-ChCl-DES as solvent and cosubstrate in ADH-catalyzed stereospecific carbonyl reduction reactions (Mourelle-Insua, Lavandera, & Gotor-Fernández, 2019). Glu-ChCl, on the one hand, serves as cosubstrate to promote the glucose-dehydrogenase (GDH)-catalyzed regeneration of NAD(P)H and as cosolvent to enable higher substrate loadings.

Choline-based DESs have recently been reported as dual-purpose solvents to also serve as stoichiometric electron donors for the reductive activation of molecular oxygen (Li et al., 2020; Ma et al., 2020, 2019). The resulting H_2O_2 can be used as oxidant to promote lipase-initiated chemoenzymatic epoxidation reactions (Ma et al., 2019) or peroxygenase-catalyzed oxyfunctionalization reactions (Scheme 6).

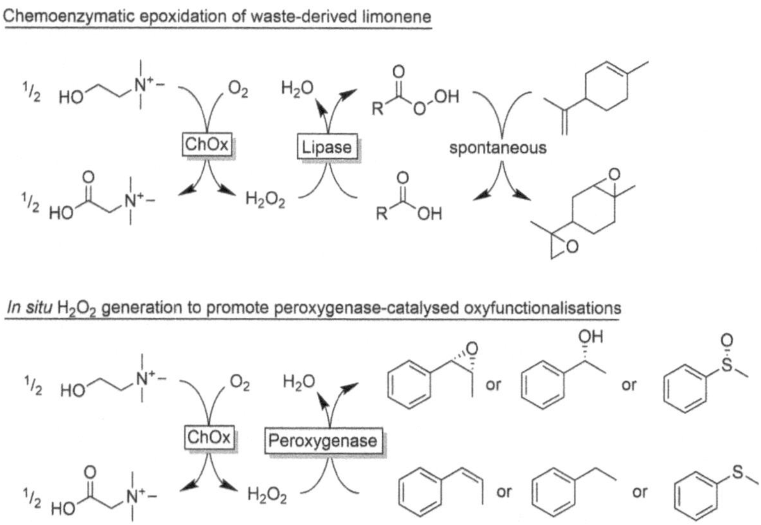

Scheme 6 Dual use of choline-based DES for the *in situ* generation of H_2O_2 catalyzed by a choline oxidase (ChOx) and the use of the H_2O_2 to promote chemoenzymatic Prilezhaev-type epoxidations and peroxygenase-catalyzed oxyfunctionalization reactions.

In case of the chemoenzymatic epoxidation of limonene, the DES (Ch-Pro) was also used to extract the terpene starting material from waste lemon peels thereby representing a triple-use (as extraction solvent, reaction solvent and sacrificial cosubstrate) for the overall process (Ma et al., 2019).

A fascinating influence of ChCl-Gly on the enantioselectivity of the Bakers' yeast-catalyzed stereospecific reduction of ketones was reported by Domínguez de María and co-workers (Scheme 7) (Maugeri & Domínguez de María, 2014). Depending on the water content of the ChCl-glycerol DES used as solvent, a marked switch of the overall enantioselectivity of the reaction was observed. Puzzling at first sight, this observation may be explained by (de)activation of enantiocomplementary ADHs within the *S. cerevisiae* cell.

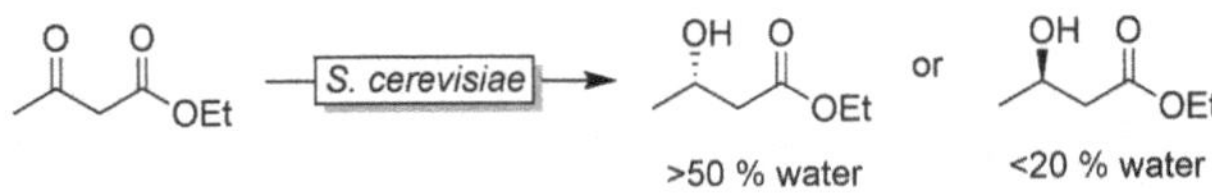

Scheme 7 Baker's yeast (*Saccharomyces cerevisiae*)-catalyzed reduction of acetoacetate in ChCl-glycerol/water mixtures. At water contents below 20% (v/v) the reaction was (*R*)-selective whereas at water contents above 50% (v/v) high (*S*)-selectivity was observed.

Similar effects have also been observed by Capriati and coworkers (Vitale et al., 2017) and Redovnikovic and coworkers (Cvjetko Bubalo, Mazur, Radošević, & Radojčić Redovniković, 2015; Panic, Delac, Roje, Redovnikovic, & Bubalo, 2019) in the Baker's yeast-catalyzed reduction of acetophenone derivates in ChCl-Gly.

Several authors have observed accelerating effects of DESs on the rate of whole cell-biotransformations. For example, the *Lysinibacillus fusiformis*-mediated transformation of isoeugenol into vanillin was markedly accelerated in the presence of various (NA)DES (Scheme 8) (Yang et al., 2017).

Generally, this is ascribed to cell wall/membrane permeabilization resulting in facilitated diffusion of the reagents into the (biocatalyst-containing) whole cells (Zhang et al., 2020).

Frequently, a stabilizing effect of DES on the biocatalysts is mentioned. Lipases have been the preferred study object for the influence of a myriad of DESs on their activity and stability (Bernardo Dias, Lucas de Carvalho, Maria Alice Zarur, & Isabel, 2019; Kim et al., 2016; Nian, Cao, & Liu, 2020; Oh et al., 2019). This vast amount of data, however, is blurred by differences in the experimental design and analysis and interpretation of the data. As a result, the sometimes contradicting findings are difficult to structure and interpret. Trends observed with one enzyme cannot be transferred easily to another (Huang et al., 2014; Wu, Wen, Xu, & Yang, 2014). Furthermore, a given DES may influence the enzyme and assay conditions in various ways, sometimes not directly obvious to the experimenter. Viscosity is frequently mentioned as a factor influencing activity assays. Some DESs, however, also exhibit emulsifying properties, which in two-liquid-phase-systems (as commonly used in lipase-catalyzed transformations) can influence the surface area of the two liquids and thereby influence the reaction rate (Lan et al., 2017).

Activity and stability data are also available for oxidoreductases such as horse liver alcohol dehydrogenase (Huang, Bittner, Domínguez de María, Jakobtorweihen, & Kara, 2020), the peroxidase from horseradish (Wu et al., 2014), versatile peroxidase (Mamashli et al., 2019), catalase

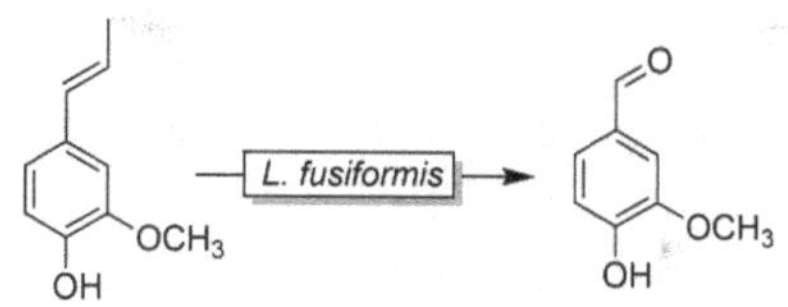

Scheme 8 *Lysinibacillus fusiformis*-catalyzed conversion of isoeugenol to vanillin.

(Harifi-Mood, Ghobadi, & Divsalar, 2017), laccase (Toledo et al., 2019) or haloalkane dehalogenases (Stepankova, Vanacek, Damborsky, & Chaloupkova, 2014).

DES have been demonstrated exhibiting a stabilizing effect on the chemoenzymatic epoxidation of C=C-double bonds (Scheme 9) (Lan et al., 2017; Ranganathan et al., 2017; Zhou, Wang, Yang, et al., 2017; Zhou, Wang, Zeng, et al., 2017).

The well-known perohydrolase activity of lipases (Björkling, Godtfredsen, & Kirk, 1990; Warwel & Klaas, 1995) allows them to accept H_2O_2 *in lieu* of water as nucleophile to hydrolyze the enzyme-acyl intermediate. The resulting peracids the mediate the Prilezhaev-type epoxidation of a broad range of C=C-double bonds. Unfortunately, the high K_M values of most lipases for H_2O_2 in water-containing media necessitate high

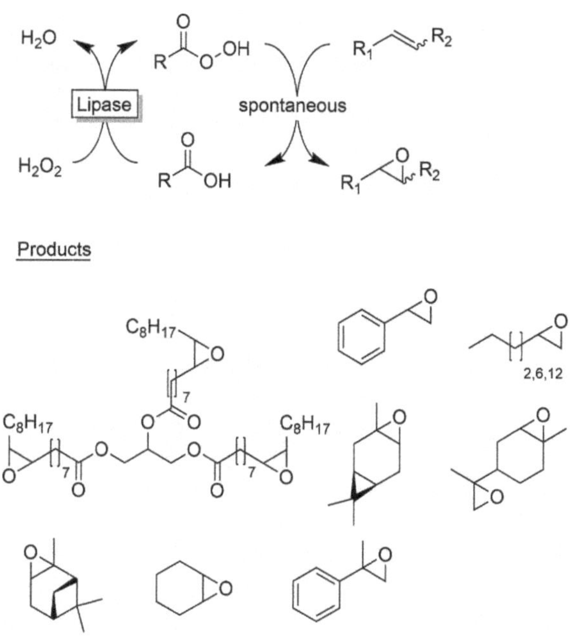

Scheme 9 Chemoenzymatic epoxidation of (non)-natural alkenes in DES.

H_2O_2-concentrations, which in turn can be detrimental to the robustness of the biocatalyst. Interestingly, DES seems to alleviate this inactivation, possibly by stabilizing the free H_2O_2 through additional H-bonding.

A unified theory rationalizing the effects of DES on enzymes is urgently needed!

Finally, also the emerging field of chemoenzymatic synthesis in DES is worth mentioning. Particularly popular are cascade reactions combining typical transition-metal catalyzed, but not known in enzyme catalysis, reactions such as cross-coupling or metathesis reactions with stereospecific enzyme-catalyzed reactions (such as the stereoselective reduction of ketones) (Scheme 10) (Cicco et al., 2018; Grabner et al., 2020; Paris et al., 2018).

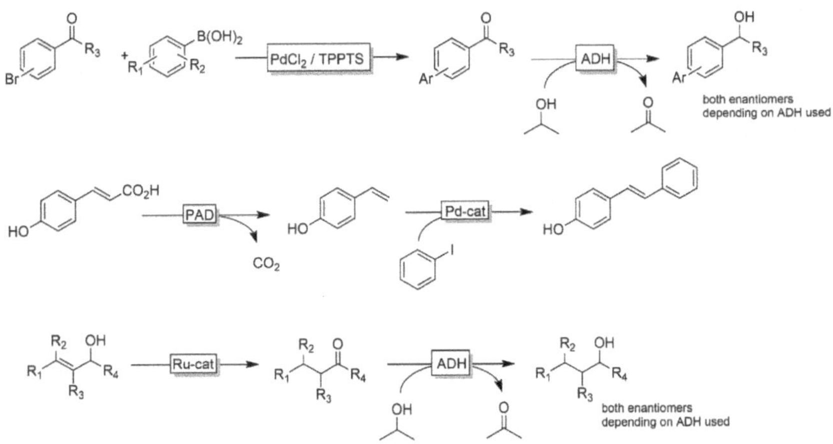

Scheme 10 Examples for chemoenzymatic cascade reactions performed in DES (Cicco et al., 2018; Grabner, Schweiger, Gavric, Kourist, & Gruber-Woelfler, 2020; Paris, Ríos-Lombardía, Morís, Gröger, & González-Sabín, 2018).

Issues of incompatibility of the two catalysis worlds, such as different requirements for the reaction conditions or mutual inactivation, are generally solved by spatial of temporal separation of the chemical and the enzymatic step.

3. Conclusions

Undoubtedly, (NA)DES represent an upcoming class of alternatives to established solvents in biocatalysis. Reactions, where too high water activities negatively influence the yield or selectivity of a reaction can benefit from using DES as (co)solvents. Also, DES can enable higher reagent solubilities than water and thereby substitute volatile, non-renewable organic solvents. Particularly interesting are those applications where the DES not only serves as solvent or enzyme stabilizer but also actively influences the reaction outcome or serves as (co)substrate itself. We are convinced that the near future will bring about many more exciting applications of (NA)DES for biocatalysis.

There are, however, certain risks we have identified: From many contributions, the specific reason behind choosing a DES as (co-)solvent is not evident and one gets the impression that many contributions are just "surfing the (NA)DES wave" without particular reason for this choice.

Frequently, terms such as "non-toxic," "biobased" and "biodegradable" are used in a prayer-wheel like fashion to underline the greenness of (NA) DES. We believe that limiting the evaluation of greenness to such terms is not sufficient. As shown in this contribution, DES have a (de)stablizing effect on proteins. Therefore, unless a broader empiric basis exists, claiming non-toxicity appears premature. Furthermore, as pointed out by Holtmann and coworkers (Pätzold, Siebenhaller, et al., 2019; Pätzold, Weimer, et al., 2019) the sometimes very high viscosity ofDES-based reaction mixtures also implies higher energy demands for pumping and stirring. As a result, unless the energy used for these processes is obtained from entirely renewable sources, also increased CO_2 emissions due to stirring and pumping can be expected.

References

Andler, S. M., Wang, L. S., Rotello, V. M., & Goddard, J. M. (2017). Influence of hierarchical interfacial assembly on lipase stability and performance in deep eutectic solvent. *Journal of Agricultural and Food Chemistry*, *65*(9), 1907–1914.

Anonymous. (2014). What is a Cell? Retrieved 29.03.2020, 2020, from https://www.nature.com/scitable/topicpage/what-is-a-cell-14023083/.

Bernardo Dias, R., Lucas de Carvalho, I., Maria Alice Zarur, C., & Isabel, M. M. (2019). Influence of betaine- and choline-based eutectic solvents on lipase activity. *Current Biochemical Engineering*, *5*(1), 57–68.

Björkling, F., Godtfredsen, S. E., & Kirk, O. (1990). Lipase-mediated formation of peroxycarboxylic acids used in catalytic epoxidation of alkenes. *Journal of the Chemical Society, Chemical Communications*, (19), 1301–1303.

Bubalo, M. C., Tusek, A. J., Vinkovic, M., Radosevic, K., Srcek, V. G., & Redovnikovic, I. R. (2015). Cholinium-based deep eutectic solvents and ionic liquids for lipase-catalyzed synthesis of butyl acetate. *Journal of Molecular Catalysis B-Enzymatic*, *122*, 188–198.

Cao, S.-L., Xu, H., Li, X.-H., Lou, W.-Y., & Zong, M.-H. (2015). Papain@magnetic nanocrystalline cellulose nanobiocatalyst: A highly efficient biocatalyst for dipeptide biosynthesis in deep eutectic solvents. *ACS Sustainable Chemistry & Engineering*, *3*(7), 1589–1599.

Chen, H., Qian, F., Lin, H., Chen, W., & Wang, P. (2020). Using choline chloride-based DESs as co-solvent for 3,5-Bis(trifluoromethyl) acetophenone bioreduction with Rhodococcus erythropolis XS1012. *Catalysts*, *10*, 30.

Cheng, Q. B., & Zhang, L. W. (2017). Highly efficient enzymatic preparation of daidzein in deep eutectic solvents. *Molecules*, *22*(1), 15.

Choi, Y. H., van Spronsen, J., Dai, Y., Verberne, M., Hollmann, F., Arends, I. W. C. E., et al. (2011). Are natural deep eutectic solvents the missing link in understanding cellular metabolism and physiology? *Plant Physiology*, *156*, 1701–1705.

Cicco, L., Ríos-Lombardía, N., Rodríguez-Álvarez, M. J., Morís, F., Perna, F. M., Capriati, V., et al. (2018). Programming cascade reactions interfacing biocatalysis with transition-metal catalysis in deep eutectic solvents as biorenewable reaction media. *Green Chemistry*, *20*(15), 3468–3475. https://doi.org/10.1039/C8GC00861B.

Craveiro, R., Meneses, L., Durazzo, L., Rocha, A., Silva, J. M., Reis, R. L., et al. (2019). Deep eutectic solvents for enzymatic esterification of racemic menthol. *ACS Sustainable Chemistry & Engineering*, 7(24), 19943–19950.

Cvjetko Bubalo, M., Mazur, M., Radošević, K., & Radojčić Redovniković, I. (2015). Baker's yeast-mediated asymmetric reduction of ethyl 3-oxobutanoate in deep eutectic solvents. *Process Biochemistry*, *50*(11), 1788–1792.

Dai, Y., Huan, B., Zhang, H. S., & He, Y. C. (2017). Effective biotransformation of ethyl 4-chloro-3-oxobutanoate into ethyl (S)-4-chloro-3-hydroxybutanoate by recombinant E-coli CCZU-T15 whole cells in ChCl Gly -water media. *Applied Biochemistry and Biotechnology*, *181*(4), 1347–1359.

Dordick, J. S., Marletta, M. A., & Klibanov, A. M. (1986). Peroxidase depolymerize lignin in organic media but not in water. *Proceedings of the National Academy of Sciences of the United States of America*, *83*(17), 6255–6257.

Durand, E., Lecomte, J., & Villeneuve, P. (2013). Deep eutectic solvents: Synthesis, application, and focus on lipase-catalyzed reactions. *European Journal of Lipid Science and Technology*, *115*(4), 379–385.

Fredes, Y., Chamorro, L., & Cabrera, Z. (2019). Increased Selectivity of Novozym 435 in the asymmetric hydrolysis of a substrate with high hydrophobicity through the use of deep eutectic solvents and high substrate concentrations. *Molecules*, *24*(4), 9.

González-Martínez, D., Gotor, V., & Gotor-Fernández, V. (2016). Application of deep eutectic solvents in promiscuous lipase-catalysed aldol reactions. *European Journal of Organic Chemistry*, *2016*(8), 1513–1519.

Gorke, J. T., Srienc, F., & Kazlauskas, R. J. (2008). Hydrolase-catalyzed biotransformations in deep eutectic solvents. *Chemical Communications*, (10), 1235–1237. https://doi.org/10.1039/B716317G.

Gotor-Fernández, V., & Paul, C. E. (2019). Deep eutectic solvents for redox biocatalysis. *Journal of Biotechnology*, *293*, 24–35.

Grabner, B., Schweiger, A. K., Gavric, K., Kourist, R., & Gruber-Woelfler, H. (2020). A chemo-enzymatic tandem reaction in a mixture of deep eutectic solvent and water in continuous flow. *Reaction Chemistry & Engineering*, *5*(2), 263–269. https://doi.org/10.1039/C9RE00467J.

Guajardo, N., Ahumada, K., & de Maria, P. D. (2020). Immobilized lipase-CLEA aggregates encapsulated in lentikats (R) as robust biocatalysts for continuous processes in deep eutectic solvents. *Journal of Biotechnology*, *310*, 97–102.

Guajardo, N., Domínguez de María, P., Ahumada, K., Schrebler, R. A., Ramírez-Tagle, R., Crespo, F. A., et al. (2017). Water as cosolvent: nonviscous deep eutectic solvents for efficient lipase-catalyzed esterifications. *ChemCatChem*, *9*(8), 1393–1396.

Harifi-Mood, A. R., Ghobadi, R., & Divsalar, A. (2017). The effect of deep eutectic solvents on catalytic function and structure of bovine liver catalase. *International Journal of Biological Macromolecules*, *95*, 115–120.

He, Y., Huang, Q., & Wang, P. (2020). Design and evaluation of novel bio-based deep eutectic solvents for highly efficient bioproduction of chiral aryl alcohol. *Journal of Chemical Technology & Biotechnology*. https://doi.org/10.1002/jctb.6386.

Huang, L., Bittner, J. P., Domínguez de María, P., Jakobtorweihen, S., & Kara, S. (2020). Modeling alcohol dehydrogenase catalysis in deep eutectic solvent/water mixtures. *ChemBioChem*, *21*(6), 811–817.

Huang, Z. L., Wu, B. P., Wen, Q., Yang, T. X., & Yang, Z. (2014). Deep eutectic solvents can be viable enzyme activators and stabilizers. *Journal of Chemical Technology and Biotechnology*, *89*(12), 1975–1981.

Hümmer, M., Kara, S., Liese, A., Huth, I., Schrader, J., & Holtmann, D. (2018). Synthesis of (-)-menthol fatty acid esters in and from (-)-menthol and fatty acids—Novel concept for lipase catalyzed esterification based on eutectic solvents. *Molecular Catalysis*, *458*, 67–72.

Ibn Majdoub Hassani, F. Z., Amzazi, S., Kreit, J., & Lavandera, I. (2020). Deep eutectic solvents as media in alcohol dehydrogenase-catalyzed reductions of halogenated ketones. *ChemCatChem*, *12*(3), 832–836.

Ibn Majdoub Hassani, F. Z., Amzazi, S., & Lavandera, I. (2019). The versatile applications of des and their influence on oxidoreductase-mediated transformations. *Molecules*, *24*(11), 2190.

Juneidi, I., Hayyan, M., Hashim, M. A., & Hayyan, A. (2017). Pure and aqueous deep eutectic solvents for a lipase-catalysed hydrolysis reaction. *Biochemical Engineering Journal*, *117*, 129–138.

Kim, S. H., Park, S., Yu, H., Kim, J. H., Kim, H. J., Yang, Y.-H., et al. (2016). Effect of deep eutectic solvent mixtures on lipase activity and stability. *Journal of Molecular Catalysis B: Enzymatic*, *128*, 65–72.

Kleiner, B., Fleischer, P., & Schorken, U. (2016). Biocatalytic synthesis of biodiesel utilizing deep eutectic solvents: A two-step-one-pot approach with free lipases suitable for acidic and used oil processing. *Process Biochemistry*, *51*(11), 1808–1816.

Kleiner, B., & Schorken, U. (2015). Native lipase dissolved in hydrophilic green solvents: A versatile 2-phase reaction system for high yield ester synthesis. *European Journal of Lipid Science and Technology*, *117*(2), 167–177.

Kourist, R., & González-Sabín, J. (2020). Non-conventional media as strategy to overcome the solvent dilemma in chemoenzymatic tandem catalysis. *ChemCatChem*. *n/a*(n/a).

Lan, D., Wang, X., Zhou, P., Hollmann, F., & Wang, Y. (2017). Deep eutectic solvents as performance additives in biphasic reactions. *RSC Advances*, 7(64), 40367–40370. https://doi.org/10.1039/C7RA06755K.

Li, Y., Ma, Y., Li, P., Zhang, X., Ribitsch, D., Alcalde, M., et al. (2020). Enantioselective sulfoxidation of thioanisole by cascading a choline oxidase and a peroxygenase in the presence of natural deep eutectic solvents. *ChemPlusChem*, *85*, 254–257.

Li, J., Wang, P., He, Y.-S., Zhu, Z.-R., & Huang, J. (2019). Toward designing a novel oligopeptide-based deep eutectic solvent: Applied in biocatalytic reduction. *ACS Sustainable Chemistry & Engineering*, 7(1), 1318–1326.

Lindberg, D., Revenga, M. D., & Widersten, M. (2010). Deep eutectic solvents (DESs) are viable cosolvents for enzyme-catalyzed epoxide hydrolysis. *Journal of Biotechnology*, *147*(3-4), 169–171.

Ma, Y., Li, Y., Ali, S., Li, P., Zhang, W., Rauch, M., et al. (2020). Natural deep eutectic solvents as performance additives for peroxygenase catalysis. *ChemCatChem*, *20*, 989–994.

Ma, Y., Li, P., Willot, S. J.-P., Zhang, W., Ribitsch, D., Choi, Y. H., et al. (2019). Natural deep eutectic solvents as multifunctional media for the valorisation of agricultural wastes. *ChemSusChem*, *12*, 1310–1315.

Mamashli, F., Badraghi, J., Delavari, B., Sabbaghian, M., Hosseini, M., & Saboury, A. A. (2019). Evaluation of versatile peroxidase's activity and conformation in the presence of a hydrated urea based deep eutectic solvent. *Journal of Solution Chemistry*, *48*(5), 689–701.

Mao, S., Li, K., Hou, Y., Liu, Y., Ji, S., Qin, H., et al. (2018). Synergistic effects of components in deep eutectic solvents relieve toxicity and improve the performance of steroid biotransformation catalyzed by Arthrobacter simplex. *Journal of Chemical Technology & Biotechnology*, *93*(9), 2729–2736.

Mao, S., Yu, L., Ji, S., Liu, X., & Lu, F. (2016). Evaluation of deep eutectic solvents as co-solvent for steroids 1-en-dehydrogenation biotransformation by Arthrobacter simplex. *Journal of Chemical Technology & Biotechnology*, *91*(4), 1099–1104.

María, P. D.d., Guajardo, N., & Kara, S. (2020). Enzyme catalysis: In DES, with DES, and in the presence of DES. *Deep Eutectic Solvents*, 257–271.

Maugeri, Z., & de Maria, P. D. (2014). Benzaldehyde lyase (BAL)-catalyzed enantioselective C-C bond formation in deep-eutectic-solvents-buffer mixtures. *Journal of Molecular Catalysis B-Enzymatic*, *107*, 120–123.

Maugeri, Z., & Domínguez de María, P. (2014). Whole-cell biocatalysis in deep-eutectic-solvents/aqueous mixtures. *ChemCatChem*, *6*(6), 1535–1537.

Maugeri, Z., Leitner, W., & Domínguez de María, P. (2013). Chymotrypsin-catalyzed peptide synthesis in deep eutectic solvents. *European Journal of Organic Chemistry*, *2013*(20), 4223–4228.

Mbous, Y. P., Hayyan, M., Hayyan, A., Wong, W. F., Hashim, M. A., & Looi, C. Y. (2017). Applications of deep eutectic solvents in biotechnology and bioengineering—Promises and challenges. *Biotechnology Advances*, *35*(2), 105–134.

Milker, S., Pätzold, M., Bloh, J. Z., & Holtmann, D. (2019). Comparison of deep eutectic solvents and solvent-free reaction conditions for aldol production. *Molecular Catalysis*, *466*, 70–74.

Mourelle-Insua, Á., Lavandera, I., & Gotor-Fernández, V. (2019). A designer natural deep eutectic solvent to recycle the cofactor in alcohol dehydrogenase-catalysed processes. *Green Chemistry*, *21*(11), 2946–2951. https://doi.org/10.1039/C9GC00318E.

Müller, C. R., Lavandera, I., Gotor-Fernández, V., & Domínguez de María, P. (2015). Performance of recombinant-whole-cell-catalyzed reductions in deep-eutectic-solvent–aqueous-media mixtures. *ChemCatChem*, 7(17), 2654–2659.

Nian, B., Cao, C., & Liu, Y. (2020). How Candida antarctica lipase B can be activated in natural deep eutectic solvents: experimental and molecular dynamics studies. *Journal of Chemical Technology & Biotechnology*, *95*(1), 86–93.

Oh, Y., Park, S., Yoo, E., Jo, S., Hong, J., Kim, H. J., et al. (2019). Dihydrogen-bonding deep eutectic solvents as reaction media for lipase-catalyzed transesterification. *Biochemical Engineering Journal*, *142*, 34–40.

Panic, M., Delac, D., Roje, M., Redovnikovic, I. R., & Bubalo, M. C. (2019). Green asymmetric reduction of acetophenone derivatives: Saccharomyces cerevisiae and aqueous natural deep eutectic solvent. *Biotechnology Letters*, *41*(2), 253–262.

Panić, M., Elenkov, M. M., Roje, M., Bubalo, M. C., & Redovniković, I. R. (2018). Plant-mediated stereoselective biotransformations in natural deep eutectic solvents. *Process Biochemistry*, *66*, 133–139.

Paris, J., Ríos-Lombardía, N., Morís, F., Gröger, H., & González-Sabín, J. (2018). Novel insights into the combination of metal- and biocatalysis: Cascade one-pot synthesis of enantiomerically pure biaryl alcohols in deep eutectic solvents. *ChemCatChem*, *10*(19), 4417–4423.

Pätzold, M., Siebenhaller, S., Kara, S., Liese, A., Syldatk, C., & Holtmann, D. (2019). Deep eutectic solvents as efficient solvents in biocatalysis. *Trends in Biotechnology*. https://doi.org/10.1016/j.tibtech.2019.03.007.

Pätzold, M., Weimer, A., Liese, A., & Holtmann, D. (2019). Optimization of solvent-free enzymatic esterification in eutectic substrate reaction mixture. *Biotechnology Reports*, *22*, e00333.

Peng, F., Chen, Q.-S., Li, F.-Z., Ou, X.-Y., Zong, M.-H., & Lou, W.-Y. (2020). Using deep eutectic solvents to improve the biocatalytic reduction of 2-hydroxyacetophenone to (R)-1-phenyl-1,2-ethanediol by Kurthia gibsonii SC0312. *Molecular Catalysis, 484*, 110773.

Peng, F., Zhao, Y., Li, F. Z., Zong, M. H., & Lou, W. Y. (2018). The effect of deep eutectic solvents on the asymmetric hydrolysis of styrene oxide by mung bean epoxide hydrolases. *Bioresources and Bioprocessing, 5*, 6.

Perna, F. M., Vitale, P., & Capriati, V. (2020). Deep eutectic solvents and their applications as green solvents. *Current Opinion in Green and Sustainable Chemistry, 21*, 27–33.

Pöhnlein, M., Ulrich, J., Kirschhöfer, F., Nusser, M., Muhle-Goll, C., Kannengiesser, B., et al. (2015). Lipase-catalyzed synthesis of glucose-6-O-hexanoate in deep eutectic solvents. *European Journal of Lipid Science and Technology, 117*(2), 161–166.

Ranganathan, S., Zeitlhofer, S., & Sieber, V. (2017). Development of a lipase-mediated epoxidation process for monoterpenes in choline chloride-based deep eutectic solvents. *Green Chemistry, 19*(11), 2576–2586. https://doi.org/10.1039/C7GC01127J.

Schweiger, A. K., Ríos-Lombardía, N., Winkler, C. K., Schmidt, S., Morís, F., Kroutil, W., et al. (2019). Using deep eutectic solvents to overcome limited substrate solubility in the enzymatic decarboxylation of bio-based phenolic acids. *ACS Sustainable Chemistry & Engineering*, 7(19), 16364–16370.

Siebenhaller, S., Kirchhoff, J., Kirschhöfer, F., Brenner-Weiß, G., Muhle-Goll, C., Luy, B., et al. (2018). Integrated process for the enzymatic production of fatty acid sugar esters completely based on lignocellulosic substrates. [Original Research]. *Frontiers in Chemistry, 6*, 421.

Siebenhaller, S., Muhle-Goll, C., Luy, B., Kirschhöfer, F., Brenner-Weiss, G., Hiller, E., et al. (2016). Sustainable enzymatic synthesis of glycolipids in a deep eutectic solvent system. *Journal of Molecular Catalysis B: Enzymatic, 133*, S281–S287.

Stepankova, V., Vanacek, P., Damborsky, J., & Chaloupkova, R. (2014). Comparison of catalysis by haloalkane dehalogenases in aqueous solutions of deep eutectic and organic solvents. *Green Chemistry, 16*(5), 2754–2761.

Tan, J.-N., & Dou, Y. (2020). Deep eutectic solvents for biocatalytic transformations: Focused lipase-catalyzed organic reactions. *Applied Microbiology and Biotechnology*.

Toledo, M. L., Pereira, M. M., Freire, M. G., Silva, J. P. A., Coutinho, J. A. P., & Tavares, A. P. M. (2019). Laccase activation in deep eutectic solvents. *ACS Sustainable Chemistry & Engineering*, 7(13), 11806–11814.

Ulger, C., & Takac, S. (2017). Kinetics of lipase-catalysed methyl gallate production in the presence of deep eutectic solvent. *Biocatalysis and Biotransformation, 35*(6), 407–416.

Vitale, P., Abbinante, V. M., Perna, F. M., Salomone, A., Cardellicchio, C., & Capriati, V. (2017). Unveiling the hidden performance of whole cells in the asymmetric bioreduction of aryl-containing ketones in aqueous deep eutectic solvents. *Advanced Synthesis & Catalysis, 359*(6), 1049–1057.

Warwel, S., & Klaas, M. R. G. (1995). Chemo-enzymatic epoxidation of unsaturated carboxylic acids. *Journal of Molecular Catalysis B-Enzymatic, 1*(1), 29–35.

Wei, P., Liang, J., Cheng, J., Zong, M. H., & Lou, W. Y. (2016). Markedly improving asymmetric oxidation of 1-(4-methoxyphenyl) ethanol with Acetobacter sp CCTCC M209061 cells by adding deep eutectic solvent in a two-phase system. *Microbial Cell Factories, 15*, 11.

Weiz, G., Braun, L., Lopez, R., de María, P. D., & Breccia, J. D. (2016). Enzymatic deglycosylation of flavonoids in deep eutectic solvents-aqueous mixtures: paving the way for sustainable flavonoid chemistry. *Journal of Molecular Catalysis B: Enzymatic, 130*, 70–73.

Wu, B. P., Wen, Q., Xu, H., & Yang, Z. (2014). Insights into the impact of deep eutectic solvents on horseradish peroxidase: Activity, stability and structure. *Journal of Molecular Catalysis B-Enzymatic, 101*, 101–107.

Wu, X., Xiong, J., Huang, Z., Cao, S., Zong, M., & Lou, W. (2019). Improving biocatalysis of cefaclor with penicillin acylase immobilized on magnetic nanocrystalline cellulose in deep eutectic solvent based co-solvent. *Bioresource Technology*, *288*, 121548.

Xu, P., Cheng, J., Lou, W.-Y., & Zong, M.-H. (2015). Using deep eutectic solvents to improve the resolution of racemic 1-(4-methoxyphenyl)ethanol through Acetobacter sp. CCTCC M209061 cell-mediated asymmetric oxidation. *RSC Advances*, *5*(9), 6357–6364. https://doi.org/10.1039/C4RA12905A.

Xu, P., Du, P.-X., Zong, M.-H., Li, N., & Lou, W.-Y. (2016). Combination of deep eutectic solvent and ionic liquid to improve biocatalytic reduction of 2-octanone with Acetobacter pasteurianus GIM1.158 cell. *Scientific Reports*, *6*(1), 26158.

Xu, P., Xu, Y., Li, X.-F., Zhao, B.-Y., Zong, M.-H., & Lou, W.-Y. (2015). Enhancing asymmetric reduction of 3-chloropropiophenone with immobilized Acetobacter sp. CCTCC M209061 cells by using deep eutectic solvents as cosolvents. *ACS Sustainable Chemistry & Engineering*, *3*(4), 718–724.

Xu, L., Zhang, L., Li, D. M., Liu, P. Z., Tan, C. P., Wang, W. F., et al. (2017). Deep eutectic solvents enable the enhanced production of n-3 PUFA-enriched triacylglycerols. *European Journal of Lipid Science and Technology*, *119*(12), 6.

Xu, P., Zheng, G.-W., Zong, M.-H., Li, N., & Lou, W.-Y. (2017). Recent progress on deep eutectic solvents in biocatalysis. *Bioresources and Bioprocessing*, *4*(1), 34.

Yang, S.-L., & Duan, Z.-Q. (2016). Insight into enzymatic synthesis of phosphatidylserine in deep eutectic solvents. *Catalysis Communications*, *82*, 16–19.

Yang, G. L., Tong, T., Yang, Y. Y., Liu, W., & Wang, X. D. (2019). Amano lipase PS-catalyzed hydrolysis of pine nut oil for the fatty acids production using deep eutectic solvent as co-solvent. *Journal of Oleo Science*, *68*(10), 977–988.

Yang, T.-X., Zhao, L.-Q., Wang, J., Song, G.-L., Liu, H.-M., Cheng, H., et al. (2017). Improving whole-cell biocatalysis by addition of deep eutectic solvents and natural deep eutectic solvents. *ACS Sustainable Chemistry & Engineering*, *5*(7), 5713–5722.

Zaks, A., & Klibanov, A. M. (1984). Enzymatic catalysis in organic media at 100-degrees-C. *Science*, *224*(4654), 1249–1251.

Zaks, A., & Klibanov, A. M. (1985). Enzyme-catalyzed processes in organic-solvents. *Proceedings of the National Academy of Sciences of the United States of America*, *82*(10), 3192–3196.

Zeng, C. X., Qi, S. J., Xin, R. P., Yang, B., & Wang, Y. H. (2015). Enzymatic selective synthesis of 1,3-DAG based on deep eutectic solvent acting as substrate and solvent. *Bioprocess and Biosystems Engineering*, *38*(11), 2053–2061.

Zhang, F., Zhu, C.-T., Peng, Q.-M., Wang, F.-Q., Sheng, S., Wu, Q.-Y., et al. (2020). Enhanced permeability of recombinant E. coli cells with deep eutectic solvent for transformation of rutin. *Journal of Chemical Technology & Biotechnology*, *95*(2), 384–393.

Zhao, H., Zhang, C., & Crittle, T. D. (2013). Choline-based deep eutectic solvents for enzymatic preparation of biodiesel from soybean oil. *Journal of Molecular Catalysis B-Enzymatic*, *85–86*, 243–247.

Zhou, P., Wang, X., Yang, B., Hollmann, F., & Wang, Y. (2017). Chemoenzymatic epoxidation of alkenes with Candida antarctica lipase B and hydrogen peroxide in deep eutectic solvents. *RSC Advances*, 7(21), 12518–12523. https://doi.org/10.1039/C7RA00805H.

Zhou, P., Wang, X., Zeng, C., Wang, W., Yang, B., Hollmann, F., et al. (2017). Deep eutectic solvents enable more robust chemoenzymatic epoxidation reactions. *ChemCatChem*, *9*, 934–936.

CHAPTER SIX

Natural deep eutectic solvents: Hypothesis for their possible roles in cellular functions and interaction with membranes and other organized biological systems

Erwann Durand[a,b,*], Pierre Villeneuve[a,b], Claire Bourlieu-lacanal[c], and Frédéric Carrière[d]

[a]CIRAD, UMR IATE, Montpellier, France
[b]IATE, Université Montpellier, CIRAD, INRAE, Institut Agro, Montpellier, France
[c]INRA, UMR IATE, Montpellier, France
[d]Aix Marseille Université, CNRS, UMR 7281 BIP, Marseille, France
[*]Corresponding author: e-mail address: erwann.durand@cirad.fr

Contents

Abstract

The involvement of natural deep eutectic solvents (NADES) in biological processes has been speculated but experimental evidence based on their observation in cells and biochemical reactions is still largely missing. Nevertheless, a sum of observations and analogies between NADES physical-chemical properties and those of peculiar cellular environments allows raising interesting hypothesis on the role they could play. To contribute to the current thinking on NADES in biological systems, an overview of these unusual liquid media is first presented with a special focus on how their occurrence in Nature (animal and plant) and hypothetical roles in cell metabolism were supported by experimental approaches. Then, a discussion pointing out the diversity of membranes formation and structures, describes, while challenging the readers, how

Advances in Botanical Research, Volume 97
ISSN 0065-2296
https://doi.org/10.1016/bs.abr.2020.09.005

NADES may be related with membrane systems. The occurrence of NADES in the cell and their possible interaction with highly organized intracellular structures and membrane-less organelles is also discussed. Observations already made with NADES metabolites in cells allow discussing the possible roles and effects of NADES on cell metabolism in connection with membranes or subcellular assemblies. Finally, suggestions for further investigations regarding the occurrence and the contribution of NADES in the regulation of cellular functions are proposed.

1. The beginning

The history of the Natural Deep Eutectic Solvent (NADES) was unlocked in the beginning of 2000s when the term Deep Eutectic Solvent (DES) was introduced in literature by the group of Prof. A. P. Abbott to name a sub-family of the well-known ionic liquids (IL) (Abbott et al., 2001; Abbott, Capper, Davies, Rasheed, & Tambyrajah, 2003). DES, also known as "deep eutectic IL," "low-melting mixtures," or "low transition temperature mixtures" are considered as a new generation of IL, with similar physical properties (low vapor pressure, chemical and thermal stability, non-inflammability, high conductivity, and high solubilizing capacity) but with beneficial attributes such as low toxicity or at least well-characterized toxicity of the individual DES components, low cost, good accessibility and sustainability. Similarly to IL, DES are mixtures of two or three components in specific molar eutectic mixture characterized by a significant (deep) decrease in the solid-liquid phase transition temperature compared to the melting point of the individual components. Abbott's group described several possibilities to prepare DES, with a high flexibility to choose the individual starting components, which are mostly small molecules having alcohols, amines and/or carboxylic acids groups in their chemical structures. These building blocks, often referred to as hydrogen-bond donor "HBD" or complexing agent, are combined with ammonium salts, often referred to as hydrogen-bond acceptor "HBA." Several groups worldwide have enriched this strategy and numerous opportunities to prepare DES are now available. Yet, most DES which have been proposed up to now have been prepared from choline-based molecules in combination with various common metabolites such as sugars, polyols or organic acids. This particularity in the DES composition held the attention of Prof. Rob Verpoorte, who had made NMR-based observations that this kind of metabolites often show high levels in various cell types (mammalian, microbial, plants, zebrafish), in concentrations too high for "reaction intermediates." He and his group linked these findings with accumulated unanswered questions over the years: How do flowers may contain up to 30% low water-soluble flavonoids? Why is a plant extract under the form of a viscous liquid? How are non-water-soluble compounds, such as terpenoids or cellulose, biosynthesized? How do plants survive in the desert? How do organisms survive at low temperatures and high salinity? How does a dry seed gets alive? Thus, they postulated that metabolites like choline, betaine, etc., in association with organic acids such as malic, succinic, citric, etc., would give a liquid phase with "DES-like" physical-chemical properties that could explain the solubilization of poorly water-soluble molecules, anti-freezing properties, resistance to high osmotic pressure and desiccation. In a preliminary assay, the combination of malic acid and choline chloride was tested in different proportions and led to a perfectly homogenous liquid mixture at 1:1 mol ratio association (Choi et al., 2011). This result was very similar to Abbot's group discovery with DES. Yet, DES have been defined as "mixtures of components that form an eutectic point in the phase diagram," but it is clear that not all NADES behave as such. Studies to characterize the molecular organization in NADES are scarce. DES were initially described as a well-organized tridimensional balance between electrostatics and H-bonding, whereas recent studies rather describe DES as disordered liquids with nanoscale heterogeneity. Moreover, it is known that simple changes in molecular structure appear to be sufficient to produce huge modifications in the nanostructure of DES. Regardless of the physical formation (melting point, glass transition) and the nanoscale organization, the most relevant information is that pure solid chemicals, including metabolites, may become liquid and stable over a large range of temperature after blending with little energy input and at ambient atmospheric pressure. In addition, NADES own distinct features from their pure components, with characteristics that do not necessarily result from the simple addition of each component properties. Since 2011 (Choi et al., 2011) when the first article on NADES was published, about hundred combinations of primary metabolites (such as sugars, amino acids or organic acids), exclusively in solid state at physiological temperature and pressure, were found to become liquid when mixed in certain molar ratios. However, several ratios are often possible for a given combination of metabolites. Indeed, even though we have pointed out that the molar ratio association between components (two or more) is fundamental to reach the eutectic in most of DES, this statement does not necessary mean that NADES has to be seen at a single combination. The researches and experiments with NADES have been frequently performed using a specific and exact ratio between constituents such as 1:1:1, 1:1, 1:2:1, 1:2, 2:2:5, etc., because assimilated (by error) to eutectic systems. In fact and, in our opinion, the term NADES should rather refer to "any association of metabolites for which the mixture is liquid at operating temperature."

2. Why "natural" in NADES?

Intuitively, the word "Natural" in NADES was thought to be due to the nature of the starting molecules involved in the NADES synthesis. In hindsight however, "Natural" should be mostly seen through the biological and physiological roles postulated for these mixtures in living organisms (Choi et al., 2011). This is the main feature that differentiates NADES from all other DES or IL made of natural components. To hypothesize about the role of NADES in Nature, we need to believe in their occurrence, since their real *in-situ* presence has not yet been recognized.

The early, and likely the strongest evidence that NADES may exist in Nature, was obtained after the investigation of highly concentrated molecules in nectars, saps, or plant exudates, for which classical theories of over-concentrated blends of oses or glassy thermodynamic state could not be applied. First, the visual appearance and physical properties of these secretions and NADES are very close; second, their qualitative analysis usually showed the presence of molecules (e.g., sugars) encountered in NADES mixtures; and third, the stoichiometry between components are very consistent and close to the ones observed in NADES. For instance, the nectar of *Cleome hassleriana* showed a sugar composition very similar to that of a fructose-glucose-sucrose NADES, with practically equimolar ratios. Similar hypothesis were made for the mixtures found in honey (glucose and fructose) or maple syrup (sucrose, glucose, fructose and malic acid). These first assumptions were enough to trigger interests of researchers, and from then, the questioning about NADES and their relationship with Nature was slightly different. NADES could then be seen as a key element to explain unclear metabolic pathways in living organisms.

For example, many plant metabolites are highly concentrated and sometimes non-water soluble, but are synthesized, stored, and transported in plants. In this perspective, NADES were hypothesized to be the medium in which, the enzyme-catalyzed reactions of poorly water-soluble molecules (e.g., secondary metabolites) or polymers such as cellulose, amylose, and lignin would occur. By a strange coincidence, this statement was strongly supported by both, an excellent solubilization capacity of these substances, and an unexpected activity/stability of enzymes in these media. For example,

NADES allowed for a much higher solubility of non-water soluble or poorly water-soluble natural products such as rutin, quercetin, cinnamic acid, carthamin, taxol. In comparison with water, the increase in solubility ranged from 18 to 460,000 (Dai, van Spronsen, Witkamp, Verpoorte, & Choi, 2013; Dai, Witkamp, Verpoorte, & Choi, 2013). Other usual macromolecules of cells, like proteins, polysaccharides or lignin, demonstrated superior solubilization capacity in NADES as compared to water (Dai, van Spronsen, et al., 2013; Dai, Witkamp, et al., 2013; Lores, Romero, Costas, Bendicho, & Lavilla, 2017). Regarding enzymatic activities, one fantastic observation lies with the denaturing properties of urea dissolved in water, whereas it does not exert significant denaturation on enzymes like the lipase B from *Candida antarctica* when it forms a NADES (e.g., with choline chloride), whereas its concentration is significantly higher than in denaturing water solutions (Durand et al., 2012). It was found that constituents of NADES form an important non-covalent complex that limit the diffusion of urea into the protein core (Monhemi, Housaindokht, Moosavi-Movahedi, & Bozorgmehr, 2014). These liquids form hydrogen bonds with the surface residues of the enzyme, which instead of enzyme denaturation, lead to a greater stability. Numerous enzymes have been studied in such systems and, unexpectedly, exhibited both high activity and stability (Xu, Zheng, Zong, Li, & Lou, 2017). Moreover, enzymes such as laccases, have been shown to be inactive in NADES but were activated upon water addition, which supports a possible role of NADES in the preservation of enzymes in plant seeds (Choi et al., 2011). In conclusion, enzyme-mediated reactions in cells spontaneously rely on water, but could NADES be the medium in which enzymes or substrates are independently stored, or simultaneously dissolved to allow biochemical reactions. For example, to conceptualize the synthesis of secondary metabolites (e.g., phenylpropanoid or flavonoid), as well as their glycosylation and deglycosylation (Le Roy, Huss, Creach, Hawkins, & Neutelings, 2016; Wagner & Hrazdina, 1984). In addition, plant exudates such as saps of carnivore plants contain high levels of typical NADES constituents such as sugars and sugar alcohols, as well as a set of digestive enzymes. These presumed sugar-based enzyme-containing NADES act as an hydrolytic mixture for the digestion of insects. NADES could also be involved in the biosynthesis of polysaccharides or lignin, by forming media in which enzymes and precursors would promote the chain elongation of building blocks. NADES could also play a fundamental role in the transfer of substrates and/or products between the enzymes (see the concept of "metabolon" later in the chapter), along with the storage and transport of water-insoluble secondary metabolites. This could explain the concentration of compounds like flavonoids and anthocyanins in cells at much higher levels than its solubility in water. For instance, the flowers of *Sophora japonica* contain up to 15% (w/w) of rutin (Paniwnyk, Beaufoy, Lorimer, & Mason, 2001). This flavonoid conjugated with sugar residues is paradoxically almost insoluble in water, because the "π-stacking" interactions of the flat aromatic hydrophobic groups strengthened by the surrounding hydrogen bonds of the hydroxyl groups (phenols), leads to the formation of aggregates with stronger interaction than the water solvation forces. In the pioneering work showing the potential application of NADES for phenolic solubilization (Dai, van Spronsen, et al., 2013; Dai, Witkamp, et al., 2013), it was found that NADES had a strong ability to dissolve rutin, with a 50- to 100-fold higher solubility than in water. This result means that, in NADES, the hydrogen interactions between the components and a solute are so strong (stronger than water with solute) that they prevail over the rest of the solute-solute electrostatic forces.

Another NADES involvement would be to allow organisms to resist or acclimate in harsh environmental conditions (cold and/or drought resistance). Nature has provided solutions to adapt living organisms to inhabit in extreme environmental conditions. The strategy to overcome and prevent dysfunctions of cell metabolisms, involves the synchronized production of diverse biological metabolites. The self-organizing structure found in NADES, allows the component to remain liquid and stable over a large temperature range (far below 0 °C, and up to 200 °C). This asset could allow both water and metabolites to remain liquid, even at extreme temperatures. Thus, NADES may be a way to prevent the crystallization of water inside living cells for winter survival in an even more effective way than classical cryoprotective effect of solubilized oses for instance. Indeed, the interactions between the NADES' network and water are strong enough to maintain water liquid even at extreme low temperatures (below −50 °C). Unlike the formation of a glass from sugars, NADES establish very singular interactions with water characterized with different thermodynamic behaviors (glass transition rather than an explicit freezing point). It has been already shown that components frequently described in the composition of NADES (small molecules such as amino acids, polyols, sugars, methylamines, or urea), are produced, stored, and used by cells organisms and tissues under specific conditions (Yancey, 2005). Some of these observations fit quite nicely with the idea of NADES as the molar concentration are close of some major compounds in the organisms. For example, in the stonefly from Alaska (*Nemoura arctica*), glycerol has been found to increase by three orders of magnitude in cold acclimation (Walters, Sformo, Barnes, & Duman, 2009). Same observation may be drawn with Beetles from central Europe, or various specimens of frogs known to survive at low temperature conditions, where resistance to winter and low temperature conditions correlate with the seasonal and high level production of sugars (trehalose and glucose among others) and polyols (glycerol), molecules which are largely recognized to form NADES (Košťál, Zahradníčková, Šimek, & Zelený, 2007). A similar observation in other insects, such as rice water weevil, codling moth, and many others was also reported. In plant, this strategy could explain how a seed can germinate after 30,000 years in permafrost, or how varieties of pea (*Pisum sativum*) can resist to intense frost. Inversely, NADES could also be involved in the survival of organisms in extreme situation such as drought (e.g., cacti, resurrection plants, lichens, prokaryotes). Indeed, it has been observed that when water is present as a part of the NADES, it is strongly retained in the liquid and cannot be easily evaporated, preventing total desiccation (Choi et al., 2011). Thus, NADES involving water molecules could be a strategy for organisms to hold water, and could explain how lichens, cacti or resurrection plants (e.g., *Selaginella lepidophylla*) that are rich in choline (a molecule abundantly present in the composition of NADES) may survive drought. Alternatively, Marine beings were found to accumulate urea, betaine and sarcosine in particular situations (Yancey, Clark, Hand, Bowlus, & Somero, 1982), and it was shown that the addition of these molecules, in specific quantities, allows mammalian cells to survive in high saline environments (Yancey & Burg, 1990). Other metabolites such as amino acids (proline, glutamine, alanine, lysine, leucine, and phenylalanine) have been reported in association with sugars and polyols as part of the extreme resistance strategy (Costanzo, Reynolds, do Amaral, Rosendale, & Lee, 2015; Overgaard et al., 2007; Slotsbo, Maraldo, Malmendal, Nielsen, & Holmstrup, 2008; Walters et al., 2009). All these metabolites are well recognized to form NADES, and we could easily imagine that their accelerated production in extreme conditions could be link with NADES-like association somewhere in the cell organism. The effects of highly concentrated polyols or sugars in water have been known for a long time (glassy state, cryoprotective effect, etc.). Yet, the particularity and the difference of sugar-based NADES would be at nanoscale from a structural view (highest concentration of sugar molecules, singular interaction with water, different coordinative and attractive forces), and from a dynamic perspective (viscoelastic state, mobility, diffusion).

Finally, another NADES' connection with nature would be to store and protect unstable essential compounds. Owing to their unique physical-chemical properties, NADES exhibit a great potential as stabilizing environment for solutes. Indeed, their strong and well-organized non-covalent tridimensional interactions decrease the movement of solute molecules, reducing contact at interfaces (e.g., with oxygen in the air) and consequently the oxidative degradation. For instance, natural pigments are stable in plants, whereas they may show very poor stability in water. The natural pigments carthamin from safflower shows a much higher stability in NADES than in water (Dai, Verpoorte, & Choi, 2014), at different temperature and light exposure, besides having the highest solubility in NADES. This implies that, in plant such as safflower, NADES may stabilize carthamin as well as other pigments. A similar result was found with cyanidin, present in flower petals of *Catharanthus roseus*, which is much more stable in NADES than in acidified ethanol due to higher interactions between cyanidin and the NADES, which improves the color stability of anthocyanin (Dai, Rozema, Verpoorte, & Choi, 2016). The sources of NADES components in living cells suggest that they may dissolve and stabilize these macromolecules intracellularly, as observed in barley seeds, where a high amount of sucrose and choline was detected in episperm during dormancy (Choi et al., 2011). Lastly, a key point in terms of putative role of NADES in cellular functions, is that it has also been reported to be a good medium to solubilize and stabilize RNA and DNA structures (Mamajanov, Engelhart, Bean, & Hud, 2010).

This review about NADES composition and possible occurrence in biological systems was essential to highlight that, (i) NADES ingredients can be found in all living organisms, (ii) under various conditions (e.g., stress) some common molecules/metabolites known to form NADES can be increased, (iii) the hypothetical involvement of NADES in living organisms is extensive, and (iv) the possible roles of NADES in cell metabolisms were supported with experimental approaches *in vitro*. However, while the presence of NADES compounds has been shown in living organisms, the direct evidence that these compounds can interact *in cellula* or *in vivo* to form NADES and contribute to cellular processes, is still missing. In addition, the assumptions on the potential roles of the NADES are based on concepts developed from static liquid systems (solvents) in a glassware, designed and scaled for laboratory experiments. Thus, one can further wonder, how NADES could be formed in a real cell?

3. How NADES may organize in living cell systems?

Most of living cell systems are composed of water (~70% w/w), therefore seeing NADES as a dynamic phase in such a dominant liquid water environment is not quite as intuitive as it could be. Indeed, although the intermolecular forces in these mixtures are relatively strong, the energy required to dissociate molecules remains weaker than for covalent bond dissociation. It is a known fact that the mixing of NADES into a large amount of a dispersible liquid leads to a loss in the singular tridimensional structure observed in bulk NADES. This result would question certain of the above-mentioned points, and thus, their relevance in nature. Actually, the liquid properties (viscosity, solubility, conductivity, etc.) can be adjusted when a NADES is dissolved with a little amount of a miscible protic liquid such as water, thus creating a ternary mixture. This is an interesting approach to design task-specific media. Indeed, self-diffusion data studies reveal that these liquids are not homogeneous and contain distinct microscopic water-rich phases (Gladden et al., 2015; Hammond, Bowron, & Edler, 2017). However, when water (or another protic liquid) becomes dominant, the structure of NADES is broken. Each component recovers its own properties, and the overall physical-chemical behavior of the medium is driven by the protic liquid. 2D NMR, FTIR and molecular dynamic simulations highlighted the rupture of the supramolecular complexes, characteristic of NADES, when the dilution reached about 50% (v/v) (Dai, Witkamp, Verpoorte, & Choi, 2015; Gutiérrez, Ferrer, Mateo, & Del Monte, 2009; Hadj-kali et al., 2015; Hammond et al., 2017). In this situation, NADES tends to become a simple solution of its individual counterparts, which is an undesired event because the above-mentioned characteristics of NADES are lost. Thus, how NADES may subsist in a dilute environment as observed with contiguous living systems? The organization of the cellular infrastructures are very complex, and rather than seeing NADES as a simple liquid phase, randomly dispersed in cellular systems, one could imagine that NADES ingredients converge to interconnect at nanometric scale, to form a layer, or a liquid cluster in different cell regions (Fig. 1). This may happen into the cytoplasm with highly organized subcellular assemblies (e.g., surrounding supramolecular protein complexes), membrane-less compartments, or membranes. One can also imagine that NADES are formed into small organelles such as plastids and vesicles. This could be the case for instance, in pyrenoids, which are membrane-less subcellular micro-compartments found in the chloroplasts of many algae (Giordano, Beardall, & Raven, 2005). Pyrenoids are associated with the carbon-concentrating mechanism (CCM) and they contain several enzymes involved in CCM such as RuBisCO, carbonic anhydrases and a dozen of other proteins (Holdsworth, 1971). Their main function would be to maintaining a high CO_2 concentration in the vicinity of these enzymes and to ensure a faster diffusion of

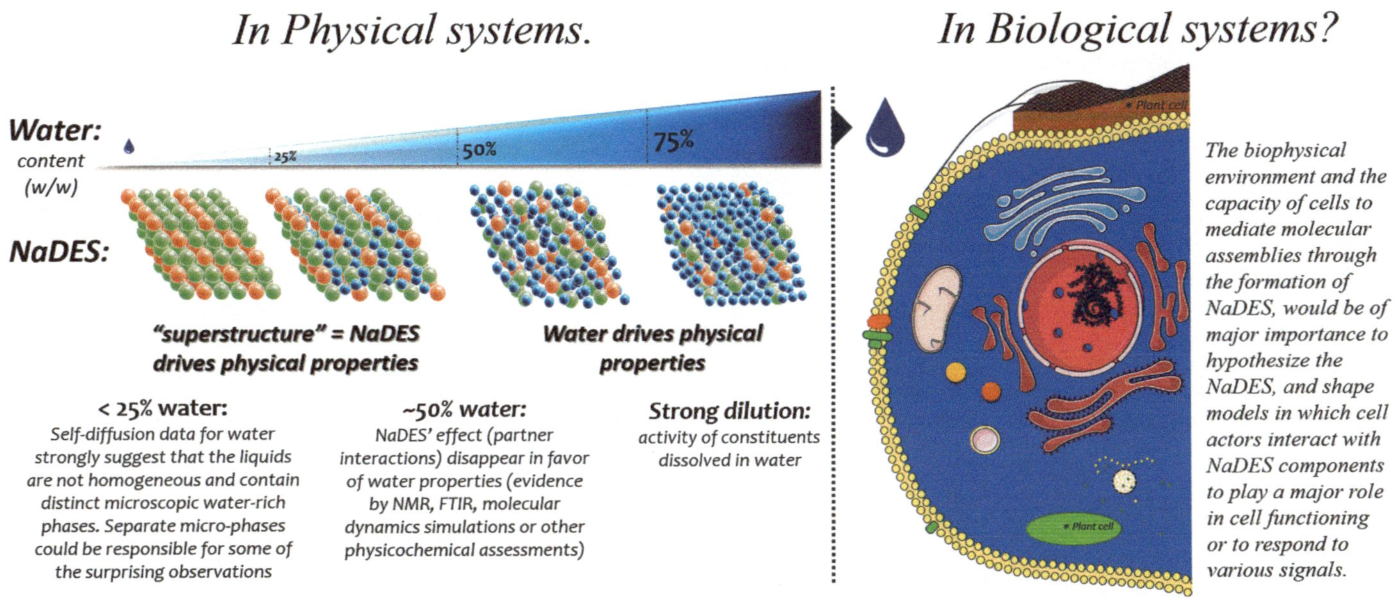

Fig. 1 NADES, from superstructure in the bulk state to a dynamic hypothetical liquid phase, layer or cluster, somewhere in biological systems.

CO_2 compared to its limited diffusion in water (Badger & Price, 1992; Meyer, Whittaker, & Griffiths, 2017). Pyrenoids are known to contain proteins and starch but their complete chemical composition is not completely known due to the difficulty in isolating these membrane-less compartments. Nevertheless, a highly organized crystalline-like structure of the pyrenoid matrix has been reported, such as a hexagonal-closed packing in the *Chlamydomonas* pyrenoid (Engel et al., 2015). In electron microscopy the pyrenoid appears as electron-dense matrix traversed by transpyrenoidal membranes (tubules), besides a surrounding starch sheath. Whether this architecture involves or results from the formation of NADES is still hypothetical. It is obvious that the demonstration of the capacity of the cell to mediate molecular assemblies through the formation of NADES, would be of major importance to conceptualize models in which cell actors interact with NADES components to form functional entities (Fig. 1).

4. NADES and membranes in living cell systems

From the identification of cells and functional cellular compartments well defined by a surrounding membrane, e.g., organelles, it has been assumed that membrane-bounded microenvironments were essential for the biosynthesis and the development of life. However, back to the primordial Earth, the exogenous extraterrestrial and endogenous synthesis molecules were suspected to provide prebiotic and biogenic mixtures that allowed the first forms of life. A scenario is that the origin of life involved the self-assembly of molecular systems within a variety of cell-sized environments (Deamer, 1997; Deamer, Dworkin, Sandford, Bernstein, & Allamandola, 2002; Segré, Ben-Eli, Deamer, & Lancet, 2001). In other words, self-assembly processes have possibly led to the emergence of membrane-bound self-reproducing molecular systems, with defined compartments sequestering units emerging in the course of natural selection and evolution. Self-assembly is a well-known process in water in which small amphiphilic molecules spontaneously associate by hydrophobic interactions into more complex structures (micelles, monolayers or bilayers). One of the simplest original prebiotic compartments could have been vesicles, which, like cells, are delimited by amphiphilic molecules (Deamer, 2017). Accordingly, it is usually considered that water and self-assembly in water were essential for the development of life. Astrobiologists studying the possible occurrence of life in other planets of the solara system have, however, some other theories according to which life could have developed in a liquid hydrocarbon environment (Shapiro & Schulze-Makuch, 2009).

What seems primordial for self-assembly and higher-order molecular organization is the transition through a liquid state. Therefore, could amphiphilic compounds spontaneously self-assemble into more complex structures, such as monolayer, bilayers and vesicles in non- or low-aqueous medium? It has been known for decades (the so-called ionothermal synthesis) that IL and DES, that share physicochemical properties with NADES, allow nanoscale growth and self-assembly conditions, by simultaneously acting as both the solvent and potential template-directing agent in the formation of solids. More recently, the self-assembly of lipids has been shown to occur in a diverse range of synthetic IL. Indeed, the formation of spherical structures from phospholipids with diameters of 200–400 nm was observed. The bilayer structure and stability of these vesicles were strongly modified by the presence of water (Gayet, Marty, Brûlet, & De Viguerie, 2011). The existence of self-assembled structures of sodium dodecyl sulfate (SDS) solubilized in choline chloride: urea DES was also observed, with considerable differences between the behavior of SDS in water (Arnold et al., 2015; Pal et al., 2014; Sanchez-Fernandez et al., 2016). The self-assembled structures in DES are larger than the equivalent structures observed in pure water, and SDS may adopt unusual cylindrical aggregate morphology (never seen in water or other polar solvents). Finally, phospholipids dissolved in DES also self-assemble spontaneously into lamellar phase, confirming that this neoteric media can mediate amphiphilic molecule self-assembly (Bryant, Atkin, & Warr, 2015). Thus, IL and DES are convenient media for self-assembly of vesicles. The synthetic origin of most IL and DES used in these experiments makes these findings difficult to rely with the formation of the early cell membranes. However, the concept of NADES and their properties that make them compatible under the conditions of the primordial earth (e.g., stable liquid under extreme temperatures), could connect NADES-like mixtures to primitive membrane origins. In addition, NADES, seen as liquid phase and therefore possibly as a "reaction medium" may provide, since the early Earth, different conditions for chemistry and thermodynamics, giving another way to figure out (i) the protection of enzyme substrates or functional macromolecules (e.g., nucleic acids), (ii) the formation or stabilization of self-assembled molecular systems (e.g., vesicles), or (iii) the stabilization of enzymes (e.g., polymerases).

In conclusion, NADES could plausibly contribute, (i) to drive self-assembly of surface-active compounds capable of membrane formation, (ii) to specific kinds of reactions in a complex prebiotic environment containing mixtures of dissolved substances, (iii) to reduce the complexity of life processes and the constraints of liquid transition temperature of mixtures and (iv) the formation of liquid phases with different physicochemical properties in comparison with water.

5. NADES (possible) interactions with membranes and highly organized subcellular biological assemblies

Cell functioning depends on the capacity of metabolites to exchange between various subcellular compartments delimited by membranes as well as between neighboring cells or extracellular environment. The plasma membrane, for instance, is a selective permeable barrier that covers the surface and separates the cell from its surrounding environment and plays a critical role in the biochemical signaling pathway. These membranes are principally composed by mixtures of proteins and lipids in dynamic motion. Lipids with different polar head groups and types of hydrophobic tails, may naturally form two leaflets in aqueous media due to their amphiphilic properties. They may be classified into different classes according to their chemical structures, namely, glycerolipids, sphingolipids and sterols. Among the glycerolipids, phospholipids predominantly occur in the plasma membranes and mitochondrial envelope, while galactolipids are characteristic of the thylakoid membrane of chloroplasts in plant cells (Van Meer, Voelker, & Feigenson, 2009). Phospholipids contain different polar head group associated with a glycerol backbone with two fatty acyl chains that regulate their physicochemical properties. Indeed, the number and position of double bonds within these hydrophobic tails, as well as the alkyl chain length, are main parameters that influence membrane fluidity. The presence of sterols and arrangement of lipids into different phases (liquid condensed vs. liquid expanded) also strongly modulate membrane fluidity and functionality. In addition, the asymmetrical lipid distribution on the membrane leaflets also conditions its physical-chemical properties and interactions with the surrounding environment. Phosphatidylcholine, glycolipids and sphingomyelin are enriched on the exoplasmic surface of the plasma membrane, whereas phosphatidylserine, phosphatidylethanolamine, and phosphatidylinositol are enriched on the cytoplasmic side. Some minor phospholipids, such as phosphatidic acid, phosphatidylinositol-4-monophosphate and phosphatidylinositol-4,5-biphosphate, may also allocate on the cytoplasmic leaflet (Dowhan, Bogdanov, & Mileykovskaya, 2016). Furthermore, cholesterol distributes preferentially on the external leaflet, although it can move easily between the two leaflets. To sum up, the lipid composition of biological membranes is highly diversified, and their organization is essential to support cell functioning, in addition to the role of membrane proteins that ensure transfers through the membrane, receptor functions and various enzyme activities. In the so-called mosaic fluid model hypothesis by Singer and Nicolson in 1972, a homogenous organization of the plasma membrane constituents with proteins embedded in the lipids layer was proposed. However, the structure and organization of biomembranes has progressively appeared as more complex with lateral heterogeneity and the identification of nanodomains (or microdomains or lipid rafts) concentrating some lipid species and proteins. The sub-compartmentalization of membranes in dynamic nanodomains is fundamental for the biochemical processes occurring at the membrane in eukaryotic and prokaryotic cells. It presents a great flexibility with respect to size, shape, and cell-type specificity. Nanoscale domains (size 10–100 nm) are small and dynamic assemblies of specific lipids and proteins, whereas microdomains (>100 nm) result of the nanodomains merging due to interactions between their individual components. These domains may be further linked to cytoskeleton and/or the cell wall depending on the cell type, which affects their mobility and dynamics, as well as their structure. For example, different lipid constituents can segregate into functionalized sub-microdomains, called lipid rafts, that are regions of plasma membranes particularly rich in sphingolipids and cholesterol, and under specific liquid-ordered physical state (Simons & Ikonen, 1997). In addition, lipid rafts can selectively segregate membrane proteins to form protein-lipid microdomains. When cell protein concentration is high, protein-protein interactions may occur to yield the formation of new microdomains, often referred to as clusters. Transmembrane or anchored protein clusters can interact with intracellular components, which can act in the diffusion as a "membrane-skeleton fence" or "anchored-protein picket." Cluster of protein and pigment molecules embedded in membrane structure, such as the thylakoid complexes (chlorophyll-protein) in chloroplasts is another example of the lateral heterogeneity of membranes that contribute to the function and structure of higher plant cells. Although non-exhaustive, this section is sufficient to point out the diversity of biophysical properties, structures and complexity of membranes, especially with the distinct subcompartments that play an important role in many cell signaling and metabolisms. This membrane heterogeneity is the result of organized interactions between molecules and macromolecules, where non-covalent interactions are predominant. Strong evidence supports the fact that these nano- or micro-scale organizations are essential for different cellular functions, from metabolic material exchange to biochemical reactions, or to regulate the interactions between membrane components (including isolated channels, receptors, other signaling complexes and specific lipids) during many physiological processes.

The highly diverse composition and architecture of membrane systems are fundamental to help conceptualizing how NADES may be bonded to lipids or proteins. This may happen with the polar head groups of membrane lipids that could bind and specifically drive metabolites associations (e.g., acids, amino acids, sugars, antioxidants, polyols, etc.) thus forming a dynamic cluster of NADES inside or around membranes (Fig. 2). Many of NADES components have already shown interesting properties in the stabilization of macromolecular structures (proteins or membranes). For example, proline, betaine, trehalose have been shown to stabilize membranes, after binding to head groups of membrane phospholipids by competition with water, thus preventing disruption of membranes (Rudolph, Crowe, & Crowe, 1986). This property could be a reason that would explain their overproduction to stabilize membranes at subzero temperatures in organisms that thrive in cold conditions. The same observation may be drawn with polyols in insects, yeast and microorganisms, that can protect proteins from denaturation at high temperatures. In addition, this postulate is supported by studies which highlight the fact that to preserve liposomes under dehydration conditions and to allow rehydration, sugars are required on both sides of the lipid bilayer membrane (Crowe, 2002). These stabilizing molecules seem to be used in nature mostly in response to stresses that destabilize macromolecules and membranes, suggesting that the stabilization is unnecessary (or minor) in the absence of external perturbation. Yet, the membrane substructures (microdomains) are highly sensitive to environmental perturbation, and subtle changes in composition can alter their interaction with specific proteins and the membrane properties. On the other hand, NADES could also be a way to localize and concentrate, being associated with membranes, antioxidants (flavonoids, vitamins, proteins, etc.) to protect the lipid bilayer against oxidation or light damage. For instance, accumulation of dense lumenal substance in the thylakoid lumen was demonstrated and proposed as a mechanism protecting the thylakoid membranes of *Haberlea rhodopensis* (Georgieva, Sárvári, & Keresztes, 2010). This compartment prevents conformational changes of membrane lipids against oxidative damage during dehydration, thus preventing recovery of photosynthesis. Although the chemical nature remains unknown, it is strongly suspected to be a dense phenolic substance. Evidence is still missing, but it is attractive to speculate that it could constitute a NADES mixture with dissolved functional substances, which could be seen as a dynamic building block that can be coordinated inside or around membranes, as well as relocated between them (Fig. 2). In addition, we have previously introduced the fact that NADES could play a fundamental role in the transfer of substrates and/or products between enzymes, as well as the synthesis, storage and transport of secondary metabolites. This characteristic reminds the concept of "metabolon," a temporary structural-functional complex formed between sequential enzymes of a metabolic pathway. In a metabolon, enzymes are held together both by non-covalent interactions and by structural elements of the cell, such as membranes. The formation of a metabolon allows the product from one enzyme to be transferred by channeling directly into the active site of the next enzyme of the metabolic pathway. The Krebs' cycle is a typical example of metabolon (Wu & Minteer, 2015; Zhang et al., 2017). Another example is the biosynthesis pathway of dhurrin, a cyanogenic glycoside produced in many plants, including sorghum. Dhurrin synthesis enzymes form a metabolon assembled in lipid membranes (Laursen et al., 2016). Other examples of metabolons are found in fatty acid (Kastritis et al., 2017) and lipid biosynthesis (Maréchal & Lupette, 2020; Parks & Casey, 1995). In metabolons, the amount of water needed to hydrate the enzymes is reduced and enzyme activity is increased (Dastmalchi & Facchini, 2016; Laursen et al., 2016). Metabolons are spatially and temporally synchronized sets of enzyme complexes (or "protein communities"). They are often defined as higher-order, dynamically associated, assemblies of macromolecules that benefit from their close proximity to each other in the cell. So far, these protein communities and their assembly have not been properly conceptualized because experimental approaches to capture this higher-order proteome organization were missing (Kastritis et al., 2017). Nevertheless, recent investigations by quantitative and cross-linking mass spectrometry, as well as by cryoEM, of the organization of cellular complexes were made using a crude cellular extract of the eukaryotic thermophile *Chaetomium thermophilum* (Kastritis et al., 2017). These extracts were found to retain basic principles of cellular organization, with for instance the identification of a metabolon involving the fatty acid synthase (FAS) complex and acetyl-coA carboxylase. These studies have allowed the identification and characterization of an intermediate layer of supramolecular assemblies between macromolecular complexes and organelles. Electron microscopy methods have allowed the observation of these higher-order assemblies of large molecular weight. The physical-chemical characterization of the solvent surrounding these assemblies still deserves detailed investigations but it is tempting to speculate that it could constitute a NADES-like environment.

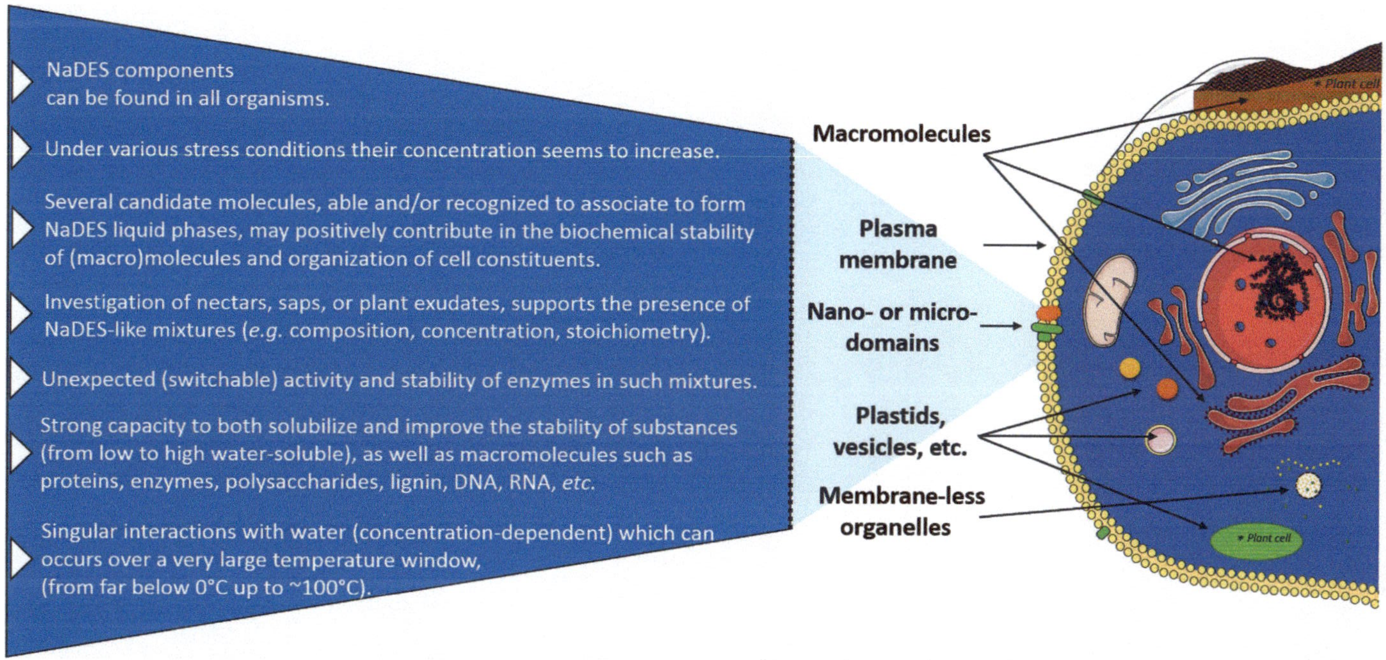

Fig. 2 Some observations that could connect the theory of NADES with a possible occurrence in cells to regulate functioning through interactions with membranes or organized assemblies.

In 1916, Neuberg invented the term "hydrotrope," to describe any compounds that increase the solubility of sparsely soluble organic substances in water (Neuberg, 1916). Although they are bearing hydrophobic and hydrophilic moieties, hydrotropes differ from conventional surfactants by some points, e.g., smaller hydrophobic part, do not form micelles, do not present a critical micellar concentration, have low cooperativity of aggregation, and work at molar concentrations. The hydrotrope formation may be seen as an action, where molecules met and interconnect through non-covalent interactions to form a cluster around a substance, creating a solubilizing microenvironment. Some measurements provide presumptive evidence for 1:1 M association to organize assemblies around solute, stoichiometry that fits with the concept of NADES. In addition, it has been recently pointed out that the IL may act as a new class of powerful hydrotropes where both the cation and the anion synergistically contribute to increase the solubility of biomolecules (vanillin and gallic acid) in water (Cláudio et al., 2015). Considering the evident similarity between IL and NADES, it seems that the driving forces behind the hydrotrope ability to dissolve a wide range of compounds, could be compatible with the NADES structure and properties. Therefore, we can suggest that NADES could act as a biological hydrotrope in cells. Recently, adenosine triphosphate (ATP), which is one of the most important and well-known molecules in the cell, was proposed to perform like a biological hydrotrope (Patel et al., 2017). Indeed, ATP was found to enhance the protein stability by preventing aggregation and dissolve previously formed protein aggregates. These results could explain why, in addition to being a source of energy for biological reactions, the ATP is observed in high concentrations (mM) in cells. It may indeed help the stability and function of macromolecules, by creating specific affinity and interaction with amino acid residues of proteins, leading to stabilized microenvironment. Similarly, we already mentioned that proline was categorized in water stressed cells to promote the thermal stability and native conformation of proteins to keep them functional. Proline is modestly surface active, but it is able to self-aggregate and provide a host system that can dissolve molecules. The microenvironmental topography found with the proline molecular assembly is somewhat different from those observed for the conventional hydrotropes, meaning that, even though the result is similar, the way to create microenvironments from binding/clustering of molecules may be different. This supramolecular chemistry, to allow cell adaptation and/or functioning in response to either internal or external factors is essentially a dynamic chemistry driven by the aptitude of the nature to create interactions of selected molecular components. In this continuum of subcellular assemblies, cells may also organize components into non-membrane-bound organelles that exist in different manner, depending on the interaction strength between constituents (Fig. 2). Indeed, the cytoplasm contains several membrane-less organelles (MLOs), called P-bodies, stress granules, or germ granules, which may be seen as condensed biological associations with well-delimited physical boundaries, but without lipid membrane barriers (Gomes & Shorter, 2019). Early works on *Caenorhabditis elegans* germ granules have shown that MLOs may behave as liquid droplets inside the cell, and are shaped via a process named liquid-liquid phase separation (LLPS). Researchers now suggest that a wide range of membrane-less structures exhibit liquid-like properties and coalesce through a LLPS mechanism. Once again, multivalent weak intermolecular interactions (such as π-π stacking, cation-π, charge-charge interactions, etc.) are fundamental to understand what makes LLPS biologically possible within the cellular environment, to balance both the entropic cost for phase separation and the interfacial free energy cost to create the phase boundary. LLPS allows cells to spatio-temporally creates essential microenvironments within the cytoplasm. For instance, intracellular phase separations and transitions by proteins have attracted much attention in the last decade in light of the multitude of physiological and pathological processes in which they are involved (Banani, Lee, Hyman, & Rosen, 2017; Bergeron-Sandoval, Safaee, & Michnick, 2016; Boeynaems et al., 2018; Holehouse & Pappu, 2018; Uversky, 2017). LLPS underlies the formation of MLOs (e.g., Cajal bodies, P-bodies) (Weidner, Wang, Prescianotto-Baschong, Estrada, & Spang, 2014), nucleolus, stress granules (Protter & Parker, 2016), centrosomes, aggresomes (Rajan, Illing, Bence, & Kopito, 2001), etc., whose function are essential for the cell and whose dysfunction is associated to various pathologies, including age-related disorders such as amyotrophic lateral sclerosis (ALS) and fronto-temporal dementia (Banani et al., 2017; Bergeron-Sandoval et al., 2016; Boeynaems et al., 2018; Holehouse & Pappu, 2018; Uversky, 2017). Many MLOs are formed during stress (Rabouille & Alberti, 2017), and most of these supramolecular assemblies are reversible and likely contribute to cell survival during stress and cell fitness upon stress relief. The proteins involved in the formation of MLOs usually possess intrinsically disordered regions (IDRs) that may be flanked by RNA binding regions (Chong & Forman-Kay, 2016; Clifford, Peter, & Rohit, 2015; Hyman, Weber, & Jülicher, 2014). Remarkably, these sequences of low compositional complexity and high conformational heterogeneity are also often at the origin of aggregation or fibrillation phenomena in amyloid-like structures via the formation of hydrogels (Boeynaems et al., 2018; Wu & Fuxreiter, 2016). By the sequestration of particular proteins, MLOs could contribute to cellular functions or perturb them (disease) subtracting proteins and metabolites from their natural pathways, or just by changing the dynamics inside the cells (Elbaum-Garfinkle & Brangwynne, 2015). LLPS mediated by IDRs also underlies the formation of viral factories (Heinrich, Maliga, Stein, Hyman, & Whelan, 2018; Nikolic et al., 2017; Zhou, Su, Samuel, & Ma, 2019). The latter are cytoplasmic inclusion bodies where the local concentration of viral proteins is increased thereby facilitating viral replication (Cifuentes-muñoz, Branttie, Slaughter, & Dutch, 2017; Heinrich, Cureton, Rahmeh, & Whelan, 2010; Hoenen et al., 2012; Lahaye et al., 2009; Rincheval et al., 2017; Zhang et al., 2013). These viral factories, which can be either double-membrane or MLOs (Heinrich et al., 2018; Nikolic et al., 2017; Zhou et al., 2019), also likely prevent the activation of host innate immunity by shielding viral components and/or impair their targeting by cellular antiviral proteins.

So doing, LLPS provide specific infrastructure, like a subcompartmentalization, that may serve to adjust and accelerate biochemical reactions, increase the concentration of certain molecules, or diffusion of biomolecules, among others. It is though that many if not all proteins can undergo LLPS in beneficial environmental conditions. In addition, LPPS has been related to the capacity of cells to resist to deleterious conditions, for example, by storing proteins and RNAs temporarily in a manner that allows their rapid recovery after the dissipation of the stress. This means that, similarly to what was obtained in NADES, the liquid microenvironment may promote the stability of macromolecules, as well as the protein folding, structuration, assemblies, etc. Thus, examining how biomolecules reorganize into MLOs to modulate activity will be an important avenue of further research, and possible connection with NADES.

6. How NADES (if they exist) may get involved in the cell functioning

The fact that NADES components can be found in all organisms, and that under various stress conditions their concentration seem to increase, shows us that they could have a role in cellular processes. Inversely and concomitantly, several candidate molecules, able and/or recognized to form NADES liquid phase when associated, may positively contribute in the biochemical stability of (macro)molecules and organization of cell constituents. In addition, we have presented different molecular assemblies driven by intermolecular interactions, either in cytoplasm or in membranes that could fit and connect with NADES technology. Since the role of membranes or organized subcellular assemblies are important in the living cell function, one may claim that NADES components may selectively interconnect and interact with membrane components or cytoplasmic macromolecules to help in the metabolism or to mitigate negative effects under stress conditions. Considering what we mentioned and pointed out in this chapter, here are some suggestion avenues for further reflections regarding the possible occurrence and the impact of NADES to regulate cell functioning through interactions with membranes, along with non-membrane-bound molecular assemblies:

- To contribute to the organization and/or stabilization of nano- or microdomains.
- To assist the dynamic of nano- or microdomains.
- Assist LLPS-mediated biological condensates formation (e.g., MLOs).
- To protect enzymes or proteins (especially after stresses including perturbing solutes, anhydrobiosis, high temperature, freezing and high hydrostatic pressure, etc.).
- To help and promote biochemical reactions of enzymes or proteins embedded in membranes.
- To stabilize macromolecules through water-solute or solute-macromolecule interactions.
- In the transport, diffusion of molecules or the circulation of fluids through membranes.
- In the reactions, or storage of molecules within, or in the vicinity, of cell membranes.
- In the plant cell conductivity.
- On membrane resistance to cold/frost.
- On permeability and membrane potential change.
- On water interaction with membrane and/or the activity of water.
- In the pH-redox of membranes.

Over the past years, an impressive collection of advanced microscopic and analytical technics have been developed to characterize the lateral heterogeneity in the plasma membrane, the organization and the mobility of components in cells. However, finding a direct approach to investigate the hypothetical roles of NADES in membrane domains or in cytoplasm with molecular assemblies *in vivo will be very challenging*. The reasons why it would be so difficult is that the involvement of NADES would likely happen in a local and dynamic way at nanoscopic scale (facing limit of diffraction, resolution, low specificity, artifacts, etc.) in complex cell systems that exhibit an enormous variety of components with diverse localizations, interactions, and functions.

References

Abbott, A. P., Capper, G., Davies, D. L., Munro, H. L., Rasheed, R. K., & Tambyrajah, V. (2001). Preparation of novel, moisture-stable, Lewis-acidic ionic liquids containing quaternary ammonium salts with functional side chains. *Chemical Communications*, *19*, 2010–2011.

Abbott, A. P., Capper, G., Davies, D. L., Rasheed, R. K., & Tambyrajah, V. (2003). Novel solvent properties of choline chloride/urea mixtures. *Chemical Communications*, *1*, 70–71.

Arnold, T., Jackson, A. J., Sanchez-Fernandez, A., Magnone, D., Terry, A. E., & Edler, K. J. (2015). Surfactant behavior of sodium dodecylsulfate in deep eutectic solvent choline chloride/urea. *Langmuir*, *31*, 12894–12902.

Badger, M. R., & Price, G. D. (1992). The CO2 concentrating mechanism in cyanobacteria and microalgae. *Physiologia Plantarum*, *84*, 606–615.

Banani, S. F., Lee, H. O., Hyman, A. A., & Rosen, M. K. (2017). Biomolecular condensates: Organizers of cellular biochemistry. *Nature Reviews. Molecular Cell Biology*, *18*, 285–298.

Bergeron-Sandoval, L. P., Safaee, N., & Michnick, S. W. (2016). Mechanisms and consequences of macromolecular phase separation. *Cell*, *165*, 1067–1079.

Boeynaems, S., et al. (2018). Protein phase separation: A new phase in cell biology. *Trends in Cell Biology*, *28*, 420–435.

Bryant, S. J., Atkin, R., & Warr, G. G. (2015). Spontaneous vesicle formation in a deep eutectic solvent. *Soft Matter*, *12*, 1645–1648.

Choi, Y. H., et al. (2011). Are natural deep eutectic solvents the missing link in understanding cellular metabolism and physiology? *Plant Physiology*, *156*, 1701–1705.

Chong, P. A., & Forman-Kay, J. D. (2016). Liquid–liquid phase separation in cellular signaling systems. *Current Opinion in Structural Biology*, *41*, 180–186.

Cifuentes-muñoz, N., Branttie, J., Slaughter, K. B., & Dutch, R. E. (2017). Human metapneumovirus induces transcription. *Journal of Virology*, *91*, 1–18.

Cláudio, A. F. M., Neves, M. C., Shimizu, K., Canongia Lopes, J. N., Freire, M. G., & Coutinho, J. P. (2015). The magic of aqueous solutions of ionic liquids: Ionic liquids as a powerful class of catanionic hydrotropes. *Green Chemistry*, *17*, 3948–3963.

Clifford, P. B., Peter, T., & Rohit, V. P. (2015). Polymer physics of intracellular phase transitions. *Nature Physics*, *11*, 899–904.

Costanzo, J. P., Reynolds, A. M., do Amaral, M. C. F., Rosendale, A. J., & Lee, R. E. (2015). Cryoprotectants and extreme freeze tolerance in a subarctic population of the wood frog. *PLoS ONE*, *10*, e0117234.

Crowe, L. M. (2002). Lessons from nature: The role of sugars in anhydrobiosis. *Comparative Biochemistry and Physiology: Part A: Molecular & Integrative Physiology*, *131*, 505–513.

Dai, Y., Rozema, E., Verpoorte, R., & Choi, Y. H. (2016). Application of natural deep eutectic solvents to the extraction of anthocyanins from Catharanthus roseus with high extractability and stability replacing conventional organic solvents. *Journal of Chromatography. A*, *1434*, 50–56.

Dai, Y., van Spronsen, J., Witkamp, G. J., Verpoorte, R., & Choi, Y. H. (2013). Natural deep eutectic solvents as new potential media for green technology. *Analytica Chimica Acta*, *766*, 61–68.

Dai, Y., Verpoorte, R., & Choi, Y. H. (2014). Natural deep eutectic solvents providing enhanced stability of natural colorants from safflower (Carthamus tinctorius). *Food Chemistry*, *159*, 116–121.

Dai, Y., Witkamp, G. J., Verpoorte, R., & Choi, Y. H. (2013). Natural deep eutectic solvents as a new extraction media for phenolic metabolites in carthamus tinctorius L. *Analytical Chemistry*, *85*, 6272–6278.

Dai, Y., Witkamp, G. J., Verpoorte, R., & Choi, Y. H. (2015). Tailoring properties of natural deep eutectic solvents with water to facilitate their applications. *Food Chemistry*, *187*, 14–19.

Dastmalchi, M., & Facchini, P. J. (2016). Plant metabolons assembled on demand. *Science*, *354*, 829–830.

Deamer, D. W. (1997). The first living systems: A bioenergetic perspective. *Microbiology and Molecular Biology Reviews*, *61*, 239–261.

Deamer, D. (2017). The role of lipid membranes in life's origin. *Lifestyles*, 7, 1–17.

Deamer, D., Dworkin, J. P., Sandford, S. A., Bernstein, M. P., & Allamandola, L. J. (2002). The first cell membranes. *Astrobiology*, *2*, 371–381.

Dowhan, W., Bogdanov, M., & Mileykovskaya, E. (2016). Functional roles of lipids in membranes. In *Biochemistry of lipids, lipoproteins and membranes* (6th ed.). Elsevier (chapter 1).

Durand, E., Lecomte, J., Baréa, B., Piombo, G., Dubreucq, E., & Villeneuve, P. (2012). Evaluation of deep eutectic solvents as new media for *Candida antarctica* B lipase catalyzed reactions. *Process Biochemistry*, *47*, 2081–2089.

Elbaum-Garfinkle, S., & Brangwynne, C. P. (2015). Liquids, fibers, and gels: The many phases of neurodegeneration. *Developmental Cell*, *35*, 531–532.

Engel, B. D., Schaffer, M., Cuellar, L. K., Villa, E., Plitzko, J. M., & Baumeister, W. (2015). Native architecture of the chlamydomonas chloroplast revealed by in situ cryo-electron tomography. *eLife*, *2015*, 1–29.

Gayet, F., Marty, J. D., Brûlet, A., & De Viguerie, N. L. (2011). Vesicles in ionic liquids. *Langmuir*, *27*(16), 9706–9710.

Georgieva, K., Sárvári, É., & Keresztes, Á. (2010). Protection of thylakoids against combined light and drought by a lumenal substance in the resurrection plant Haberlea rhodopensis. *Annals of Botany*, *105*, 117–126.

Giordano, M., Beardall, J., & Raven, J. A. (2005). CO_2 concentrating mechanisms in algae: Mechanisms, environmental modulation, and evolution. *Annual Review of Plant Biology*, *56*, 99–131.

Gladden, L. F., Mantle, M. D., Abbott, A. P., Ahmed, I. M., Al-Murshedi, A. Y., & Harris, R. C. (2015). Molecular and ionic diffusion in aqueous—Deep eutectic solvent mixtures: Probing inter-molecular interactions using PFG NMR. *Physical Chemistry Chemical Physics*, *17*, 15297–15304.

Gomes, E., & Shorter, J. (2019). The molecular language of membraneless organelles. *The Journal of Biological Chemistry*, *294*, 7115–7127.

Gutiérrez, M. C., Ferrer, M. L., Mateo, C. R., & Del Monte, F. (2009). Freeze-drying of aqueous solutions of deep eutectic solvents: A suitable approach to deep eutectic suspensions of self-assembled structures. *Langmuir*, *25*, 5509–5515.

Hadj-kali, M. K., Alkhidir, K. E., Wazeer, I., El-blidi, L., Mulyono, S., & Alnashef, I. M. (2015). Application of deep eutectic solvents and their individual constituents as surfactants for enhanced oil recovery. *Colloids and Surfaces A: Physicochemical and Engineering Aspects*, *487*, 221–231.

Hammond, O. S., Bowron, D. T., & Edler, K. J. (2017). The effect of water upon deep eutectic solvent nanostructure: An unusual transition from ionic mixture to aqueous solution. *Angewandte Chemie, International Edition*, *56*, 9782–9785.

Heinrich, B. S., Cureton, D. K., Rahmeh, A. A., & Whelan, S. P. J. (2010). Protein expression redirects vesicular stomatitis virus RNA synthesis to cytoplasmic inclusions. *PLoS Pathogens*, *6*, 1–14.

Heinrich, B. S., Maliga, Z., Stein, D. A., Hyman, A. A., & Whelan, S. P. J. (2018). Phase transitions drive the formation of vesicular stomatitis virus replication compartments. *MBio*, *9*, 1–10.

Hoenen, T., et al. (2012). Inclusion bodies are a site of Ebolavirus replication. *Journal of Virology*, *86*, 11779–11788.

Holdsworth, R. H. (1971). The isolation and partial characterization of the pyrenoid protein of eremosphaera viridis. *The Journal of Cell Biology*, *51*, 499–513.

Holehouse, A. S., & Pappu, R. V. (2018). Functional implications of intracellular phase transitions. *Biochemistry*, *57*, 2415–2423.

Hyman, A. A., Weber, C. A., & Jülicher, F. (2014). Liquid-liquid phase separation in biology. *Annual Review of Cell and Developmental Biology*, *30*, 39–58.

Kastritis, P. L., et al. (2017). Capturing protein communities by structural proteomics in a thermophilic eukaryote. *Molecular Systems Biology*, *13*, 936–950.

Koštál, V., Zahradníčková, H., Šimek, P., & Zelený, J. (2007). Multiple component system of sugars and polyols in the overwintering spruce bark beetle, Ips typographus. *Journal of Insect Physiology*, *53*, 580–586.

Lahaye, X., et al. (2009). Functional characterization of Negri bodies (NBs) in rabies virus-infected cells: Evidence that NBs are sites of viral transcription and replication. *Journal of Virology*, *83*, 7948–7958.

Laursen, T., et al. (2016). Characterization of a dynamic metabolon producing the defense compound dhurrin in sorghum. *Science*, *354*, 890–893.

Le Roy, J., Huss, B., Creach, A., Hawkins, S., & Neutelings, G. (2016). Glycosylation is a major regulator of phenylpropanoid availability and biological activity in plants. *Frontiers in Plant Science*, 7. article 735.

Lores, H., Romero, V., Costas, I., Bendicho, C., & Lavilla, I. (2017). Natural deep eutectic solvents in combination with ultrasonic energy as a green approach for solubilisation of proteins: Application to gluten determination by immunoassay. *Talanta*, *162*, 453–459.

Mamajanov, I., Engelhart, A. E., Bean, H. D., & Hud, N. V. (2010). DNA and RNA in anhydrous media: Duplex, triplex, and G-quadruplex secondary structures in a deep eutectic solvent. *Angewandte Chemie, International Edition*, *49*, 6310–6314.

Maréchal, E., & Lupette, J. (2020). Relationship between acyl-lipid and sterol metabolisms in diatoms. *Biochimie*, *169*, 3–11.

Meyer, M. T., Whittaker, C., & Griffiths, H. (2017). The algal pyrenoid: Key unanswered questions. *Journal of Experimental Botany*, *68*, 3739–3749.

Monhemi, H., Housaindokht, M. R., Moosavi-Movahedi, A. A., & Bozorgmehr, M. R. (2014). How a protein can remain stable in a solvent with high content of urea: Insights from molecular dynamics simulation of Candida antarctica lipase B in urea:choline chloride deep eutectic solvent. *Physical Chemistry Chemical Physics*, *16*, 14882–14893.

Neuberg, C. (1916). Hydrotropic phenomena. *Biochemische Zeitschrift*, *76*, 107.

Nikolic, J., et al. (2017). Negri bodies are viral factories with properties of liquid organelles. *Nature Communications*, *8*, 1–12.

Overgaard, J., et al. (2007). Metabolomic profiling of rapid cold hardening and cold shock in Drosophila melanogaster. *Journal of Insect Physiology*, *53*, 1218–1232.

Pal, M., Rai, R., Yadav, A., Khanna, R., Baker, G. A., & Pandey, S. (2014). Self-aggregation of sodium dodecyl sulfate within (choline chloride plus urea) deep eutectic solvent. *Langmuir*, *30*, 13191–13198.

Paniwnyk, L., Beaufoy, E., Lorimer, J. P., & Mason, T. J. (2001). The extraction of rutin from flower buds of Sophora japonica. *Ultrasonics Sonochemistry*, *8*, 299–301.

Parks, L. W., & Casey, W. M. (1995). Physiological implications of sterol biosynthesis in yeast. *Annual Review of Microbiology, 49*, 95–116.

Patel, A., et al. (2017). Biochemistry: ATP as a biological hydrotrope. *Science, 356*, 753–756.

Protter, D. S. W., & Parker, R. (2016). Principles and properties of stress granules. *Trends in Cell Biology, 26*, 668–679.

Rabouille, C., & Alberti, S. (2017). Cell adaptation upon stress: The emerging role of membrane-less compartments. *Current Opinion in Cell Biology, 47*, 34–42.

Rajan, R. S., Illing, M. E., Bence, N. F., & Kopito, R. R. (2001). Specificity in intracellular protein aggregation and inclusion body formation. *Proceedings of the National Academy of Sciences of the United States of America, 98*, 13060–13065.

Rincheval, V., et al. (2017). Functional organization of cytoplasmic inclusion bodies in cells infected by respiratory syncytial virus. *Nature Communications, 8*, 1–11.

Rudolph, A. S., Crowe, J. H., & Crowe, L. M. (1986). Effects of three stabilizing agents-Proline, betaine, and trehalose-on membrane phospholipids. *Archives of Biochemistry and Biophysics, 245*, 134–143.

Sanchez-Fernandez, A., et al. (2016). Micelle structure in a deep eutectic solvent: A small-angle scattering study. *Physical Chemistry Chemical Physics, 18*, 14063–14073.

Segré, D., Ben-Eli, D., Deamer, D. W., & Lancet, D. (2001). The lipid world. *Origins of Life and Evolution of the Biosphere, 31*, 119–145.

Shapiro, R., & Schulze-Makuch, D. (2009). The search for alien life in our solar system: Strategies and priorities. *Astrobiology, 9*, 335–343.

Simons, K., & Ikonen, E. (1997). Functional rafts in cell membranes. *Nature, 387*, 569–572.

Slotsbo, S., Maraldo, K., Malmendal, A., Nielsen, N. C., & Holmstrup, M. (2008). Freeze tolerance and accumulation of cryoprotectants in the enchytraeid Enchytraeus albidus (Oligochaeta) from Greenland and Europe. *Cryobiology, 57*, 286–291.

Uversky, V. N. (2017). Intrinsically disordered proteins in overcrowded milieu: Membrane-less organelles, phase separation, and intrinsic disorder. *Current Opinion in Structural Biology, 44*, 18–30.

Van Meer, G., Voelker, D. R., & Feigenson, G. W. (2009). Membrane lipids: Where they are. *Nature Reviews. Molecular Cell Biology, 101*, 1–4.

Wagner, G. J., & Hrazdina, G. (1984). Endoplasmic reticulum as a site of phenylpropanoid and flavonoid metabolism in hippeastrum. *Plant Physiology, 74*, 901–906.

Walters, K. R., Sformo, T., Barnes, B. M., & Duman, J. G. (2009). Freeze tolerance in an arctic Alaska stonefly. *The Journal of Experimental Biology, 212*, 305–312.

Weidner, J., Wang, C., Prescianotto-Baschong, C., Estrada, A. F., & Spang, A. (2014). The polysome-associated proteins Scp160 and Bfr1 prevent P body formation under normal growth conditions. *Journal of Cell Science, 127*, 1992–2004.

Wu, H., & Fuxreiter, M. (2016). The structure and dynamics of higher-order assemblies: Amyloids, signalosomes, and granules. *Cell, 165*, 1055–1066.

Wu, F., & Minteer, S. (2015). Krebs cycle metabolon: Structural evidence of substrate channeling revealed by cross-linking and mass spectrometry. *Angewandte Chemie, International Edition, 54*, 1851–1854.

Xu, P., Zheng, G. W., Zong, M. H., Li, N., & Lou, W. Y. (2017). Recent progress on deep eutectic solvents in biocatalysis. *Bioresources and Bioprocessing, 4*, 1–18.

Yancey, P. H. (2005). Organic osmolytes as compatible, metabolic and counteracting cytoprotectants in high osmolarity and other stresses. *The Journal of Experimental Biology, 208*, 2819–2830.

Yancey, P. H., & Burg, M. B. (1990). Counteracting effects of urea and betaine in mammalian cells in culture. *The American Journal of Physiology, 258*, 198–204.

Yancey, P. H., Clark, M. E., Hand, S. C., Bowlus, R. D., & Somero, G. N. (1982). Living with water stress: Evolution of osmolyte systems. *Science*, *217*, 1214–1222.

Zhang, S., et al. (2013). An amino acid of human Parainfluenza virus type 3 nucleoprotein is critical for template function and cytoplasmic inclusion body formation. *Journal of Virology*, *87*, 12457–12470.

Zhang, Y., et al. (2017). Protein-protein interactions and metabolite channelling in the plant tricarboxylic acid cycle. *Nature Communications*, *8*, 1–11.

Zhou, Y., Su, J. M., Samuel, C. E., & Ma, D. (2019). Measles virus forms inclusion bodies with properties of liquid organelles. *Journal of Virology*, *93*. e00948–19.

CHAPTER SEVEN

Natural deep eutectic solvents in plants and plant cells: In vitro evidence for their possible functions☆

Yuntao Dai[a,*], Eleni Maria Varypataki[b], Elena A. Golovina[c], Wim Jiskoot[b], Geert-Jan Witkamp[d], Young Hae Choi[a], and Robert Verpoorte[a]

[a]Natural Products Laboratory, Institute of Biology, Leiden University, Leiden, The Netherlands
[b]Division of BioTherapeutics, Leiden Academic Centre for Drug Research, Leiden University, Leiden, The Netherlands
[c]Laboratory of Plant Physiology, Department of Plant Sciences, Wageningen University, Wageningen, The Netherlands
[d]Division of Biological and Environmental Science and Engineering, King Abdullah University of Science and Technology, Thuwal, Saudi Arabia
[*]Corresponding author: e-mail address: dai_yuntao@live.cn

Contents

☆This chapter is an update from the nonpublished Chapter 8 of the PhD thesis of Dr. Yuntao Dai.

Advances in Botanical Research, Volume 97
ISSN 0065-2296
https://doi.org/10.1016/bs.abr.2020.09.012

Abstract

The components of natural deep eutectic solvents (NADES) are abundant in plants. This led to our hypothesis that NADES may play an important role in solubilizing, storing, and transporting poorly water-soluble metabolites in living cells, adjusting the water content of plants, and protecting cells when in harsh conditions. In order to test these hypothetical roles, diverse plant materials were analyzed, including leaves, petals, plant secretions and seeds. Comparatively high amounts of ingredients of NADES are observed in those organs. In particular, resurrection plants in dry state contain a higher amount of NADES components than fresh ones, and the level of NADES components is specifically higher in the outside layer (aleurone and seed cover) of barley, than in the inside (endosperm and embryo) layer. A high accumulation of sugars, sugar alcohols, amines, amino acids, and organic acids dominate plant secretions such as sap and nectar, often in typical molar ratios of NADES. This strongly supports the hypothesis of the existence of NADES in plants. For the roles, experimentally, NADES and water were mixed resulting in liquids with different compositions and properties. In the case of plants, NADES and water co-exist in the cells and may form ideal solvents for metabolites of diverse polarities and macromolecules. Some NADES are hygroscopic, providing evidence for possible water level controlling effects of NADES in plants. Most importantly, NADES may accumulate around the lipid bilayers, form intermolecular bonds with the polar heads of lipids, and stabilize the membrane, as revealed in experiments with liposomes. This study gives in vitro evidence for the different roles NADES may play in living organisms, and opens perspectives for further exploring the existence and functions of NADES in plants cells. The omics allows now to identify all molecules in an organism or even in a cell. The challenge for future research will be to understand how there molecules interact in the dynamic cellular processes and their compartmentation on a nanoscale. In other words the challenge is to unravel the molecular interactions in the three dimensions of space and the one of time, which will require a true multidisciplinary collaboration.

1. Introduction

Based on the observations made in metabolomics that all kind of microbial, plant and animal cells have similar patterns of some abundant primary metabolites in certain molar ratios, Choi et al. proposed the occurrence of a third liquid phase in living organisms. This liquid consists of two or more common solid (crystalline) primary metabolites that form an ionic liquid, or a deep eutectic solvent (Choi et al., 2011; Dai, van Spronsen, Witkamp, Verpoorte, & Choi, 2013; Dai, Witkamp, Verpoorte, & Choi, 2015). They called these liquids natural deep eutectic solvents (NADES). In fact, one may consider NADES as liquid supermolecules, composed of natural compounds in certain molar ratios connected by intermolecular interactions with each other, particularly H-bonding, i.e., NADES consist of a hydrogen bond donor and acceptor (Choi et al., 2011). The strong hydrogen bonding between these compounds is the crux for the specific characteristics of the NADES such as non-volatile, viscous and liquid in a wide temperature range. Their polarity is in the range of water and alcohols, making them particularly optimal solvents for medium polar compounds, which often have poor solubility in water, like most secondary metabolites.

When adding water to a NADES it is clear from the NMR spectra that until a dilution with about 50% of water, there is still an interaction between the NADES components. At higher dilutions the NMR spectra are like an aqueous solution of the NADES ingredients (Dai et al., 2015). In our view NADES are thus more a concept of the dynamics of different liquid phases in cells and organisms than a strict static physicochemical phenomenon of a deep eutectic mixture of two compounds. The components of NADES are metabolites that are found in high amounts in living cells (sugars, sugar alcohols, organic acids, amino acids, amines) as well as water. NADES possess some properties that make them very interesting solvents (Dai, van Spronsen, et al., 2013; Dai, Witkamp, Verpoorte, & Choi, 2013, 2015), e.g., negligible volatility, low melting point (liquid even far below −20 °C), broad polarity range and high solubilization power of a wide range of compounds (Choi et al., 2011; Dai, van Spronsen, et al., 2013, 2015). Poorly water-soluble low molecular weight metabolites and macromolecules (e.g., DNA, proteins, and polysaccharides) have shown good solubility in NADES (Mamajanov, Engelhart, Bean, & Hud, 2010; Dai, van Spronsen, et al., 2013; Dai, 2013; Dai, Witkamp, et al., 2013; El Achkar, Fourmentin, & Greige-Gerges, 2019; Núñez-Pertíñez & Wilks, 2020, see chapter "Proteins in deep eutectic solvents: Structure, dynamics and interactions with the solvent" by Sanchez-Fernandez and Jackson). Moreover NADES are suitable for enzymatic reactions (Zhao, Baker, & Holmes, 2011; Choi et al., 2011; chapter "Natural deep eutectic solvents as performance additives for biocatalysis" by Erol and Hollmann of this issue) and biotransformations (Gutiérrez, Ferrer, Yuste, Rojo, & del Monte, 2010).

The hypothesis of NADES being a third liquid phase in organisms still needs more solid evidence. The indirect evidence comes from plant tolerance against, e.g., drought, cold, and salinity, in which tolerance has been found to be correlated with high levels of some of the mentioned common cell metabolites such as sugars (e.g., sucrose, trehalose, raffinose), sugar alcohols (e.g., inositol, mannitol), organic acids (e.g., ascorbic acid), amino acids (e.g., proline), and amines (e.g., betaine, choline) (Bartels & Sunkar, 2005, chapters "NADES formation in vegetative desiccation tolerance: Prospects and challenges" by du Toit et al. and "Natural deep eutectic solvents—A new era of cryopreservation" by Jesus et al. of this issue). For example, the content of proline increased more than 10 times to nearly the same molar concentrations as sucrose during the natural desiccation of the resurrection plant *Selaginella bryopteris* (L.) Baker (Pandey et al., 2010). Whereas in other resurrection plants accumulation of high levels of glucose, sucrose and amino acids was observed, e.g., *Sporobolus stapfianus* Gand. (Whittaker, Martinelli, Farrant, Bochicchio, & Vazzana, 2007), and of sucrose and raffinose in *Xerophyta viscosa* Baker (Peters, Mundree, Thomson, Farrant, & Keller, 2007). Moreover, higher amounts of sugars (fructose, glucose, raffinose), proline, and galactinol were observed in *Arabidopsis* during cold acclimatization (Kaplan et al., 2007). In fact, in cryopreservation of tissues the common cryoprotectants are sugars, sugar alcohols, and proline (Mustafa, de Winter, van Iren, & Verpoorte, 2011, see chapter "Natural deep eutectic solvents—A new era of cryopreservation" by Jesus et al. of this issue). Proline is important for the cold acclimatization of plants as well as for other organisms (Kovács et al., 2011). All these metabolites that correlate with drought, cold, or salinity have been classified as compatible solutes for organisms, i.e., non-toxic molecules which do not interfere with normal metabolism and accumulate predominantly in the cytoplasm at high concentration under various stress conditions (Bartels & Sunkar, 2005; Yancey, Clark, Hand, Bowlus, & Somero, 1982).

In our view, these compatible substances might be part of NADES that could be formed in various cellular compartments. Drought and freezing may result in the formation of a NADES phase containing proteins and low molecular weight compounds, whereas the water evaporates or freezes, respectively. This process keeps all the macro- and micro-molecules in different cellular compartments stored in a non-aqueous liquid phase, stabilized by a compartment specific NADES. Dilution with water, i.e., melting in case of frozen material or uptake in case of resurrection plants, restores the cellular structures and cells resulting in a living cell or organism.

Thus, the indirect or chemical evidence for the roles of NADES in living systems exists in the form of their high abundance in organisms, often in certain molar ratios like 1:1 or 2:1. However, real in situ occurrence of NADES in any type of organism still needs solid experimental proof. To prove their in situ presence will, first of all, require microscopy approaches. Assuming the presence of NADES in cells, however, one may predict certain physical characteristics of NADES that enables the interaction with other cellular components such as membranes, cell walls, and macromolecules including

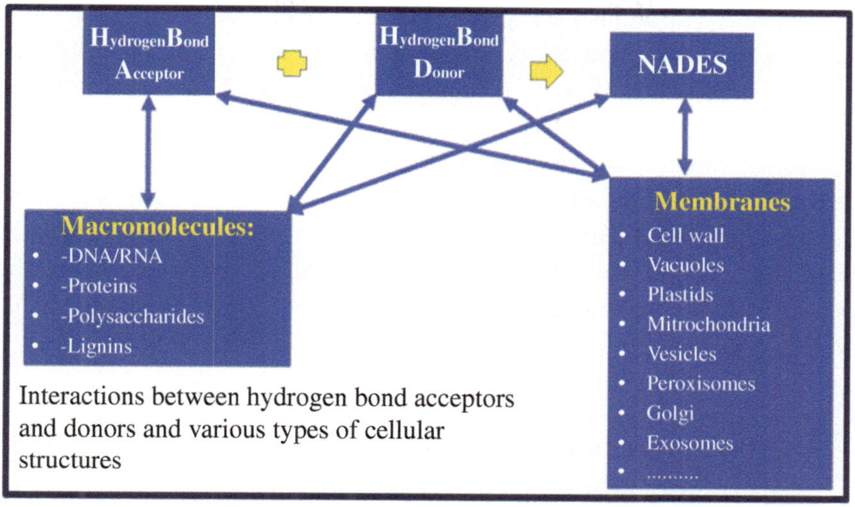

Fig. 1 Potential interactions between the hydrogen bond acceptor and hydrogen bond donor and various cellular structures.

polysaccharides, proteins, DNA, RNA and lignin. Particularly, the strong hydrogen bonding capacities of NADES components are an important characteristic in this context. A scheme of possible interactions of proton donor and proton acceptor molecules is shown in Fig. 1. Possibly all these interactions may occur in cells on a nanoscale, in a dynamic system. Some of these processes may play a role in an organism to overcome various forms of stress, like saline conditions, freezing temperatures, drought, wounding, and germination. Some NADES are part of static systems, like in seeds for long-term survival as well as in plants at low temperatures and in resurrection plants.

The aim of this chapter is to give an overview of the work we performed for the characterization of the biological roles of NADES found in plants. That includes both some previously reported and unpublished experimental data from the PhD thesis of Dr. Dai. Different plant materials and secretions have been collected and analyzed to obtain more evidence for the natural occurrence of NADES (Choi et al., 2011, see also chapters "Natural deep eutectic solvents present in plant exudates? A case study on the saps of Drosera species" by Vanda et al. and "Honey in traditional Chinese medicine: A guide to future applications of NADES to medicines" by Dai et al. of this issue). An interesting phenomenon of NADES is their potential to bind water, i.e., hygroscopicity could play a role in connection with drought-, and cold-stress as well as storage of seeds. As mentioned above, even up to a dilution with about 50% of water the components of NADES still interact with each other. This raises the question if a phase separation of NADES and water could occur, e.g., a layer of NADES adsorbed on membranes, proteins, polysaccharides, DNA or RNA being in equilibrium with an aqueous phase containing a host of different small molecules. Therefore, the diffusion process in a unstable two-phase water—NADES system was investigated to get information about the dynamics of such a system. Finally, to learn more about the interaction between NADES and membranes, the behavior of liposome bilayers in the presence of NADES in the media was studied.

2. Occurrence of NADES in plants

2.1 Plants and stress

To study a possible role of NADES in resurrection plants, *Selaginella pallescens* (C. Presl) Spring was analyzed by ^{1}H NMR spectroscopy. These experiments showed that a comparatively high amount of the NADES components sugar (glucose), organic acids (succinic acid, tartaric acid, acetic acid, γ-aminobutyric acid), and amino acids (alanine, threonine, arginine) are present in dry *S. pallescens* if compared with fresh plant material (Dai, van Spronsen, et al., 2013). The increased levels of these metabolites as a reaction to drought was also reported for other plants, such as *Arabidopsis*, which showed increased levels of sugar (sucrose), amino acids (proline, alanine, arginine), organic acids (succinic acid, fumaric acid, malic acid), and an amine (choline) in water depleted conditions if compared with normal growing conditions (Dai, van Spronsen, et al., 2013). So, a relatively high amount of components of NADES exists in plants under dry conditions. However, to find direct evidence of the existence of NADES in plants to survive various harsh conditions, in situ analysis of plant materials on the levels of tissues, cells and cellular compartments would be required instead of the indirect evidence from the presence of NADES components in crude extracts of plant material.

2.2 Flower petals containing pigments

In cells, NADES might help to dissolve and preserve the non-water-soluble or non-water-stable compounds (e.g., many hydrophobic colorants) due to the fact that NADES solubilize non-water-soluble phenolic compounds (Dai, Verpoorte, & Choi, 2014, 2015; Dai, Rozema, Verpoorte, & Choi, 2016).

Extremely high amount of rutin in the flower buds of *Sophora japonica* L. were reported to contain up to 22.9% (DW) of rutin (Couch, Naghski, & Krewson, 1952). Paniwnyk, Beaufoy, Lorimer, and Mason (2001) optimized the extraction method and found about 12% of rutin in the flower bud material they analyzed. Considering its very low water solubility of rutin, the high rutin level in flower buds is a mystery, but a similar case has been reported for anthocyanins. Considering their solubility in water, a higher than expected concentration of anthocyanins was found in vacuolar inclusions (Markham et al., 2000). This might be explained by the presence of a high level of NADES in anthocyanoplasts and anthocyanic vacuoles. The co-existing NADES could dissolve high amounts of anthocyanins, and flavonoids, or even these phenolics may be part of a NADES itself. It is noteworthy that small amounts of water in NADES may improve the solubility of phenolics (Dai et al., 2015).

The types of NADES are expected to differ depending on tissue, organ, and even cellular compartment, connected with different roles of the NADES. In a metabolomics analysis of plant materials, one will thus find the ingredients for different NADES, which in the tissues, cells and cellular compartments could be clearly compartmentalized. Metabolomics analysis of extracts of plant materials does not give any information about the localization of the NADES ingredients. Only in situ analysis of the metabolites might be able to show that components of a NADES are together in the same space. For example, analyzing the extracts flowers of *Catharanthus roseus* (L.) G. Don a number of potential NADES components were found (Pan et al., 2014). To get insight into the molecular organization of the small molecules in the flowers, flower petals were directly measured by high-resolution magic angle spinning (HR-MAS) ^{1}H NMR spectroscopy, a fast NMR spectroscopy method that is non-destructive and can be used to analyze biological materials without the need of extraction. It was expected that NADES components would be visible, as well as the interactions between these components, and even the interaction between NADES and the phenolics. However, the ^{1}H NMR spectrum of petals is dominated by sugars and organic acids (Fig. 2) and no signals of the phenolics were observed. Also, rutin, a major compound in the dry flower buds of *Sophora japonica* (upto 22.9%, DW), was hardly detected in HR-MAS ^{1}H NMR using optimized conditions for small molecules (Dai et al. unpublished results). Further experiments are needed to also visualize the phenolics, which might be present in a bound form in a sort of polymeric state that is not visible with the applied HR-MAS NMR conditions for small molecules. Pohjala and

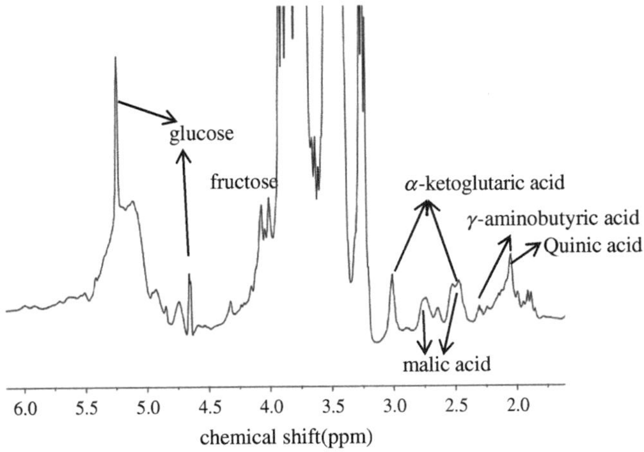

Fig. 2 High-resolution-magic angle spinning ^{1}H NMR spectrum (400 MHz) of the red petals of *Catharanthus roseus.*

Tammela (2012) reported that flavonoids aggregate easily, which may affect, for example, various bioassays. Scheidt, Pampel, Nissler, Gebhardt, and Huster (2004) showed that flavonoids interact with artificial membranes in a freeze dried preparation of liposomes and flavonoids by using MAS ^{1}H NOESY NMR spectroscopy. These observations might be an interesting model for studying the effect of NADES on the interaction of the phenolics and membranes, e.g., in cryopreservation experiments and in resurrection plants.

It is difficult to detect NADES itself in plants. The dynamic changes of NADES in cells and the presence of different potential ingredients of NADES in tissues, cells or cellular compartments, such as vacuoles, vesicles, exosomes, peroxisomes, mitochondria, or plastids point to the possibility that different NADES may occur in different cellular compartments. That means on a nanolevel. This requires further study on the cellular localization of NADES and NADES ingredients, i.e., on nanolevel.

2.3 Plants secretions

Various secretions, such as maple syrup, nectar, and mucilages are present extracellularly in plants. For example, sucrose:fructose:glucose with a molar ratio around 1:1:1 was detected in the nectar of *Cleome houtteana* Schltdl. (synonym *Cleome hassleriana* Chodat), and honey is composed of glucose and fructose (1:1, molar ratio) (Choi et al., 2011). In chapter "Natural deep eutectic solvents present in plant exudates? A case study on the saps of Drosera species" by Vanda et al. of this issue some further studies on various exudates of plants are discussed.

2.4 Plant seeds

A seed, as a temporary static system, is an interesting model to study the presence of NADES. In our viewpoint, NADES could be involved in the cold and drought resistance of seeds and keeping seeds viable in dormant stage, i.e., germinate when under the right conditions. In this context, NADES could keep essential enzymes dissolved and upon addition of water these enzymes are activated (Choi et al., 2011). The ^{1}H NMR of extracts from the outside and inside parts of barley seeds revealed the presence of, among others, sucrose, organic acids (acetic acid, succinic acid), amino acids (threonine, alanine), alcohols (isobutanol, ethanol), and amines (choline, betaine) (Fig. 3) (Choi et al., 2011). Interestingly, the outside layer (aleurone and seed cover) showed about equal molar amounts of sucrose and choline as major components. These may be present as a NADES in which some of the germination related enzymes are dissolved. These enzymes may become active after dilution of the NADES (Choi et al., 2011). Schuurink, Sedee, and Wang (1992) emphasized the need for protection of enzymes stored in the outside layer of the seed that must survive in extremely dry conditions during the seeds dormant period. NADES have been shown to well dissolve

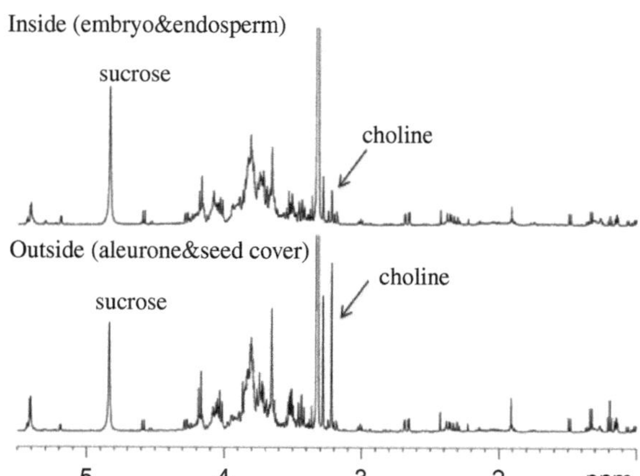

Fig. 3 ^{1}H NMR spectra (500 MHz) of extracts made with 50% $CH3OH\text{-}d_4$ in buffer (90 mM KH_2PO_4 in D_2O) containing 0.05% TMSP (trimethylsilyl propionic acid sodium salt (w/v)) (A) the inside and (B) outside parts of barley seeds (aleurone).

proteins and to stabilize them (Arakawa & Timasheff, 1982; Choi et al., 2011, see also chapters "Proteins in deep eutectic solvents: Structure, dynamics and interactions with the solvent" by Sanchez-Fernandez and Jackson and "Natural Deep Eutectic solvents as performance additives for biocatalysis" by Erol and Hollmann in this issue).

3. The physicochemical properties of NADES

3.1 Liquid characteristics NADES

Several physicochemical properties of NADES were measured: the density of NADES lies in the range of 1.08–1.36 g/cm^3 (Dai, van Spronsen, et al., 2013) and their viscosity between 37 and 720 mm^2/s, which is much higher than water (1 mm^2/s). The viscosity substantially increases at low temperature or with low water contents. The NADES have a low T_g (temperature of glass formation) far below −50 °C, thus they are stable liquids over a wide temperature range.

Several studies have reported the physicochemical properties of cytoplasm in cells before the intracellular molecular glass formation in plant anhydrobiosis. The cytoplasmic glass in dry cells results from a complex of sugars and other cytoplasmic components such as organic acids, amino acids, and salts (Buitink & Leprince, 2004). The viscosity of cytoplasm increases during drying-out conditions and eventually reaches a glass state. In general, a glass has the following physical properties: high density, temperature dependent mobility and slow mobility in solid state, high viscosity, and low T_g (Buitink & Leprince, 2008; Dijksterhuis, Nijsse, Hoekstra, & Golovina, 2007), which are also typical for the physical properties of NADES (Dai, van Spronsen, et al., 2013). The increasing levels of several NADES constituents in plants during anhydrobiosis attest a role of NADES in protecting the cells and tissues, e.g., in stabilizing the essential proteins and the various cellular membranes (see below). In fact, freezing of cells or plants would result in a similar process, in which water will freeze as pure water, leaving all compounds, both of small and high molecular weight, in the cellular liquids as a rest phase like in case of anhydrobiosis. Freezing resistance could thus be built upon the production of high levels of NADES ingredients. In fact, in cryopreservation of plant cells, typical cryoprotectants added to improve the chances of survival of the plant cells are all potential NADES components (Mustafa et al., 2011, see also chapter "Natural deep eutectic solvents—A new era of cryopreservation" by Jesus et al. in this issue).

3.2 The diffusion of NADES in water

Most NADES are hydrophilic and are miscible with water since the individual components are polar compounds. The high viscosity of NADES slows down the diffusion of water in a NADES. An example of a high viscosity NADES is sucrose–choline chloride–water (1:4:4) (SuCH), whereas 1,2-propanediol–choline chloride–water (1:1:1) (PpCH) is an example of a low viscosity NADES (Dai, van Spronsen, et al., 2013). To study the diffusion in a two-phase system consisting of NADES and water, experiments were performed with either of these NADES as the lower layer, on top of which water was carefully pipetted. In case of equal diffusion rates of the constituents in both layers, one would expect the disappearance of the interface when both layers have become identical. If the diffusion rates in both layers are not equal, the interface would move up (water diffuses faster into lower NADES layer, than NADES into the water layer) or down (NADES components have faster diffusion in upper layer, than water in the lower layer). To be able to follow the diffusion process over time, carthamin was added as dye in either phase. The diffusion between the NADES and water behaved differently than expected, as instead of two phases, three phases (upper, middle and lower) were observed (Fig. 4, day 10 SuCH;

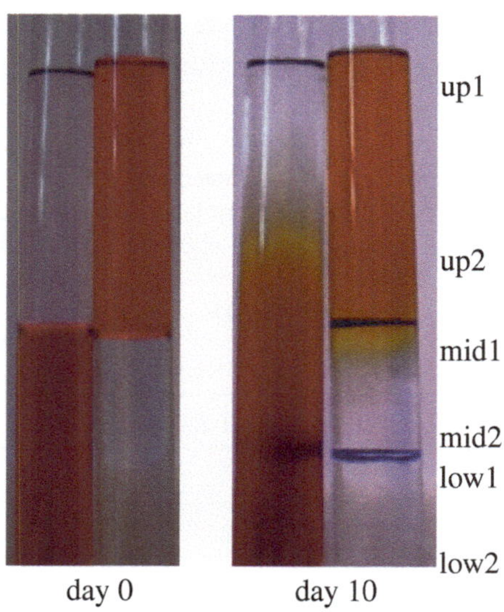

Fig. 4 A picture of diffusion tests between sucrose–choline chloride–water (1:4:4) and water on day 0 and 10, and 3 layers (upper, middle, and lower) were observed with two interfaces labeled in blue on day 10 (left: carthamin in the sucrose–choline chloride layer; right: carthamin in the water layer).

the two interfaces marked with blue color to improve the visibility). Apparently, the dye diffuses from the lower layer to the middle and from there to the top layer, whereas little diffusion of the dye occurs from the top layer to the lower layers, though the solubility of the dye is higher in the NADES than in water (Dai, 2013; Dai et al., 2014).

The same results were observed with the low viscosity PpCH, though the intermediate layer disappeared already in 2 days. So, the dye diffusion between PpCH and water was much faster than that from SuCH and water. Faster diffusion was also observed when the starting phase of SuCH was diluted with 10% (v/v) water. Viscosity clearly plays a complex role in the phase behavior between NADES and water.

To learn more about the composition of the intermediate phase, samples were collected from the upper and the lower phase, as well as of two different levels of the middle layer. The water content of each sample was determined and the organic components were quantified by means of ^{1}H NMR (Table 1). At day 10, the water content in the water layer showed a gradient from about 90% to 60% (up to down), in the middle layer 50%–20% and more or less unchanged in the lower phase, showing that instead of complete mixing, a new metastable NADES was formed as an intermediate phase (Table 1). The ratio of the two components in NADES is different in the layers. The molar ratio of choline chloride to sucrose in SuCH is 4 when prepared. During the diffusion process, the ratio was around 15 in the upper layer, 3 in the middle layer, and remained at the same value of 4 in the lower

Table 1 The water weight percentage and molecular ratio of components of sucrose-choline chloride in two parts of three different layers (upper, middle, and lower) after 10 days' diffusion between water and sucrose–choline chloride–water (1:4:4) ($n=3$) (see Fig. 4).

		up1	up2	mid1	mid2	low1	low2
Water w%	1	90 ± 2.2	60.1 ± 3.3	50.0 ± 4.5	18.0 ± 3.7	9.0 ± 2.8	6.4 ± 1.5
	2	91.9 ± 1.7	64.2 ± 5.7	42.1 ± 4.5	15.3 ± 2.7	6.4 ± 1.0	6.3 ± 0.9
	3	90	62	49	15	7.9	7.2
Molar ratio (choline: sucrose)	1	14.5	3.2	2.8	3.1	3.5	3.7
	2	16.6	3.0	3.0	3.5	3.5	3.9
	3	10.3	3.0	2.7	3.1	3.5	3.7

NADES components were quantified by means of ^{1}HNMR measurements.

layer. Apparently, in the interface of SuCH and water a new phase is formed from which choline can move into the water layer whereas the sugar is kept in the middle layer by the hydrogen bonding.

This mixing behavior indicates the complexity of the physicochemical properties of the cellular contents. A change in pH, concentration, temperature, etc., may cause the formation of separate phases. In a dynamic system like the endoplasmic reticulum, conditions for the biosynthesis of poorly water-soluble compounds may be created transiently via the formation of a third NADES phase in which enzymes are dissolved and where poorly water-soluble substrates are concentrated from the cytosol. Metabolons could thus be formed in this way (see chapter "Metabolons and biocondensates: The essence of plant plasticity and the key elements in development of green production systems" by Møller and Laursen). Layers of NADES might even be formed on membranes in a dynamic equilibrium with the liquid phase surrounding the membrane.

3.3 Water content in cells and plants

Some NADES we noted are hygroscopic. That raised the question if NADES in plants may play a role in controlling water levels, e.g., by absorbing water from the surroundings. Hygroscopicity measurements of glucose–choline chloride–water (2:5:5) (GCH) showed gradual increases of water content with an increasing humidity level in the surroundings from 0% to 80%. The process is reversible when the humidity goes down (Fig. 5A). The water percentage in GCH reached equilibrium in about 10h for a certain condition (Fig. 5B). So, the NADES may function in adjusting the water level in a plant through interaction with water vapor in the air.

To explore a hygroscopicity effect in plants, the water level of a barley seed was tested at a relative humidity level of 80% at 25 °C (Fig. 6). As can be seen, most water was absorbed during the first 48h and the equilibrium water content for 80% external relative humidity was about 9.1% (w/w). When the humidity level decreased to zero, the amount of absorbed water in the barley seed returned to the initial level within the same time interval. The observation fits our hypothesis of NADES being involved in water loss and uptake, e.g., several NADES are clearly hygroscopic and at the same time are able to strongly retain certain minimum amounts of water. The barley aleurone has been shown to contain sucrose and choline in an about

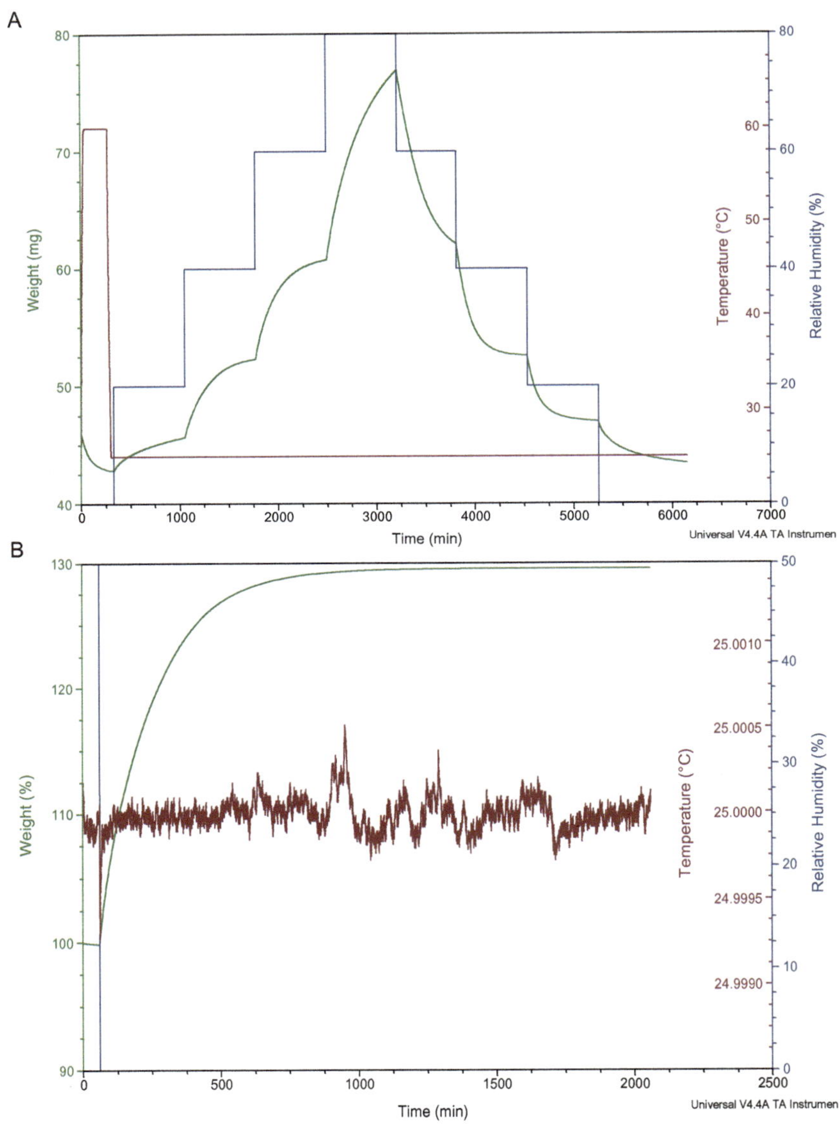

Fig. 5 Hygroscopicity experiment of glucose:choline chloride:water (1:2.5:2.5) at 25 °C with relative humidity steps (A) and 50% relative humidity level (B). The green line indicates the weight of glucose:choline chloride:water and the blue line refers to the relative humidity of test conditions.

1:1 M ratio. The NADES sucrose–choline chloride was found to be hygroscopic (Dai et al., unpublished results). The real in situ molecular organization of the aleurone is not known, so we cannot conclude yet if indeed these components are present as a NADES. But, apparently, plants can lose water

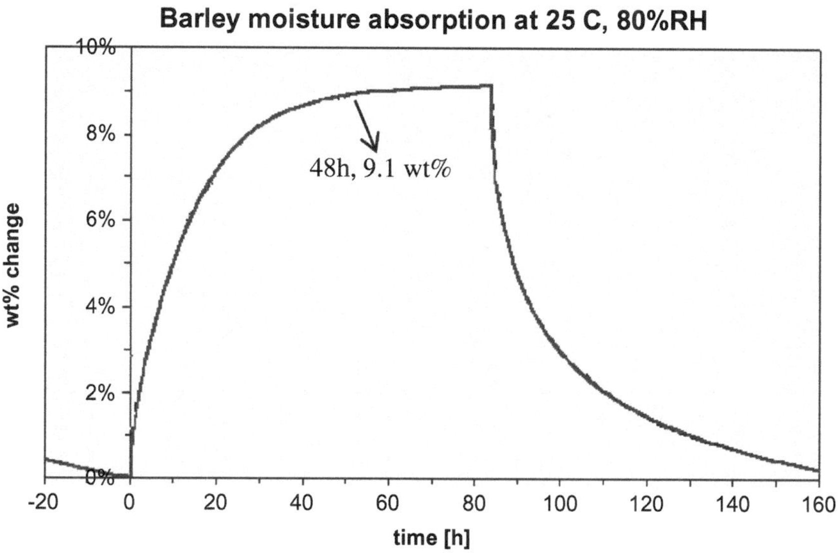

Fig. 6 The hygroscopicity experiment with intact barley seeds at 25 °C with 20 h drying, followed by 84 h 80% relative humidity level, and then back to 0% relative humidity level.

in relative dry conditions and absorb water in relative humid conditions. The hygroscopicity may explain how mosses can live on rocks, and how cacti can survive in deserts. Cacti open their stomata during the night which allows the exchange of oxygen and carbon dioxide. But at the same time with increasing relative humidity of the air due to low night temperatures, cacti may also catch the water molecules via a hygroscopic NADES. That NADES could be formed during the day time from sugars produced from the carbon dioxide and water that were absorbed during the night. The water uptake may activate the enzymes of photosynthesis that subsequently produce sugars again during the day thus restoring a hygroscopic NADES for nightly water harvesting.

3.4 The effect of NADES on the size, zeta potential and stability of liposomes

As mentioned above, our hypothesis is that NADES could be attached to cellular membranes by hydrogen bonding and hydrogen accepting interactions with the polar head groups in the lipid membranes. Therefore, the effect of some NADES on liposomes was tested as a model for the interaction of various cellular membranes with NADES. In these experiments artificial liposomes were made with 7.8 mmol 1,2-dioleoyl-3-trimethylammonium-propane (chloride salt), 3.9 mmol soybean L-α-phosphatidylcholine, and

Table 2 The size (hydrodynamic diameter, nm) and *zeta* potential (ZP) of liposome in three media ($n=3$): phosphate buffer (pH = 7.0), buffer with 18.0% (w/w) sucrose:malic acid (SM), buffer with 27.8% (w/w) sucrose:fructose:glucose (SFG) and buffer with 18.2% (w/w) sucrose-proline (SPr)

	Size (d, nm)			ZP	
	Average	**PDI[a]**	**RSD%**	**Average**	**RSD%**
Buffer + SM	117.2	0.124	0.60	40.5	0.65
buffer + SFG	192.2	0.175	4.85	20.4	12.70
Buffer + SPr	790	0.369	25.90	15.2	34.30
Buffer	162.5	0.092	1.07	27.4	1.03

[a]PDI: polydispersity index, the width of molecular size distribution.

3.9 mmol cholesterol. The liposomes were kept in a phosphate buffer of pH 7. When NADES was added, the size (hydrodynamic diameter) of liposomes was increased by 30 nm in a buffer with 27.8% sucrose:fructose:glucose (1:1:1) (SFG), increased by 600 nm in buffer with 18.2% sucrose:proline (1:1) SPr, and decreased by 40 nm in buffer with 18.0% sucrose:malic acid (1:1) (SM), compared with phosphate buffer (Table 2). The levels of the NADES are such that the liquid behaves like an aqueous solution, i.e., have a similar viscosity as the buffer (Dai, van Spronsen, et al., 2013). It means that NADES do affect liposomes, they may attach to them, or vacuoles may aggregate which increases the size of liposomes in the case of SFG and SPr, whereas SM decreases the liposome size.

The zeta potential decreased in buffer with SFG, SPr and increased in buffer with SM, compared with phosphate buffer. The different zeta potentials confirm interactions between NADES and liposomes and also imply that the interactions between liposomes and various NADES are different. For SFG, the sugar components may hydrogen bond to the head group of the lipids and form a film around the liposome through a hydrogen bonding network, increasing the size of liposome (Crowe, Hoekstra, & Crowe, 1992). Moreover the covering of the liposome with a neutral NADES may explain the lower zeta potential. For SM, the negatively charged acid group of malic acid may have ionic interactions with the positively charged head group of the lipids and this interaction may shrink the size of liposome due to the attraction effect. The size of liposomes with SPr increased but they were physically unstable probably due to the zwitterionic form of proline. The preparation of liposomes with sucrose: choline chloride (SC) in buffer (pH 7.0) was also attempted but the hydration of the lipid film failed and the lipid precipitated in the buffer containing SuC. This might be because of the repulsive forces between choline and the head group of the lipid (1,2-dioleoyl-3-trimethylammonium-propane). In all, the different behavior of liposomes in buffers containing NADES with neutral (SFG), acid (SM), basic (SC) and zwitterionic (SPr) components indicates the existence of interactions between liposomes and the molecules of NADES components.

The stability of liposomes in different media was investigated with the size and zeta potential as parameters. The size, polydispersity index, and zeta potential of liposomes were practically stable for 1 month at 4 °C in buffer and buffer with SFG or SM, but unstable in SPr. This indicates that liposomes are stable for shape, homogeneity, and interaction between NADES and liposomes for at least 1 month in buffer, buffer with SFG or SM. However, after 5 months at 4 °C, the liposomes had precipitated in both buffer and SPr, while they were still in suspension in SFG and SM. This proves a stabilizing effect of some NADES on liposomes. The stabilizing effects of sugars on membranes and liposomes have been known already for many years (Crowe et al., 1988, 1992; Crowe, Crowe, Carpenter, & Aurell Wistrom, 1987; Crowe, Leslie, & Crowe, 1994). Further studies should establish if the NADES effects are different from the sugar effect. The strong proton donor/proton acceptor interactions in the NADES may lead to the formation of a NADES layer on membranes, in which even selectivity may play a role, e.g., a different binding with an acid (e.g., malic acid) or a base (e.g., choline) containing NADES, depending on the character of the membranes.

3.5 The effect of NADES on the membrane dynamics of liposomes

The effect of NADES on membrane dynamics was studied with the EPR spin probe technique. A spin label 5-doxylstearic acid (5-DS) was used to probe the membrane interface in liposomes. In the 5-DS molecule the nitroxide doxyl group (a stable radical) is attached in a rigid, stereospecific manner to stearic acid at the fifth carbon from the COOH group. In phospholipid membranes the COOH group of spin-labeled stearic acid is inserted between the polar heads. Therefore, the nitroxide (doxyl group) of 5-DS resides in a polar area of the bilayer, which is called the bilayer surface or bilayer interface. The EPR spectral shape of 5-DS depends on the motion and angular orientation of the nitroxide group with respect to the membrane lipid-water interface (Marsh, 1981). This spin label allows the probing of the motional freedom in membranes at the lipid-water interface of the bilayer.

The anisotropic character of the spectral shape of 5-DS in liposomes in normal buffer (Fig. 7A) results from the restricted angular freedom of the radical group of 5-DS in the bilayer interface. The spectral parameters A_{max} and A_{min} indicate the outer and inner hyperfine splitting in an experimental spectrum as shown in Fig. 7A. The membrane-order parameter S_{zz} can be calculated as the ratio between the observed hyperfine anisotropy ($A_{max} - A_{min}$) to the maximum theoretically obtainable value, which corresponds to the completely rigid orientation of 5-DS (Knowles, Marsh, & Rattle, 1976). With membrane fluidization outer splitting (A_{max}) decreases and inner splitting (A_{min}) increases, so that the order parameter decreases. Later on, some corrections have been proposed to account for differences in polarity and for the range of membrane-order parameter. Here the order parameter is calculated according to the formula proposed by Marsh and Schorn (1998). This formula takes into account the principle splitting values for 5-DS and is corrected for polarity of the spin label environment. $S_{zz} = 1/7\ (A_{max} + 2A_{min}) - \text{sqrt}\{[1/7(A_{max} + 2A_{min})]^2 - 0.46\ (A_{max} - A_{min}) + 0.6\}$.

The EPR spectra of the 5-DS incorporated into the liposome in four media were compared: phosphate buffer, phosphate buffer with SFG (27.8%), phosphate buffer with SM (18.0%), and phosphate buffer with SPr (18.1%). The data on $2A_{max}$ and S_{zz} are presented in Table 3. Outer splitting $2A_{max}$ considerably decreases in the presence of 18.0% SM and decreases insignificantly in the presence of SFG and SPr. The order parameter S_{zz} of the EPR spectra of 5-DS in liposomes in the presence of 18.0% SM is considerably reduced (Table 3). In the presence of SFG and SPr the order parameter slightly increases.

The decrease of outer splitting $2A_{max}$ and order parameter S_{zz} in the presence of SM result from the fluidization of the membrane interface. This data is also supported by a decrease in the width of the central line (Table 3). The results are in accordance with a considerable decrease of liposome size from 162.5 to 117.2 nm (Table 2). Obviously, the decrease of liposome size inevitably causes the fluidization of the liposome bilayer outer surface due to an increased bilayer curvature and the spacing between lipids. The effect of bilayer curvature on phospholipid behavior has been demonstrated by molecular dynamics simulations (Risselada & Marrink, 2009). The size of liposomes in the presence of SFG and SPr considerably increases (Table 3). However,

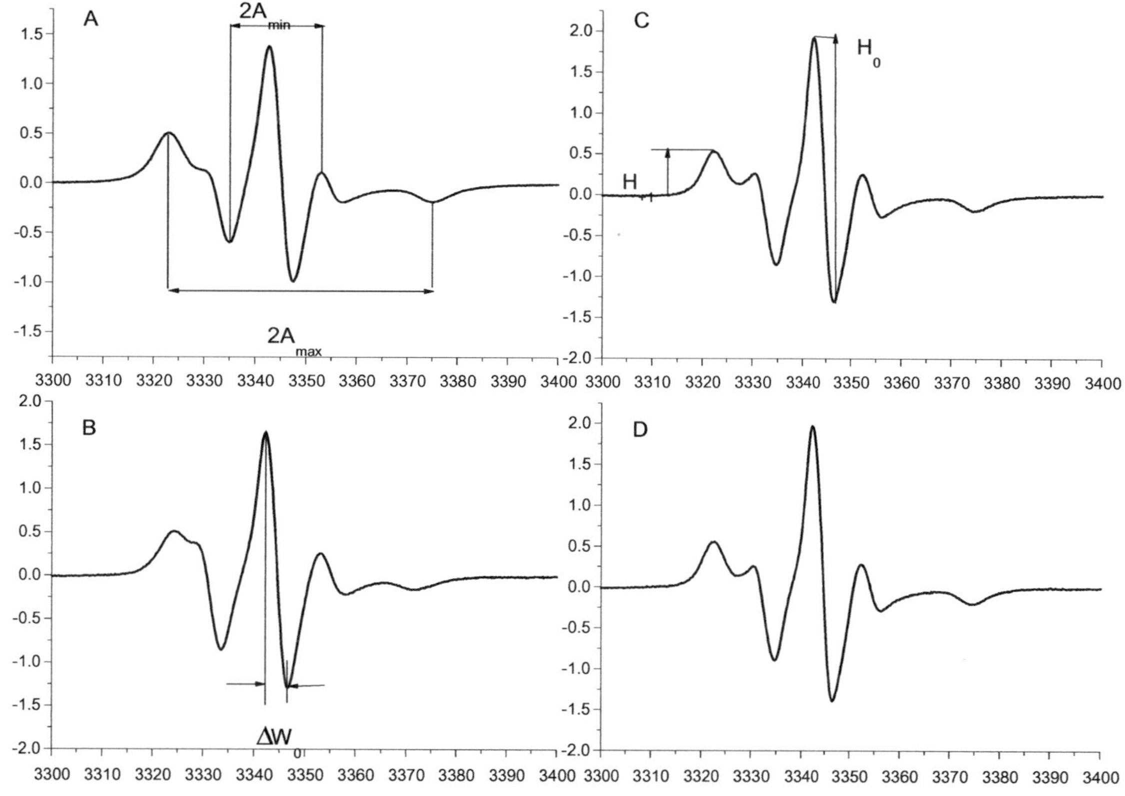

Fig. 7 Electron paramagnetic resonance spectra of 5-doxyl stearic acid in liposomes in PBS phosphate buffer (pH 7.0) (A); in PBS phosphate buffer with 18.0% (w/w) sucrose:malic acid: water (SM) (B); in PBS phosphate buffer with 27.8% (w/w) sucrose:fructose:glucose (SFG) (C); and in PBS phosphate buffer with 18.1% (w/w) sucrose:proline (SPr) (D).

Table 3 Parameters of electron paramagnetic resonance spectra of 5-doxyl stearic acid in liposomes in the presence of three different natural different deep eutectic solvents (buffer with 18.0 (w/w) sucrose:malic acid (SM), buffer with 27.8% (w/w) sucrose: fructose:glucose (SFG), and buffer with 18.1% (w/w) sucrose:proline (SPr), compared with phosphate buffer (pH=7.0)).

Media composition	$2A_{max}$, (Gauss)	S_{zz} (Rel. units)	ΔW_0 (Gauss)	H_0/H_{+1} (Rel. units)
Buffer (pH=7.0)	52.08	0.617	4.6919	2.71391
Buffer+27.8 (w/w) SFG	51.99	0.629	4.30102	3.67734
Buffer+18.0% (w/w) SM	47.45	0.507	4.30103	3.26143
Buffer+18.1% (w/w) SPr	51.87	0.622	4.30102	3.55936

The parameters include membrane-order parameter (S_{zz}), the outer hyperfine splitting ($2A_{max}$, in gauss), the width of the central line (ΔW_0, in gauss) and the ratio between the heights of the central (H_0) and low-field (H_{+1}) lines (H_0/H_{+1}, relative units).

this causes only a slight ordering of the membrane interface as determined by order parameter (Table 3). If the ordering/immobilization increases slightly in the presence of SFG and SPr, an increase of the outer splitting $2A_{max}$ would also be expected. However, this was not the case. The slight decrease of the $2A_{max}$ for these cases could be explained by the decrease of the polarity in the vicinity of spin label moiety.

The width of the central line decreases in the presence of all NADES (Table 3, Fig. 7), while the ratio between the heights of the central (H_0) and low-field (H_{+1}) lines (as indicated in Fig. 6C) increases in the presence of all NADES (Table 3, Fig. 8). There are many factors that determine the line width in the EPR spectra. Line broadening can be paramagnetic and motional. Line narrowing can only be explained in the case of SM by fluidization of the membrane interface, because in this case the order parameter of the spectrum also decreases. In the case of SFG and SPr the situation is more complicated. Both spectra are identical, thus the mechanism of line narrowing for these two NADES must be the same. In these samples, line narrowing occurs without the decrease of the order parameter. The order parameter relates to the average angle of the departure of the molecular axis of the acyl chain from the normal to the bilayer surface (angular orientation or angular freedom). When the system is disordered, this angle increases and the spin label moiety experiences more freedom to move. This results in line narrowing. If order parameter does not change, this can only mean that the rate of rotation around the molecular axis of the acyl chain increases without system disordering. Usually such effect is observed in the membrane

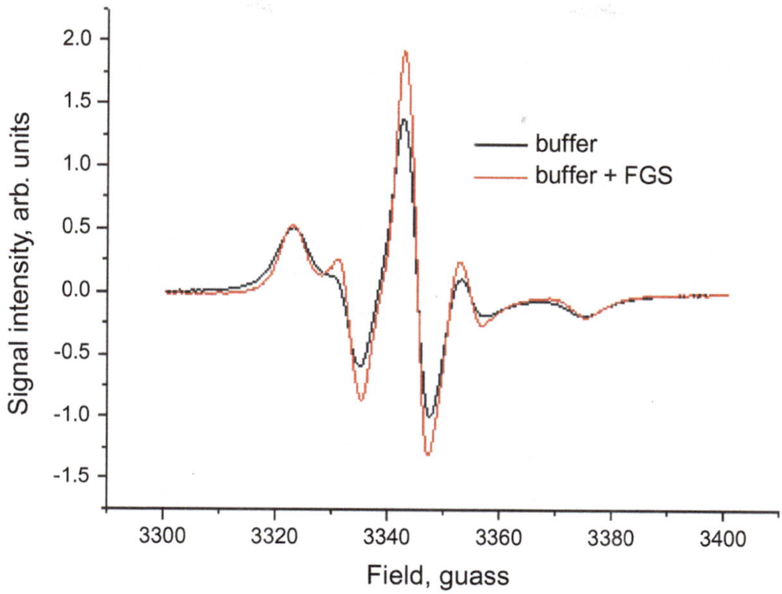

Fig. 8 Comparison of the electron paramagnetic resonance spectral shape of 5-doxyl stearic acid in liposomes in buffer (black lines) and in buffer with SFG (red lines).

interface at higher temperature. The fact that the same phenomenon is observed in case of NADES at room temperature, is very intriguing.

If line narrowing originates from a decreased probability of interaction between the spin label moieties at the liposomal membrane (paramagnetic broadening due to dipole-dipole interaction or exchange), then such an effect may result from decreased membrane lateral diffusion (Sachse, King, & Marsh, 1987). The restriction of lateral diffusion of lipids within bilayers in the presence of NADES would be one other possible mechanism for their stabilizing effect. All the parameters in EPR show an unexpected behavior of 5-DS which may indicate the establishment of a NADES layer around the liposomes. Generally spin label experiments are in aqueous surroundings, with the -COOH group of 5-DS at the polar heads of the lipid bilayer and the doxyl group in a polar area of the bilayer. If NADES builds up a shell around the liposomes, the label will be in NADES, a less polar but viscous environment, in which the 5-DS might bind more loosely with the polar heads of the bilayer as it dissolves better in the NADES than in water. In other words, the 5-DS will have a different behavior in a water-lipid system than in a water-NADES-lipid system due to a different interface at the bilayer. The decrease of the line width and increase of the H_0/H_{+1} ratio in the spectra of 5-DS in liposomes in the presence of SFG and SPr are a clear indication of the presence of another fraction of 5-DS, which demonstrates liquid-type behavior. These spectra may be considered as the sum of two spectra—one originates from the immobilized doxyl group (anisotropic spectrum), another—from the freely rotating spin label (isotropic spectrum with narrow lines). Superposition of two spectra results in the central line narrowing together with almost non-changed order parameter. The interaction of 5-DS with the bilayer is not homogenous, the differences may be caused by different localizations of the spin-label moiety in the bilayer. One localization might allow the free rotation of the doxyl group, the other location might be one more embedded in the bilayer. The isotropic spectrum might also result from any local spacing in the bilayer due to insertions of some molecular groups between phospholipid molecules, created by the insertion local spacing allows faster and less restricted rotation of the doxyl group. These measurements show that the addition of a NADES to liposome in a buffer results in changes in the liposome membranes. Different NADES have different effects. The effects include for some NADES an increase of the liposome stability. To translate these results to living cells is at this point pure speculation, but at least it is clear that different NADES may interact with different membranes in cells and thus may have a host of functions in the cell. For example one may envisage that the vacuolar membrane on the inside has a sugar-malic acid NADES, because of the acidic nature of the vacuolar liquid, and on the outside a sugar–choline chloride NADES, thus creating a kind of buffer to stabilize the tonoplast potential.

The most important lesson form these experiments is that the addition of NADES to an aqueous liposome suspension, changes the character of the liposomes. This fits our thoughts about the role of NADES in biological systems, NADES is a concept in which complex interactions between various molecules need to be considered to better understand various cellular functions, in which the NADES can be considered as a third liquid phase.

4. Conclusion

The aim of this chapter was to summarize and assess the in vitro evidence we accumulated for the existence of NADES in biological systems and to discuss some of their possible functions. NADES were proposed as a third liquid phase in cellular metabolism (Choi et al., 2011). Based on that hypothesis several aspects have been studied.

A number of flowers have been reported to have much higher levels of flavonoids and anthocyanins than water solubility would enable. These compounds are very well soluble in NADES, containing a certain percentage of water. High levels of these compounds could be found in the extracts of such flowers, but solid phase MAS NMR did not detect the phenolics using a small molecules targeted method. That means that probably in flowers these compounds are bound to large molecules, or form aggregates that behave as high molecular weight compounds and thus remain undetected in the solid phase NMR targeted at small molecules.

NADES may be involved in controlling the water level in plants and even trap water from humid air. NADES components are present in all cells and some are even the major compounds. It is well known that the levels of some NADES components are increased in plants as a defense against freezing and against drought. The use of NADES components in cryopreservation is good evidence for the possible role of NADES against freezing. Further studies in this field maybe important for breeding frost resistant plants and to further optimize cryopreservation of various organisms. Desert plants are obvious subjects for further studies of the role of NADES in resurrection plants as well as studying how plants like cacti can use NADES to hold water or even capture water from the atmosphere when at low temperatures the relative humidity gets very high and hygroscopic sugars based NADES may capture water from the air.

If NADES can be formed in cells, e.g., in vesicles, metabolons or as a layer on cellular membranes, the question is how diffusion would work at the interface between NADES and water in, e.g., the vacuole or the cytosol. An experiment in which a water layer was brought on top of a NADES it was observed that slowly a three-phase system was formed, in which the middle layer was a new NADES with a higher level of water. A dye added to either upper or lower layer diffused upward, but not downward. This shows the complexity of NADES in all kinds of cellular processes.

The stabilizing effects of sugars on membranes and proteins are known already for many years. The experiments we did with NADES showed also clear effects on liposome characteristics, including an increased stability. Assuming that sugars can bind to both membranes and proteins, the NADES open the way to build a whole sheet around the membranes or a protein. Strong reversible hydrogen bonding may result in transiently putting together a series of proteins capable of catalyzing a biosynthetic pathway, i.e., creating a metabolon (Bassard & Halkier, 2018; see chapter "Metabolons and bio-condensates: The essence of plant plasticity and the key elements in development of green production systems" by Møller and Laursen of this issue). The combination of this concept with membranes, leads to vesicles, which creates an environment where non-water-soluble intermediates are dissolved and may be transferred from one enzyme to the next in the sequence of the biosynthesis pathway. The characteristics of an enzyme in a NADES might be quite different from the same enzyme in water, among others because of a much higher concentration of substrate in the NADES liquid phase than in water. To fully address this hypothesis on the central role of NADES in many cellular processes, further experiments are required that can do in situ measurements on the interactions of molecules. Eventually that might revolutionize our views on cell physiology, as well as the physiology of whole organisms.

In this chapter we have brought together the results of a number of separate experiments we did in the past years in which we explored various models for possible roles of NADES in living organisms. Most of the results only give indirect evidence for possible roles, but for the final proof the occurrence, compartmentation and interactions of the NADES molecules in living cells must be mapped, in space and in time, i.e., full characterization of the dynamic system of all molecules in a living cell. Present day omics technologies can characterize all the molecules in an organism and even a cell, but how they interact in a dynamic system and how they are localized on a nanoscale is now the last step into the full understanding of life. A major challenge! A challenge that requires a multidisciplinary approach to be able to unravel all the molecular interactions in the three dimensions of space and the one of time.

Acknowledgments

We are grateful to Dr. Jieun Shin (Plant Developmental Biology Department, Max Planck Institute, Germany) for providing *Arabidopsis* plants. We thank KMB Jansen (Mechanics of Materials, Delft University of Technology, the Netherlands) for hydroscopicity tests.

References

Arakawa, T., & Timasheff, S. N. (1982). Stabilization of protein structure by sugars. *Biochemistry*, *21*, 6536–6544.

Bartels, D., & Sunkar, R. (2005). Drought and salt tolerance in plants. *Critical Reviews in Plant Sciences*, *24*, 23–58.

Bassard, J. E., & Halkier, B. A. (2018). How to prove the existence of metabolons? *Phytochemistry Reviews*, *17*, 211–227.

Buitink, J., & Leprince, O. (2004). Glass formation in plant anhydrobiotes: Survival in the dry state. *Cryobiology*, *48*, 215–228.

Buitink, J., & Leprince, O. (2008). Intracellular glasses and seed survival in the dry state. *Comptes Rendus Bbiologies*, *331*, 788–795.

Choi, Y. H., van Spronsen, J., Dai, Y., Verberne, M., Hollmann, F., Arends, I. W., et al. (2011). Are natural deep eutectic solvents the missing link in understanding cellular metabolism and physiology? *Plant Physiology*, *156*, 1701–1705.

Couch, J. F., Naghski, J., & Krewson, C. F. (1952). Rutin content of Sophora japonica L. *Journal of the American Chemical Society*, *74*, 424–425.

Crowe, J. H., Crowe, L. M., Carpenter, J. F., & Aurell Wistrom, C. (1987). Stabilization of dry phospholipid bilayers and proteins by sugars. *The Biochemical Journal*, *242*, 1–10.

Crowe, J. H., Crowe, L. M., Carpenter, J. F., Rudolph, A. S., Wistrom, C. A., Spargo, B. J., et al. (1988). Interactions of sugars with membranes. *Biochimica et Biophysica Acta*, *947*, 367–384.

Crowe, J. H., Hoekstra, F. A., & Crowe, L. M. (1992). Anhydrobiosis. *Annual Review of Physiology*, *54*, 579–599.

Crowe, J. H., Leslie, S. B., & Crowe, L. M. (1994). Is vitrification sufficient to preserve liposomes during freeze-drying? *Cryobiology*, *31*, 355–366.

Dai, Y. (2013). *Natural deep eutectic solvents and their application in natural products research and development.* PhD thesis Leiden University, ISBN:978-90-821172-0-2. Chapter 8.

Dai, Y., Rozema, E., Verpoorte, R., & Choi, Y. H. (2016). Application of natural deep eutectic solvents to the extraction of anthocyanins from *Catharanthus roseus* with high extractability and stability replacing conventional organic solvents. *Journal of Chromatography. A*, *1434*, 50–56.

Dai, Y., van Spronsen, J., Witkamp, G. J., Verpoorte, R., & Choi, Y. H. (2013). Natural deep eutectic solvents as new potential media for green technology. *Analytica Chimica Acta*, *766*, 61–68.

Dai, Y., Verpoorte, R., & Choi, Y. H. (2014). Natural deep eutectic solvents providing enhanced stability of natural colorants from safflower (*Carthamus tinctorius*). *Food Chemistry*, *159*, 116–121.

Dai, Y., Witkamp, G. J., Verpoorte, R., & Choi, Y. H. (2013). Natural deep eutectic solvents as a new extraction media for phenolic metabolites in *Carthamus tinctorius* L. *Analytical Chemistry*, *85*, 6272–6278.

Dai, Y., Witkamp, G. J., Verpoorte, R., & Choi, Y. H. (2015). Tailoring properties of natural deep eutectic solvents with water to facilitate their applications. *Food Chemistry*, *187*, 14–19.

Dijksterhuis, J., Nijsse, J., Hoekstra, F. A., & Golovina, E. A. (2007). High viscosity and anisotropy characterize the cytoplasm of fungal dormant stress-resistant spores. *Eukaryotic Cell*, *6*, 157–170.

El Achkar, T., Fourmentin, S., & Greige-Gerges, H. (2019). Deep eutectic solvents: An overview on their interactions with water and biochemical compounds. *Journal of Molecular Liquids.*, *288*. article number UNSP 111028.

Gutiérrez, M. C., Ferrer, M. L., Yuste, L., Rojo, F., & del Monte, F. (2010). Bacteria incorporation in deep-eutectic solvents through freeze-drying. *Angewandte Chemie (International Ed. in English)*, *49*, 2158–2162.

Kaplan, F., Kopka, J., Sung, D. Y., Zhao, W., Popp, M., Porat, R., et al. (2007). Transcript and metabolite profiling during cold acclimation of *Arabidopsis* reveals an intricate relationship of cold-regulated gene expression with modifications in metabolite content. *The Plant Journal*, *50*, 967–981.

Knowles, P. F., Marsh, D., & Rattle, H. W. E. (1976). *Magnetic resonance of biomolecules: An introduction to the theory and practice of NMR and ESR in biological systems.* United Kingdom: John Wiley & Sons.

Kovács, Z., Simon-Sarkadi, L., Sovány, C., Kirsch, K., Galiba, G., & Kocsy, G. (2011). Differential effects of cold acclimation and abscisic acid on free amino acid composition in wheat. *Plant Science*, *180*, 61–68.

Mamajanov, I., Engelhart, A. E., Bean, H. D., & Hud, N. V. (2010). DNA and RNA in anhydrous media: Duplex, triplex, and G-quadruplex secondary structures in a deep eutectic solvent. *Angewandte Chemie (International Ed. in English)*, *49*, 6310–6314.

Markham, K. R., Gould, K. S., Winefield, C. S., Mitchell, K. A., Bloor, S. J., & Boase, M. R. (2000). Anthocyanic vacuolar inclusions-their nature and significance in flower colouration. *Phytochemistry, 55*, 327–336.

Marsh, D. (1981). Electron spin resonance: Spin labels. *Molecular Biology, Biochemistry, and Biophysics, 31*, 51–142.

Marsh, D., & Schorn, K. (1998). Correction for anisotropically averaged hypefine splittings and order parameters from pseudopowder electron paramagnetic resonance (EPR) line shapes. An update for slow-motion contributions to lipid spin label spectra from membranes. In L. Berliner (Ed.), *Biological Magnetic Resonance* (pp. 405–410). New York: Plenum Press.

Mustafa, N. R., de Winter, W., van Iren, F., & Verpoorte, R. (2011). Initiation, growth and cryopreservation of plant cell suspension cultures. *Nature Protocols, 6*, 715–742.

Núñez-Pertíñez, S., & Wilks, T. R. (2020). Deep eutectic solvents as media for the prebiotic DNA-templated synthesis of peptides. *Frontiers in Chemistry*. https://doi.org/10.3389/fchem.2020.00041.

Pan, Q., Dai, Y., Nuringtyas, T. R., Mustafa, N. R., Schulte, A. E., Verpoorte, R., et al. (2014). Investigation of the chemomarkers correlated with flower colour in different organs of *Catharanthus roseus* using NMR-based metabolomics. *Phytochemical Analysis: PCA, 25*, 66–74.

Pandey, V., Ranjan, S., Deeba, F., Pandey, A. K., Singh, R., Shirke, P. A., et al. (2010). Desiccation-induced physiological and biochemical changes in resurrection plant, *Selaginella bryopteris*. *Journal of Plant Physiology, 167*, 1351–1359.

Paniwnyk, L., Beaufoy, E., Lorimer, J. P., & Mason, T. J. (2001). The extraction of rutin from flower buds of *Sophora japonica*. *Ultrasonics Sonochemistry, 8*, 299–301.

Peters, S., Mundree, S. G., Thomson, J. A., Farrant, J. M., & Keller, F. (2007). Protection mechanisms in the resurrection plant *Xerophyta viscosa* (Baker): Both sucrose and raffinose family oligosaccharides (RFOs) accumulate in leaves in response to water deficit. *Journal of Experimental Botany, 58*, 1947–1956.

Pohjala, L., & Tammela, P. (2012). Aggregating behavior of phenolic compounds—A source of false bioassay results? *Molecules, 17*, 10774–10790.

Risselada, H. J., & Marrink, S. J. (2009). Curvature effects on lipid packing and dynamics in liposomes revealed by coarse grained molecular dynamics simulations. *Physical Chemistry Chemical Physics, 11*, 2056–2067.

Sachse, J. H., King, M. D., & Marsh, D. J. (1987). ESR determination of lipid translational diffusion coefficients at low spin-label concentrations in biological membranes, using exchange broadening, exchange narrowing, and dipole-dipole interactions. *Journal of Magnetic Resonance, 71*, 385–404.

Scheidt, H. A., Pampel, A., Nissler, L., Gebhardt, R., & Huster, D. (2004). Investigation of the membrane localization and distribution of flavonoids by high-resolution magic angle spinning NMR spectroscopy. *Biochimica et Biophysica Acta, 1663*, 97–107.

Schuurink, R. C., Sedee, N. J., & Wang, M. (1992). Dormancy of the barley grain is correlated with gibberellic acid responsiveness of the isolated aleurone layer. *Plant Physiology, 100*, 1834–1839.

Whittaker, A., Martinelli, T., Farrant, J. M., Bochicchio, A., & Vazzana, C. (2007). Sucrose phosphate synthase activity and the co-ordination of carbon partitioning during sucrose and amino acid accumulation in desiccation-tolerant leaf material of the C4 resurrection plant *Sporobolus stapfianus* during dehydration. *Journal of Experimental Botany, 58*, 3775–3787.

Yancey, P. H., Clark, M. E., Hand, S. C., Bowlus, R. D., & Somero, G. N. (1982). Living with water stress: Evolution of osmolyte systems. *Science (New York, N.Y.), 217*, 1214–1222.

Zhao, H., Baker, G. A., & Holmes, S. (2011). New eutectic ionic liquids for lipase activation and enzymatic preparation of biodiesel. *Organic & Biomolecular Chemistry, 9*, 1908–1916.

CHAPTER EIGHT

Metabolons and bio-condensates: The essence of plant plasticity and the key elements in development of green production systems

Birger Lindberg Møller and Tomas Laursen*

Plant Biochemistry Laboratory, Department of Plant and Environmental Sciences, University of Copenhagen, Copenhagen, Denmark

*Corresponding author: e-mail address: tola@plen.ku.dk

Contents

Abstract

An insufficient vocabulary and lack of a proper molecular grammar tends to simplify the exquisite organization of the cellular environment in which plant metabolism is orchestrated. Chaos at the cellular level would not have enabled plants to evolve as the chemists par excellence in nature. Certain molecules such as sugars, amino acids and organic acids, are always present in considerable amounts in plant cells. In the proper stoichiometric ratios, they may form natural deep eutectic solvents (NADES) and thereby provide a third intracellular phase in addition to oil and water. With a negligible vapor pressure and superior ability to solubilize natural products, they may fine tune the localization and activity of enzymes and guide storage of natural products in dense bio-condensates through liquid-liquid phase separation. In the absence of a stringent lipid

Advances in Botanical Research, Volume 97
ISSN 0065-2296
https://doi.org/10.1016/bs.abr.2020.09.006

bilayer border, such dynamic compartments may allow for swift mixing and demixing depending on metabolic demands. In plants, aromatic amino acids serve as precursors of an impressive diversity of bioactive natural compounds including alkaloids, cyanogenic glucosides and phenylpropanoids. This product diversity is largely achieved through combinatorial assembly of sequential enzymes into efficient enzyme complexes, metabolons, that facilitate substrate channeling and dependent on their configuration establish metabolic highways towards formation of specific end products. Based on specific experimental systems, we discuss how new technologies with improved spatial and temporal resolution are now making it possible to study the dynamics of plant metabolism based on metabolon and bio-condensate formation making these substantive words in the glossary.

1. Introduction

We are moving into The Planthropocene Era where petrochemicals are to be replaced by use of photosynthetic organisms (higher plants, ferns, moss, algae or cyanobacteria) for production of food, biomaterials and energy and as a rich source of nutraceuticals, condiments, flavors, colorants as well as medicinal compounds (Møller, 2017). In contrast to primary metabolites, bioactive natural products are produced by a specific subset of plant species, which often cannot be grown at a commercial scale or are red-listed. Typically, bioactive natural products are produced in low amounts. In this context, it would be important to understand the mechanisms that enable a small number of selected plant species to orchestrate synthesis of high and sometimes extremely high amounts of natural products. In this review, we highlight recent literature on the possible involvement of enzyme complexes (metabolons) and Natural Deep Eutectic Solvents (NADES) in the biosynthesis, accumulation and storage processes enabling formation of dense bio-condensates of plant natural products.

On demand biosynthesis of plant natural products may involve combinatorial organization of both membrane anchored and soluble enzymes in metabolons facilitating the formation of metabolic highways. Numerous pathways involved in both general and specialized metabolism have been proposed to organize in metabolons on the endoplasmic reticulum surface (Jørgensen et al., 2005; Laursen, Møller, & Bassard, 2015; Møller & Conn, 1980) that facilitate substrate channeling and prevents leakage of potentially toxic intermediates (Møller & Conn, 1980; Srere, 1985). The potential for directing metabolism through organization of the biosynthetic enzymes in metabolons is well-illustrated in nature, which allows for accumulation of natural products up to 30% of the dry mass in specialized tissues

(Halkier & Møller, 1989) despite the biosynthetic enzymes constituting less than 1% of the total protein content (Laursen et al., 2016). To date only a handful of metabolons involved in specialized metabolism have been experimentally proven with the requirements to involve both substrate channeling and protein-protein interactions. Hence, it remains unknown how relatively low abundant sequential enzymes (Bassard, Møller, & Laursen, 2017) succeed to organize on-demand in the seemingly chaotic lipid membrane "ocean." One may speculate that the intracellular milieu comprises a mosaic of micro environments and domains rather than well-mixed phases. Such mosaic structures are likely to be involved in organization of biosynthetic enzymes thereby circumventing bulk equilibria and undesired release of intermediates.

On March 14th 2018, Nature had a feature article entitled "Cell biology's new phase" (Dolgin, 2018) (Fig. 1). As an equivalent to cellular compartments confined by lipid bilayers, liquid-liquid phase separation within living animal cells was discussed as a means to concentrate some molecules in bio-condensates and exclude others and thereby orchestrate order in the crowded chaos of the cell (Banani, Lee, Hyman, & Rosen, 2017; Choi et al., 2011; Courchaine, Lu, & Neugebauer, 2016). In animal cells, numerous membrane-less droplet organelles have been defined to co-exist with canonical organelles (Aguilera-Gomez & Rabouille, 2017; Aumiller & Keating, 2017; Banani et al., 2017; Courchaine et al., 2016; Kuchler, Yoshimoto, Luginbühl, Mavelli, & Walde, 2016; Mitrea & Kriwacki, 2016) with the best-studied examples being the nucleolus, the site of ribosome biogenesis in the nucleus in eukaryotic cells. *In vitro* experimental models to study liquid-liquid phase coexistence using polymer solutions have been established (Aumiller & Keating, 2017; Berry, Brangwynne, & Haataja, 2018; Courchaine et al., 2016).

In plants, bio-active natural products have classically been thought to be stored in the central vacuole, in specific chromoplasts or in trichomes. This view of compartments as clearly separated entities encapsulated by membranes is far too simplistic and has been challenged by Verpoorte and Choi and their co-workers (Choi et al., 2011). They propose, that in plants bio-condensates of largely undefined origin and nature are stabilized by and composed of **NA**tural **D**eep **E**utectic **S**olvents (**NADES**) (Fig. 2). This NADES phase would then represent a yet-uncharacterized third liquid phase in the live plant cell and function as a crucial solvent of bioactive natural products. The concept of NADES was proposed based on NMR studies of the ubiquitous presence of sugars, protein amino acids, choline and

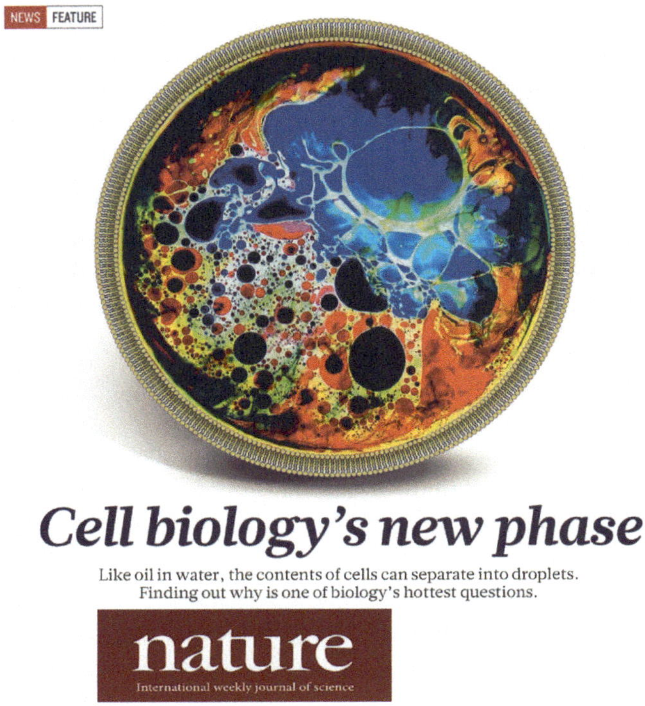

Fig. 1 Cell biology's new phase. Illustration from a Feature Article in Nature March 14th, 2018, on the emerging importance of aqueous phase separation as a tool used in biology to create order in the crowed space of the cell. The biological droplets formed by aqueous-aqueous phase partitioning were hypothesized to speed up some reactions by concentrating substrates while preventing undesired reactions to take place. The focus was on the content of proteins and RNAs in the droplets as exemplified by the nucleolus being a dense condensate of proteins and genetic material in the cell nucleus with a clear drop-like behavior. In this review, we discuss the possible roles of biological droplets in plant cells as storage sites for dense bio-condensates of natural products.

organic acids in plant cells (Choi et al., 2011; Kim, Choi, & Verpoorte, 2010) (Fig. 2). In laboratory experiments, different types of NADES (Dai, Witkamp, Verpoorte, & Choi, 2013; Grimplet et al., 2009; Zhao et al., 2015) were demonstrated to be excellent solvents for sparsely soluble bio-active natural products such as vanillin glucoside, rutin and other flavonoids. Today, NADES are used as green solvents for industrial scale isolation of bioactive natural products from a large number of plants (Choi et al., 2011; Dai, Witkamp, et al., 2013; González, Mustafa, Erica, Verpoorte, & Choi, 2018; Zhao et al., 2015).

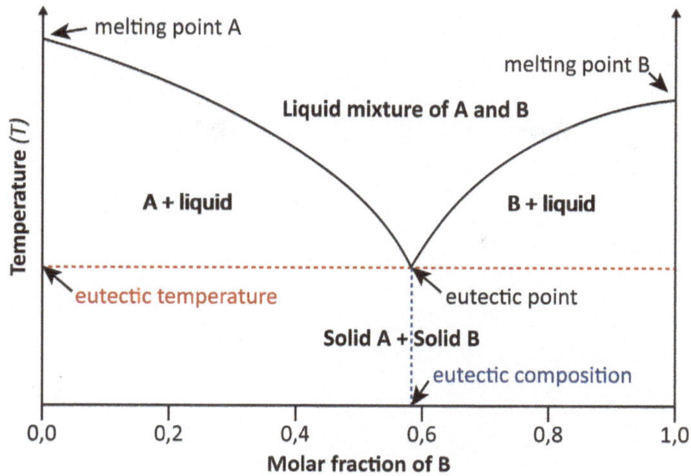

Fig. 2 Binary eutectic phase diagram. A *eutectic mixture* is defined as a *mixture* of two or more components which when present in a specific stoichiometric ratio (the eutectic composition) inhibit each other's crystallization to form a solid. As a result, this eutectic mixture has a lower melting point (the eutectic point) than either of its components. Sugars, protein amino acids, choline and organic acids that are ubiquitously present constituents in plants and are typical components of eutectic mixtures. When formed in plants, they are therefore termed Natural Deep Eutectic Solvents (NADES). The term eutectic is derived from the Greek "εύ" (eu = well) and "τῆξις" (tēxis = melting). The eutectic phase diagram shown is composed of two components, A and B.

A common denominator for NADES is that they are composed of crystalline components that form a highly viscous liquid when mixed in the right stoichiometric proportions, dissolved in water and lyophilized until the mass remains constant (approx. 8% water remaining). This applies for a mixture of glucose: tartaric acid in a 1:1 M ratio as found in raisins (Fig. 3). Raisins maintain a liquid phase despite an almost complete removal of water. The NADES form a liquid crystal through formation of inter- and intramolecular hydrogen bonds. This results in a high melting point depression causing the solids to form viscous oils and, in many cases, to remain fluid at room temperature (Choi et al., 2011). In this way, the NADES might function as super-molecules mimicking the polymers known to stabilize liquid-liquid phase separation in animal cells. The unparalleled solubilization capacity of NADES would suggest that they possess the ability to establish hydrogen bonds to specific bioactive natural products. Upon addition of water, the unique properties of the NADES gradually abolish. However, at present

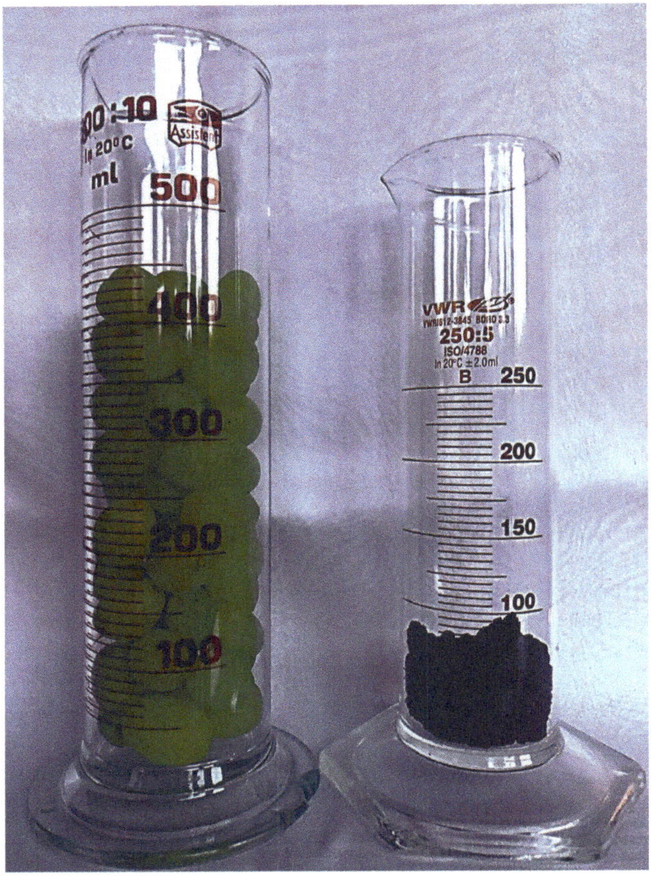

Fig. 3 Raisins are a good representative of a NADES. Raisins are obtained from dried grapes and retains an oily liquid phase formed by the presence of glucose and tartaric acid which forms a Natural Deep Eutectic Solvent. The picture shows the same number of grapes and dried raisins illustrating the low water content of the latter while avoiding the formation of sugar cubes.

no direct evidence for the *in planta* existence or possible mode of formation of NADES based bio-condensates has been reported.

In this review, we discuss the establishment of high-flux metabolic highways mediated by metabolon formation and selected plant systems where the formation and predicted functionality of NADES would appear to explain some of the experimental results obtained on the formation and storage of natural products. In these plant systems, bio-active natural products with sparse water solubility are kept in soluble form at surprisingly high molar concentrations in confined cellular domains.

2. Bio-condensates provides a suiting explanation for how plants biosynthesize and store natural products

All plant species produce a number of natural products with different complexity and in different amounts. Remarkable examples of plants containing tissues with extraordinary high levels of phytochemicals are the almost black flower petals of the gentiana "flower of death" (*Lisianthius nigrescens*) (Fig. 4). In these flower petals, the anthocyanin content constitutes 24% of the dry mass (Markham et al., 2004). The buds and flowers of sophora (*Sophora japonica*, Pagoda tree) contain the flavonoid rutin that constitutes up to 30% of their dry mass (Choi et al., 2011; Horosanskaia, Nguyen, Vu, Seidel-Morgenstern, & Lorenz, 2017; Zhao et al., 2015). Transcriptome sequencing has provided gene candidates encoding the enzymes catalyzing rutin biosynthesis (Zhang et al., 2017). In the upper part of sorghum (*Sorghum bicolor*) seedlings, the cyanogenic glucoside dhurrin represents 30% of the dry mass (Halkier & Møller, 1989; Kojima, Poulton, Thayer, & Conn, 1979) equivalent to molar concentrations at the storage sites. The enzymes catalyzing dhurrin synthesis have all been isolated and characterized (Bak, Olsen, Halkier, & Møller, 2000; Kahn, Fahrendorf, Halkier, & Møller, 1999; Koch, Sibbesen, Halkier, Svendsen, & Møller, 1995; Kristensen et al., 2005; Møller & Conn, 1979; Nielsen, Tattersall, Jones, & Møller, 2008; Sibbesen, Koch, Halkier, & Møller, 1994, 1995; Thorsoe et al., 2005). In vanilla (*Vanilla planifolia*) pods, vanillin glucoside is present in 4.7 M concentrations stored in senescent chloroplasts (Brillouet et al., 2014; Gallage et al., 2018) (Fig. 5). The vanillin synthase, which

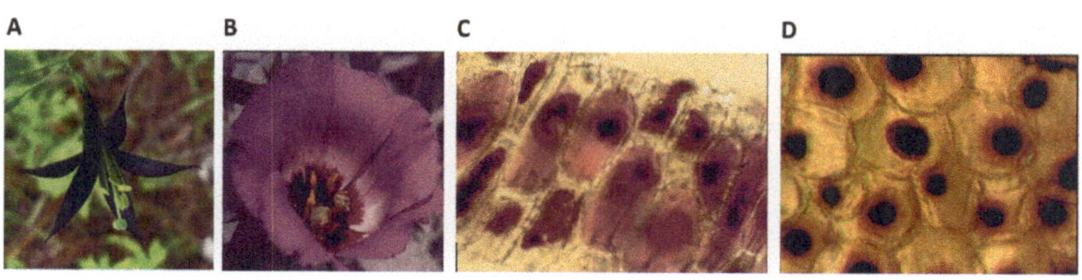

Fig. 4 Flowers containing bio-condensates of anthocyanins. Flowers of (A) *Lisianthius nigrescens* ("flower of death") and (B) *Eustoma grandiflorum* (purple lisianthus) and transverse light-microscopy sections showing anthocyanin bio-condensates from (C) purple lisianthus and (D) blue-gray carnation (*Dianthus caryophyllus*). *Illustration from Markham, K. R., Bloor, S. J., Nicholson, R., Rivera, R., Shemluck, M., Kevan, P. G., et al. (2004). Black flower coloration I wild Lisianthius nigrescens: Its chemistry and ecological consequences.* Zeitschrift für Naturforschung C, *59, 624–630.*

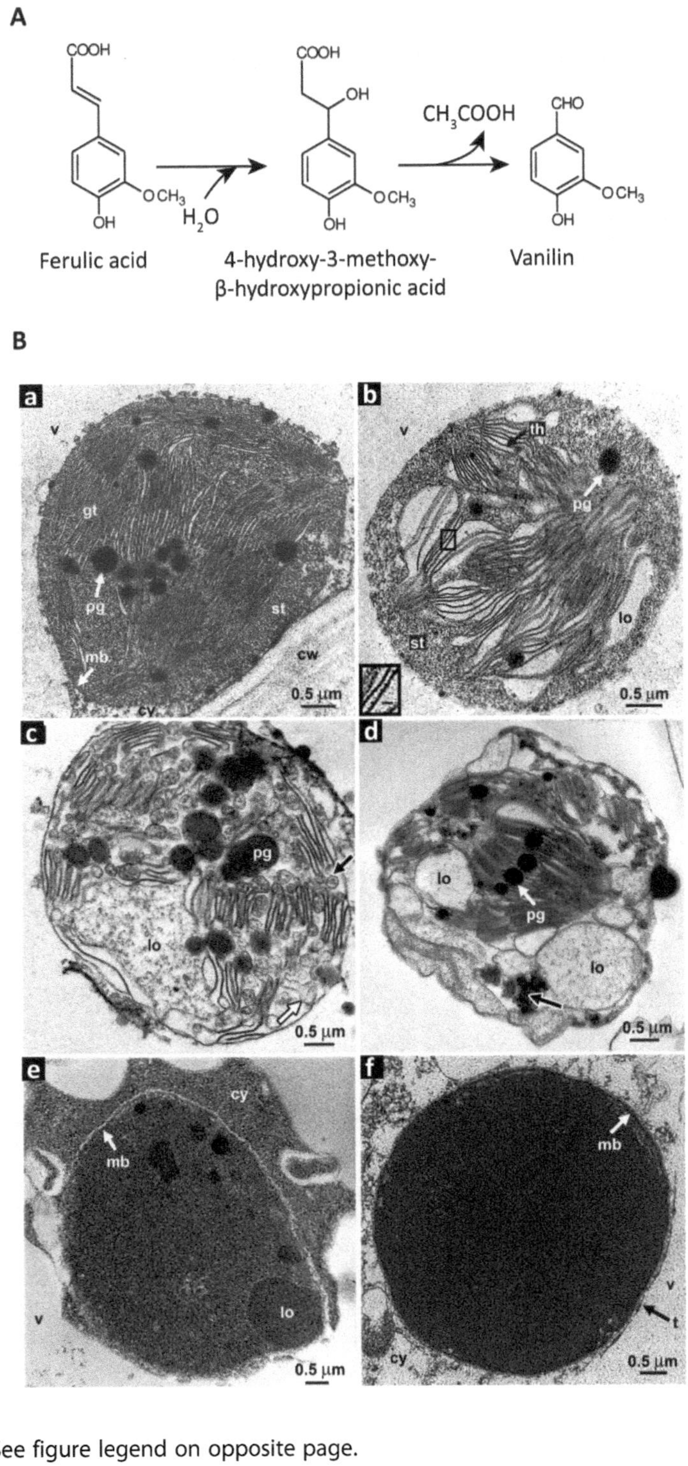

Fig. 5 See figure legend on opposite page.

catalyzes the conversion of ferulic acid into vanillin has also been isolated and characterized (Gallage et al., 2014, 2018; Gallage & Møller, 2015). Indian coleus (*Coleus forskohlii*) accumulates the manoyl oxide derived diterpenoid forskolin in specialized cells in the outer root bark (Pateraki et al., 2014). The biosynthetic pathway for forskolin has also been elucidated and the genes and enzymes involved identified (Pateraki et al., 2017).

Specific knowledge on each of the above discussed experimental plant systems with respect to the organization and localization of the enzyme systems involved and the manner in which the different types of natural products are stored in specific plant tissues, cells or organelles are outlined below, followed by a description of the intriguing opportunities and challenges offered by each system.

2.1 Vanillin hijacks plastids and transform them into phenyloplasts

The main aroma compound in vanilla extracts is vanillin. In the pod, vanillin is produced as vanillin glucoside. The biosynthesis takes off approximately three months after pollination when the pods have reached their full-grown size and continues until pod maturity at 7 to 8 months. After the first three months, the pods have few requirements for *de novo* cell wall synthesis. This may guide the shift of the use of phenylpropanoid precursors including ferulic acid from lignin production to synthesis of vanillin glucoside catalyzed

Fig. 5 The conversion of ferulic acid to vanillin catalyzed by the enzyme vanillin synthase and accumulation of vanillin glucoside in senescent chloroplasts re-differentiating into phenyloplasts in mature vanilla pods. (A) Ferulic acid is converted to vanillin in a retro-aldol elimination reaction catalyzed by the enzyme vanillin synthase which has high sequence similarity to cysteine proteases; (B) Progression of re-differentiation of a chloroplast into a phenyloplast in a maturing vanilla pod. (a) Functionally active chloroplast in immature vanilla pod showing grana thylakoids and plastoglobules; (b) Initial phase of chloroplast re-differentiation with appearance of granular stroma and grana thylakoid membranes with loculi formed between them; (c and d) Progression of re-differentiation showing budding of the thylakoid membranes into pseudocircular vesicles and an increasing number of loculi; (e) A senescent plastid showing its twin membranes and a locule filled with vanillin glucoside; (f) Completed re-differentiation of chloroplasts into phenyloplasts harboring a bio-condensate of vanillin glucoside visualized as a highly osmiophilic matrix surrounded by a membrane system. *Illustration from Brillouet, J. M., Verdeil, J. L., Odoux, E., Lartaud, M., Grisoni, M., Conéjéro, G. (2014). Phenol homeostasis is ensured in vanilla fruit by storage under solid form in a new chloroplast-derived organelle, the phenyloplast.* Journal of Experimental Botany, 65, *2427–2435.*

by vanillin synthase (*Vp*VAN) (Gallage et al., 2014, 2018; Gallage & Møller, 2015) (Fig. 5). *Vp*VAN converts ferulic acid or ferulic acid glucoside into vanillin or vanillin glucoside, respectively. The conversion proceeds as a retro-aldol elimination reaction resulting in the two-carbon cleavage of the side chain. The vanillin glucoside concentration reaches more than 300 mM in the aqueous phase of the mesocarp (Gallage et al., 2018). Biosynthetic studies with radiolabeled precursors demonstrated that the biosynthesis takes place in the inner part of the mesocarp. In agreement with this observation, immunohistological studies showed that *Vp*VAN is also localized in the inner part of the mesocarp where it specifically localizes to the chloroplasts. During pod ontogeny and approaching pod maturity, the chloroplasts senesce (Brillouet et al., 2014) (Fig. 5). This transition is accompanied by loss of their photosynthetic capacity, thylakoid lamella and chlorophyll. Using confocal microscopy combined with fluorescence spectral analysis, vanillin glucoside was found to stockpile within the stroma of the senesced chloroplasts as amorphous material at concentrations above 4 M. The re-differentiated chloroplasts are therefore termed phenyloplasts (Fig. 5).

Isolated intact chloroplasts and phenyloplasts are able to biosynthesize vanillin glucoside from ^{14}C-phenylalanine (Gallage et al., 2018). Thus, the site of synthesis and storage of vanillin glucoside are identical. This implies that the entire set of enzymes required for vanillin glucoside synthesis is available in the chloroplasts or provided by metabolic continuities obtained by transorganellar complementation between chloroplasts and the endoplasmic reticulum (Mehrshahi, Johnny, & DellaPenna, 2014; Mehrshahi et al., 2013). No knowledge on the organization of the enzymes involved and their possible direct interactions in such a system is available. The synthesis of vanillin glucoside proceeds over a 3–4 months period offering ample time for vanillin glucoside production even though the *Vp*VAN synthase may not be present in high quantities or not be highly active or the catalytic rate be limited by low levels of ferulic acid substrate. In this context, it remains to be resolved how extremely high molar concentrations of vanillin glucoside are stored in the lumen of the phenyloplasts without causing unbalanced homeostasis (Brillouet et al., 2014) (Fig. 5). This is essential to avoid autotoxicity. An obvious way to avoid this is aqueous-aqueous phase partitioning enabling formation of NADES-based dense bio-condensates, in which the vanillin glucoside is stored. Although no knowledge on the accumulation of NADES in the phenyloplasts is available, it is noticeable that NADES composed of lactic acid: 1,2-propanediol in 1:1 stoichiometric ratio is used for "green" extraction of vanillin from vanilla pods (González et al., 2018). It would be of great interest to analyze the possible content of NADES in the lumen of the isolated chloroplasts and phenyloplasts. Alternatively, micro-syringe or laser-capture based micro-dissection might be used for characterization of their lumen content. If indeed, the lumen of the chloroplasts and phenyloplasts are charged with NADES, it is to be elucidated how these are synthesized and imported.

The molecular mechanisms that triggers initiation of the re-differentiation process of the chloroplasts in the vanilla pod into phenyloplasts remain unknown. But in this context, it is worth noting that *Vp*VAN shows high amino acid sequence similarity to enzymes within the cysteine protease family (Gallage et al., 2014; Garcia-Lorenzo, Sjodin, Jansson, & Funk, 2006). Some plant species contain more than 150 different cysteine proteinases. The cysteine proteases present in plants are functionally diverse and the function of the individual cysteine proteinases is typically not well characterized. In general, cysteine proteases are expressed as pre-proteins with an N-terminal endoplasmic reticulum (ER)—targeting signal peptide as part of the pro-peptide domain, that comprises 130–160 amino acid residues (Cambra, Hernandez, Diaz, & Martinez, 2012). This also applies for *Vp*VAN (Gallage et al., 2014). The pro-peptide sequence may be removed either by a processing enzyme or by auto catalytical processing, thereby generating the mature cysteine proteinase (Turk et al., 2012). Two putative protease cleavage sites were identified in *Vp*VAN (Gallage et al., 2014). *In vitro* transcription/translation experiments showed no evidence of autocatalytic processing of the *Vp*VAN protein formed indicating that removal of the pro-peptide requires the action of a separate processing enzyme (Gallage et al., 2014). The pro-peptide sequence may control intracellular targeting, promote proper folding of the mature enzyme, and/or serve to maintain the enzyme in an inactive form in the cell to balance its function. A putative sumoylation site is located near the C-terminal end of *Vp*VAN (Gallage et al., 2014). It is interesting that SUMO modification of a tomato cysteine proteinase targeted the cysteinase to the nucleus, which resulted in activation of a gene in the ethylene biosynthetic pathway (Matarasso, Schuster, & Avni, 2005). If a similar coordinated process takes place in vanilla early in pod maturation, the ancestral role of *Vp*VAN synthase may have been to induce pod senescence. In the course of evolution, natural mutations may then have arisen enabling the enzyme to catalyze vanillin synthesis from ferulic acid.

2.2 Anthocyanins condensates in vacuolar inclusions

As already mentioned, *L. nigrescens* and *S. japonica* are two remarkable examples of plants containing tissues with extraordinary high levels of anthocyanins (Fig. 4). A literature search did not reveal any specific details on the cellular or subcellular accumulation of anthocyanins in these species (Zhang et al., 2017). Nevertheless, the biosynthesis and storage of anthocyanins have been studied in numerous other plant species.

The phenylpropanoid pathway constitutes one of the most versatile pathways in the plant kingdom (Vogt, 2010). Anthocyanin pigments are part of the flavonoid branch of the phenylpropanoid pathway that colors most flowers, fruits and seeds (Koes, Vorweij, & Quattrocchio, 2005) including *Petunia* (Fig. 6). The general precursor for anthocyanins, naringenin, is derived from the core phenylpropanoid pathway (Fig. 6). The colorless naringenin is converted into the simplest anthocyanin, pelargonidin, by a reduction at the 4-position and hydroxylation at the 3-position of the central ring. The orange pigment pelargonidin may be further modified by hydroxylations and methylations of the B-ring and glycosylation at the 3 position of the central ring resulting in formation of a palette of colors ranging from orange to purple (Fig. 6). In the plant, dynamic assembly of metabolons with different configurations is thought to facilitate the formation of metabolic highways for the production of specific anthocyanins (Jørgensen et al., 2005; Winkel-Shirley, 2001, 2004) (Fig. 6).

The plant vacuole is a multifunctional compartment and is considered the major subcellular storage compartment for flower pigments of the anthocyanin type. Early reports describe the storage of anthocyanins in anthocyanoplasts (Pecket & Small, 1980). Anthocyanins accumulate inside the vacuole, either as soluble compounds or as dense droplets with intense coloration termed anthocyanic vacuolar inclusions (AVIs) (Tohge, de Souza, & Fernie, 2017; Pourcel et al., 2010; Markham et al., 2000) (Fig. 4). The AVIs may or may not be membrane encapsulated (Chanoca et al., 2015; Kallam et al., 2017; Markham et al., 2000; Pecket & Small, 1980) and some membrane encapsulated condensates reported may be artifacts formed during sample preparation (Chanoca et al., 2015; Kallam et al., 2017; Pecket & Small, 1980). The inclusions vary in size and structure depending on the plant species, ranging from 0.5 to 15 μm in grapevine (*Vitis vinifera*) (Conn, Franco, & Zhang, 2010) and from 11 to 15 μm in transgenic *N. benthamiana* (Kallam et al., 2017). A vacuolar localization of anthocyanins together with glutathione may serve to prevent oxidation (Pradedova, Nimaeva, Rakevich, & Salyaev, 2019) and the solubility of the anthocyanins

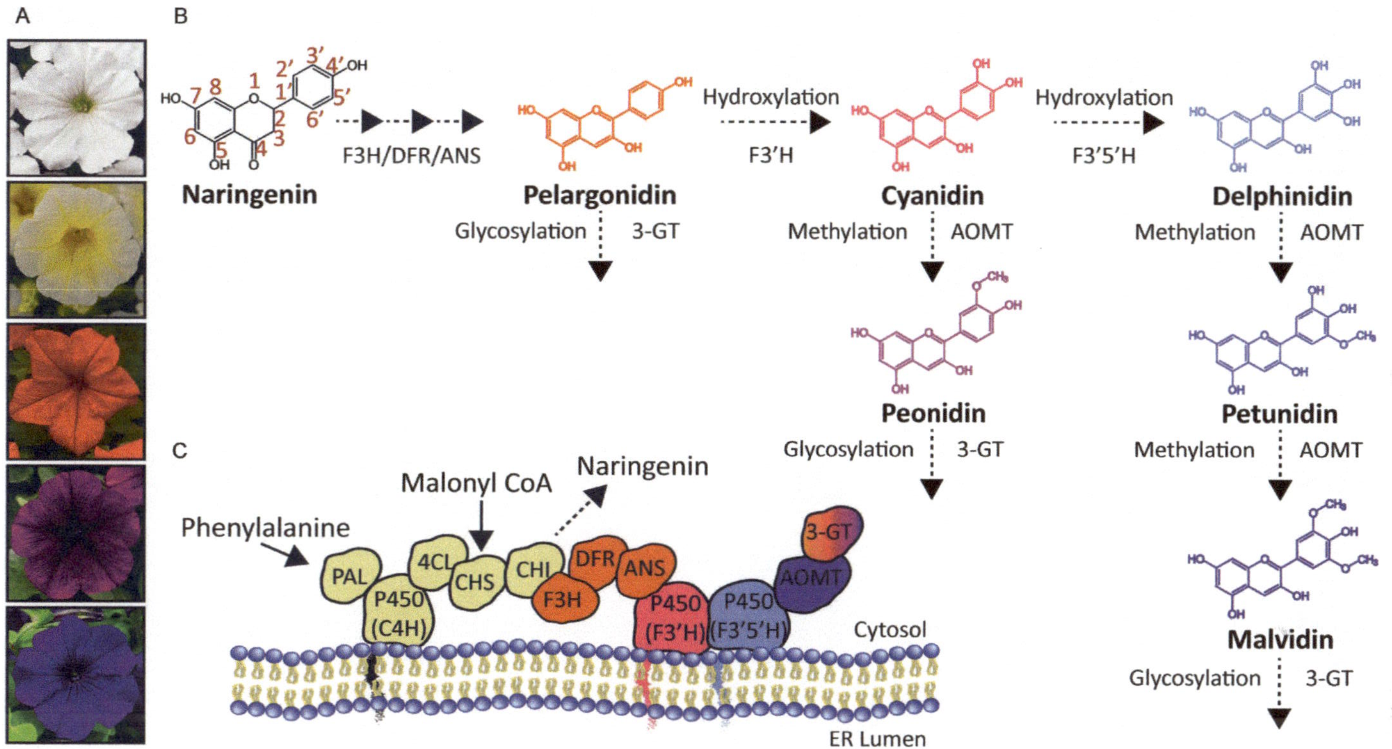

Fig. 6 Anthocyanin biosynthesis in petunia. (A) Petunia hybrids accumulating different anthocyanins. (B) Biosynthesis of the six major anthocyanins found in petunia derived from the colorless naringenin precursor. (C) Metabolon formation involved in biosynthesis of anthocyanins tethered to the endoplasmic reticulum membrane system *via* membrane anchored cytochrome P450 enzymes. The core biosynthetic pathway involves, PAL: phenylalanine ammonia lyase, C4H: cinnamate 4-hydroxylase, 4CL: 4-coumaric acid-CoA ligase, CHS: chalcone synthase, CHI: chalcone isomerase, F3H: flavanone 3-hydroxylase, DFR: dihydroflavonol 4-reductase, ANS: anthocyanin synthase, F3′H: flavonoid 3′-hydroxylase, F3′5′H: flavonoid 3′5′-hydroxylase, AOMT: anthocyanin *O*-methyltransferase and 3-GT: anthocyanin 3-*O*-glucosyltransferase.

may be improved by the low pH (Markham et al., 2000; Pourcel et al., 2010). Several possible routes to AVI formation have been proposed including vesicular trafficking, vacuolar membrane transporter-mediated import and microautophagy of cytoplasmic anthocyanin condensates (Chanoca et al., 2015; Knudsen, Gallage, Hansen, Møller, & Laursen, 2018; Tohge et al., 2017). Likewise, the factors controlling aggregation of the anthocyanins into droplet-like structures remain unknown. *In vitro* experiments simulating AVI formation demonstrated, that high salt concentrations, pH as well as aromatic acylation of the cyanidin-backbone are key parameters controlling stacking (Kallam et al., 2017). At high ionic strength and at pH values above 4.5, coumaroylation of the 5-OH-position instead of glycosylation facilitates intramolecular stacking and condensation (Kallam et al., 2017). *In vivo*, AVIs may form inside vesicles in the cytosol followed by fusion with the vacuole in a process mimicking micro-autophagy (Kallam et al., 2017). A common denominator for the AVI formation models is that aromatic acylated anthocyanins are able to pack tightly together in aggregates due to intra- and inter-molecular associations (Fernandes, Brás, Mateus, & de Freitas, 2015). It is possible that the high ionic strength observed to facilitate AVI formation *in vitro* (Dai, van Spronsen, Witkamp, Verpoorte, & Choi, 2013; Laursen et al., 2015), may *in vivo* be represented by a specific NADES that facilitates organized self-stacking of the anthocyanins and maintaining them in a soluble state while also securing lipid-lipid phase separation to avoid unbalanced osmo homeostasis. Anthocyanins are easily extracted from plant tissues using NADES in which they display increased stability comparable extraction into acidified organic solvents (Dai, Rozema, Verpoorte, & Choi, 2016).

Sorghum produces phytoalexins of the flavonoid type as a site-specific response to infection by the fungus *Colletotrichum graminicola* (Fig. 7) (Nicholson, Kollipara, Vincent, Lyons, & Cadena-Gomez, 1987a, 1987b; Nielsen et al., 2004; Snyder & Nicholson, 1990). The 3-deoxyanthocyanidins produced migrate as part of inclusions to surround the host epidermal cell being subject to fungal hyphae penetration and are visible as orange inclusions. No evidence for encapsulation of the inclusions within a membrane was found. With time, the vesicles move directly towards the cell harboring the point of attachment of the fungal appressorium and fuse to produce 20 μm inclusions and change color to dark red reflecting acidification to pH 3 in the course of infection (Nielsen et al., 2004). Upon physical contact between the 3-deoxyanthocyanidin inclusions and the plasma membrane, the infected epidermal cell collapses

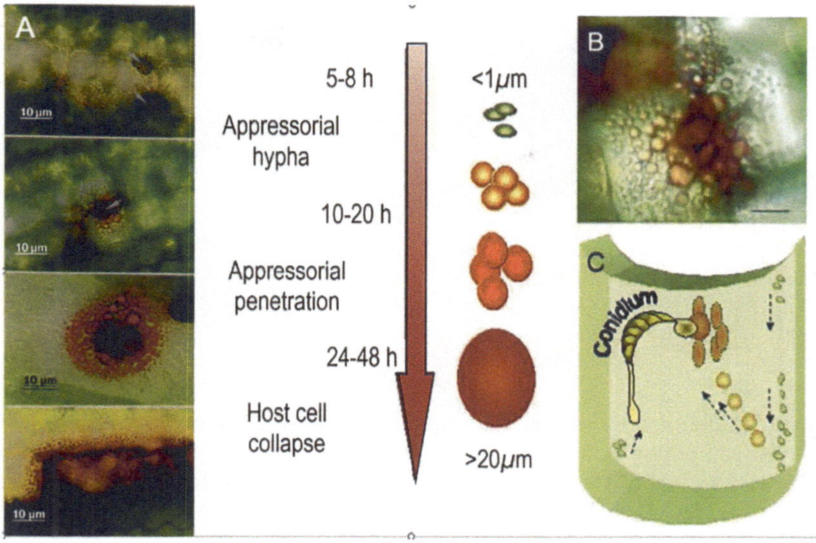

Fig. 7 Time-course showing development of a *C. graminicola* infection on a sorghum leaf. (A) The surface views show how epidermal cells next to the fungal appressoria infection site react by producing inclusions that change from almost colorless to dark red and finally release their content of 3-deoxyanthocyanidin flavonoids into the infected cell. *The illustrations are from Nielsen, K. A., Gotfredsen, C. H., Buch-Pedersen, M. J., Ammitzbøll, H., Mattsson, O., Duus, J. Ø., et al. (2004). Inclusions of flavonoid 3-deoxyanthocyanidins in Sorghum bicolor self-organize into spherical structures.* Physiological and Molecular Plant Pathology, 65, *187–196. and Snyder, B. A., & Nicholson, R. L. (1990). Synthesis of phytoalexins in sorghum as a site-specific response to fungal ingress.* Science, 248, *1637–1639.*

(Nielsen et al., 2004) (Fig. 7). Further studies are needed to elucidate factors promoting this cytosolic demixing. Interestingly, the biosynthetic machinery of flavonoid 3-deoxyanthocyanidins, localized on the cytosolic side of the ER, is in close proximity to the initiation of condensation adding to the importance of ER dynamics and interactions between cellular compartments (Phillips & Voeltz, 2016).

2.3 Cyanogenic glucosides may accumulate in invisible condensates

Sorghum produces the tyrosine derived cyanogenic glucoside dhurrin (Kahn et al., 1999; Koch et al., 1995; Kristensen et al., 2005; Nielsen et al., 2008; Sibbesen et al., 1995). Dhurrin is part of a two component defense system (Gleadow & Møller, 2014; Sibbesen et al., 1995).

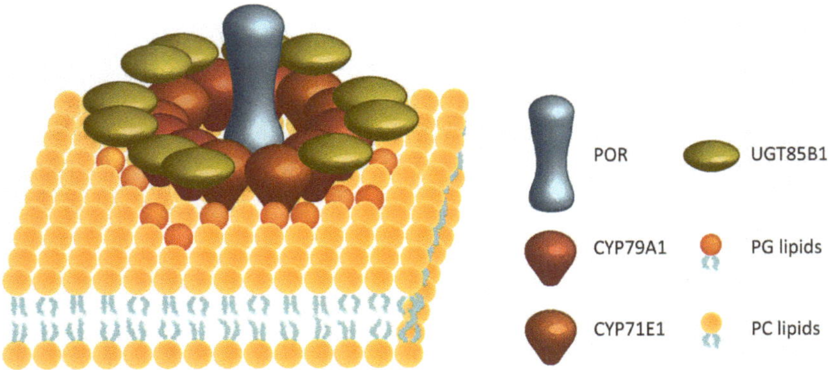

Fig. 8 The dhurrin metabolon. The dhurrin metabolon organizes as a multienzyme complex comprising homo- and hetero oligomerization of CYP79A1, CYP71E1 and UGT85B1. Enrichment of negatively charged phosphatidylglycerol (PG) lipids stimulate the activity of CYP71E1 and thereby substrate channeling. *Illustration is from Laursen, T., Borch, J., Knudsen, C., Bavishi, K., Torta, F., Martens, H. J., et al. (2016). Characterization of a dynamic metabolon producing the defense compound dhurrin in sorghum.* Science, 354, *890–893.*

Recently the dynamic and membrane bound dhurrin metabolon catalyzing dhurrin synthesis was published in Science (Dastmalchi & Facchini, 2016; Laursen et al., 2016) (Fig. 8). The dhurrin pathway is a well-established experimental system: all pathway genes have been identified, heterologous expression and reconstitution into various bio-mimetic membrane systems has been established (Bavishi, Laursen, Martinez, Møller, & Della Pia, 2016; Bavishi et al., 2018; Bertram et al., 2015; Hamann, Laursen, & Møller, 2009; Laursen, Naur, & Møller, 2013; Laursen et al., 2014; Wadsäter et al., 2012), free flow electrophoresis separation systems established (Justesen et al., 2013), specific antibodies to all enzymes involved have been prepared (Laursen et al., 2016; Schmidt et al., 2018) and routes for chemical synthesis of all dhurrin pathway intermediates and dhurrin have been established (Møller, Mcfarlane, & Conn, 1977; Møller, Olsen, & Motawia, 2016).

In Raman spectra, cyanogenic glucosides including dhurrin give rise to a prominent and distinct vibration band at 2245 cm^{-1} derived from the triple bond in the nitrile moiety. This band constitutes an unmistakable marker of cyanogenic glucosides, as the band occurs in an otherwise "quiet" region of the biological spectrum (Thygesen, Jørgensen, Møller, & Engelsen, 2004). Using label-free Raman hyperspectral imaging analysis and etiolated sorghum seedlings, dhurrin was demonstrated to be absent in the central vacuole (Heraud et al., 2018). Considering the high amounts of dhurrin present, a likely option is therefore that dhurrin is stored in dense bio-condensates that may be surrounded by a membrane or formed without a membrane by aqueous-aqueous phase partitioning. Isolation of such bio-condensates using technologies like free-flow electrophoresis, density gradient fractionation, or direct extraction from the cell using micropipettes or laser-capture micro-dissection might reveal all constituents present in the bio-condensates using NMR, metabolomics and proteomics. Dhurrin is produced in both etiolated and light-grown seedlings (Halkier & Møller, 1989; Kojima et al., 1979) offering two independent experimental systems. If dhurrin turns out to be localized in chloroplasts, etiolated seedlings could be exposed to light to clarify the ontogeny of the bio-condensate formation during the transition of prolamellar bodies into fully developed chloroplasts using metabolomics, transcriptomics and proteomics. The sorghum *tcd1* mutant which cannot produce dhurrin was obtained using a combined biochemical screen and TILLING approach (Blomstedt et al., 2012). Thus, a genetic control system is available enabling studies with wild-type sorghum seedlings to focus on features relevant to dhurrin synthesis, transport and storage. Dhurrin had earlier been reported to accumulate in isolated vacuoles (Saunders & Conn, 1978). The isolated vacuoles later turned out to be protoplasts *i.e.*, to represent the entire cell content (Kojima et al., 1979).

As discussed above, sorghum produces phytoalexins of the flavonoid type as a site-specific response to infection by the fungus *Colletotrichum graminicola* (Fig. 7) (Nicholson et al., 1987a, 1987b; Nielsen et al., 2004; Snyder & Nicholson, 1990). This response also happens in young seedlings where dhurrin content is very high. In the study from 2004, no data is provided on simultaneous changes in dhurrin synthesis and the possible co-accumulation of dhurrin in the 3-deoxyanthocyanidin containing inclusions. Repetition of the fungal infection experiments and isolation of the vesicles formed and characterization of all constituents present would reveal whether this distinct defense response also involves changes in the accumulation of dhurrin and detonation of a dhurrin based hydrogen cyanide bomb. It would also be interesting to analyze whether the high density of natural products in the inclusions formed was obtained and stabilized by the presence of NADES.

2.4 Cannabinoid loaded trichomes

Cannabis sativa produces a number of cannabinoids of the alkyl type carrying a monoterpenolic isoprenyl moiety (C10) and a pentyl side-chain (C5) (Hanuš, Meyer, Muñoz, Taglialatela-Scafati, & Appendino, 2016).

The enzyme catalyzed biosynthesis of cannabinoids produces acidic cannabinoids (C22, "pre-cannabinoids") with Δ^9-THCA, CBDA, CBCA and CBGA being the main products. These are labile compounds that following non-enzymatic oxidation, decarboxylation and cyclization reactions convert into Δ^9-THC, CBD, CBC and CBG (ElSohly & Slade, 2005; Flores-Sanchez & Verpoorte, 2008; Happyana & Kayser, 2016; Sirikantaramas et al., 2005). The phytocannabinoids and other terpenoids accumulate in glandular trichomes located all over the above-ground parts of the plant and in highest density on the female flowers (Flores-Sanchez & Verpoorte, 2008; Happyana et al., 2013). Within the glandular trichomes, the cannabinoids are secreted into a balloon-shaped secretory cavity positioned at the tip of the capitate-stalked trichome (Hammond & Mahlberg, 1977; Happyana et al., 2013; Livingston et al., 2019; Mahlberg & Kim, 2004, 2017). When the trichome ruptures under influence of high ambient temperatures or as a result of herbivory, the released cannabinoids form a viscous coating on the plant's surface (Garrett & Hunt, 1974; Mahlberg & Kim, 2017). The sticky surface coating glues together mandibles and legs of potential herbivores while also preventing desiccation (Mauseth, 2006).

The pathway for phytocannabinoid biosynthesis in *C. sativa* has recently been elucidated (Gülck et al., 2020; Luo et al., 2019). The individual biosynthetic steps take place in different cell types and organelles: the cytosol of the gland cells, the plastids and the extracellular storage cavity (Fellermeier, Eisenreich, Bacher, & Zenk, 2001; Hammond & Mahlberg, 1977; Livingston et al., 2019; Mahlberg & Kim, 2004, 2017). The operation of the biosynthetic enzymes in different cell types and organelles obviously prevents formation of an enzyme metabolon. Nevertheless, in hop (*Humulus lupulus*) a related species in the Cannabaceae family, it is worth noticing that the aromatic prenyl transferases *Hl*PT1L and *Hl*PT2 localize to the plastid membranes, where they form heterodimeric complexes catalyzing three sequential prenylations in the biosynthesis of bitter acids (Li et al., 2015). This catalytic activity is influenced by non-catalytic chalcone isomerase-like proteins (CHIL proteins), which serve as chaperones and improve the production of chalcone-derived metabolites (Ban et al., 2018). This serves to demonstrate that in spite of the splitting of the biosynthetic steps between different cell types and organelles, enzyme interactions at the local level is important. It remains to be resolved how the pathway intermediates are transported between the different compartments. The tetrahydrocannabinolic acid synthase and cannabidiolic acid synthase catalyzing the final step in the synthesis of specific cannabinoids are localized and catalytically active in the hydrophobic phase of the balloon-shaped extracellular storage cavity (Rodziewicz, Loroch, Marczak, Sickmann, & Kayser, 2019; Sirikantaramas et al., 2005). This storage cavity was also shown to contain hydrophilic, amphiphilic and osmoprotective compounds that possibly could represent NADES preserving and stabilizing the activity of the enzymes present as well as keeping the cannabinoids present in solution (Rodziewicz et al., 2019). Semi-quantitative assessment of the relative concentrations of different cannabinoids in the extracellular storage cavity has been carried out using laser-microdissection (Happyana et al., 2013). The content of cannabinoids on the bracts of *C. sativa* inflorescences is in the range of 5–15 mg per 100 mg dry mass dependent on strain and ontogeny (Turner, Hemphill, & Mahlberg, 1980). The total cannabinoid content per capitate-stalked gland is typically in the range of 30–100 ng per gland (Lanyon, Turner, & Mahlberg, 1981). With an approximate gland diameter of 75 μm (Happyana et al., 2013), the total content of cannabinoids may reach around 450 mg/mL in the gland lumen. At this concentration, the presence of NADES could serve to maintain the cannabinoids in solution. In this context, the reported presence of hydrophilic and amphiphilic compounds in the glands are interesting (Rodziewicz et al., 2019).

2.5 Diterpenoid Forskolin in specialized root cork oil bodies

In *Coleus forskohlii*, the structurally complex diterpenoid forskolin is produced and accumulate in specialized cells in the outer root cortex (Pateraki et al., 2014, 2017). Within each of these cells, the forskolin is contained in a single spherical body (5–9 μm diameter) with histochemical properties like those of an oil body. The forskolin-containing bodies can be isolated by gradient centrifugation (Pateraki et al., 2014). Their structural composition is unknown. Based on comparative transcriptomic analyses of cork, steele, stem, flower and leaves and transient expression of candidate genes in *N. benthamiana*, the genes encoding the enzymes catalyzing forskolin synthesis from geranylgeranyl diphosphate (GGPP) have been identified (Pateraki et al., 2017). The conversion of the terpene 13R-manoyl oxide into forskolin requires six regio- and stereospecific consecutive monooxygenation reactions and a final regiospecific acetylation. Of the tested CYPs, six were found to convert 13R-Manoyl oxide into oxygenated derivatives. These P450s all belong to the CYP76AH subfamily and show somewhat overlapping oxygenation capacities. Three of these P450s and a single acetyltransferase catalyze the entire conversion of 13R manoyl oxide into forskolin when transiently expressed in *N. benthamiana*. Interestingly, CYP76AH15 also catalyzes conversion of miltiradiene to ferruginol. The function of the remaining three CYP76AHs not essential for forskolin synthesis may be to catalyze monooxygenation reactions required in the formation of abietane diterpenopids which are also present in the *C. forskohlii* roots. The partly overlapping catalytic activities of the CYP76AHs expressed in the roots indicate that these P450s participate in formation of dynamic metabolons of which the composition is defined in response to metabolic demands or environmental challenges to offer on-demand production opportunities for different diterpenoids.

Upon expression in yeast, forskolin titers around 0.4g forskolin/L culture are obtained (Pateraki et al., 2017). Protein engineering of the substrate recognition sites of the involved CYP76AHs improved their efficiency toward forskolin synthesis as accentuated by CYP76AG15 where modification of a single amino acid residue has a major impact on the catalytic properties (Forman, Bjerg-Jensen, Dyekjær, Møller, & Pateraki, 2018).

3. Metabolons; nature's conveyor belts for production of natural products

In plants, the shikimate pathway channels approximately 20% of the carbon flux towards chorismate, the precursor from which the three aromatic amino acids phenylalanine, tyrosine and tryptophan are produced (Herrmann, 1995). A significant proportion of these amino acids are further metabolized towards an impressive diversity of specialized metabolites including, alkaloids, glucosinolates, cyanogenic glucosides and phenylpropanoids. Such biosynthetic pathways can be highly branched sharing a few common steps. This provides a dynamic tool-box for combinatorial assembly of biosynthetic enzymes enabling plants to respond on-demand to environmental and developmental cues. Assembly of sequential enzymes into efficient complexes, termed metabolons, allows for orchestration of the metabolic grid by guiding the metabolism towards specific branches or metabolites. A key example of such a metabolic highway is the phenylpropanoid pathway, which may redirect phenylalanine towards the formation of monolignol units used in lignin biosynthesis or towards formation of bioactive natural products required for defense depending on the needs (Knudsen et al., 2018; Laursen et al., 2015). The latter part of the phenylpropanoid pathway further differentiates into sub-branches leading to biosynthesis of a broad spectrum of metabolites. Assembly of metabolons is thought to facilitate the channeling of phenylalanine *via* *p*-coumaroyl-CoA (Bassard, Mutterer, Duval, & Werck-Reichhart, 2012) toward the different branches for biosynthesis of monolignols (Bassard, Mutterer, et al., 2012; Bassard, Richert, et al., 2012), flavonoids (Dastmalchi, Bernards, & Dhaubhadel, 2016; Fujino et al., 2018), sporopollenin (Lallemand, Erhardt, Heitz, & Legrand, 2013; Morant et al., 2007) and vanilloids such as vanillin glucoside (Gallage et al., 2014).

P450 enzymes catalyze key reactions in the biosynthesis of bioactive compounds and play an important role in recruitment of soluble enzymes, such as lyases, transferases and reductases during metabolon formation (Jørgensen et al., 2005). Their assembly is commonly considered to involve transient protein-protein interactions between globular domains (Bassard, Mutterer, et al., 2012; Bassard, Richert, et al., 2012). However, the evolution of complex biosynthetic pathways in plants has favored the synthesis of membrane anchored P450s and their redox partner (NADPH-dependent cytochrome P450 oxidoreductase, POR) as opposed to their soluble bacterial ancestors, which points to the question: *What is the role of the membrane in guiding the assembly of metabolons?* It is notable that choline, which is known to efficiently take part as a NADES component, is found in the head group of a significant proportion of the membrane phospholipids.

Besides the core biosynthetic enzymes, additional proteins have been postulated to be involved in guiding the metabolic flux of specific pathways and possibly the assembly of metabolons. Most notable are the small C-terminal anchored cytochrome b_5 (b_5) redox proteins. Co-expression of b_5 from *Artemisia* in yeast together with the biosynthetic enzymes for production of artemisinic acid resulted in an almost 20-fold increase in yield (Paddon et al., 2013). Likewise, a b_5 from *Petunia* was shown to be required for the activity of the P450 F3'5'H (CYP75A1) leading to the formation of purple anthocyanins (Fig. 6). More recently, a non-catalytic Membrane Steroid Binding Protein (MSBP) was shown to act as scaffolding protein for production of specific mono lignol units in the phenylpropanoid pathway (Gou, Ran, Martin, & Liu, 2018). The possible role of the local lipid environment and non-catalytic proteins on the assembly of specific metabolon configurations is supported by a recent study using a detergent free approach to isolate the intact dhurrin metabolon (Fig. 9) (Laursen et al., 2016). Following isolation, an enrichment of negatively charged lipids and co-purification of an MSBP and several b_5 isoforms was observed. This suggests that plants utilize an arsenal of biological modules to regulate the assembly of metabolons composed of different configurations.

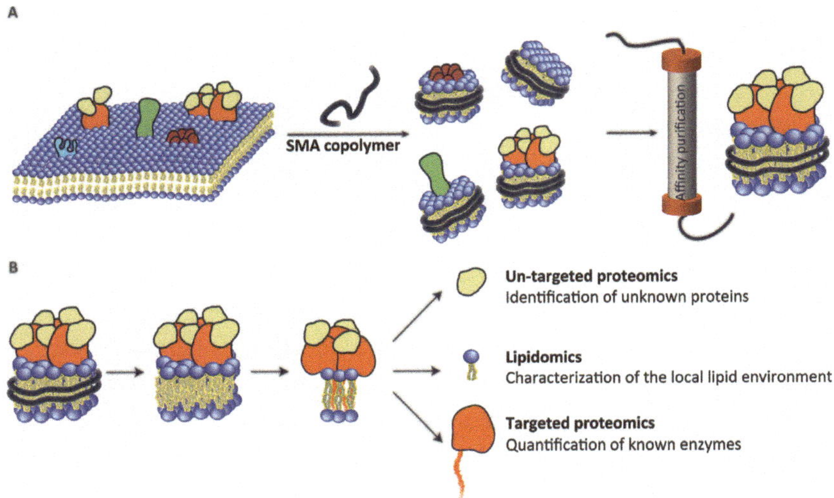

Fig. 9 Isolation and characterization of different metabolon configurations. (A) Detergent free isolation of metabolons directly from biological membranes using the Styrene Maleic Acid copolymer to carve out discrete Lipid Particles (SMALPs) combined with affinity chromatography. (B) Disassembly of intact SMALPs into their individual components and mass spectrometry-based characterization of proteins and lipids.

Such microdomains renders the cellular interior a continuum of dynamic mosaic substructures that readily mix and demix in a demand dependent fashion.

3.1 Dynamic metabolons

Although multiple studies have suggested the existence of metabolon formation in various specialized metabolic pathways, the requirements for obtaining "metabolon" status as stipulated by Srere (1985) is so far only met by a handful of pathways. The two best studied metabolons in plant specialized metabolism are the cyanogenic glucoside producing dhurrin metabolon (Laursen et al., 2016; Møller & Conn, 1980) and the core phenylpropanoid metabolon (Achnine, Blancaflor, Rasmussen, & Dixon, 2004). The dynamic nature of metabolons has hampered their isolation using classical detergent based methodologies. This renders it difficult to provide solid experimental proofs for their existence, which for decades have made metabolons a topic of intense debate. Most of the research on metabolons in the phenylpropanoid metabolic grid has involved binary interactions between sequential enzymes of the different branches. These studies are mainly based on classical experimental approaches such as split-ubiquitin or yeast-two-hybrid, Förster resonance energy transfer (FRET), bimolecular fluorescence complementation (BiFC) and co-immunoprecipitation (Co-IP), for detailed reviews see (Bassard & Halkier, 2018; Obata, 2019). These studies have mostly focused on interactions between enzymes with known catalytic functions. Factors stimulating metabolon assembly in the phenylpropanoid pathway remains rather elusive. Furthermore, there is limited evidence for substrate channeling beyond the core phenylpropanoid pathway, which is a key feature and determinant of metabolons.

Phenylpropanoid metabolon: Although multiple studies support the hypothesis for metabolon assembly in the phenylpropanoid pathway, only the core pathway has been experimentally proven to organize in metabolons. Since the concept of enzyme complexes in the phenylpropanoid pathway was proposed in the 1970s (Stafford, 1974), research was focusing on substrate channeling and protein-protein interaction between enzymes of the core pathway, phenylalanine ammonia-lyase (PAL) and cinnamate-4-hydroxylase (C4H, CYP73A) (Achnine et al., 2004). More recently, the role of P450s as nucleation site for complex assembly and recruitment of soluble proteins has been documented (Bassard, Mutterer, et al., 2012; Bassard, Richert, et al., 2012; Dastmalchi et al., 2016; Fujino et al., 2018). As demonstrated for the lignin branch of the phenylpropanoid pathway, co-expression of the membrane anchored *p*-coumaroylshikimate 3′-hydroxylase (C3′H, CYP83A3) together with the soluble hydroxycinnamoyl transferase (HCT) or 4-coumaric acid CoA ligase (4-CL) results in relocation of the soluble proteins towards the ER membrane (Bassard, Mutterer, et al., 2012; Bassard, Richert, et al., 2012). The upstream P450 (C4H, CYP73A5) does not affect the localization of the soluble enzymes alone, however co-expression together with C3′H and 4-CL improves the association between HCT and C3′H. This demonstrates synergistic effects on enzyme complex formation by the presence of additional sequential enzymes.

In soybean (*Glycine* max), the chalcone isomerase (CHI) and chalcone synthase (CHS) associates with the P450 isoflavone synthase (IFS, CYP93C) in the isoflavanoid branch (Dastmalchi et al., 2016; Fujino et al., 2018; Waki et al., 2016). Chalcone reductase (CHR) interacts specifically with one of the IFS isoforms (Dastmalchi et al., 2016). This serves to demonstrate how plants use a relatively simple tool-box of biosynthetic enzymes to generate a complex metabolic grid. In the phenylpropanoid branch leading to biosynthesis of flavonoids in snapdragon (*Antirrhinum majus*), CHS and CHI interact specifically with flavone synthase II (FNSII, CYP93B) and

CHI interacts with the downstream P450 flavonoid 3′-hydroxylase (F3′H, CYP75B) and thereby guides the metabolism towards production of anthocyanin pigments (Fujino et al., 2018) (Fig. 6). The exact basic mechanisms regulating the metabolic grid of the phenylpropanoid pathways remain to be elucidated, however, it is likely that metabolon assembly plays a key role. Although protein-protein interaction studies suggest that several branches are organized in complexes, it has been challenging to provide evidence for substrate channeling, mainly because this single pathway gives rise to a plethora of products and therefore most likely constitutes a continuum of conformations.

Dhurrin metabolon: Dhurrin metabolism in sorghum (*Sorghum bicolor*) has been studied for half a century and remains one of the best characterized metabolons involved in specialized metabolism in plants. Dhurrin is synthesized from the substrate L-tyrosine by the action of two P450 enzymes (CYP79A1 and CYP71E1), POR and a soluble UDP-glucosyltransferase (UGT85B1) (Andersen, Hansen, Laursen, Weitzel, & Simonsen, 2016; Bak, Kahn, Nielsen, Møller, & Halkier, 1998; Bavishi et al., 2018; Bertram et al., 2015; Jones, Møller, & Høj, 1999; Kahn, Bak, Svendsen, Halkier, & Møller, 1997; Kahn et al., 1999; Koch et al., 1995; Kristensen et al., 2005; Laursen, Jensen, & Møller, 2011; Laursen et al., 2014; Wadsäter et al., 2012). The efficiency of the dhurrin pathway has been studied since the 1980s demonstrating tight channeling based on simultaneous administration of substrates and intermediates radiolabelled with different isotopes to *Sorghum bicolor* microsomes (Møller & Conn, 1980; Halkier & Møller, 1989). This prompted the hypothesis that the enzymes would assemble in metabolons on demand. Since then and following intense studies of the pathway, experimental proof for the existence of the dhurrin metabolon was reported in 2016 in a publication in Science (Fig. 8) (Laursen et al., 2016). Using a combination of multiple techniques, the metabolon was isolated and functionally characterized. A major turning point in the studies of dynamic membrane anchored metabolons has been the application of a *co*-polymer of styrene and maleic acid (SMA). This polymer spontaneously integrates into biological membranes and form discrete lipid particles (SMALPs) (Fig. 9) (Dörr et al., 2016; Jamshad et al., 2011; Lee et al., 2016). The SMALP technology enabled isolation of the dhurrin metabolon directly from *Sorghum bicolor* microsomal fractions with a single affinity chromatography step using POR as bait (Laursen et al., 2016). Quantitative proteomics revealed strong enrichment of all the known biosynthetic enzymes as well as potentially non-catalytic proteins such as b_5´s and an MSBP.

The latter proteins could provide the regulatory mechanism governing on demand assembly. *In planta* protein-protein interactions were evaluated by FRET using a transient expression platform in *Nicotiana benthamiana* with fluorescent- tagged proteins. In short, CYP79A1, CYP71E1 and UGT85B1 form homo and hetero oligomers, with an enhancement of UGT85B1 oligomerization when CYP79A1 or CYP71E1 is co-expressed. *In vitro* functional characterization of reconstituted purified enzymes in proteo-liposomes revealed that recruitment of the soluble UGT85B1 stimulates both flux and channeling between CYP79A1 and CYP71E1. This effect is independent of the presence of UDP-glucose signifying that physical interaction between the enzymes is required for efficient turnover and independent on the ability of UGT85B1 to carry out a catalytic reaction (Laursen et al., 2016). Finally, the lipid composition of the SMALP isolated dhurrin metabolon reveals an enrichment of negatively charged phospholipids (Fig. 9). In subsequent *in vitro* experiments, the presence of negatively charged phospholipids was shown to stimulate the activity of CYP71E1 and to improve substrate channeling. Despite half a decade's research on the dhurrin pathway, we are still only just starting to understand how plants are able to orchestrate their metabolic grids ultimately leading to biosynthesis of molar concentrations of bioactive natural products, like dhurrin.

3.2 Dynamic assembly of bio-condensates

One of the major questions for which an answer remains to be provided is how relatively low abundant proteins succeed to organize and synthesize molar concentrations of bioactive natural products. Indeed, metabolon formation facilitates substrate channeling and improved flux, however, is this really enough. One might speculate whether substrates accumulate at the site of synthesis through liquid—liquid phase separation giving rise to increased local concentrations. Likewise, concentration of potentially volatile, toxic and labile end products in bio-condensates may prevent auto toxicity and promote the ability to sequester enormous concentrations of small molecules in bio-condensates.

It is to be noted that amino acids, like phenylalanine, tyrosine and tryptophan which are the precursors for formation of a multitude of aromatic natural products in plants, are excellent NADES constituents (Dai, van Spronsen, et al., 2013; Dai, Witkamp, et al., 2013). The aromatic natural products as well as their derived glycosides show a high solubility in NADES as demonstrated by the solubility of vanillin (620 mg/mL) in

NADES composed of lactic acid: 1,2-propanediol in 1:1 stoichiometric ratio, a level comparable to its solubility in methanol (González et al., 2018). It remains unresolved whether NADES components accumulate in the phenyloplasts, the storage site of vanillin glucoside in the vanilla pods. However, it is tempting to envision that nature has developed similar systems to co-store and concentrate both substrates and products of plant specialized metabolism in NADES based bio-condensates.

In addition to providing storage capacity for phytochemicals within a cell, formation of NADES-based assemblies may also serve to mediate and fine-tune numerous biological processes. A higher order of metabolic control may be achieved by limiting or improving enzyme access and activity towards their substrates or by embedding entire pathways in NADES based assemblies. Dehydration of bio-active compounds and/or enzymes stored in NADES might prevent some enzymatic conversions. This points to the possibility that in NADES assemblies, enzymes may be stored together with their substrates and co-factors without catalyzing any reactions. This would *e.g.*, offer an entirely different and much more dynamic orchestration of the detonation of two component plant defense systems (Morant et al., 2008). Classically, these two component defense systems are reported to be based on membrane-based compartmentalized storage of a glucosylated defense molecule (like vanillin glucoside or a cyanogenic glucoside) and a β-glucosidase catalyzing its hydrolysis. Upon cell and organelle disruption as caused by a chewing insect larva, the defense system would be detonated by the hydrolytic action of the β-glucosidase (Morant et al., 2008). If the substrate and hydrolytic enzyme are co-localized within a NADES assembly, a simple dilution resulting from cell disruption would detonate the defense bomb.

An intracellular NADES phase could serve to preserve the integrity of organelles and proteins upon abiotic stresses like dehydration, heat and cold. As example, *Sorghum bicolor* is a highly drought resistant plant, which is able to withstand long periods without water. However, the biosynthetic machineries are maintained awaiting the moment the rain comes. The dhurrin pathway from *Sorghum bicolor* was recently shown to be highly stabilized in a glucose: tartrate based NADES (Knudsen et al., 2020). The otherwise labile P450 enzymes (CYP79A1 and CYP71E1) maintained functional activity upon storage at room temperature for a month in NADES compared to less than a week in aqueous buffer. Likewise, all the dhurrin biosynthetic enzymes were stabilized towards heat treatment maintaining activity upon treatment of up to 70–80 °C for 30 min. This demonstrates yet another functional property of NADES as a third solvent for stabilization of cellular integrity.

Plants are sessile organisms that needs to retain high metabolic plasticity to adapt to rapid changes in their growth environment. A continuum of phase separation and dissolution could be essential elements in providing the necessary control functions to optimize the plants responses towards both abiotic and biotic challenges. Localized high concentrations of substrates would serve to circumvent bulk equilibria whereas storage of enzymes and products would render the plant ready to act on demand (Knudsen et al., 2020). With the emergence of technical platforms with temporal and spatial resolution future studies shall reveal the dynamic behavior of both metabolons and NADES in real time with single molecule resolution (Bavishi et al., 2016; Laursen et al., 2014; Stella, Mesa, Thomsen, et al., 2018) and non-disruptive techniques of native biological samples.

Bio-condensates observed in plants may have different origins. Some may be membrane-encapsulated, some may arise as a result of liquid-liquid phase separation within an organelle (*e.g.,* vacuole, plastid or cytosol) (Markham et al., 2000; Pecket & Small, 1980). In animals, the membrane-less droplets accumulate proteins and nucleic acids (Aguilera-Gomez & Rabouille, 2017; Kallam et al., 2017). In plants they may have an important function by accumulating bio-active natural products (Mitrea & Kriwacki, 2016). In plants, the organization of NADES as super-molecules would mimic the observed stabilizing function of proteins with long intrinsically disordered regions (IDRs) in animal cells (Aguilera-Gomez & Rabouille, 2017; Brady et al., 2017; Courchaine et al., 2016; van der Lee et al., 2014). The inherent properties of NADES could enhance plant robustness because a NADES has almost no vapor pressure and thus retains water during desiccation caused by drought stress and heat waves. When used as nutraceuticals and pharmaceuticals, the bioavailability of bio-active natural products may be improved when orally administered dissolved in non-toxic NADES (Faggian et al., 2016).

4. Future research based on technical advancement of spatial and temporal resolution

We envision that the knowledge gained will guide targeted formation of high-density bio-condensates harboring desired high value bio-active natural products in engineered heterologous photosynthetic hosts in high amounts. Design of more efficient biological production systems facilitates the transition towards a bio-based society also from an economical point of view (Møller, 2017).

NMR based analyses (Choi et al., 2011; Kim et al., 2010) should be carried out in additional plant species known to accumulate high amounts of natural products in specific tissues to identify possible presence of specific type of NADES constituents present in the plant tissue and cell types in which the target natural product accumulates. If NADES are shown to be present, this may reveal tribe or family specific patterns or patterns dependent on the class and specific structural properties of the accumulated natural product. *In vitro* solubility assays of the natural product under study could be used to guide the search for proper NADES candidates. In other studies where a NADES has been demonstrated to be involved in establishing a bio-condensate, *in vitro* solubility studies with the identified NADES could be used to determine its solvent potential in comparison, *e.g.*, to water and methanol.

Mass spectrometry based bioimaging of tissue sections known to contain interesting natural products with high resolution mass accuracy is a rapidly developing technology (Bjarnholt, Li, & D'Alvise, & Janfelt, 2014; Li, Bjarnholt, Hansen, & Janfelt, 2011; Li et al., 2013; Montini et al., 2020) enabling determination of the cellular or sub-cellular localization of natural products with respect to specific cell types (*e.g.*, epidermal cells, vascular bundles, root cork, specific bud tissues, trichomes) or organelles (*e.g.*, chloroplasts, etioplasts, chromoplasts, vacuole, cytosol). This technology would enable simultaneous monitoring of co-localization of the natural product with possible NADES components. When the MS-bioimaging approach is based on approaches eliminating component diffusion or hydrolysis issues encountered using liquid sprays, definitive results on possible colocalization should be achievable. At present, the technology offers five μm spatial resolution. This methodology enables precise recording (positive or negative mode) of most targeted natural products and of the different types of NADES constituents possibly present as identified by an initial NMR study.

Free flow electrophoresis-based fractionation of organelles and other sub-structures present in tissue homogenates may also be used to identify specific cellular sub-compartments in the tissue to which the natural product is associated (Justesen et al., 2013). Co-partitioning of NADES and the natural product under study could be analyzed using NMR and LC-MS.

Monitoring of possible loss or redistribution of the natural product or NADES during fractionation by free flow electrophoresis will be critical to observe and may provide clues whether the dynamics of the experimental system results in artifact formation. Alternatively, separation may be achieved by density gradient centrifugation as previously used to isolate intact chloroplasts and phenyloplasts from mature vanilla pods (Gallage et al., 2018). (Fig. 5) and to isolate forskolin bodies from the outer root bark tissue of *Coleus forskohli* (Pateraki et al., 2014, 2017). Micro-syringe or laser-capture based micro-dissection may also be used for isolation of specific vesicles/droplets. Possible spatial distribution of the target natural product, putative derivatives (esters, glucosides) (Koja et al., 2018) and viscosity differences within the isolated substructures/droplets or tissue slices may be analyzed by multiphoton fluorescence lifetime imaging (Heskes et al., 2012; Heskes, Lincoln, Goodger, Woodrow, & Smith, 2012). In general, specific antibodies may also be relevant to prepare against all the proteins catalyzing the biosynthesis of the studied compounds. Localization of the target natural product within a canonical organelle or other membrane encapsulated substructures may be investigated using specific stains, fluorescence and immunochemistry in combination with electron microscopy. Such studies would benefit by support by state-of-the-art lipidomics and proteomics (Laursen et al., 2016). In cases where the cellular subcompartment is membrane-less, proteomics may be used to identify proteins present and the presence of natural products and other small molecules which could be NADES constituents including identified by NMR and unbiased metabolomics. The dynamics of small molecules within the phase separated droplets can then be monitored using NMR. Proteins present should be analyzed for the presence of large intrinsically disordered regions (IDRs) (Brady et al., 2017) and if present IDRs functionally classified using available database resources (van der Lee et al., 2014).

Inter- and intramolecular interactions of natural products and NADES mediating increased solubility and interplay among other constituents present may be possible to study *in vitro* and *in vivo* using a variety of *spectrometric techniques*. These include ^{1}H NMR and ^{1}H–^{1}H-nuclear Overhauser enhancement spectroscopy (NOESY), FT-IR spectrometry and multimodal vibrational imaging of cells to monitor formation of fully organized structures controlled by extensive hydrogen bonding (Dai, Verpoorte, & Choi, 2014; Fernandes et al., 2015; Nielsen et al., 2004; Perez-Guaita et al., 2017). Effect of dilution with water will then reveal the sequential loss of specific interactions.

Co-localization of NADES and specific natural products in bio-condensates with their respective biosynthetic enzymes or in case of two component defense systems their hydrolytic β-glucosidases could be analyzed by *immunochemical methods and confocal microscopy* (Laursen et al., 2016). Such technologies have already been used to document co-localization of vanillin glucoside and vanillin synthase (Gallage et al., 2018) and to demonstrate vesiculate structures harboring a prunasin β-glucosidase in almonds (Sánchez-Pérez, Jørgensen, Motawia, Dicenta, & Møller, 2009). Observed co-localizations may be followed by *in vitro* experiments to study the enzyme activities in NADES. In *in vitro* exploratory experiments in sorghum producing the cyanogenic glucoside dhurrin, the biosynthetic enzymes (CYP79A1 and CYP71E1) were found to be remarkable more tolerant to high temperatures and long-term storage when kept in a NADES composed of a 1:1 stoichiometric ratio of glucose and dipotassium tartrate (Knudsen et al., 2020). In contrast, the β-glucosidase was completely inhibited in the NADES. If these *in vitro* experiments mimic an *in vivo* situation, this offers the possibility of co-storage of a substrate and its hydrolytic enzyme and thus immediate activation of the defense system upon cell disruption and dilution of the NADES. The ability of NADES to improve the chemical stability of natural products and enzymes *e.g.*, in relation to long distance shipments as commercial products may also be of interest to investigate (Choi et al., 2011; Dai, Witkamp, et al., 2013; Zhao et al., 2015).

In systems where NADES are shown to play a major role in storage of natural products, studies should be initiated to identify the *metabolic origin of the NADES constituents*. What are their metabolic precursors and where are the enzymes catalyzing NADES production localized? If the bio-condensates are encapsulated by membranes, do the membranes then harbor transporters? Or are the natural products in NADES based bio-condensates in some cases simply acquired by liquid-liquid phase separation? How are auto-toxic effects of the bio-active natural products on the host cells avoided? Unbiased metabolomics and transcriptomics guided by biosynthetic radiotracer studies, enzyme assays and MS-bio-imaging may be the approaches used to study NADES constituent formation.

Dependent on the different experimental systems, plant ontogeny, exposure to light, drought or fungal attack are expected to be main drivers of metabolic re-organizations of production and storage of natural products. Physiological studies in this context are key to transition the research aims into studies of the mechanisms behind plant plasticity (Grotewold & Davies, 2008; Knudsen et al., 2018; Nicholson et al., 1987a, 1987b;

Nielsen et al., 2004). The knowledge on NADES based bio-condensate formation may also be used to increase the yields and storage capacity of high value compounds in synthetic biology approaches using heterologous production systems.

5. Concluding remarks

The many entangled global challenges related to the use of fossil fuels, climate change, food security, rapid spread of diseases, environmental preservation and financial inequality are well known. Overcoming these challenges and transitioning to a bio-based society will require interdisciplinary research and development, marketing of new and innovative products manufactured using renewable resources, and novel green technologies that are economically feasible and have transformative power (Møller, 2017). Plants and plant science have a key role in making these aspirations come through. The plasticity of photosynthetic organisms as observed in nature constitutes a treasure box. In collaboration with nature and based on concerted long-term efforts, we envision that the inherent ability of photosynthetic organisms to produce food, bio-materials, energy and a plethora of key lead compounds for the pharmaceutical industries based on carbon dioxide from the air and solar light may be a decisive advantage with respect to establishing robust, efficient and environmentally benign global production systems for the future. Knowledge of the molecular mechanisms controlling the establishment of metabolons and bio-condensates enabling formation and storage of high amounts of plant natural products is key to optimize such green production systems for the future.

References

Achnine, L., Blancaflor, E. B., Rasmussen, S., & Dixon, R. A. (2004). Colocalization of L-phenylalanine ammonia-lyase and cinnamate 4-hydroxylase for metabolic channeling in phenylpropanoid biosynthesis. *Plant Cell*, *16*, 3098–3109.

Aguilera-Gomez, A., & Rabouille, C. (2017). Membrane-bound organelles versus membrane-less compartments and their control of anabolic pathways in Drosophila. *Developmental Biology*, *428*, 310–317.

Andersen, T. B., Hansen, N. B., Laursen, T., Weitzel, C., & Simonsen, H. T. (2016). Evolution of NADPH-cytochrome P450 oxidoreductases (POR) in Apiales—POR 1 is missing. *Molecular Phylogenetics and Evolution*, *98*, 21–28.

Aumiller, W. M., & Keating, C. D. (2017). Experimental models for dynamic compartmentalization of biomolecules in liquid organelles: Reversible formation and partitioning in aqueous biphasic systems. *Advance in Colloid and Interface Science*, *239*, 75–87.

Bak, S., Kahn, R. A., Nielsen, H. L., Møller, B. L., & Halkier, B. A. (1998). Cloning of three A-type cytochromes p450, CYP71E1, CYP98, and CYP99 from Sorghum bicolor (L.)

Moench by a PCR approach and identification by expression in Escherichia coli of CYP71E1 as a multifunctional cytochrome p450 in the biosynthesis of the cyanogenic glucoside dhurrin. *Plant Molecular Biology, 36*, 393–405.

Bak, S., Olsen, C. E., Halkier, B. A., & Møller, B. L. (2000). Transgenic tobacco and Arabidopsis plants expressing the two multifunctional sorghum cytochrome P450 enzymes, CYP79A1 and CYP71E1, are cyanogenic and accumulate metabolites derived from intermediates in Dhurrin biosynthesis. *Plant Physiology, 123*, 1437–1448.

Ban, Z., Qin, H., Mitchell, A. J., Liu, B., Zhang, F., Weng, J.-K., et al. (2018). Noncatalytic chalcone isomerase-fold proteins in Humulus lupulus are auxiliary components in prenylated flavonoid biosynthesis. *PNAS, 115*, E5223–E5232.

Banani, S. F., Lee, H. O., Hyman, A. A., & Rosen, M. K. (2017). Biomolecular condensates: Organizers of cellular biochemistry. *Nature Reviews. Molecular Cell Biology, 18*, 285–298.

Bassard, J.-E., & Halkier, B. A. (2018). How to prove the existence of metabolons? *Phytochemistry Reviews, 17*, 211–227.

Bassard, J.-E., Møller, B. L., & Laursen, T. (2017). Assembly of dynamic P450-mediated metabolons—Order versus chaos. *Current Molecular Biology Reports, 3*, 37–51.

Bassard, J. E., Mutterer, J., Duval, F., & Werck-Reichhart, D. (2012). A novel method for monitoring the localization of cytochromes P450 and other endoplasmic reticulum membrane associated proteins: A tool for investigating the formation of metabolons. *FEBS Journal, 279*, 1576–1583.

Bassard, J. E., Richert, L., Geerinck, J., Renault, H., Duval, F., Ullmann, P., et al. (2012). Protein-protein and protein-membrane associations in the lignin pathway. *Plant Cell, 24*, 4465–4482.

Bavishi, K., Laursen, T., Martinez, K. L., Møller, B. L., & Della Pia, E. A. (2016). Application of nanodisc technology for direct electrochemical investigation of plant cytochrome P450s and their NADPH P450 oxidoreductase. *Scientific Reports, 6*, 29459.

Bavishi, K., Li, D., Eiersholt, S., Hooley, E. N., Petersen, T. C., Møller, B. L., et al. (2018). Direct observation of multiple conformational states in cytochrome P450 oxidoreductase and their modulation by membrane environment and ionic strength. *Scientific Reports, 8*, 6817.

Berry, J., Brangwynne, C. P., & Haataja, M. (2018). Physical principles of intracellular organization via active and passive phase transitions. *Reports on Progress in Physics, 81*, 046601.

Bertram, N., Laursen, T., Barker, R., Bavishi, K., Møller, B. L., & Cárdenas, M. (2015). Nanodisc films for membrane protein studies by neutron reflection: Effect of the protein scaffold choice. *Langmuir, 31*, 8386–8391.

Bjarnholt, N., Li, B., D'Alvise, J., & Janfelt, C. (2014). Mass spectrometry imaging of plant metabolites principles and possibilities. *Natural Product Reports, 31*, 818–837.

Blomstedt, C. K., Gleadow, R. M., O'Donnell, N., Naur, P., Jensen, K., Laursen, T., et al. (2012). A combined biochemical screen and TILLING approach identifies mutations in *Sorgum bicolor* L. Moench resulting in acyanogenic forage production. *Plant Biotechnology Journal, 10*, 54–66.

Brady, J. P., Farber, P. J., Sekhar, A., Lin, Y. H., Huang, R., Bah, A., et al. (2017). Structural and hydrodynamic properties of an intrinsically disordered region of a germ cell-specific protein on phase separation. *Proceedings of the National Academy of Sciences of the United States of America, 114*, E8194–E8203.

Brillouet, J. M., Verdeil, J. L., Odoux, E., Lartaud, M., Grisoni, M., & Conéjéro, G. (2014). Phenol homeostasis is ensured in vanilla fruit by storage under solid form in a new chloroplast-derived organelle, the phenyloplast. *Journal of Experimental Botany, 65*, 2427–2435.

Cambra, I., Hernandez, D., Diaz, I., & Martinez, M. (2012). Structural basis for specificity of propeptide-enzyme interaction in barley C1A cysteine peptidases. *PLoS One*, *7*, e37234.

Chanoca, A., Kovinich, N., Burkel, B., Stecha, S., Bohorquez-Restrepo, A., Ueda, T., et al. (2015). Anthocyanin vacuolar inclusions form by a microautophagy mechanism. *The Plant Cell*, *27*, 2545–2559.

Choi, Y. H., van Spronsen, J., Dai, Y., Verberne, M., Hollmann, F., Arends, et al. (2011). Are natural deep eutectic solvents the missing link in understanding cellular metabolism and physiology? *Plant Physiology*, *156*, 1701–1705.

Conn, S., Franco, C., & Zhang, W. (2010). Characterization of anthocyanic vacuolar inclusions in Vitis vinifera L. cell suspension cultures. *Planta*, *231*, 1343–1360.

Courchaine, E. M., Lu, A., & Neugebauer, K. M. (2016). Droplet organelles? *The EMBO Journal*, *35*, 1603–1612.

Dai, Y., Rozema, E., Verpoorte, R., & Choi, Y. H. (2016). Application of natural deep eutectic solvents to the extraction of anthocyanins from Catharanthus roseus with high extractability and stability replacing conventional organic solvents. *Journal of Chromatography A*, *1434*, 50–56.

Dai, Y., van Spronsen, J., Witkamp, G. J., Verpoorte, R., & Choi, Y. H. (2013). Ionic liquids and deep eutectic solvents in natural products research: Mixtures of solids as extraction solvents. *Journal of Natural Products*, *76*, 2162–2173.

Dai, Y., Verpoorte, R., & Choi, Y. H. (2014). Natural deep eutectic solvents providing enhanced stability of natural colorants from safflower (*Carthamus tinctorius*). *Food Chemistry*, *159*, 116–121.

Dai, Y., Witkamp, G.-J., Verpoorte, R., & Choi, Y. H. (2013). Natural deep eutectic solvents as new extraction media for phenolic metabolites in *Carthamus tinctorius* L. *Analytical Chemistry*, *85*, 6272–6278.

Dastmalchi, M., Bernards, M. A., & Dhaubhadel, S. (2016). Twin anchors of the soybean isoflavonoid metabolon: Evidence for tethering of the complex to the endoplasmic reticulum by IFS and C4H. *The Plant Journal*, *85*, 689–706.

Dastmalchi, M., & Facchini, P. J. (2016). Plant metabolons assembled on demand. *Science*, *354*, 829–830.

Dolgin, E. (2018). Cell biology's new phase: Like oil in water, the contents of cells can separate into droplets. Finding out why is one of biology's hottest questions. *Nature*, *555*, 300–302.

Dörr, J. M., Scheidelaar, S., Koorengevel, M. C., Dominguez, J. J., Schafer, M., van Walree, C. A., et al. (2016). The styrene-maleic acid copolymer: A versatile tool in membrane research. *European Biophysics Journal*, *45*, 3–21.

ElSohly, M. A., & Slade, D. (2005). Chemical constituents of marijuana: The complex mixture of natural cannabinoids. *Life Sciences*, *78*, 539–548.

Faggian, M., Sut, S., Perissutti, B., Baldan, V., Grabnar, I., & Dall'Acqua, S. (2016). Natural deep eutectic solvents (NADES) as a tool for bioavailability improvement: Pharmacokinetics of rutin dissolved in proline/glycine after oral administration in rats: Possible application in nutraceuticals. *Molecules*, *21*, 1531.

Fellermeier, M., Eisenreich, W., Bacher, A., & Zenk, M. H. (2001). Biosynthesis of cannabinoids incorporation experiments with 13C-labeled glucoses. *European Journal of Biochemistry*, *268*, 1596–1604.

Fernandes, A., Brás, N. F., Mateus, N., & de Freitas, V. (2015). A study of anthocyanin self-association by NMR spectroscopy. *New Journal of Chemistry*, *39*, 2602–2611.

Flores-Sanchez, I. J., & Verpoorte, R. (2008). Secondary metabolism in cannabis. *Phytochemistry Reviews*, 7, 615–639.

Forman, V., Bjerg-Jensen, N., Dyekjær, J. D., Møller, B. L., & Pateraki, I. (2018). Engineering of CYP76AH15 can improve activity and specificity towards forskolin biosynthesis in yeast. *Microbial Cell Factories*, *17*, 181.
Fujino, N., Tenma, N., Waki, T., Ito, K., Komatsuzaki, Y., Sugiyama, K., et al. (2018). Physical interactions among flavonoid enzymes in snapdragon and torenia reveal the diversity in the flavonoid metabolon organization of different plant species. *The Plant Journal*, *94*, 372–392.
Gallage, N. J., Hansen, E. H., Kannangara, R., Olsen, C. E., Motawia, M. S., Jørgensen, K., et al. (2014). Vanillin formation from ferulic acid in Vanilla planifolia is catalysed by a single enzyme. *Nature Communications*, *5*, 4037.
Gallage, N. J., Jørgensen, K., Janfelt, C., Nielsen, A. J. Z., Naake, T., Duński, E., et al. (2018). The intra-cellular localization of the vanillin biosynthetic machinery in pods of *Vanilla planifolia*. *Plant Cell & Physiology*, *59*, 304–318.
Gallage, N. J., & Møller, B. L. (2015). Vanillin – Bioconversion and bioengineering of the most popular plant flavor and its *de novo* biosynthesis in the vanilla orchid. *Molecular Plant*, *8*, 40–57.
Garcia-Lorenzo, M., Sjodin, A., Jansson, S., & Funk, C. (2006). Protease gene families in Populus and Arabidopsis. *BMC Plant Biology*, *6*, 30.
Garrett, E. R., & Hunt, C. A. (1974). Physicochemical properties, solubility, and protein binding of Δ9-tetrahydrocannabinol. *Journal of Pharmaceutical Sciences*, *63*, 1056–1064.
Gleadow, R. M., & Møller, B. L. (2014). Cyanogenic glucosides: Synthesis, physiology, and plant plasticity. *Annual Review of Plant Biology*, *65*, 155–185.
González, C. G., Mustafa, N. R., Erica, G. W., Verpoorte, R., & Choi, Y. H. (2018). Application of natural deep eutectic solvents for the "green" extraction of vanillin from vanilla pods. *Flavour and Fragrance Journal*, *33*, 91–96.
Gou, M., Ran, X., Martin, D. W., & Liu, C. J. (2018). The scaffold proteins of lignin biosynthetic cytochrome P450 enzymes. *Nature Plants*, *4*, 299–310.
Markham, K. R., Gould, K. S., Winefield, C. S., Mitchell, K. A., Bloor, S. J., & Boase, M. R. (2000). Anthocyanin vacuolar inclusions-their nature and significance in flower colouration. *Phytochemistry*, *55*, 327–336.
Grimplet, J., Wheatley, M. D., Jouira, H. B., Deluc, L. G., Cramer, G. R., & Cushman, J. C. (2009). Proteomics and selected metabolite analysis of grape berry tissue under well-watered and water-deficit stress conditions. *Proteomics*, *9*, 2503–2528.
Grotewold, E., & Davies, K. (2008). Trafficking and sequestration of anthocyanins. *Natural Product Communication*, *3*, 1251–1258.
Gülck, T., Booth, J. K., Carvalho, Â., Crocol, C., Motawia, M. S., Møller, B. L., et al. (2020). Synthetic biology of cannabinoids and cannabinoid glycosides in *Nicotiana benthamiana* and *Saccharomyces cerevisiae*. *Journal of Natural Products*. preliminary accepted.
Halkier, B. A., & Møller, B. L. (1989). Biosynthesis of the cyanogenic glucoside dhurrin in seedlings of *Sorghum bicolor* (L.) Moench and partial purification of the enzyme system involved. *Plant Physiology*, *90*, 1552–1559.
Hamann, T., Laursen, T., & Møller, B. L. (2009). Functional expression of N-terminally tagged membrane bound cytochrome P450. *Protein Expression and Purification*, *68*, 18–21.
Hammond, C. T., & Mahlberg, P. G. (1977). Morphogenesis of capitate glandular hairs of Cannabis sativa (Cannabaceae). *American Journal of Botany*, *64*, 1023–1031.
Hanuš, L. O., Meyer, S. M., Muñoz, E., Taglialatela-Scafati, O., & Appendino, G. (2016). Phytocannabinoids: A unified critical inventory. *Natural Product Reports*, *33*, 1357–1392.
Happyana, N., Agnolet, S., Muntendam, R., Van Dam, A., Schneider, B., & Kayser, O. (2013). Analysis of cannabinoids in laser-microdissected trichomes of medicinal Cannabis sativa using LCMS and cryogenic NMR. *Phytochemistry*, *87*, 51–59.
Happyana, N., & Kayser, O. (2016). Monitoring metabolite profiles of Cannabis sativa L. Trichomes during flowering period using 1H NMR-based metabolomics and real-time PCR. *Planta Medica*, *82*, 1217–1223.

Heraud, P., Cowan, M. F., Marzec, K. M., Møller, B. L., Blomstedt, C. K., & Gleadow, R. (2018). Label-free Raman hyperspectral imaging analysis localizes the cyanogenic glucoside dhurrin to the cytoplasm in sorghum cells. *Scientific Reports*, *8*, 2691.
Herrmann, K. M. (1995). The shikimate pathway as an entry to aromatic secondary metabolism. *Plant Physiology*, *107*, 7–12.
Heskes, A. M., Goodger, J. Q., Tsegay, S., Quach, T., Williams, S. J., & Woodrow, I. E. (2012). Localization of Oleuropeyl glucose esters and a flavanone to secretory cavities of Myrtaceae. *PLoS One*, *7*, e40856.
Heskes, A. M., Lincoln, C. N., Goodger, J. Q., Woodrow, I. E., & Smith, T. A. (2012). Multiphoton fluorescence lifetime imaging shows spatial segregation of secondary metabolites in *Eucalyptus* secretory cavities. *Journal of Microscopy*, *247*, 33–42.
Horosanskaia, E., Nguyen, T. M., Vu, T. D., Seidel-Morgenstern, A., & Lorenz, H. (2017). Crystallization-based isolation of pure rutin from herbal extract of Sophora japonica L. *Organic Process Research & Development*, *21*, 1769–1778.
Jamshad, M., Lin, Y. P., Knowles, T. J., Parslow, R. A., Harris, C., Wheatley, M., et al. (2011). Surfactant-free purification of membrane proteins with intact native membrane environment. *Biochemical Society Transactions*, *39*, 813–818.
Jones, P. R., Møller, B. L., & Høj, P. B. (1999). The UDP-glucose:P-hydroxymandelonitrile-O-glucosyltransferase that catalyzes the last step in synthesis of the cyanogenic glucoside dhurrin in Sorghum bicolor. Isolation, cloning, heterologous expression, and substrate specificity. *The Journal of Biological Chemistry*, *274*, 35483–35491.
Jørgensen, K., Rasmussen, A. V., Morant, M., Nielsen, A. H., Bjarnholt, N., Zagrobelny, M., et al. (2005). Metabolon formation and metabolic channeling in the biosynthesis of plant natural products. *Current Opinion in Plant Biology*, *8*, 280–291.
Justesen, B. H., Laursen, T., Weber, G., Fuglsang, A. T., Møller, B. L., & Pomorski, T. G. (2013). Isolation of monodisperse nanodisc-reconstituted membrane proteins using free flow electrophoresis. *Analytical Chemistry*, *85*, 3497–3500.
Kahn, R. A., Bak, S., Svendsen, I., Halkier, B. A., & Møller, B. L. (1997). Isolation and reconstitution of cytochrome P450ox and in vitro reconstitution of the entire biosynthetic pathway of the cyanogenic glucoside dhurrin from sorghum. *Plant Physiology*, *115*, 1661–1670.
Kahn, R. A., Fahrendorf, T., Halkier, B. A., & Møller, B. L. (1999). Substrate specificity of the cytochrome P450 enzymes CYP79A1 and CYP71E1 involved in the biosynthesis of the cyanogenic glucoside dhurrin in Sorghum bicolor (L.) Moench. *Archives of Biochemistry and Biophysics*, *363*, 9–18.
Kallam, K., Appelhagen, I., Luo, J., Albert, N., Zhang, H., Deroles, S., et al. (2017). Aromatic decoration determines the formation of anthocyanic vacuolar inclusions. *Current Biology*, *27*, 945–957.
Kim, H. K., Choi, Y. H., & Verpoorte, R. (2010). NMR-based metabolomic analysis of plants. *Nature Protocols*, *5*, 536–549.
Knudsen, C., Bavishi, K., Viborg, K. M., Drew, D. P., Simonsen, H. T., Motawia, M. S., et al. (2020). Stabilization of dhurrin biosynthetic enzymes from *Sorghum bicolor* using a natural deep eutectic solvent. *Phytochemistry*, *170*, 112214.
Knudsen, C., Gallage, N. J., Hansen, C. C., Møller, B. L., & Laursen, T. (2018). Dynamic metabolic solutions to the sessile life style of plants. *Natural Product Report*, *35*, 1140–1155.
Koch, B. M., Sibbesen, O., Halkier, B. A., Svendsen, I., & Møller, B. L. (1995). The primary sequence of cytochrome P450tyr, the multifunctional N-hydroxylase catalyzing the conversion of L-tyrosine to p-Hydroxyphenylacetaldehyde oxime in the biosynthesis of the cyanogenic glucoside Dhurrin in Sorghum bicolor (L.) Moench. *Archives of Biochemistry and Biophysics*, *323*, 177–186.
Koes, R., Vorweij, W., & Quattrocchio, F. (2005). Flavonoids: A colorful model for the regulation of biochemical pathways. *Trends in Plant Science*, *10*, 236–242.

Koja, E., Ohata, S., Maruyama, Y., Suzuki, H., Shimosaka, M., & Taguchi, G. (2018). Identification and characterization of a rhamnosyltransferase involved in rutin biosynthesis in *Fagopyrum esculentum* (common buckwheat). *Bioscience, Biotechnology, and Biochemistry*, *82*, 1790–1802.

Kojima, M., Poulton, J. E., Thayer, S. S., & Conn, E. E. (1979). Tissue distributions of dhurrin and of enzymes involved in its metabolism in leaves of *Sorghum bicolor*. *Plant Physiology*, *63*, 1022–1028.

Kristensen, C., Morant, M., Olsen, C. E., Ekstrøm, C. T., Galbraith, D. W., Møller, B. L., et al. (2005). Metabolic engineering of dhurrin in transgenic Arabidopsis plants with marginal inadvertent effects on the metabolome and transcriptome. *Proceedings of the National Academy of Sciences of the United States of America*, *102*, 1779–1784.

Kuchler, A., Yoshimoto, M., Luginbühl, S., Mavelli, F., & Walde, P. (2016). Enzymatic reactions in confined environments. *Nature Nanotechnology*, *11*, 409–420.

Lallemand, B., Erhardt, M., Heitz, T., & Legrand, M. (2013). Sporopollenin biosynthetic enzymes interact and constitute a metabolon localized to the endoplasmic reticulum of tapetum cells. *Plant Physiology*, *162*, 616–625.

Lanyon, V. S., Turner, J. C., & Mahlberg, P. G. (1981). Quantitative analysis of cannabinoids in the secretory product from capitate-stalked glands of Cannabis sativa L. (Cannabaceae). *Botanical Gazette*, *142*, 316–319.

Laursen, T., Borch, J., Knudsen, C., Bavishi, K., Torta, F., Martens, H. J., et al. (2016). Characterization of a dynamic metabolon producing the defense compound dhurrin in sorghum. *Science*, *354*, 890–893.

Laursen, T., Jensen, K., & Møller, B. L. (2011). Conformational changes of the NADPH-dependent cytochrome P450 reductase in the course of electron transfer to cytochromes P450. *Biochimica Et Biophysica Acta-Proteins and Proteomics*, *1814*, 132–138.

Laursen, T., Møller, B. L., & Bassard, J. E. (2015). Plasticity of specialized metabolism as mediated by dynamic metabolons. *Trends in Plant Science*, *20*, 20–32.

Laursen, T., Naur, P., & Møller, B. L. (2013). Amphipol trapping of a functional CYP system. *Biotechnology and Applied Biochemistry*, *60*, 119–127.

Laursen, T., Singha, A., Rantzau, N., Tutkus, M., Borch, J., Hedegård, P., et al. (2014). Single molecule activity measurements of cytochrome P450 oxidoreductase reveal the existence of two discrete functional states. *ACS Chemical Biology*, *9*, 630–634.

Lee, S. C., Knowles, T. J., Postis, V. L., Jamshad, M., Parslow, R. A., Lin, Y. P., et al. (2016). A method for detergent-free isolation of membrane proteins in their local lipid environment. *Nature Protocols*, *11*, 1149–1162.

Li, H., Ban, Z., Qin, H., Ma, L., King, A. J., & Wang, G. (2015). A heteromeric membrane-bound Prenyltransferase complex from hop catalyzes three sequential aromatic prenylations in the bitter acid pathway. *Plant Physiology*, *167*, 650–659.

Li, B., Bjarnholt, N., Hansen, S. H., & Janfelt, C. (2011). Characterization of barley leaf tissue using direct and indirect desorption electrospray ionization imaging mass spectrometry. *Journal of Mass Spectrometry*, *46*, 1241–1246.

Li, B., Knudsen, C., Hansen, N. K., Jørgensen, K., Kannangara, R., Bak, S., et al. (2013). Visualizing metabolite distribution and enzymatic conversion in plant tissues by desorption electrospray ionization mass spectrometry imaging. *The Plant Journal*, *74*, 1059–1071.

Livingston, S. J., Quilichini, T. D., Booth, J. K., Wong, D. C. J., Rensing, K. H., Laflamme-Yonkman, J., et al. (2019). Cannabis glandular trichomes alter morphology and metabolite content. *The Plant Journal*, *101*, 37–56.

Luo, X., Reiter, M. A., d'Espaux, L., Wong, J., Denby, C. M., Lechner, et al. (2019). Complete biosynthesis of cannabinoids and their unnatural analogues in yeast. *Nature*, *567*, 123–126.

Mahlberg, P., & Kim, E. (2004). Accumulation of cannabinoids in glandular trichomes of Cannabis (Cannabaceae). *Journal of Industrial Hemp*, *9*, 15–36.

Mahlberg, P. G., & Kim, E. (2017). Cuticle development on glandular Trichomes of Cannabis sativa (Cannabaceae). *American Journal of Botany*, *78*, 1113–1122.

Markham, K. R., Bloor, S. J., Nicholson, R., Rivera, R., Shemluck, M., Kevan, P. G., et al. (2004). Black flower coloration I wild *Lisianthius nigrescens*: Its chemistry and ecological consequences. *Zeitschrift für Naturforschung C*, *59*, 624–630.

Matarasso, N., Schuster, S., & Avni, A. (2005). A novel plant cysteine protease has a dual function as a regulator of 1-aminocyclopropane-1-carboxylic acid synthase gene expression. *Plant Cell*, *17*, 1205–1216.

Mauseth, J. D. (2006). Structure-function relationships in highly modified shoots of cactaceae. *Annals of Botany*, *98*, 901–926.

Mehrshahi, P., Johnny, C., & DellaPenna, D. (2014). Redefining the metabolic continuity of chloroplasts and ER. *Trends in Plant Science*, *19*, 501–507.

Mehrshahi, P., Stefano, G., Andaloro, J. M., Brandizzi, F., Froehlich, J. E., & DellaPenna, D. (2013). Transorganellar complementation redefines the biochemical continuity of endoplasmic reticulum and chloroplasts. *PNAS*, *110*, 12126–12131.

Mitrea, D. M., & Kriwacki, R. W. (2016). Phase separation in biology; functional organization of a higher order. *Cell Communication and Signaling*, *14*, 1.

Møller, B. L. (2017). Solar-driven synthesis of bioactive natural products. *Chemical Engineering Progress*, *113*, 46–54.

Møller, B. L., & Conn, E. E. (1979). The biosynthesis of cyanogenic glucosides in higher plants. N-Hydroxytyrosine as an intermediate in the biosynthesis of dhurrin by *Sorghum bicolor* (Linn) Moench. *Journal of Biological Chemistry*, *254*, 8575–8583.

Møller, B. L., & Conn, E. E. (1980). The biosynthesis of cyanogenic glucosides in higher plants. Channeling of intermediates in dhurrin biosynthesis by a microsomal system from *Sorghum bicolor* (linn) Moench. *Journal of Biological Chemistry*, *255*, 3049–3056.

Møller, B. L., Mcfarlane, I. J., & Conn, E. E. (1977). Chemical synthesis and disproportionation of *N*-hydroxytyrosine. *Acta Chemica Scandinavica*, *B31*, 343–344.

Møller, B. L., Olsen, C. E., & Motawia, M. S. (2016). General and stereocontrolled approach to the chemical synthesis of naturally occurring cyanogenic glucosides. *Journal of Natural Products*, *79*, 1198–1202.

Montini, L., Crocoll, C., Gleadow, R., Motawia, M. S., Janfeldt, C., & Bjarnholt, N. (2020). Matrix-assisted laser desorption/ionization-mass spectrometry imaging of metabolites during sorghum germination. *Plant Physiology*, *183*, 925–942.

Morant, A. V., Jørgensen, K., Paquette, S. M., Sánchez-Pérez, R., Møller, B. L., & Bak, S. (2008). β-Glucosidases as detonators of plant chemical defense. *Phytochemistry*, *69*, 1795–1813.

Morant, M., Jørgensen, K., Schaller, H., Pinot, F., Møller, B. L., Werck-Reichhart, D., et al. (2007). CYP703 is an ancient cytochrome P450 in land plants catalyzing in-chain hydroxylation of lauric acid to provide building blocks for sporopollenin synthesis in pollen. *Plant Cell*, *19*, 1473–1487.

Nicholson, R. L., Kollipara, S. S., Vincent, J. R., Lyons, P. C., & Cadena-Gomez, G. (1987a). Phytoalexin synthesis by the sorghum mesocotyl in response to infection by pathogenic and nonpathogenic fungi. *Proceedings of the National Academy of Sciences of the United States of America*, *84*, 5520–5524.

Nicholson, R. L., Kollipara, S. S., Vincent, J. R., Lyons, P. C., & Cadena-Gomez, G. (1987b). Phytoalexin synthesis by the sorghum mesocotyl in response to infection by pathogenic and nonpathogenic fungi. *Proceedings of the National Academy of Sciences of the United States of America*, *84*, 5520–5524.

Nielsen, K. A., Gotfredsen, C. H., Buch-Pedersen, M. J., Ammitzbøll, H., Mattsson, O., Duus, J.Ø., et al. (2004). Inclusions of flavonoid 3-deoxyanthocyanidins in *Sorghum bicolor* self-organize into spherical structures. *Physiological and Molecular Plant Pathology, 65*, 187–196.
Nielsen, K. A., Tattersall, D. B., Jones, P. R., & Møller, B. L. (2008). Metabolon formation in dhurrin biosynthesis. *Phytochemistry, 69*, 88–98.
Obata, T. (2019). Metabolons in plant primary and secondary metabolism. *Phytochemistry Reviews, 18*, 1483–1507.
Schmidt, F. B., Cho, S. K., Olsen, C. E., Yang, S. W., Møller, B. L., & Jørgensen, K. (2018). Diurnal regulation of cyanogenic glucoside biosynthesis and endogenous turnover in cassava. *Plant Direct, 2*, 1–9.
Paddon, C. J., Westfall, P. J., Pitera, D. J., Benjamin, K., Fisher, K., McPhee, D., et al. (2013). High-level semi-synthetic production of the potent antimalarial artemisinin. *Nature, 496*, 528–532.
Pateraki, I., Andersen-Ranberg, J., Hamberger, B., Heskes, A. M., Martens, H. J., Zerbe, P., et al. (2014). Manoyl oxide (13R), the biosynthetic precursor of forskolin, is synthesized in specialized root cork cells in *Coleus forskohlii*. *Plant Physiology, 164*, 1222–1236.
Pateraki, I., Andersen-Ranberg, J., Jensen, N. B., Wubshet, S. G., Heskes, A. M., Forman, et al. (2017). Total biosynthesis of the cyclic AMP booster forskolin from *Coleus forskohlii*. *eLife, 6*, e23001.
Pecket, R. C., & Small, C. J. (1980). Occurrence, location and development of anthocyanoplasts. *Phytochemistry, 19*, 2571–2576.
Perez-Guaita, D., Kochan, K., Miguela, M., Andrew, D. W., Heraud, P., Richards, et al. (2017). Multimodal vibrational imaging of cells. *Vibrational Spectroscopy, 91*, 46–58.
Phillips, M. J., & Voeltz, G. K. (2016). Structure and function of ER membrane contact sites with other organelles. *Nature Reviews Molecular Cell Biology, 17*, 69–82.
Pourcel, L., Irani, N. G., Lu, Y. H., Riedl, K., Schwartz, S., & Grotewold, E. (2010). The formation of Anthocyanic vacuolar inclusions in *Arabidopsis thaliana* and implications for the sequestration of anthocyanin pigments. *Molecular Plant, 3*, 78–90.
Pradedova, E. V., Nimaeva, O. D., Rakevich, A. L., & Salyaev, R. K. (2019). Comparative analyses of glutathione system of vacuoles and leucoplasts isolated from the storage parenchyma cells of dormant red beetroots (Beta vulgaris L.). *Plant Physiology and Biochemistry, 145*, 52–63.
Rodziewicz, P., Loroch, S., Marczak, Ł., Sickmann, A., & Kayser, O. (2019). Cannabinoid synthases and osmoprotective metabolites accumulate in the exudates of Cannabis sativa L. glandular trichomes. *Plant Science, 284*, 108–116.
Sánchez-Pérez, R., Jørgensen, K., Motawia, M. S., Dicenta, F., & Møller, B. L. (2009). Tissue and cellular localization of individual ß-glycosidases using a substrate specific sugar reducing assay. *The Plant Journal, 60*, 894–906.
Saunders, J. A., & Conn, E. E. (1978). Presence of the cyanogenic glucoside dhurrin in isolated vacuoles from *Sorghum*. *Plant Physiology, 61*, 154–157.
Sibbesen, O., Koch, B., Halkier, B. A., & Møller, B. L. (1994). Isolation of the heme-thiolate enzyme cytochrome P-450TYR, which catalyzes the committed step in the biosynthesis of the cyanogenic glucoside dhurrin in Sorghum bicolor (L.) Moench. *Proceedings of the National Academy of Sciences of the United States of America, 91*, 9740–9744.
Sibbesen, O., Koch, B., Halkier, B. A., & Møller, B. L. (1995). Cytochrome P-450TYR is a multifunctional Heme-thiolate enzyme catalyzing the conversion of L-tyrosine to *p*-Hydroxyphenylacetaldehyde oxime in the biosynthesis of the cyanogenic glucoside Dhurrin in *Sorghum bicolor* (L.) Moench. *The Journal of Biological Chemistry, 270*, 3506–3511.

Sirikantaramas, S., Taura, F., Tanaka, Y., Ishikawa, Y., Morimoto, S., Shoyama, Y., et al. (2005). Tetrahydrocannabinolic acid synthase, the enzyme controlling marijuana psychoactivity, is secreted into the storage cavity of the glandular trichomes. *Plant and Cell Physiology*, *46*, 1578–1582.
Snyder, B. A., & Nicholson, R. L. (1990). Synthesis of phytoalexins in sorghum as a site-specific response to fungal ingress. *Science*, *248*, 1637–1639.
Srere, P. A. (1985). The metabolon. *Trends in Biochemical Sciences*, *10*, 109–110.
Stafford, H. (1974). The metabolism of aromatic compounds. *Annual Review of Plant Physiology*, *25*, 459–486.
Stella, S., Mesa, P., Thomsen, J., et al. (2018). Conformational activation promotes CRISPR-Cas12a catalysis and resetting of the endonuclease activity. *Cell*, *175*, 1856–1871.
Thorsoe, K. S., Bak, S., Olsen, C. E., Imberty, A., Breton, C., & Lindberg Møller, B. (2005). Determination of catalytic key amino acids and UDP sugar donor specificity of the cyanohydrin glycosyltransferase UGT85B1 from Sorghum bicolor. Molecular modeling substantiated by site-specific mutagenesis and biochemical analyses. *Plant Physiology*, *139*, 664–673.
Thygesen, L. G., Jørgensen, K., Møller, B. L., & Engelsen, S. B. (2004). Raman spectroscopic analysis of cyanogenic glucosides in plants. Development of a flow injection surface enhanced Raman scatter (FI-SERS) method for determination of cyanide. *Applied Spectroscopy*, *58*, 212–217.
Tohge, T., de Souza, L. P., & Fernie, A. R. (2017). Current understanding of the pathways of flavonoid biosynthesis in model and crop plants. *Journal of Experimental Botany*, *68*, 4013–4028.
Turk, V., Stoka, V., Vasiljeva, O., Renko, M., Sun, T., Turk, B., et al. (2012). Cysteine cathepsins: From structure, function and regulation to new frontiers. *Biochimica et Biophysica Acta*, *1824*, 68–88.
Turner, J. C., Hemphill, J. K., & Mahlberg, P. G. (1980). Trichomes and cannabinoid content of developing leaves and bracts of Cannabis sativa L. (Cannabaceae). *American Journal of Botany*, *67*, 1397–1406.
van der Lee, R., Buljan, M., Lang, B., Weatheritt, R. J., Daughdrill, G. W., Dunker, A. K., et al. (2014). Classification of intrinsically disordered regions and proteins. *Chemical Reviews*, *114*, 6599–6631.
Vogt, T. (2010). Phenylpropanoid biosynthesis. *Molecular Plant*, *3*, 2–20.
Wadsäter, M., Laursen, T., Singha, A., Hatzakis, N., Stamou, D., Barker, R., et al. (2012). Monitoring shifts in the conformation equilibrium of the membrane protein cytochrome P450 reductase (POR) in nanodiscs. *Journal of Biological Chemistry*, *287*, 34596–34603.
Waki, T., Yoo, D., Fujino, N., Mameda, R., Denessiouk, K., & Yamashita, S. (2016). Identification of protein-protein interactions of isoflavonoid biosynthetic enzymes with 2-hydroxyisoflavanone synthase in soybean (Glycine max (L.) Merr.). *Biochemical and Biophysical Research Communications*, *469*, 546–551.
Winkel-Shirley, B. (2001). Flavonoid biosynthesis. A colorful model for genetics, biochemistry, cell biology, and biotechnology. *Plant Physiology*, *126*, 485–493.
Winkel-Shirley, B. (2004). Metabolic channeling in plants. *Annual Review of Plant Biology*, *55*, 85–107.
Zhang, F.-U., Wang, Q.-Y., Pu, Y.-J., Chen, T.-Y., Qin, X.-M., & Gao, J. (2017). Identification of genes involved in flavonoid biosynthesis in Sophora japonica through transcriptome sequencing. *Chemistry and Biodiversity*, *14*, e1700369.
Zhao, B.-Y., Xu, P., Yang, F.-X., Wu, H., Zong, M.-H., & Lou, W.-Y. (2015). Biocompatible deep eutectic solvents based on choline chloride: Characterization and application to the extraction of rutin from *Sophora japonica*. *ACS Sustainable Chemistry & Engineering*, *3*, 2746–2755.

CHAPTER NINE

NADES formation in vegetative desiccation tolerance: Prospects and challenges

S. Francois du Toit, Joanne Bentley, and Jill M. Farrant*
Department of Molecular and Cell Biology, University of Cape Town, Rondebosch, South Africa
*Corresponding author: e-mail address: jill.farrant@uct.ac.za

Contents

Abstract

Vegetative desiccation tolerance is rare, occurring in approximately only 0.04% of angiosperms. Such species survive extreme water loss to the air-dry state, remaining quiescent until rehydration, upon which metabolism is rapidly recovered in existing tissues. In many desiccation-tolerant species, sucrose levels increase considerably during dehydration, and this extensive accumulation has typically been attributed to cytoplasmic vitrification. However, an equally compelling explanation for the role of large-scale sugar accumulation during desiccation could be attributed to NADES formation. Recent high-resolution metabolomics studies on desiccation-tolerant plants have noted increased organic acids and amino acids during dehydration, which together with sugar accumulation implies the existence of NADES. In this chapter we explore the circumstantial evidence for cytoplasmic and organellar NADES formation during dehydration, evaluating the role of this mechanism in the concentration and stability of crucial cellular components, which would ultimately contribute to both survival at low cellular water contents and subsequent revival of tissues.

1. Introduction

As an incompressible fluid, water is an ideal "solvent of life" that occupies the vast majority of intra- and intercellular space within cells and tissues. It thus comes as no surprise that exceptionally few organisms

Advances in Botanical Research, Volume 97
ISSN 0065-2296
https://doi.org/10.1016/bs.abr.2020.09.007

are able to survive even minor loss thereof. In plants, water plays a structural role, pushing against the cell membrane and cell wall, producing the turgor pressure necessary for maintaining not only the structural integrity of the cell, but the organism as a whole (Iljin, 1957; Levitt, 1980). At the molecular level, water plays multiple roles, being the fluid matrix facilitating the diffusion of substrates to the active sites of enzymes and being the reactant and by-product of many important metabolic reactions. In most plants, loss of water (and thus turgor) results in progressive cell shrinkage and ultimately plasmolysis and cytorrhysis. This is accompanied by ionic imbalances, destabilization of macromolecules and membrane structures, and uncontrolled generation of reactive oxygen species (ROS), particularly from photosynthetic metabolism and other free radicals, which collectively lead to plant death.

The seeds of the majority of plants are desiccation tolerant (DT), this being defined in the seed literature as the ability to survive drying to 0.1 g H_2O/g dry weight or 10% water content on a wet-weight basis (Sun, 2002) for prolonged periods of time. In such seeds, a variety of protective mechanisms are laid down during the late stages of embryological development to withstand this extreme water loss. However, the ability to tolerate prolonged desiccation is lost upon germination, except in a very small number of angiosperms (135 species or 0.04% of flowering plants) that display vegetative desiccation tolerance (Farrant, Cooper, Dace, Bentley, & Hilgart, 2017; Gaff & Oliver, 2013). Collectively these are termed "resurrection plants" due to their ability to revive from a seemingly dead state when supplied with water. Like seeds, resurrection plants prepare for quiescence and subsequent revival during the dehydration phase by accumulating the proteins, metabolites, and other components necessary for maintaining cellular integrity during desiccation, with some repair being effected upon rehydration (reviewed *inter alia* in Blomstedt et al., 2010; Dinakar & Bartels, 2013; Farrant, Brandt, & Lindsey, 2007; Farrant et al., 2017; Gaff, 1989; Gaff & Oliver, 2013; Liu, Moyankova, Djilianov, & Deng, 2019; Oliver et al., 2020). This is in contrast to nontracheophytes, where protection has been proposed to be largely constitutive and where repair is actuated upon rehydration (Oliver, Wood, & O'Mahony, 1998; Stark, 2017). Importantly, metabolic activity is maintained during tissue dehydration in resurrection plants, even to water contents below *ca.* 30% of that at full turgor value. The use of relative water content (RWC; Barr & Weatherley, 1962), which refers to the water content of a tissue relative to that at full turgor, is generally used for comparisons among species and will thus be utilized in our commentary here.

The tolerance of desiccation ultimately implies stable metabolic arrest and quiescence together with the orderly revival of the tissues upon rehydration. Our observation of significant metabolic changes at RWCs at and below *ca.* 30% RWC in resurrection plants, signifying the continuation of metabolic activity in water-scarce cellular environments, led us to explore the possible molecular scenarios that might facilitate this. At these water contents, cytoplasmic transitions are believed to move from an amorphous phase to a glass (Oliver et al., 2020; Walters & Koster, 2007). However, in this chapter, we propose a greater, but not mutually exclusive, role for NADES in this phase transition in resurrection plants.

2. The cellular environment under desiccation

Upon imposition of water deficit stress due to a lack of available water at the root level, plants first attempt to reduce leaf water loss by accumulating or concentrating osmotically-active compounds and closing the stomata, thus limiting water loss from transpiration. Resurrection plants do the same, but unlike desiccation-sensitive plants, they are able to prevent or minimize the damage associated with ongoing photosynthetic ROS production and further turgor loss as tissues dehydrate to below 90%–50% RWC (depending on the species) (reviewed in Gaff & Oliver, 2013; Oliver et al., 2020; Zhang & Bartels, 2018). At water contents below this, desiccation-sensitive species experience increased plasmolysis and excessive levels of unmitigated ROS formation, resulting in aberrant metabolism, damage to nucleic acids, proteins, lipids, and carbohydrates, and ultimately, tissue senescence (Farooq, Wahid, Lee, Ito, & Siddique, 2009; Rout & Shaw, 2001).

In resurrection plants, dehydration results in changes in gene expression and metabolite abundance, which appears to occur in two successive phases, namely a mid-response and late-response to dehydration (Costa, Cooper, Hilhorst, & Farrant, 2017; Djilianov et al., 2013; Gechev et al., 2013; Ingle, Schmidt, Farrant, Thomson, & Mundree, 2007; Kuang et al., 1995; Oliver, Guo, et al., 2011, Oliver, Jain, et al., 2011; Radermacher, du Toit, & Farrant, 2019; Van Buren et al., 2018; Xiao et al., 2015; Zhang & Bartels, 2018). While the exact water contents at which these phases might start and end is complicated by the inconsistencies among studies in terms of the actual water contents sampled for tissue analysis, it has generally been observed that entry into the mid-response stage correlates with the onset of the controlled down-regulation of photosynthetic activity, usually initiating at *ca.* 60% RWC depending on the species. There are two strategies used by resurrection plants in this process, termed homoiochlorophylly and poikilochlorophylly. The former is typified by retention of chlorophyll and maintenance and protection of the photosynthetic apparatus during desiccation, allowing rapid institution of photosynthesis upon rehydration. In the latter, chlorophyll is degraded and thylakoids are dismantled during desiccation, requiring their regeneration during rehydration. Despite these differences (which imply different mechanisms of subcellular protection), it is during this stage that resurrection plants universally begin to accumulate various protective proteins and compounds such as sugars, amino acids, and organic acids (Table 1; Fig. 1A). While the exact origin of these molecules remains contested–whether they result from de novo synthesis, catabolism of larger molecules, or are mobilized from other organs or tissues–the accumulation thereof appears consistent even across distantly-related species. We propose that the accumulation of these metabolites serves to enable the ongoing metabolism observed at low water contents, ultimately contributing to the creation of a stable subcellular environment during the late stages of dehydration.

Late-stage responses in gene transcription and metabolite accumulation are initiated at RWCs between 40% and 20% (species- and study-dependent) (reviewed in Farrant et al., 2017; Zhang & Bartels, 2018). Of particular interest is that even in this extreme state of dehydration, respiration is ongoing (Farrant, 2000; Oliver et al., 2020; Tuba, Lichtenthaler, Csintalan, Nagy, & Szente, 1996) and de novo transcription and biochemical synthesis is apparent (Costa et al., 2017; Liu et al., 2018; Oliver, Guo, et al., 2011, Oliver, Jain, et al., 2011; Radermacher et al., 2019; Xiao et al., 2015; van Buren et al., 2018). Certain antioxidant enzymes, particularly glutathione reductase (GR) and superoxide dismutase (SOD), retain the ability for activity (demonstrated by in vitro assays) during late and early rehydration in resurrection plants, whereas these enzymes are denatured in desiccation-sensitive species (Farrant et al., 2007; Illing et al., 2005). While this implies their protection during desiccation, it is also possible that these enzymes may remain active or resume activity as needed at low RWCs provided that the nature of their cellular environment allows for this. Finally, it is also evident that in some species, such as *Xerophyta humilis* (Baker) T. Durand and Schinz and *X. schlechteri* (Baker) N.L. Menezes, many of the transcripts accumulated during this late stage are not translated but are required for recovery of metabolic competency upon early rehydration (Costa et al., 2017; Dace, Sherwin, Illing, & Farrant, 1998), implying not only their stable storage but also early initiation of translation prior to full hydration.

Table 1 Potential NADES-forming metabolites present in resurrection plant species.

Metabolite	*Barbacenia purpurea*	*Boea hygroscopica*	*Haberlea rhodopensis*	*Craterostigma plantagineum*	*Craterostigma pumilim*	*Eragrostis nindensis*	*Myrothamnus flabellifolia*	*Ramonda serbica*
Sugars								
Fructose	Yes	Yes	Yes*	Yes	Yes	Yes	Yes*	Yes*
Sucrose	Yes**	Yes	Yes**	Yes**	Yes**	Yes	Yes	Yes
Raffinose	Yes**	Yes*	Yes*	Yes**	Yes**	Yes	Yes	Yes*
Stachyose	Yes**	Yes*	–	Yes**	Yes**	Yes	Yes*	–
Trehalose	Yes**	–	Yes	–	–	Yes	Yes	–
Organic acids								
Citric acid	–	–	Yes**	–	Yes	–	–	–
Fumaric acid	Yes**	–	–	–	–	–	–	–
Malic acid	–	–	Yes**	–	Yes	–	–	–
Amino acids								
y-Aminobutyric acid	Yes	–	–	–	–	–	–	–
1,2-Aminobutyric acid	–	–	Yes**	–	–	–	–	–
2-Aminoadipate	–	–		–	–	–	–	–
Arginine	Yes**	–	–	–	–	–	–	–

Continued

Table 1 Potential NADES-forming metabolites present in resurrection plant species.—cont'd

Metabolite	***Barbacenia purpurea***	***Boea hygroscopica***	***Haberlea rhodopensis***	***Craterostigma plantagineum***	***Craterostigma pumilim***	***Eragrostis nindensis***	***Myrothamnus flabellifolia***	***Ramonda serbica***
Asparagine	–	–	Yes**	–	Yes	–	–	–
Aspartate	Yes	–	–	–	–	–	–	–
B-alanine	–	–	–	–	–	–	–	–
Cyanoalanine	–	–	–	–	–	–	–	–
L-alanine	–	–	Yes**	–	–	–	–	–
B-Aminoisobutyric acid	–	–	Yes	–	–	–	–	–
Betaine	–	–	–	–	–	–	–	–
Cystatheonine	–		Yes*	–	–	–	–	–
Cysteine	–		Yes*	–	–	–	–	–
Cysteine-glutathione disulfide	–	–	–	–	–	–	–	–
Dimethylglycine	–	–	–	–	–	–	–	–
Gamma-aminobutyrate (GABA)	–	–	–	–	–	–	–	–
Glutamic acid	Yes	–	Yes**	–	–	–	–	–
Glycine	Yes	–	–	–	–	–	–	–

Histidine	Yes	–	–	–	–	–	–	–
Homostachydrine	–	–	–	–	–	–	–	–
Hydroxyproline	Yes	–	Yes	–	–	–	–	–
Isoleucine	–	–	–	–	–	–	–	–
Leucine	Yes	–	Yes*	–	Yes	–	–	–
Lysine	Yes	–	–	–	–	–	–	–
3-Methyl-2-oxovalerate	–	–	–	–	–	–	–	–
4-Methyl-2-oxopentanoate	–	–	–	–	–	–	–	–
N-acetylproline	–	–	–	–	–	–	–	–
N-acetylthreonine	–	–	–	–	–	–	–	–
N-acetylvaline	–	–	–	–	–	–	–	–
Ornithine	–	–	–	–	–	–	–	–
Phenylalanine	–	–	Yes*	–	–	–	–	–
Pipecolic acid	–	–	–	Yes	–	–	–	–
Proline	Yes**	–	–	Yes*	–	Yes	Yes	Yes
Serine	Yes	–	–	–	–	–	–	–

Continued

Table 1 Potential NADES-forming metabolites present in resurrection plant species.—cont'd

Metabolite	*Barbacenia purpurea*	*Boea hygroscopica*	*Haberlea rhodopensis*	*Craterostigma plantagineum*	*Craterostigma pumilim*	*Eragrostis nindensis*	*Myrothamnus flabellifolia*	*Ramonda serbica*
Stachydrine	–	–	–	–	–	–	–	–
Threonine	–	–	Yes	–	Yes	–	–	–
Tryptophan	Yes	–	–	–	–	–	–	–
Tyrosine	Yes	–	Yes	–	–	–	–	–
Urocanate	–	–	–	–	–	–	–	–
Valine	Yes**	–	–	–	–	–	–	–
Other								
Glycerol	–	–	Yes	–	–	–	–	–
References	Suguiyama, da Silva, Meirelles, Centeno, and Braga (2014)	Albini et al. (1999)	Moyankova et al. (2014)	Tymms and Gaff (1979), Bianchi, Gamba, Murelli, Salamini, and Bartels (1991), Egert, Eicher, Keller, and Peters (2015)	Egert et al. (2015), Ricco, Hilhorst, Farrant, and Dace (2018)	Tymms and Gaff (1979), Ghasempour, Gaff, Williams, and Gianello (1998)	Tymms and Gaff (1979), Moore et al. (2007)	Živković, Quartacci, Stevanović, Marinone, and Navari-Izzo (2005)

Metabolite	*Ramonda nathaliae*	*Reaumuria soongorica*	*Sporobolus stapfianus*	*Xerophyta humilis*	*Xerophyta schlecteri*	*Xerophyta villosa*	*Xerophyta viscosa*
Sugars							
Fructose	Yes*	–	Yes*	Yes*	Yes**	Yes	Yes*
Sucrose	Yes	Yes	Yes**	Yes	Yes**	Yes	Yes**
Raffinose	Yes*	–	Yes**	–	–	Yes	Yes**
Stachyose	–	–	Yes**	–	–	Yes	Yes**
Trehalose	–	–	–	–	Yes	No	–
Organic acids							
Citric acid	–	Yes	Yes**	Yes	Yes**	–	–
Fumaric acid	–	–	Yes**	–	–	–	–
Malic acid	–	Yes*	Yes**	Yes*	Yes	–	–
Amino acids							
y-Aminobutyric acid	–	–		–	–	–	–
1,2-Aminobutyric acid	–	–		–	–	–	–
2-Aminoadipate	–	–	Yes**	–	–	–	–
Arginine	–	–	Yes**	–	–	–	–
Asparagine	–	–		–	Yes	–	–

Continued

Table 1 Potential NADES-forming metabolites present in resurrection plant species.—cont'd

Metabolite	*Ramonda nathaliae*	*Reaumuria soongorica*	*Sporobolus stapfianus*	*Xerophyta humilis*	*Xerophyta schlecteri*	*Xerophyta villosa*	*Xerophyta viscosa*
Aspartate	–	–	Yes	Yes	–	–	–
B-alanine	–	–	Yes**	–	Yes	–	–
Cyanoalanine	–	–	Yes	–	–	–	–
L-alanine	–	–	Yes**	–	Yes	–	–
B-Aminoisobutyric acid	–	–		–	–	–	–
Betaine	–	–	Yes**	–	–	–	–
Cystatheonine	–	–		–	–	–	–
Cysteine	–	–	Yes**	–	No	–	–
Cysteine-glutathione disulfide	–	–	Yes	–	–	–	–
Dimethylglycine	–	–	Yes**	–	–	–	–
Gamma-aminobutyrate (GABA)	–	–	Yes**	–	–	–	–
Glutamic acid	–	–	Yes	–	Yes**	–	–
Glycine	–	–	Yes**	Yes	Yes	–	–

Histidine	–	–	Yes**	–	–	–	–
Homostachydrine	–	–	Yes**	–	–	–	–
Hydroxyproline	–	–	–	–	–	–	–
Isoleucine	–	–	Yes**	–	Yes	–	–
Leucine	–	–	Yes**	–	Yes	–	–
Lysine	–	–	Yes**	–	Yes	–	–
3-Methyl-2-oxovalerate	–	–	Yes	–	–	–	–
4-Methyl-2-oxopentanoate	–	–	Yes	–	–	–	–
N-acetylproline	–	–	Yes**	–	–	–	–
N-acetylthreonine	–	–	Yes**	–	–	–	–
N-acetylvaline	–	–	Yes	–	–	–	–
Ornithine	–	–	Yes	–	Yes	–	–
Phenylalanine	–	–	Yes	Yes	Yes**	–	–
Pipecolic acid	–	–	Yes	–	Yes**	–	–
Proline	–	Yes	Yes**	–	No	Yes	–
Serine	–	–	Yes**	–	Yes	–	–

Continued

Table 1 Potential NADES-forming metabolites present in resurrection plant species.—cont'd

Metabolite	***Ramonda nathaliae***	***Reaumuria soongorica***	***Sporobolus stapfianus***	***Xerophyta humilis***	***Xerophyta schlecteri***	***Xerophyta villosa***	***Xerophyta viscosa***
Stachydrine	–	–	Yes	–	–	–	–
Threonine	–	–	Yes**	Yes	–	–	–
Tryptophan	–	–	Yes**	Yes	Yes**	–	–
Tyrosine	–	–	Yes**	Yes	No	–	–
Urocanate	–	–	Yes	–	–	–	–
Valine	–	–	Yes**	–	No	–	–
Other							
Glycerol	–	–	Yes*	–	Yes**	–	–
References	Müller, Sprenger, Bortlik, Boller, and Wiemken (1997)	Liu et al. (2007)	Oliver, Guo, et al. (2011), Oliver, Jain, et al. (2011), Yobi et al. (2017)	Dace (2014)	Radermacher et al. (2019)	Tymms and Gaff (1979), Ghasempour et al. (1998)	Peters, Mundree, Thomson, Farrant, and Keller (2007)

"Yes" indicates that the metabolite is present in the plant; Yes* indicates that the metabolite is present in the plant and is more abundant at RWCs >35% than at RWCs of ≤35%; "Yes**" indicates that the metabolite is higher in abundance at RWCs of ≤35% than at RWCs of >35%; "No" indicates that the metabolite has been detected at other RWCs but not at RWC ≤35%; and "–" indicates that we could not find information on the metabolite.

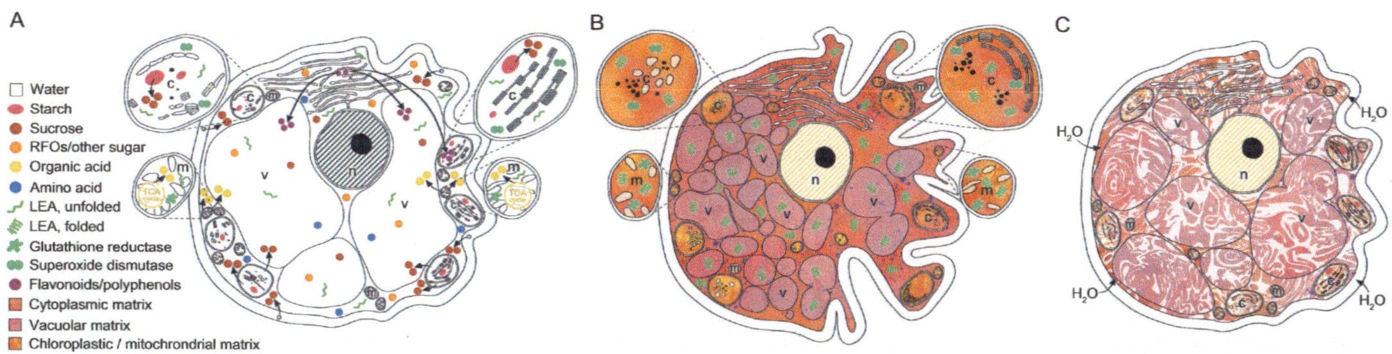

Fig. 1 Schematic representation of subcellular changes and metabolite production and distribution at the initiation of the mid-phase dehydration, (A), in the desiccated state, (B), and upon early rehydration, (C). In each instance the left hand side of the cell is typical of that of poikilochlorophyllous species and the right hand side typical of that of homoiochlorophyllous species. Enlargements of chloroplasts (c) and mitochondria (m) are shown, further depicting the respective strategies used by such species. The presence of NADES forming metabolites are shown in various colors, with arrows showing potential movement of these during the mid to late stages of drying. All white areas represent water. Presence of LEA proteins and the antioxidants glutathione reductase and superoxide dismutase are also depicted. v = vacuole, n = nucleus.

These observations suggest that the nature of the cellular environment at the late-response must be such to allow the movement and functioning of polymerases, chaperone proteins, respiratory and antioxidant enzymes, and others responsible for metabolite accumulation. Considering that the bulk of this environment would consist of the compounds produced during the mid-late responses, it follows that those metabolites allow the aforementioned movement and functioning of large molecules without being depleted and with limited dependency on molecular water. Finally, this cellular environment must provide a degree of chemical, structural, and thermal stability and act in a buffering capacity to the cell and its organelles during dehydration, as well as the maintenance of a minimal RWC of between 10% and 5%.

Sucrose has been implicated as being vitally important in providing cellular protection to resurrection plants during dehydration and the quiescent state, with suggested roles in cellular stabilization and signaling (reviewed in Farrant et al., 2007, 2017; Oliver et al., 2020; Zhang, Song, & Bartels, 2016). One of the prevailing theories of how this stabilization is achieved invokes the formation of sugar-based glasses in a process termed vitrification (Crowe, Carpenter, & Crowe, 1998; Hoekstra, Golovina, & Buitink, 2001; Vicré, Farrant, & Driouich, 2004). Vitrification in the context of plant sciences refers to the transition of a sugar-saturated biological solution from a liquid state to an extremely viscous glass-like state that takes on the properties of a solid (Vanhal & Blond, 1999). This implication is fundamentally based on the observation that resurrection plants, like DT seeds, accumulate large quantities of sugars, particularly sucrose and to a lesser extent the raffinose family oligosaccharides (RFOs) raffinose and stachyose (Berjak, Farrant, & Pammenter, 2007; Dinakar & Bartels 2013; Farrant et al., 2017; Leopold & Vertucci, 1986; Whittaker, Bochicchio, Vazzana, Lindsey, & Farrant, 2001; Zhang et al., 2016; see Table 1). However, this theory does not adequately explain the ongoing metabolism and molecular movement apparent in resurrection plants at low water contents.

While the existence of glass-like structures in desiccated seed tissues has been ably demonstrated by the use of techniques such as Differential Scanning Calorimetry, Fourier-Transform Infrared Analysis and Dynamic Mechanical Analysis (e.g., Ballesteros & Walters, 2019; Sakurai et al., 2008; Vertucci, 1990), this has not yet been confirmed for DT vegetative tissues. While it is highly plausible that sugar-based glasses are formed, we propose that additional methods of subcellular and organellar stabilization occur in such tissues. Unlike seeds, leaf tissues do not accumulate the complex storage reserves that contribute toward the nature of mechanical stabilization during desiccation. In seeds, large proportions of the subcellular environment are occupied by starch-containing plastids, protein bodies and/or lipid bodies, considerably reducing the area constituting the "cytoplasm," which is presumably protected by sugar glasses (Bewley, Bradford, Hilhorst, & Nonogaki, 2013; Oliver et al., 2020). In resurrection plant cells, mechanical stabilization is variably achieved among species by a combination of increased vacuolation (the vacuoles being proposed to contain "compatible solutes," the nature of which is likely to vary among species) and by wall-folding, usually in inverse proportions (Farrant et al., 2007; Oliver et al., 2020). Should the formation of sugar glasses be the sole manner of subcellular stabilization during desiccation, a far higher proportion of sucrose and other glass-forming sugars should be present relative to other metabolites. From the few studies that have yielded absolute metabolite concentrations, this does not appear to be the case, with some organic acids and amino acids existing in high ratios relative to sucrose by the late response (Liu et al., 2007; Mladenov et al., 2015). Furthermore, resurrection plants often survive for longer in the desiccated, quiescent state under harsh environmental conditions (elevated temperatures, varying relative humidities, etc.), which facilitate the destabilization of sugar glasses in vitro, than is predicted possible should sugar glasses be the sole mechanism of cytoplasmic stabilization in the desiccated state (Bajic, 2006; Farrant & Kruger, 2001; Vanhal & Blond, 1999; Van Scoik & Carstensen, 1990). This then raises the question: what is the role of the other metabolites seen to accumulate in the late stages of dehydration in providing the stabilization and longevity of quiescence? While longevity could in part be related to the nature and relative contributions of sugars comprising the subcellular glasses formed, which may in turn contribute to their degrees of fragility (Crowe et al., 1998; Wright, Baez, & Vanderkooi, 2002), we propose that additional biochemical players are involved that contribute to subcellular stabilization, and which perhaps, *inter alia,* act in a thermal and mechanical buffering capacity, facilitating longevity under such harsh and variable environmental conditions.

3. A role for NADES?

As outlined elsewhere in this volume, many plant metabolites can form and act as deep eutectic solvents. The question remains: do NADESs form during the later stages of drying in resurrection plants, and if so where and to what purpose? Despite the fact that absolute concentrations of individual metabolites present in the dry state are rarely reported, all comprehensive metabolomics studies reveal the presence and accumulation of NADES-forming compounds (as proposed by Dai, van Spronsen, Witkamp, Verpoorte, & Choi, 2013) in both the late-response and the quiescent state (Table 1). While Dai et al. (2013) have provided a comprehensive list of NADES combinations, it should be noted that, to the best of our knowledge, many of the compounds do not occur in the discrete ratios indicated as necessary for NADES formation in vitro. However, this does not negate the possible existence of NADES as, unlike an in vitro combination of pure compounds, the cytoplasm is a heterogeneous environment comprising various structures, metabolic compounds, organelles, and membraneless bodies, to which the concept of NADES can be applied. Indeed, the universally high presence of sucrose in the late-response indicates that a hydrogen-bond acceptor is available for the myriad of potential hydrogen-bond donors, such as some organic acids and amino acids, which could form complex mixtures and gradients of NADES and NADES-like sugar glasses with chemophysical properties tailored to the requirements of the organelles and local intracellular spaces. To this end, we provide a summary of typical NADES-forming metabolites that have been detected in various resurrection plants using high-resolution mass spectrometry-based metabolomics (Table 1). Interestingly, many of these metabolites accumulate at higher relative amounts between water contents of 30% and the air-dry state in comparison to other water contents, lending support to the hypothesis that they are functioning in facilitating some metabolism at these low water contents and providing an environment for the storage of important metabolites for early rehydration. In particular, sucrose, citric acid, and malic acid are consistently identified as being differentially abundant at RWCs at and below 30% (Table 1). Furthermore, most species accumulate particular amino acids during late dehydration, with proline being the consistent among them. Despite the increasing use of technologies to aid our understanding of metabolism under water-limited conditions, there is as yet a limited understanding of the fine-scale molecular interactions in the subcellular and cellular environment of dehydrating, desiccated, and rehydrating tissues of resurrection plants. Thus, we can only speculate on how NADES may explain some of the observations emanating from research on resurrection plants. These are summarized in Fig. 1.

In the first instance, it is possible that in various organelles custom NADES may form in order to suit the specific needs of the organelle/

compartment in extreme states of dehydration and early rehydration. Indeed it has been proposed by Radermacher et al. (2019) that the accumulation of citrate and sucrose in the mitochondria during the late stages of drying in *X. schlechteri* may result in NADES formation in that organelle, this possibly enabling the ongoing respiration recorded at low water contents in this species (Mundree & Farrant, 2000). Oliver et al. (2020) suggested that the accumulation of citrate observed in numerous resurrection plants at low water contents could be the result of the shutdown of the tricarboxylic acid (TCA) cycle, which could contribute to NADES formation and thus respiratory activity and ATP formation at these water contents.

In another study, Bentley, Olsen, Moore, and Farrant (2020) showed that a NADES comprising sucrose and citric acid enabled superior extractability (relative to other NADES combinations) of anthocyanins (flavonoids) from desiccated tissues (<10% RWC) of the resurrection plant *Myrothamnus flabellifolia* Welw. Anthocyanins are stored in specialized anthocyanic vacuoles in most plants, at concentrations higher than their solubility in water would allow (Markham et al., 2000), and play a role in preventing oxidation. Choi et al. (2011) demonstrated that the flavonoid rutin had a solubility of up to 100 times that in water, in various NADES combinations. Flavonoids and anthocyanins are thought to be synthesized in the cytoplasm, near the exterior surface of the endoplasmic reticulum, before likely being relocated to various storage vacuoles and organelles where they behave in an antioxidant capacity and provide protection (Hrazdina & Jensen, 1992). In this regard, Moore et al. (2005) reported a 50% increase (per gram dry mass of leaf tissue) in the potent antioxidant polyphenol 3,4,5-tri-*O*-galloylquinic acid in *M. flabellifolia*, this being proposed to accumulate predominantly in the vacuoles and acting as a redox buffer. Enhanced extraction of this same compound by a sucrose-citric acid NADES relative to other NADES (Bentley et al., 2020) suggests that the presence of such a NADES in the vacuoles might enable the concentration and functioning of this compound at low water contents. As a further potential role for NADES in polyphenol accumulation and functions, Georgieva, Sárvári, and Keresztes (2010) reported the presence of a dense luminal substance, thought to be phenolic in nature (Choi et al., 2011; van Steveninck & Van Steveninck, 1980), in the thylakoid lumen of desiccated tissues of the homoiochlorophyllous resurrection plant *Haberlea rhodopensis* Friv. Considering that the bulk appearance of various phenolics and anthocyanins occurs in the late-response in many species of resurrection plants (reviewed in Farrant et al., 2007), this strongly suggests that the presence of an adequate solvent for their biosynthesis, transport, and concentration is required. Thus in accordance with Choi et al. (2011), we speculate that an organic acid-sucrose NADES in the vacuoles, cytoplasm, and organelles at low water contents and during early rehydration might facilitate the concentration of, and protective functions of, these important antioxidants (Fig. 1B, C).

As indicated variously above, one of the most remarkable observations is the apparent ongoing metabolism, including enzyme activity, at RWCs below 30%. The reported ability of some NADES to maintain enzymes in solution and even facilitate their functioning in low hydration conditions in vitro (Huang, Wu, Wen, Yang, & Yang, 2014) could explain this phenomenon when applied to intracellular sites of metabolic activity and active organelles. Indeed, the mitochondrial antioxidant enzyme GR has been shown to have an increased efficiency in vitro in NADES (Fan, 2014), indirectly confirming the suggestion that NADES may enable mitochondrial functioning and antioxidant activity at low water contents. Similarly, it is possible that the presence of NADES in chloroplasts could account for the protection and functioning of SOD observed in many resurrection species (reviewed in Farrant et al., 2007; Liu et al., 2019). In a study of the nuclear proteome of *Xerophyta viscosa* Baker, Abdalla and Rafudeen (2012) report the consistent expression of citrate synthase and an increase in malate dehydrogenase during desiccation, which might suggest the existence of NADES in the nucleus of this species, possibly enabling the late transcription of genes observed in this and many other resurrection plants (Costa et al., 2017; Pardo et al., 2020; Rodriguez, 2010; Xiao et al., 2015).

Some NADES combinations have been reported to have better protein solvation properties than water (Dai et al., 2013; Huang et al., 2014). This characteristic could account for the protection of certain enzymes as discussed above but could also account for the protection of other proteins required for the resumption of metabolism upon rehydration. Protein concentrations increase considerably during dehydration, due to both loss of solvent volume and de novo synthesis, and these risk precipitation as water is lost from the cell (Ingle et al., 2007; Jiang et al., 2006; Schneider, Wells, Schmelzer, Salamini, & Bartels, 1993). In particular, resurrection plants are known to accumulate a large variety of late embryogenesis abundant (LEA) proteins during dehydration (reviewed in Artur, Costa, Farrant, & Hilhorst, 2019; Oliver et al., 2020). These proteins are mostly intrinsically disordered in aqueous solutions and thus to date their roles remain largely unknown. However, it has been shown, using circular dichroism spectroscopy, that such proteins do change structure, in many instances to a predominantly alpha helical form, in non-aqueous solutions (Artur et al., 2019). This, in part, confirms the postulate that these proteins adopt configurations suitable to their protective function in the late stages of drying and quiescence (Ingram & Bartels, 1996; Hincha & Thalhammer, 2012; Hundertmark, Popova, Rausch, Seckler, & Hincha, 2012; Popova, Hundertmark, Seckler, & Hincha, 2011). While NADES have not yet been shown to influence LEA protein structure and function, we propose that they might play a role in this regard (illustrated in Fig. 1).

As a further potential role, some NADES have been shown to allow the dissolution of otherwise water-insoluble biological compounds (Choi et al., 2011; Dai et al., 2013). Cao, Nian, Li, Wu, and Liu (2019) demonstrated that a malic acid-glucose NADES allows starch to enter solution, and the slight dilution of the starch-NADES solution results in the reprecipitation/appearance of starch. A feature universal of resurrection plants is the apparent loss of starch during desiccation, followed by its reappearance during early rehydration (Dinakar, Djilianov, & Bartels, 2012; Farrant, 2000; Ingle et al., 2008; Radermacher et al., 2019; Tan et al., 2017). While it has been suggested that starch breakdown contributes toward sucrose formation, it is clear that this alone cannot account for the amounts of sucrose observed in the dry tissues, and thus other means of sucrose production are likely (Dinakar et al., 2012; Farrant et al., 2017). Of particular interest is the observation that starch granules reappear in chloroplasts well before photosynthesis is operating at full capacity, this being particularly evident in poikilochlorophyllous species, which at this stage have not regenerated thylakoid membranes (Ingle et al., 2008; Radermacher et al., 2019). Assuming that a NADES-like environment is created below a certain level of hydration and is then dissipated again above a similar level of rehydration due to dilution, the observed disappearance and reappearance of starch granules could be possible (illustrated in Fig. 1B and C). Whether this plays any important metabolic roles, aids in supplying volume to the chloroplast, or is merely an artefactual property of the cellular environment is unclear.

Since NADES are highly customizable, and different species of resurrection plants accumulate similar compounds in varying proportions, it may be that species-specific NADES combinations are variably sectioned within their respective cells. For instance, the vacuoles are utilized to various extents by different species to facilitate mechanical stabilization (discussed above). As there is no bulk water in desiccated tissues, one can assume that the vacuole is comprised of metabolites important for recovery upon rehydration, but that exist in a NADES-like state. In a targeted study on the vacuolar and cytoplasmic content of the leaves of *Eragrostis nindensis* Ficalho & Hiern during desiccation, it was found that proline content increased and preferentially located to the vacuole (Vander Willigen, Pammenter, Mundree, & Farrant, 2004). Sucrose was also shown to increase in the vacuoles of the dry leaves, although the authors concluded that the latter was more predominant in the cytoplasm. This study was limited, however, in that it did not examine other potential NADES-forming biochemicals, and the authors did conclude that proline and sucrose alone could not account for the vacuolar content. We propose that species-specific accumulation of amino acids may further contribute to NADES formation in this compartment, *inter alia*.

Finally, differences among species in terms of their longevity in the desiccated state could be explained by the types and amounts of NADES formed therein. We base this suggestion on the observation that different NADES and different molecular ratios thereof differ in their chemophysical properties, such as viscosity and response to temperature and humidity fluctuations (Dai, Witkamp, Verpoorte, & Choi, 2015; Savi et al., 2019).

Entertaining the concept of a NADES-like cellular environment during quiescence provides an additional explanation for an ordered return to metabolic competence during rehydration. An often-overlooked phenomenon is the danger associated with uncontrolled rehydration. In most resurrection plants, water uptake upon rehydration occurs predominantly through the root tissues. Initially, this occurs passively due to the high osmotic potential of the dry roots. Once in the root xylem, capillary action enables water movement to the leaf tissues where it moves apoplastically at first, finally entering the cytoplasm. The preferential rehydration of cell walls prior to water movement into cells could cause shearing of the plasmalemma, especially at the plasmodesmatal sites (Pammenter & Berjak, 1999). Yet this is not seen in ultrastructural studies of rehydrating tissue of resurrection plants, despite these using aqueous fixation, suggesting some degree of cytoplasmic plasticity (Farrant, 2000; Farrant et al., 2007; Radermacher et al., 2019; Vecchia, El Asmar, Calamassi, Rascio, & Vazzana, 1998; Vicré et al., 2004). Given a NADES-like cellular environment in the quiescent state, this environment must too be the first to interact with the influx of water during rehydration, acting as a sort of hydration buffer and enforcing some passive control on the way water moves into the cell and is dispersed in the cellular environment. This at least until the biological system restores sufficient metabolic activity to proceed with the rehydration process. In a comprehensive review on the chemophysical changes occurring in solution, Ma, Laaksonen, Liu, Lu, and Ji (2018) provide a model for how constituents of a DES solution interact with molecules of water at increasing dilution. In contrast with a pure ionic salt, such as NaCl, the constituents of a DES are proposed to not immediately dissociate as free ions/solutes upon the introduction of water. Rather, water molecules are suggested to first be absorbed into the DES matrix and form strong molecular interactions with the DES solution and become evenly dispersed through it (Ma et al., 2018; Zhekenov, Toksanbayev, Kazakbayeva, Shah, & Mjalli, 2017). Following further addition of water, the DES constituents begin to disperse until they eventually dissociate as free solutes once the amount of water increases sufficiently. Importantly, in the context of the current work, increasing the hydration of some DES results in changes in solution viscosity, with the DES becoming increasingly less viscous and progressively resembling the properties of a dilute solution (de la Parra et al., 2015, Ma et al., 2018; Illustrated in Fig. 1C).

Superimposing these properties on our NADES-like cellular environment, bearing in mind that it is likely to be a very complex one, it becomes plausible that it may indeed act as a buffer during rehydration, providing a transition of properties as the solvent shifts from concentrated to dilute and allowing the system to respond accordingly. The transition in solvent properties may in fact act as a regulator of biological processes, allowing stored enzymes to become active as their immediate cellular environment becomes favorable, thus allowing for the systematic activation of rehydration processes such as translation, repair, and organelle reassembly. In contrast, the high hygroscopicity of a largely amorphous sucrose glass would not provide the same means of hydration buffering, as it will likely behave much like an ionic salt would, rapidly attracting and dissolving in water and potentially triggering catastrophic nucleation events (Van Scoik & Carstensen, 1990).

4. The way forward

Resurrection plants provide an excellent model for the investigation of NADES in vivo. We have shown the accumulation of NADES-forming metabolites that are also highly likely to be locally concentrated during desiccation. Given their concentration at appropriate ratios in organelles and/or cytoplasmic compartments, NADES formation is a definite possibility.

While the observations referred to in this chapter are largely speculative and do not supply definitive evidence of a NADES-like cellular environment in the late-response and quiescent states, they do provide an attractive and plausible explanation for some of the yet not well-explained phenomena observed in resurrection plants. Given that the study of NADES is still in its infancy and that resurrection plant research has yet to investigate the possible in vivo occurrences of NADES, it is difficult to define a clear path forward. While more in vitro work is needed in order to understand the properties of potential NADES, both simple and complex, more comprehensive metabolomic studies of DT species are also required to identify likely NADES candidates. Furthermore, non-aqueous isolation of organelles and cytoplasmic compartments, and identification of their molecular constituents (as attempted by vander Willigen et al., 2004), will enable the determination of their specific contents, perhaps confirming the postulation of organellar-specific as well as species-specific accumulation of NADES-forming compounds. The biological molecules present in the late-response and quiescent state, such as LEA proteins, can also be screened in vitro for changes in conformation and action in a NADES-based environment.

However, it is in the sentiment of the authors that the largest hurdle that remains is providing empirical evidence for the presence of NADES-like cellular environments. For this, insight into the fine-scale molecular interactions present in the extreme dehydration states is needed, potentially with the creative use of technologies such as Nuclear Magnetic Resonance and Raman Spectroscopy on intact tissue. Nonaqueous fractionation techniques can also be used to isolate different organelles for targeted assessments of their metabolite compositions (such as in Vander Willigen et al., 2004). Additionally, information will need to be gathered on the mechanical and structural properties of the cellular environment at these extreme states, perhaps with aid from Brillouin spectroscopy or dynamic mechanical analysis and molecule tracking with careful manipulation of the system.

Acknowledgments

J.M.F. acknowledges and thanks the South African Department of Science and Innovation and the National Research Foundation for her research Chair (grant number 98406), from which Scholarships were provided for S.F. du Toit and J.B. We thank Keren Cooper, Department of Molecular and Cell Biology, University of Cape Town, for the artwork provided in Fig. 1.

References

Abdalla, K. O., & Rafudeen, M. S. (2012). Analysis of the nuclear proteome of the resurrection plant *Xerophyta viscosa* in response to dehydration stress using iTRAQ with 2DLC and tandem mass spectrometry. *Journal of Proteomics*, *75*, 2361–2374. https://doi.org/10.1016/j.jprot.2012.02.006.

Albini, F. M., Murelli, C., Finzi, P. V., Ferrarotti, M., Cantoni, B., Puliga, S., et al. (1999). Galactinol in the leaves of the resurrection plant *Boea hygroscopica*. *Phytochemistry*, *51*(4), 499–505.

Artur, M. A. S., Costa, M. C. D., Farrant, J. M., & Hilhorst, H. W. (2019). Genome-level responses to the environment: Plant desiccation tolerance. *Emerging Topics in Life Sciences*, *3*(2), 153–163.

Artur, M. A. S., Rienstra, J., Dennis, T. J., Farrant, J. M., Ligterink, W., & Hilhorst, H. (2019). Structural plasticity of intrinsically disordered LEA proteins from *Xerophyta schlechteri* provides protection in vitro and in vivo. *Frontiers in Plant Science*, *10*, 1272.

Bajic, J. (2006). *Exploring the longevity of dry Craterostigma wilmsii (homoiochlorophyllous) and Xerophyta humilis (poikolichlorophyllous) under simulated field conditions (Doctoral dissertation)*. University of Cape Town.

Ballesteros, D., & Walters, C. (2019). Solid-state biology and seed longevity: A mechanical analysis of glasses in pea and soybean embryonic axes. *Frontiers in Plant Science*, *10*, 920.

Barr, H. D., & Weatherley, P. E. (1962). A re-examination of the relative turgidity technique for estimating water deficit in leaves. *Australian Journal of Biological Sciences*, *15*(3), 413–428.

Bentley, J., Olsen, E. K., Moore, J. P., & Farrant, J. M. (2020). The phenolic profile extracted from the desiccation-tolerant medicinal shrub *Myrothamnus flabellifolia* using Natural Deep Eutectic Solvents varies according to the solvation conditions. *Phytochemistry*, *173*, 112323.

Berjak, P., Farrant, J. M., & Pammenter, N. W. (2007). Seed desiccation-tolerance mechanisms. *Plant Desiccation Tolerance*, *1*, 151–192.

Bewley, J. D., Bradford, K. J., Hilhorst, H. W. M., & Nonogaki, H. (2013). *Seeds. Physiology of development, germination and dormancy* (3rd ed.). New York: Springer.

Bianchi, G., Gamba, A., Murelli, C., Salamini, F., & Bartels, D. (1991). Novel carbohydrate metabolism in the resurrection plant *Craterostigma plantagineum*. *The Plant Journal*, *1*(3), 355–359.

Blomstedt, C. K., Griffiths, C. A., Fredericks, D. P., Hamill, J. D., Gaff, D. F., & Neale, A. D. (2010). The resurrection plant *Sporobolus stapfianus*: An unlikely model for engineering enhanced plant biomass? *Plant Growth Regulation*, *62*(3), 217–232.

Cao, C., Nian, B., Li, Y., Wu, S., & Liu, Y. (2019). Multiple hydrogen-bonding interactions enhance the solubility of starch in natural deep eutectic solvents: Molecule and macroscopic scale insights. *Journal of Agricultural and Food Chemistry*, *67*(45), 12366–12373.

Choi, Y. H., van Spronsen, J., Dai, Y., Verberne, M., Hollmann, F., Arends, I. W., et al. (2011). Are natural deep eutectic solvents the missing link in understanding cellular metabolism and physiology? *Plant Physiology*, *156*(4), 1701–1705.

Costa, M. C. D., Cooper, K., Hilhorst, H. W., & Farrant, J. M. (2017). Orthodox seeds and resurrection plants: Two of a kind? *Plant Physiology*, *175*(2), 589–599.

Crowe, J. H., Carpenter, J. F., & Crowe, L. M. (1998). The role of vitrification in anhydrobiosis. *Annual Review of Physiology*, *60*(1), 73–103.

Dace, H. (2014). *Metabolomics of desiccation tolerance in Xerophyta humilis (PhD diss.)*. University of Cape Town.

Dace, H., Sherwin, H. W., Illing, N., & Farrant, J. M. (1998). Use of metabolic inhibitors to elucidate mechanisms of recovery from desiccation stress in the resurrection plant *Xerophyta humilis*. *Plant Growth Regulation*, *24*(3), 171–177.

Dai, Y., van Spronsen, J., Witkamp, G. J., Verpoorte, R., & Choi, Y. H. (2013). Natural deep eutectic solvents as new potential media for green technology. *Analytica chimica acta*, *766*, 61–68.

Dai, Y., Witkamp, G. J., Verpoorte, R., & Choi, Y. H. (2015). Tailoring properties of natural deep eutectic solvents with water to facilitate their applications. *Food Chemistry*, *187*, 14–19.

de la Parra, C. J., Zambrano, J. R., Bermejo, M. D., Martín, Á., Segovia, J. J., & Cocero, M. J. (2015). Influence of water concentration in the viscosities and densities of cellulose dissolving ionic liquids. Correlation of viscosity data. *The Journal of Chemical Thermodynamics*, *91*, 8–16.

Dinakar, C., & Bartels, D. (2013). Desiccation tolerance in resurrection plants: New insights from transcriptome, proteome, and metabolome analysis. *Frontiers in Plant Science*, *4*, 1–14.

Dinakar, C., Djilianov, D., & Bartels, D. (2012). Photosynthesis in desiccation tolerant plants: Energy metabolism and antioxidative stress defense. *Plant Science*, *182*, 29–41.

Djilianov, D. L., Dobrev, P. I., Moyankova, D. P., Vankova, R., Georgieva, D. T., Gajdošová, S., et al. (2013). Dynamics of endogenous phytohormones during desiccation and recovery of the resurrection plant species *Haberlea rhodopensis*. *Journal of Plant Growth Regulation*, *32*(3), 564–574.

Egert, A., Eicher, B., Keller, F., & Peters, S. (2015). Evidence for water deficit-induced mass increases of raffinose family oligosaccharides (RFOs) in the leaves of three Craterostigma resurrection plant species. *Frontiers in Physiology*, *6*, 206.

Fan, C. (2014). *In vitro investigation into the possible role of natural deep eutectic solvents (NADES) in resurrection plants*. BSc Honours research thesis South Africa: University of Cape Town.

Farooq, M., Wahid, A., Lee, D. J., Ito, O., & Siddique, K. H. (2009). Advances in drought resistance of rice. *Critical Reviews in Plant Sciences*, *28*(4), 199–217.

Farrant, J. M. (2000). Comparison of mechanisms of desiccation tolerance among three angiosperm resurrection plants. *Plant Ecology*, *151*, 29–39.

Farrant, J. M., Brandt, W., & Lindsey, G. G. (2007). An overview of mechanisms of desiccation tolerance in selected angiosperm resurrection plants. *Plant Stress Journal*, *1*, 72–84.

Farrant, J. M., Cooper, K., Dace, H. J. W. S., Bentley, J., & Hilgart, A. (2017). Desiccation tolerance. Chapter 9 In S. Shabala (Ed.), *Plant stress physiology* (2nd ed., pp. 217–252). Wallingford, UK: CAB International. ISBN-13:978 1 78064 729 6.

Farrant, J. M., & Kruger, L. A. (2001). Longevity of dry *Myrothamnus flabellifolius* in simulated field conditions. *Plant Growth Regulation*, *35*(2), 109–120.

Gaff, D. F. (1989). Responses of desiccation tolerant 'resurrection' plants to water stress. In *Structural and functional responses to environmental stresses* (pp. 255–268).

Gaff, D. F., & Oliver, M. (2013). The evolution of desiccation tolerance in angiosperm plants: A rare yet common phenomenon. *Functional Plant Biology*, *40*(4), 315–328.

Gechev, T. S., Benina, M., Obata, T., Tohge, T., Sujeeth, N., Minkov, I., et al. (2013). Molecular mechanisms of desiccation tolerance in the resurrection glacial relic *Haberlea rhodopensis*. *Cellular and Molecular Life Sciences*, *70*(4), 689–709.

Georgieva, K., Sárvári, É., & Keresztes, A. (2010). Protection of thylakoids against combined light and drought by a lumenal substance in the resurrection plant *Haberlea rhodopensis*. *Annals of Botany*, *105*(1), 117–126.

Ghasempour, H. R., Gaff, D. F., Williams, R. P. W., & Gianello, R. D. (1998). Contents of sugars in leaves of drying desiccation tolerant flowering plants, particularly grasses. *Plant Growth Regulation*, *24*(3), 185–191.

Hincha, D. K., & Thalhammer, A. (2012). *LEA proteins: IDPs with versatile functions in cellular dehydration tolerance*.

Hoekstra, F. A., Golovina, E. A., & Buitink, J. (2001). Mechanisms of plant desiccation tolerance. *Trends in Plant Science*, *6*(9), 431–438.

Hrazdina, G., & Jensen, R. A. (1992). Spatial organization of enzymes in plant metabolic pathways. *Annual Review of Plant Biology*, *43*(1), 241–267.

Huang, Z. L., Wu, B. P., Wen, Q., Yang, T. X., & Yang, Z. (2014). Deep eutectic solvents can be viable enzyme activators and stabilizers. *Journal of Chemical Technology & Biotechnology*, *89*(12), 1975–1981.

Hundertmark, M., Popova, A. V., Rausch, S., Seckler, R., & Hincha, D. K. (2012). Influence of drying on the secondary structure of intrinsically disordered and globular proteins. *Biochemical and Biophysical Research Communications*, *417*(1), 122–128.

Iljin, W. S. (1957). Drought resistance in plants and physiological processes. *Annual Review of Plant Physiology*, *8*(1), 257–274.

Ingle, R. A., Collett, H., Cooper, K., Takahashi, Y., Farrant, J. M., & Illing, N. (2008). Chloroplast biogenesis during rehydration of the resurrection plant *Xerophyta humilis*: Parallels to the etioplast–chloroplast transition. *Plant, Cell & Environment*, *31*(12), 1813–1824.

Ingle, R. A., Schmidt, U. G., Farrant, J. M., Thomson, J. A., & Mundree, S. G. (2007). Proteomic analysis of leaf proteins during dehydration of the resurrection plant *Xerophyta viscosa*. *Plant, Cell & Environment*, *30*(4), 435–446.

Ingram, J., & Bartels, D. (1996). The molecular basis of dehydration tolerance in plants. *Annual Review of Plant Biology*, *47*(1), 377–403.

Jiang, G., Wang, Z., Shang, H., Yang, W., Hu, Z., Phillips, J., et al. (2006). Proteome analysis of leaves from the resurrection plant *Boea Hygrometrica* in response to dehydration and rehydration. *Planta*, *225*, 1405–1420. https://doi.org/10.1007/s00425-006-0449-z.

Leopold, A. C., & Vertucci, C. W. (1986). Physical attributes of desiccated seeds. In *Membranes, metabolism and dry organisms* (pp. 22–34). Ithaca, NY: Cornell University Press.

Levitt, J. (1980). Responses of plants to environmental stresses. In *Volume II. Water, radiation, salt, and other stresses* (2nd ed.). Academic Press.

Liu, J., Moyankova, D., Djilianov, D., & Deng, X. (2019). Common and specific mechanisms of desiccation tolerance in two Gesneriaceae resurrection plants. Multiomics evidences. *Frontiers in Plant Science*, *10*, 1067.

Liu, J., Moyankova, D., Lin, C. T., Mladenov, P., Sun, R. Z., Djilianov, D., et al. (2018). Transcriptome reprogramming during severe dehydration contributes to physiological and metabolic changes in the resurrection plant *Haberlea rhodopensis*. *BMC Plant Biology*, *18*(1), 1–16.

Liu, Y. B., Wang, G., Liu, J. Y., Zhao, X., Tan, H. J., & Li, X. R. (2007). Anatomical, morphological and metabolic acclimation in the resurrection plant *Reaumuria soongorica* during dehydration and rehydration. *Journal of Arid Environments*, *70*(2), 183–194.

Kuang, J., Gaff, D. F., Gianello, R. D., Blomstedt, C. K., Neale, A. D., & Hamill, J. D. (1995). Changes in in vivo protein complements in drying leaves of the desiccation-tolerant grass *Sporobolus stapfianus* and the desiccation-sensitive grass *Sporobolus pyramidalis*. *Functional Plant Biology*, *22*(6), 1027–1034.

Ma, C., Laaksonen, A., Liu, C., Lu, X., & Ji, X. (2018). The peculiar effect of water on ionic liquids and deep eutectic solvents. *Chemical Society Reviews*, *47*(23), 8685–8720.

Markham, K. R., Gould, K. S., Winefield, C. S., Mitchell, K. A., Bloor, S. J., & Boase, M. R. (2000). Anthocyanic vacuolar inclusions—Their nature and significance in flower colouration. *Phytochemistry*, *55*(4), 327–336.

Mladenov, P., Finazzi, G., Bligny, R., Moyankova, D., Zasheva, D., Boisson, A. M., et al. (2015). In vivo spectroscopy and NMR metabolite fingerprinting approaches to connect the dynamics of photosynthetic and metabolic phenotypes in resurrection plant *Haberlea rhodopensis* during desiccation and recovery. *Frontiers in Plant Science*, *6*, 564.

Moore, J. P., Hearshaw, M., Ravenscroft, N., Lindsey, G. G., Farrant, J. M., & Brandt, W. F. (2007). Desiccation-induced ultrastructural and biochemical changes in the leaves of the resurrection plant *Myrothamnus flabellifolia*. *Australian Journal of Botany*, *55*(4), 482–491.

Moore, J. P., Westall, K. L., Ravenscroft, N., Farrant, J. M., Lindsey, G. G., & Brandt, W. F. (2005). The predominant polyphenol in the leaves of the resurrection plant *Myrothamnus flabellifolius*, 3, 4, 5 tri-O-galloylquinic acid, protects membranes against desiccation and free radical-induced oxidation. *Biochemical Journal*, *385*(1), 301–308.

Moyankova, D., Mladenov, P., Berkov, S., Peshev, D., Georgieva, D., & Djilianov, D. (2014). Metabolic profiling of the resurrection plant *Haberlea rhodopensis* during desiccation and recovery. *Physiologia Plantarum*, *152*(4), 675–687.

Müller, J., Sprenger, N., Bortlik, K., Boller, T., & Wiemken, A. (1997). Desiccation increases sucrose levels in Ramonda and Haberlea, two genera of resurrection plants in the Gesneriaceae. *Physiologia Plantarum*, *100*(1), 153–158.

Mundree, S. G., & Farrant, J. M. (2000). Some physiological and molecular insights into the mechanisms of desiccation tolerance in the resurrection plant *Xerophyta viscosa* Baker. In *Plant tolerance to abiotic stresses in agriculture: Role of genetic engineering* (pp. 201–222). Dordrecht: Springer.

Oliver, M. J., Guo, L., Alexander, D. C., Ryals, J. A., Wone, B. W., & Cushman, J. C. (2011). A sister group contrast using untargeted global metabolomic analysis delineates the biochemical regulation underlying desiccation tolerance in Sporobolus stapfianus. *The Plant Cell*, *23*(4), 1231–1248.

Oliver, M. J., Jain, R., Balbuena, T. S., Agrawal, G., Gasulla, F., & Thelen, J. J. (2011). Proteome analysis of leaves of the desiccation-tolerant grass, *Sporobolus stapfianus*, in response to dehydration. *Phytochemistry*, *72*(10), 1273–1284.

Oliver, M. J., Farrant, J. M., Hilhorst, H. W., Mundree, S., Williams, B., & Bewley, J. D. (2020). Desiccation tolerance: Avoiding cellular damage during drying and rehydration. *Annual Review of Plant Biology*, *71*.

Oliver, M. J., Wood, A. J., & O'Mahony, P. (1998). To dryness and beyond-preparation for the dried state and rehydration in vegetative desiccation-tolerant plants. *Plant Growth Regulation*, *24*, 193–201.

Pammenter, N. W., & Berjak, P. (1999). A review of recalcitrant seed physiology in relation to desiccation-tolerance mechanisms. *Seed Science Research*, *9*(1), 13–37.

Pardo, J., Wai, C. M., Chay, H., Madden, C. F., Hilhorst, H. W., Farrant, J. M., et al. (2020). Intertwined signatures of desiccation and drought tolerance in grasses. *Proceedings of the National Academy of Sciences*, *117*(18), 10079–10088.

Peters, S., Mundree, S. G., Thomson, J. A., Farrant, J. M., & Keller, F. (2007). Protection mechanisms in the resurrection plant *Xerophyta viscosa* (Baker): Both sucrose and raffinose family oligosaccharides (RFOs) accumulate in leaves in response to water deficit. *Journal of Experimental Botany*, *58*(8), 1947–1956.

Popova, A. V., Hundertmark, M., Seckler, R., & Hincha, D. K. (2011). Structural transitions in the intrinsically disordered plant dehydration stress protein LEA7 upon drying are modulated by the presence of membranes. *Biochimica et Biophysica Acta (BBA)-Biomembranes*, *1808*(7), 1879–1887.

Radermacher, A. L., du Toit, S. F., & Farrant, J. M. (2019). Desiccation-driven senescence in the resurrection plant Xerophyta schlechteri (Baker) NL Menezes: Comparison of anatomical, ultrastructural and metabolic responses between senescent and non-senescent tissues. *Frontiers in Plant Science*, *10*, 1396.

Ricco, C. R., Hilhorst, H., Farrant, J. M., & Dace, H. (2018). *Metabolite analysis of resurrection plants (Master thesis in Plant Sciences)*. Wageningen University.

Rodriguez, M. C. S. (2010). Transcriptomes of the desiccation-tolerant resurrection plant *Craterostigma plantagineum*. *The Plant Journal*, *63*, 212–228.

Rout, N. P., & Shaw, B. P. (2001). Salt tolerance in aquatic macrophytes: Possible involvement of the antioxidative enzymes. *Plant Science*, *160*(3), 415–423.

Sakurai, M., Furuki, T., Akao, K. I., Tanaka, D., Nakahara, Y., Kikawada, T., et al. (2008). Vitrification is essential for anhydrobiosis in an African chironomid, *Polypedilum vanderplanki*. *Proceedings of the National Academy of Sciences*, *105*(13), 5093–5098.
Savi, L. K., Dias, M. C. G. C., Carpine, D., Waszczynskyj, N., Ribani, R. H., & Haminiuk, C. W. I. (2019). Natural deep eutectic solvents (NADES) based on citric acid and sucrose as a potential green technology: A comprehensive study of water inclusion and its effect on thermal, physical and rheological properties. *International Journal of Food Science & Technology*, *54*(3), 898–907.
Schneider, K., Wells, B., Schmelzer, E., Salamini, F., & Bartels, D. (1993). Desiccation leads to the rapid accumulation of both cytosolic and chloroplastic proteins in the resurrection plant *Craterostigma plantagineum* Hochst. *Planta*, *189*(1), 120–131.
Stark, L. R. (2017). Ecology of desiccation tolerance in bryophytes: A conceptual framework and methodology. *Bryologist*, *120*, 130–166.
Suguiyama, V. F., da Silva, E. A., Meirelles, S. T., Centeno, D. D. C., & Braga, M. R. (2014). Leaf metabolite profile of the Brazilian resurrection plant *Barbacenia purpurea* Hook. (Velloziaceae) shows two time-dependent responses during desiccation and recovering. *Frontiers in Plant Science*, *5*, 96.
Sun, W. Q. (2002). Methods for the study of water relations under desiccation stress. In *Desiccation and survival in plants: Drying without dying* (pp. 47–91).
Tan, T., Sun, Y., Luo, S., Zhang, C., Zhou, H., & Lin, H. (2017). Efficient modulation of photosynthetic apparatus confers desiccation tolerance in the resurrection plant *Boea hygrometrica*. *Plant and Cell Physiology*, *58*(11), 1976–1990.
Tuba, Z., Lichtenthaler, H. K., Csintalan, Z., Nagy, Z., & Szente, K. (1996). Loss of chlorophylls, cessation of photosynthetic CO_2 assimilation and respiration in the poikilochlorophyllous plant *Xerophyta scabrida* during desiccation. *Physiologia Plantarum*, *96*(3), 383–388.
Tymms, M. J., & Gaff, D. F. (1979). Proline accumulation during water stress in resurrection plants. *Journal of Experimental Botany*, 165–168.
Van Buren, R., Wai, C. M., Pardo, J., Giarola, V., Ambrosini, S., Song, X., et al. (2018). Desiccation tolerance evolved through gene duplication and network rewiring in Lindernia. *The Plant Cell*, *30*(12), 2943–2958.
Van Scoik, K. G., & Carstensen, J. T. (1990). Nucleation phenomena in amorphous sucrose systems. *International Journal of Pharmaceutics*, *58*(3), 185–196.
van Steveninck, M. E., & Van Steveninck, R. F. M. (1980). Plastids with densely staining thylakoid contents in Nymphoides indica. *Protoplasma*, *103*(4), 343–360.
Vander Willigen, C., Pammenter, N. W., Mundree, S. G., & Farrant, J. M. (2004). Mechanical stabilization of desiccated vegetative tissues of the resurrection grass *Eragrostis nindensis*: Does a TIP 3; 1 and/or compartmentalization of subcellular components and metabolites play a role? *Journal of Experimental Botany*, *55*(397), 651–661.
Vanhal, I., & Blond, G. (1999). Impact of melting conditions of sucrose on its glass transition temperature. *Journal of Agricultural and Food Chemistry*, *47*(10), 4285–4290.
Vecchia, F. D., El Asmar, T., Calamassi, R., Rascio, N., & Vazzana, C. (1998). Morphological and ultrastructural aspects of dehydration and rehydration in leaves of *Sporobolus stapfianus*. *Plant Growth Regulation*, *24*(3), 219–228.
Vertucci, C. W. (1990). Calorimetric studies of the state of water in seed tissues. *Biophysical Journal*, *58*(6), 1463.
Vicré, M., Farrant, J. M., & Driouich, A. (2004). Insights into the cellular mechanisms of desiccation tolerance among angiosperm resurrection plant species. *Plant, Cell & Environment*, *27*(11), 1329–1340.

Walters, C., & Koster, K. L. (2007). *Structural dynamics and desiccation damage in plant reproductive organs* (pp. 251–280). Iowa, USA: Blackwell Publishing.

Whittaker, A., Bochicchio, A., Vazzana, C., Lindsey, G., & Farrant, J. (2001). Changes in leaf hexokinase activity and metabolite levels in response to drying in the desiccation-tolerant species *Sporobolus stapfianus* and *Xerophyta viscosa*. *Journal of Experimental Botany*, *52*(358), 961–969.

Wright, W. W., Baez, J. C., & Vanderkooi, J. M. (2002). Mixed trehalose/sucrose glasses used for protein incorporation as studied by infrared and optical spectroscopy. *Analytical Biochemistry*, *307*(1), 167–172.

Xiao, L., et al. (2015). The resurrection genome of *Boea hygrometrica*: A blueprint for survival of dehydration. *Proceedings of the National Academy of Sciences of the United States of America*, *112*, 5833–5837.

Yobi, A., Schlauch, K. A., Tillett, R. L., Yim, W. C., Espinoza, C., Wone, B. W., et al. (2017). *Sporobolus stapfianus*: Insights into desiccation tolerance in the resurrection grasses from linking transcriptomics to metabolomics. *BMC Plant Biology*, *17*(1), 67.

Zhang, Q., & Bartels, D. (2018). Molecular responses to dehydration and desiccation in desiccation-tolerant angiosperm plants. *Journal of Experimental Botany*, *69*(13), 3211–3222.

Zhang, Q., Song, X., & Bartels, D. (2016). Enzymes and metabolites in carbohydrate metabolism of desiccation tolerant plants. *Proteomes*, *4*(4), 40.

Zhekenov, T., Toksanbayev, N., Kazakbayeva, Z., Shah, D., & Mjalli, F. S. (2017). Formation of type III Deep Eutectic Solvents and effect of water on their intermolecular interactions. *Fluid Phase Equilibria*, *441*, 43–48.

Živković, T., Quartacci, M. F., Stevanović, B., Marinone, F., & Navari-Izzo, F. (2005). Low-molecular weight substances in the poikilohydric plant *Ramonda serbica* during dehydration and rehydration. *Plant Science*, *168*(1), 105–111.

CHAPTER TEN

Natural deep eutectic solvents present in plant exudates? A case study on the saps of *Drosera* species

Henni Vanda[a,b], Natali Rianika Mustafa[a], Robert Verpoorte[a], Peter G.L. Klinkhamer[c], and Young Hae Choi[a,]*

[a]Natural Products Laboratory, Institute of Biology, Leiden University, Leiden, The Netherlands
[b]Faculty of Veterinary Medicine, Universitas Syiah Kuala, Banda Aceh, Indonesia
[c]Plant Ecology and Phytochemistry, Institute of Biology, Leiden University, Leiden, The Netherlands
*Corresponding author: e-mail address: y.h.choi@biology.leidenuniv.nl

Contents

Abstract

Many plants produce exudates as defense mechanism to hamper pests and pathogens. Plant exudates include latexes, resins, gums, root exudates, and also biofluids produced by carnivorous plants. Plants exudates consist of, among others, polysaccharides, lipids, proteins, organic acids and bases, sugars, and amino acids. All these small molecules are NADES components. In this chapter, the chemical compositions of plant exudates, especially leaf exudates, and *Drosera* plant biofluids will be elaborated. Metabolic profiling approaches based on ^{1}H NMR and GC–MS were performed on the biofluids of the *Drosera* species. The ^{1}H NMR spectra showed a high level of *myo*-inositol and sugars in the biofluids as well as some organic acids. To further confirm the identity of the sugars and to identify some minor components, GC–MS was employed. This gave similar results as the ^{1}H NMR spectroscopy. All the tested exudates were found to contain sugars, such as fructose, glucose, arabinose, sucrose, and xylose; organic acids, such as glucuronic acid and ascorbic acid; and *myo*-inositol. Those *Drosera* ingredients are

Advances in Botanical Research, Volume 97
ISSN 0065-2296
https://doi.org/10.1016/bs.abr.2020.09.014

known to be strong hydrogen-bond donors. Based on the information obtained from the ^{1}H NMR and GC–MS results, new NADESs compositions were proposed and they were tested in vitro, of which indeed five combinations formed liquids. However, considering that these *Drosera* biofluids consist mainly of water and other materials, it may indicate that the stable high viscosity droplets typical for the *Drosera* species must be due in part to other physical-chemical forces, such as gel formation by polysaccharides with calcium ions. A possible role of the NADES-like liquids could be in stabilizing proteins, or the formation of biofilms enabling a faster digesting of the prey. The *myo*-inositol that was clearly a major compound in the exudate might be involved in protecting the leaves and the sugar containing exudate from herbivory. The occurrence of some common products found in fermentation processes might indicate that microorganisms are present in the biofluid.

1. Overview

Exudates are produced by various plant species, and have defensive roles against herbivores and environmental stress. These fluids are produced by special cells and released to the surface through interconnected tubes or canals. Plant exudates are, among others in the form of latexes, resins, gums or root exudates, which are locally oozed out where the plant is damaged (Agrawal & Konno, 2009; Lambert, Wu, & Santiago-Blay, 2005; Moutim, Silva, Lopes, Fernandes, & Salas, 1999).

As an example of plant exudates, latex is a creamy milky fluid which is synthesized and stored in laticifer cells and distributed in roots, stems, petioles, and leaves of plants. Plant species that produce latex mainly belong to family *Papaveraceae* (e.g., *Papaver somniferum* L.), *Euphorbiaceae* (e.g., *Jathropa curcas* L., *Hevea brasiliensis* (Willd. ex A.Juss.) Müll.Arg., *Apocynaceae* (e.g., *Asclepias syriaca* L.), and *Moraceae* (e.g., *Ficus carica* L.). Another type of exudate are plant resins which are translucent solutions, insoluble in water, but soluble in various organic solvents. Gums, on the other hand, are water based and contain high molecular weight polysaccharides, some of which are used widely in cosmetics, pharmaceuticals, and food industry (Lambert et al., 2005).

There are also root exudates which consist of organic substances released into the rhizosphere. These exudates serve as an energy source for the microbiome, and to regulate the growth of various microorganisms, to benefit the beneficial microorganisms and reduce levels of potential pathogens (Badri & Vivanco, 2009; Jaeger, Lindow, Miller, Clark, & Firestone, 1999).

Another kind of plant exudate consists of the mucilages or biofluids produced by carnivorous plants. This type of biofluids are released to the surface of the plant as a trap to catch small animals as their prey. *Drosera* species are probably the best-known carnivorous plants due to the sticky exudate produced by the glands on the tentacles as trapping mechanism. The glue droplets look like dew on the surface of the tentacles, from which these plants got their name sundew. Once the prey is captured, the leaves will roll-up and perform an outer stomach to cover the prey for enzymatic processing (Barthlott, Porembski, Seine, & Theisen, 2007).

The enzymatic processing in *Drosera* species starts when the prey is captured, the sessile glands produce digestive enzymes that will smother the prey, digest it, and provide small nutrient molecules for assimilation in the plant (Barthlott et al., 2007; Matušíková et al., 2005; Takeuchi et al., 2011). The enzymatic processes will take approximately 2 h up to 2 days (Barthlott et al., 2007; Pate & Dixon, 1978; Schaefer & Ruxton, 2008).

The fascinating fact about these biofluid is that they remain liquid and do not evaporate easily. It is hypothesized that NADESs may play a role as an alternative liquid to water or lipids. Previous experiments conducted by Choi et al. (2011) showed that many combinations of ubiquitous natural compounds present at high concentrations in all living cells can form a liquid. In this chapter we probe our hypothesis that plant exudates such as latexes, gums, and *Drosera* mucilages are natural deep eutectic solvents (NADESs) which have the necessary physiological properties such as to protect the plants against various forms of stress (Choi et al., 2011), as well as to dissolve and stabilize enzymes for essential digestive enzymatic processes as in case of *Drosera*.

2. Characteristics of plant exudates

Plant exudates have specific physical and biological characteristics depending on the species. The roles of plant exudates are diverse, and include defense against pests and diseases and allelopathic activity.

Latex is a milky emulsion composed of small organic compounds suspended in a liquid dispersion medium. Not all latexes are whitish, there are some which are transparent or have a distinct color, such as yellow, orange, brown, or red (Abarca, Klinkhamer, & Choi, 2019; Konno, 2011). Most latexes are sticky and viscous after excretion, and even coagulate and form clots. However, latexes excreted from leaves of the mulberry tree (*Morus* spp.) and the oleander tree (*Nerium oleander* L.) are not viscous (Konno, 2011). The chemical properties of latex are related to their functions. Latexes act as the first line of defense against herbivores and pathogens, and it also plays a role in environmental stress conditions (Abarca et al., 2019; Agrawal & Konno, 2009; Konno, 2011). The endophytes living inside the host plant are also believed to contribute via the latex to the production and biotransformation of diverse metabolites related to the plant defense against pests and diseases (Abarca et al., 2019; Kusari, Zühlke, & Spiteller, 2009; Yang, Rogers, Song, Guo, & Kolattukudy, 2005).

Resins are exudates produced in specialized surface glands such as glandular hairs or internal ducts. They serve the protection against insects and pathogens. The characteristics of resins are: stable, inert, amorphous, nonvolatile, insoluble in water, but soluble in organic solvents.

Plant root exudates are produced as attractants and repellants in the rhizosphere. The compounds also may regulate the soil microbial community, control the herbivores, facilitate beneficial symbioses, change the chemical and physical properties of the soil, and inhibit the growth of competing plant species (Bais, Weir, Perry, Gilroy, & Vivanco, 2006; Estabrook & Yoder, 1998; Nardi et al., 2000).

Plant gums are present in large quantities in a variety of plants, with different structural and metabolic functions. Gums are considered to be pathological products as a result of plant injury or due to drought conditions. Some highly viscous gums, the mucilages, on the other hand, are water-insoluble, and produced as normal metabolites in epidermal cells of leaves, seed coats, and barks (Bhosale, Osmani, & Moin, 2014). Both gums and mucilages are translucent amorphous substances, consisting of polymers of a monosaccharide or mixed monosaccharides combined with uronic acids, they may also contain hydrophilic molecules, and when combined with water will form gels (Bhosale et al., 2014; Malsawmtluangi et al., 2014). The gums and mucilages also play a role in the response to the various forms of stress for the plant or plant cells.

Drosera exudates fit best the class of gums. The exudate is transparent, just like drops of glue on the surface of tentacles, and act as trap to catch insects. The trapping mechanism involves two types of glands: stalked glands and sessile glands. Stalked glands specifically secrete a sweet mucilage to attract, capture, and digest the prey, while sessile glands particularly absorb the nutrients. There are different opinions about how the insects are attracted to the plants. Some researchers believe that it is due to the red color of the tentacles that contrast with the background (Schaefer & Ruxton, 2008), while other researchers believe that the insects are interested in nectar as a source of nutrients, which supports the idea that sugars play an important role in the trapping mechanism (Bennett & Ellison, 2009).

Sugars found in the sticky fluid have a particular ratio that may form a NADES that act as a solvent of biomaterials (Choi et al., 2011; Gowda, Reuter, & Schauer, 1982).

3. Chemical compositions of plant exudates

Plant exudates consist of diverse chemicals, including primary metabolites and secondary metabolites. Compared to other plant exudates, latexes have more complex chemical compositions with high concentrations of secondary metabolites, the concentration is even higher than in leaves. Many of these compounds provide resistance to insects and other herbivores by their toxicity, antinutritive effects, or sticky character (Agrawal & Konno, 2009).

Latexes consist of alkaloids (Itenov, Mølgaard, & Nyman, 1999; Konno et al., 2006), terpenoids (Rees & Harborne, 1985; Sessa, Bennett, Lewis, Mansfield, & Beale, 2000), rubber (Bushman et al., 2006), cardenolides (Dussourd & Hoyle, 2000; Rasmann, Johnson, & Agrawal, 2009), and also enzymes such as proteases (John, Bhat, & Rao, 2003; Rasmann et al., 2009), chitinases (Ramos et al., 2010), and oxidases (Sethi, McAuslane, Rathinasabapathi, Nuessly, & Nagata, 2009).

Terpenoids are found abundantly in latex, for example, in Euphorbiaceae, phorbol, diterpenes and their derivatives are the major components, which are toxic for insects (Konno, 2011). In the Asteraceae family, several sesquiterpene lactones, including lactucin, lactucopicrin, and 8-deoxylactucin were found (Rees & Harborne, 1985; Sessa et al., 2000). Cardenolides such as G-strophanthin, voruscharin, ushcharidin, and alotropagenin were found in the Apocynaceae family (Dussourd & Hoyle, 2000; Rasmann et al., 2009; Samuelsson & Bohlin, 2009), while toxicariosides was reported in Moraceae family (Carter et al., 1997).

Alkaloids derived from latex are morphine (Hartmann, Schmid, Van Tuinen, & Berg, 2009; Itenov et al., 1999) and sanguinarine (Tomè & Colombo, 1995) in Papaveraceae, lobeline in Campanulaceae (Oppel, Dussourd, & Garimella, 2009), and sugar-mimicking alkaloids in Moraceae (Konno et al., 2006). Most alkaloids are poorly soluble in water, yet they present in numerous plant latex.

Besides secondary metabolites, latexes also contain proteins, such as various types of proteases, protease inhibitors, oxidases, and chitinases. The defensive role of these enzymes on herbivores are unclear, however, it is assumed that they exhibit a toxic effect or hamper the digestion of the plan material by targeting to essential proteins in the insects (Zhu-Salzman, Bi, & Liu, 2005). In some latexes, these enzymes are bound to carbohydrates, as reported by Odani, Yokokawa, Takeda, Abe, and Odani (1996). The protease inhibitor in latex of *Carica papaya* L. was bound to two carbohydrate chains, composed of mannose, xylose, fucose, and *N*-acetylglucosamine residues. A study reported by Hullar and Smith (1966) stated that rubber latex contains *myo*- and L-inositol, quebrachitol (1-0-methyl-L-inositol), galactinol, and raffinose. Latex of *Jatropha neopauciflora* Pax was reported to contain 8.52 mg/mL carbohydrates, and 79% of this was fructose (Hernandez-Hernandez et al., 2017). Latex of *Jatropha multifida* L. was reported to contain multifidol attached to a glucose moiety (Kosasi, Van Der Sluis, & Labadie, 1989). Barbieri et al. (1983) also reported glucose, galactose, xylose, and fucose in latex of *Hura crepitans* L. and *Euphorbia characias* L. All of these sugars are potential candidates for NADESs and might play a role in enzyme preservation, and drought and cold resistance.

Plant resins are composed of mono-, sesqui-, di- and triterpenes (Leonhardt, Schmitt, & Blüthgen, 2011). Plant gums are usually heteropolysaccharides, consisting of arabinose, galactose, glucose, xylose, mannose, and organic acids (Amid, Mirhosseini, & Kostadinović, 2012; Amin, Ahmad, Yin, Yahya, & Ibrahim, 2007; Mirhosseini & Amid, 2012). Root exudates comprise amino acids, organic acids, phenolics, secondary metabolites, polysaccharides, and also proteins. Though such general qualitative information is available on chemical composition of root exudates from diverse plant species, real in depth studies are lacking for most exudates (Walker, Bais, Grotewold, & Vivanco, 2003).

4. *Drosera* plant exudate

To learn more about the constituents of *Drosera* trapping liquids they were collected from seven species of *Drosera* from the Hortus Botanicus in Leiden. The exudates of *Drosera adelae* F. Muell., *D. regia* Stephens, *D. capensis* L. variety Alba, *D. capencis* L. variety Rubra, *D. capensis* L. variety Giant, *D. binata* Labill., and *D. slackii* Cheek were collected by microtubes which were prefilled with two solvents, MeOH-aqueous KH_2PO_4 buffer pH 6 (4:1) (v/v) or KH_2PO_4 buffer pH 6. All the samples were dried and further

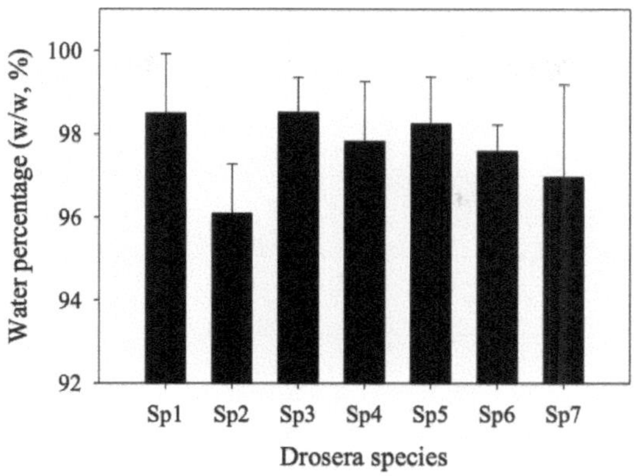

Fig. 1 Water percentage (w/w) of seven *Drosera* exudates in triplicate collected in June 2013. Sp1 *Drosera adelae*, Sp2 *Drosera regia*, Sp3 *Drosera capensis* Alba, Sp4 *Drosera capensis* Giant, Sp5 *Drosera capensis* Rubra, Sp6 *Drosera binata*, Sp7 *Drosera slackii.*

analyzed by ^1H NMR spectroscopy (Kim, Choi, & Verpoorte, 2010) and GC–MS to acquire comprehensive qualitative and quantitative data about the exudate's composition.

A water content measurement showed that the average percentage ranged from 96% to 98.5% (w/w) (Fig. 1), which makes the exudate suitable for enzymatic processes.

4.1 ^1H NMR analysis of *Drosera* exudate

The analysis of the exudates by means of ^1H NMR spectroscopy showed that all the *Drosera* species had *myo*-inositol as main ingredient. Six species contained sucrose, and five contained glucose (Figs. 2–3). The exudate of *D. capensis* Rubra was found to have a high level of oligosaccharides. Ethanol was also detected as a major ingredient in all exudates. Acetic-, acetoacetic-, lactic, and formic acid, were detected in all the species in varying concentrations. These compounds are typical products of various fermentation processes and may point to the presence of microorganisms in the biofluid.

Organic acids found in the exudate were citric acid, fumaric acid, and malic acid. Malic acid in *D. capensis* and *D. slackii* was higher than in other species, at a molar concentration similar to sucrose (ca. 1:1). The chemical compositions of the exudates are listed in Table 1, and the average concentration of *Drosera* biofluids compounds is presented in Fig. 4.

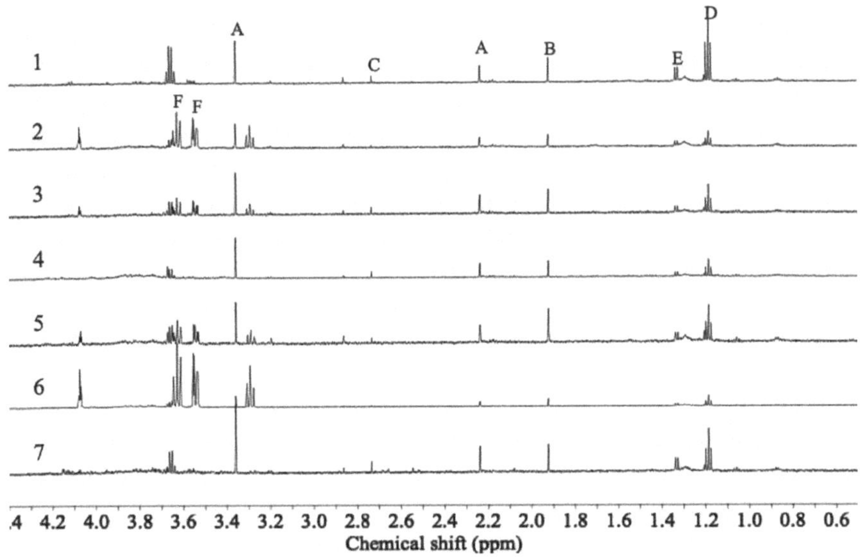

Fig. 2 ^1H NMR spectra of seven *Drosera* species (600 MHz, KH_2PO_4 buffer containing 0.01% trimethylsilylpropanoic acid (TMSP, w/w) in the range of δ 0.3–4.5. 1. *Drosera adelae*, 2. *Drosera regia*, 3. *Drosera capensis* Alba, 4. *Drosera capensis* Giant, 5. *Drosera capensis* Rubra, 6. *Drosera binata*, 7. *Drosera slackii*. A: acetoacetic acid, B: acetic acid, C: citric acid, D: ethanol, E: lactic acid, F: *myo*-inositol.

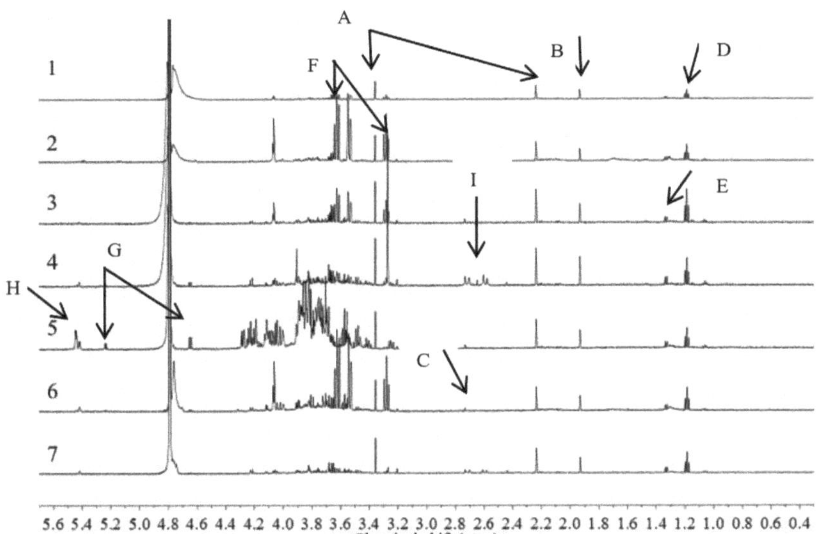

Fig. 3 ^1H NMR spectra of seven *Drosera* species (600 MHz, CH_3OH-d_4-KH_2PO_4 buffer (8:2, v/v) containing 0.01% trimethylsilylpropanoic acid (TMSP, w/w) in the rage of δ 0.3–4.5. 1. *Drosera adelae*, 2. *Drosera regia*, 3. *Drosera capensis* Alba, 4. *Drosera capensis* Giant, 5. *Drosera capensis* Rubra, 6. *Drosera binata*, 7. *Drosera slackii*. A: acetoacetic acid, B: acetic acid, C: citric acid, D: ethanol, E: lactic acid, F: *myo*-inositol, G: glucose, H: sucrose, I: malic acid.

Table 1 Detected metabolites of *Drosera* exudates in ^{1}H NMR spectrum [600 MHz, CH_3OH-d_4-KH_2PO_4 buffer (8:2, v/v) containing 0.01% TMSP (w/w)].

Compound	Characteristic chemical shift in ppm (splitting pattern and coupling constant in Hz)[a]	*Drosera* species						
		Sp1[b]	Sp 2[c]	Sp3[d]	Sp4[e]	Sp 5[f]	Sp 6[g]	Sp 7[h]
Acetoacetic acid	3.36 (s), 2.24 (s)	+	+	+	+	+	+	+
Acetic acid	1.93 (s)	+	+	+	+	+	+	+
Citric acid	2.71 (d, J=16.3), 2.59 (d, J=16.3)				+			+
Ethanol	3.63 (q, J=7.1), 1.19 (t, J=7.1)	+	+	+	+	+	+	+
Formic acid	8.5 (s)	+	+	+	+	+	+	+
Fumaric acid	6.76 (s)		+					
Glucose	5.23 (d, J=3.9), 4.65 (d, J=8.0)			+	+	+	+	+
myo-Inositol	3.63 (t, J=9.54), 3.54 (dd, J=10.1, 3.5),	+	+	+	+	+	+	+
Sucrose	5.42 (d, J=3.9), 4.64 (d, J=8.0)		+	+	+	+	+	+
Malic acid	2.6 (d, J=16.9), 2.7 (d, J=16.9)				+			+
Lactic acid	1.3 (d, J=5.64)	+	+	+	+	+	+	+
Methanol	3.35 (s)	+	+	+	+	+	+	+

[a]All the chemical shifts were calculated based on 0.00 ppm of TMSP signal, s: singlet, d: doublet, dd: double doublet, t: triplet, m: multiplet.
[b]*Drosera adelae*.
[c]*Drosera regia*.
[d]*Drosera capensis* Alba.
[e]*Drosera capensis* Giant.
[f]*Drosera capensis* Rubra
[g]*Drosera binata*.
[h]*Drosera slackii*. ($n=3$).

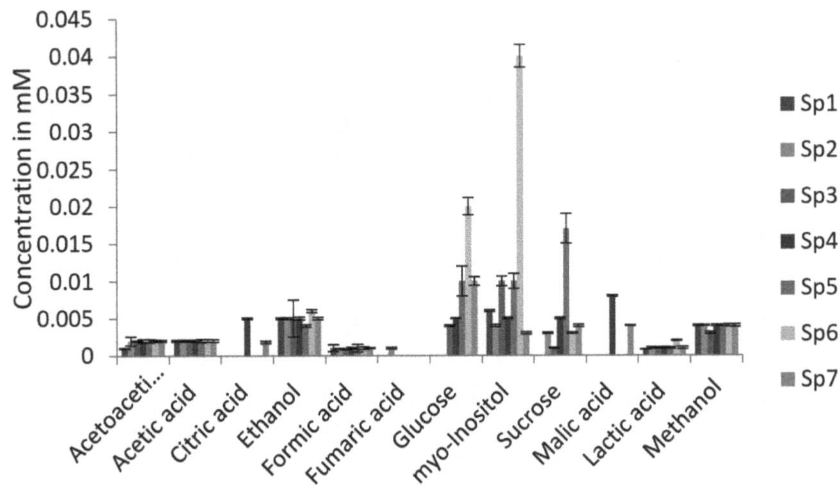

Fig. 4 ^{1}H NMR average quantification of compounds of *Drosera* biofluids. Sp1: *Drosera adelae*, Sp2: *Drosera regia*, Sp3: *Drosera capensis* Alba, Sp4: *Drosera capensis* Giant, Sp5: *Drosera capensis* Rubra, Sp6: *Drosera binata*, Sp7: *Drosera slackii* ($n=3$ with standard deviation).

4.2 GC–MS analysis *Drosera* exudates

^{1}H NMR spectroscopy showed that the main metabolites in *Drosera* species are organic acids and sugars, which might be components of NADESs. To analyze the chemical composition of the exudates in more detail, gas chromatography–mass spectrometry (GC–MS) was employed after derivatization of non-volatiles. This resulted in the identification of arabinose, fructose, mannose, galactose, sucrose, glucose, xylose, *myo*-inositol, ascorbic acid, glucuronic acid, and sorbose. In all the exudates, *myo*-inositol was detected as the main carbohydrate in both buffer and MeOH extract by GC–MS. Mannose and glucose were found in all MeOH extracts from the leaves. In buffer extract, fewer compounds were detected, as most of the sugars bind to the resin produced by the glandular hairs of the plants. The non-soluble resin fraction possibly helps keeping up the water content in the exudate and prevents evaporation (Dell, 1977). These results show that all the exudates of the seven species of *Drosera* have very similar sugar compositions. GC–MS confirmed the presence of *myo*-inositol in the exudate of all seven *Drosera* species (Table 2).

Previous experiments on *D. capensis* (Gowda et al., 1982; Rost & Schauer, 1977) reported that the exudates contained arabinose, mannose, galactose, xylose, and glucuronic acid, which is similar to our experiment.

Table 2 Detected metabolites of MeOH and buffer extracts of *Drosera* exudates in GC–MS analysis.

		MeOH extract							Buffer extract						
Compound	**Retention time**	**Sp1**	**Sp2**	**Sp3**	**Sp4**	**Sp5**	**Sp6**	**Sp7**	**Sp1**	**Sp2**	**Sp3**	**Sp4**	**Sp5**	**Sp6**	**Sp7**
Mannose	13.15	+	+	+		+	+	+			+				+
myo-inositol	17.26	+	+	+	+	+	+	+	+	+	+	+	+	+	++
Glucose	15.30; 16.16	+	+	+	+	+	+	+						+	
Fructose	14.37		+	+	+	+		+			+				
Sorbose	14.50		+	+		+	+	+							
Arabinose	14.80	+		+	+	+	+	+		+		+		+	
Galactose	15.40	+	+				+	+						+	+
Ascorbic acid	15.85		+							+					
Glucuronic acid	16.25					+			+					+	+
Xylose	16.39		+	+	+	+	+	+	+	+					++
Sucrose	21.7	+	+	+	+	+	+	+	+	+			+		

Sp1: Drosera adelae, Sp2: Drosera regia, Sp3: Drosera capensis Alba, *Sp4: Drosera capensis* Giant, *Sp5: Drosera capensis* Rubra, *Sp6: Drosera binata, Sp7: Drosera slackii.*

A study by Kokubun (2017) confirmed that the mucilage of *Drosera capensis* contain a significant amount of *myo*-inositol together with polysaccharide, and *myo*-inositol might act as a cross-linker (hydrogellator) between the polysaccharide strands, probably through the hydrogen bond-network between the hydrogel groups. Huang, Wang, Sun, Agrawal, and Zhang (2015) also reported that *Drosera* mucilage is a naturally occurring hydrogel, composed of nano particles assembled with polysaccharide, driven by electrostatic interactions mediated with divalent cations (Ca^{2+} and M^{g2+}).

5. Natural deep eutectic solvents formation with the ingredients detected in plant exudates

From *Drosera* plant exudate, it is clear that the exudates consist of more than 95% of water, the very strong hydrogen bonding capacities of the sugars and some of the acids might play role in producing nanoparticles that contain the various enzymes present in the exudate, to encapsulate proteins that might be more stable than in normal water solutions, or to adsorb on the prey to help to dissolve the prey for further digesting by the enzymes. To see if any of the ingredients may form NADESs we tested some potential NADESs, which could be formed by the small molecules present in the *Drosera* biofluids.

The NADESs were prepared with *myo*-inositol combined with various sugars and acids, because *myo*-inositol was found in all *Drosera* species. However, none of the combinations tested was able to form a liquid, therefore a non-organic ingredient, Ca^{2+} was added, in a form of $CaCl_2 \cdot 2H_2O$, as this component was the main cation found in *D. capensis* and *D. binata* (Huang et al., 2015; Rost & Schauer, 1977). All of the components were mixed together according to Choi et al. (2011) and with certain molar ratios, NADESs were formed. The tested combinations are shown in Table 3.

Other combinations were made of a number of sugars in combination with organic acids, which refers to previous experiments by Gowda et al. (1982, 1983). The combinations included arabinose, xylose, galactose, mannose, glucuronic acid, and water with a molar ratio of 3.6:1:4.9:8.4:8.2:80, with a water concentration of about 23% were all a liquid. A different molar ratio 8.4:1:9.6:18.3:17.1:60, with 10% of water also gave a liquid. The combinations which gave liquids stayed liquid even after freeze-drying. The viscosity of the fluid decreased when more water was added.

Table 3 Composition and ratios of compounds tested for the formation of NADESs (based on the analysis of the *Drosera* samples: Gowda et al., 1982; Gowda, Reuter, & Schauer, 1983, this chapter).

Composition	Mole ratio	Result
Myo-inositol:sucrose:water	1:2:5	Solid
Myo-inositol:sucrose:$CaCl_2 \cdot 2H_2O$:water	1:2:2:20	Liquid
D-mannose:glucose:choline chloride:water	1:2:1:2	Liquid
D-mannose:D-glucuronic acid:betaine:water	1:1:1:4	Liquid
Myo-inositol:malic acid:water	1:2:2	Solid
Myo-inositol:glucose:malic acid:choline chloride: water	1:1:1:2	Solid
Arabinose:D-mannose:*Myo*-inositol:water	2:1:1:5	Solid
Myo-inositol:acetic acid:water	1:1:5	Solid
Myo-inositol:formic acid:water	1:2:5	Solid
D-arabinose:D-xylose:D-galactose:D-mannose: D-glucuronic acid:water	8.4:1:9.6:18.3:17.1:60	Liquid
D-arabinose:D-xylose:D-galactose:D-mannose: D-glucuronic acid:water	3.6:1:4.9:8.4:8.2:80	Liquid

Sugar-based NADESs usually have thicker, more viscous consistency. Using the ingredients found in the biofluids plus choline chloride or betaine were tested for the formation of less viscous NADESs.

6. Conclusions

Exudates are thought to protect the producing plants against herbivores, insects, and pathogens. They have similar physical and chemical characteristics; however, the roles of plant exudates are diverse, and depends on the plant species. Plant exudates are mainly composed of sugars, amino acids, and organic acids, which may constitute a NADES. They also contain secondary metabolites, water, and minerals. In case of insect catching *Drosera* species, ^1H NMR and GC–MS were used to analyze metabolites in the exudates of seven *Drosera* species. The NMR results were confirmed by the results of GC–MS, which also resulted in the identification of some minor sugars in all exudates. The role of the compounds found could be for stabilization of the proteins or the formation of biofilms around the caught preys enabling a faster enzymatic digesting of the prey. The *Drosera* biofluids consist of 95–99% water, and their viscosity and stability are preserved without water loss over prolonged periods. As NADESs seem not to be a primary factor in this special behavior of the *Drosera* biofluids, other explanations need to be explored in future research. The role of the compounds found in the *Drosera* biofluid could be an antifeedant activity against herbivores, the attraction of their preys, stabilization of the proteins or the formation of biofilms around the caught preys enabling a faster enzymatic digesting of the prey. Calcium alginate is a well-known example of a gel formed from a polysaccharide, which is used to immobilize enzymes or cells in various biotechnological applications. Taking that as an example the combination of polysaccharide and/or proteins with cations might play a role in keeping the *Drosera* biofluids in a highly viscous liquid state. Finally, the occurrence of ethanol, acetoacetic acid, acetic acid and lactic acid as typical products of various fermentation processes may point to the presence of one or more microorganisms in the biofluid.

References

Abarca, L. F. S., Klinkhamer, P. G., & Choi, Y. H. (2019). Plant latex, from ecological interests to bioactive chemical resources. *Planta Medica*, *85*, 856–868.

Agrawal, A. A., & Konno, K. (2009). Latex: A model for understanding mechanisms, ecology, and evolution of plant defense against herbivory. *Annual Review of Ecology, Evolution, and Systematics*, *40*, 311–331.

Amid, B. T., Mirhosseini, H., & Kostadinović, S. (2012). Chemical composition and molecular structure of polysaccharide-protein biopolymer from *Durio zibethinusseed*: Extraction and purification process. *Chemistry Central Journal*, *6*, 117.

Amin, A. M., Ahmad, A. S., Yin, Y. Y., Yahya, N., & Ibrahim, N. (2007). Extraction, purification and characterization of durian (*Durio zibethinus*) seed gum. *Food Hydrocolloids*, *21*, 273–279.

Badri, D. V., & Vivanco, J. M. (2009). Regulation and function of root exudates. *Plant, Cell & Environment*, *32*, 666–681.

Bais, H. P., Weir, T. L., Perry, L. G., Gilroy, S., & Vivanco, J. M. (2006). The role of root exudates in rhizosphere interactions with plants and other organisms. *Annual Review of Plant Biology*, *57*, 233–266.

Barbieri, L., Falasca, A., Franceschi, C., Licastro, F., Rossi, C. A., & Stirpe, F. (1983). Purification and properties of two lectins from the latex of the euphorbiaceous plants *Hura crepitans* L. (sand-box tree) and *Euphorbia characias* L. (Mediterranean spurge). *Biochemical Journal*, *215*, 433–439.

Barthlott, W., Porembski, S., Seine, R., & Theisen, I. (2007). *The curious world of carnivorous plants: A comprehensive guide to their biology and cultivation*. Portland, Or: Timber Press.

Bennett, K. F., & Ellison, A. M. (2009). Nectar, not colour, may lure insects to their death. *Biology Letters*, *5*, 469–472.

Bhosale, R. R., Osmani, R. A. M., & Moin, A. (2014). Natural gums and mucilages: A review on multifaceted excipients in pharmaceutical science and research. *International Journal of Pharmacognosy and Phytochemical Research*, *15*(4), 901–912.

Bushman, B. S., Scholte, A. A., Cornish, K., Scott, D. J., Brichta, J. L., Vederas, J. C., et al. (2006). Identification and comparison of natural rubber from two *Lactuca* species. *Phytochemistry*, *67*, 2590–2596.

Carter, C. A., Forney, R. W., Gray, E. A., Gehring, A. M., Schneider, T. L., Young, D. B., et al. (1997). Toxicarioside a. a new cardenolide isolated from *Antiaris toxicaria* latex-derived dart poison. Assignment of the 1H-and 13C-NMR shifts for an antiarigenin aglycone. *Tetrahedron*, *53*, 13557–13566.

Choi, Y. H., van Spronsen, J., Dai, Y., Verberne, M., Hollmann, F., Arends, I. W. C. E., et al. (2011). Are natural deep eutectic solvents the missing link in understanding cellular metabolism and physiology? *Plant Physiology*, *156*, 1701–1705.

Dell, B. (1977). Distribution and function of resins and glandular hairs in Western Australian plants. *Journal of the Royal Society of Western Australia*, *59*, 119–123.

Dussourd, D. E., & Hoyle, A. M. (2000). Poisoned plusiines: Toxicity of milkweed latex and cardenolides to some generalist caterpillars. *Chemoecology*, *10*, 11–16.

Estabrook, E. M., & Yoder, J. I. (1998). Plant-plant communications: Rhizosphere signaling between parasitic angiosperms and their hosts. *Plant Physiology*, *116*, 1–7.

Gowda, D. C., Reuter, G., & Schauer, R. (1982). Structural features of an acidic polysaccharide from the mucin of Drosera binata. *Phytochemistry*, *21*, 2297–2300.

Gowda, D. C., Reuter, G., & Schauer, R. (1983). Structural studies of an acidic polysaccharide from the mucin secreted by Drosera capensis. *Carbohydrate Research*, *113*, 113–124.

Hartmann, A., Schmid, M., Van Tuinen, D., & Berg, G. (2009). Plant-driven selection of microbes. *Plant and Soil*, *321*, 235–257.

Hernandez-Hernandez, A. B., Alarcon-Aguilar, F. J., Almanza-Perez, J. C., Nieto-Yañez, O., Olivares-Sanchez, J. M., Duran-Diaz, A., et al. (2017). Antimicrobial and anti-inflammatory activities, wound-healing effectiveness and chemical characterization of the latex of *Jatropha neopauciflora* Pax. *Journal of Ethnopharmacology*, *204*, 1–7.

Huang, Y., Wang, Y., Sun, L., Agrawal, R., & Zhang, M. (2015). Sundew adhesive: A naturally occurring hydrogel. *Journal of the Royal Society Interface*, *12*, 20150226.

Hullar, T. L., & Smith, F. (1966). The neutral carbohydrates of ammoniated rubber latex. *Archives of Biochemistry and Biophysics*, *115*, 505–509.

Itenov, K., Mølgaard, P., & Nyman, U. (1999). Diurnal fluctuations of the alkaloid concentration in latex of poppy *Papaver somniferum* is due to day–night fluctuations of the latex water content. *Phytochemistry*, *52*, 1229–1234.

Jaeger, C. H., Lindow, S. E., Miller, W., Clark, E., & Firestone, M. K. (1999). Mapping of sugar and amino acid availability in soil around roots with bacterial sensors of sucrose and tryptophan. *Applied and Environmental Microbiology*, *65*, 2685–2690.

John, K. S., Bhat, S. G., & Rao, U. P. (2003). Biochemical characterization of sap (latex) of a few Indian mango varieties. *Phytochemistry*, *62*, 13–19.

Kim, H. K., Choi, Y. H., & Verpoorte, R. (2010). NMR-based metabolomic analysis of plants. *Nature Protocols*, *5*, 536–549.

Kokubun, T. (2017). Occurrence of *myo*-inositol and alkyl-substituted polysaccharide in the prey-trapping mucilage of *Drosera capensis*. *The Science of Nature*, *104*, 83.

Konno, K. (2011). Plant latex and other exudates as plant defense systems: Roles of various defense chemicals and proteins contained therein. *Phytochemistry*, *72*, 1510–1530.

Konno, K., Ono, H., Nakamura, M., Tateishi, K., Hirayama, C., Tamura, Y., et al. (2006). Mulberry latex rich in antidiabetic sugar-mimic alkaloids forces dieting on caterpillars. *Proceedings of the National Academy of Sciences*, *103*, 1337–1341.

Kosasi, S., Van Der Sluis, W. G., & Labadie, R. (1989). Multifidol and multifidol glucoside from the latex of Jatropha multifida. *Phytochemistry*, *28*, 2439–2441.

Kusari, S., Zühlke, S., & Spiteller, M. (2009). An endophytic fungus from *Camptotheca acuminata* that produces camptothecin and analogues. *Journal of Natural Products*, *72*, 2–7.

Lambert, J. B., Wu, Y., & Santiago-Blay, J. A. (2005). Taxonomic and chemical relationships revealed by nuclear magnetic resonance spectra of plant exudates. *Journal of Natural Products, 68*, 635–648.
Leonhardt, S. D., Schmitt, T., & Blüthgen, N. (2011). Tree resin composition, collection behavior and selective filters shape chemical profiles of tropical bees (Apidae: Meliponini). *PLoS One, 6*, e23445.
Malsawmtluangi, C., Thanzami, K., Lalhlenmawia, H., Selvan, V., Palanisamy, S., Kandasamy, R., et al. (2014). Physicochemical characteristics and antioxidant activity of *Prunus cerasoides* D. Don gum exudates. *International Journal of Biological Macromolecules, 69*, 192–199.
Matušíková, I., Salaj, J., Moravčíková, J., Mlynárová, L., Nap, J.-P., & Libantová, J. (2005). Tentacles of in vitro-grown round-leaf sundew (*Drosera rotundifolia.* L.) show induction of chitinase activity upon mimicking the presence of prey. *Planta, 222*, 1020–1027.
Mirhosseini, H., & Amid, B. T. (2012). A review study on chemical composition and molecular structure of newly plant gum exudates and seed gums. *Food Research International, 46*, 387–398.
Moutim, V., Silva, L. G., Lopes, M. T. P., Fernandes, G. W., & Salas, C. E. (1999). Spontaneous processing of peptides during coagulation of latex from *Carica papaya. Plant Science, 142*, 115–121.
Nardi, S., Concheri, G., Pizzeghello, D., Sturaro, A., Rella, R., & Parvoli, G. (2000). Soil organic matter mobilization by root exudates. *Chemosphere, 41*, 653–658.
Odani, S., Yokokawa, Y., Takeda, H., Abe, S., & Odani, S. (1996). The primary structure and characterization of carbohydrate chains of the extracellular glycoprotein proteinase inhibitor from latex of *Carica papaya. European Journal of Biochemistry, 241*, 77–82.
Oppel, C. B., Dussourd, D. E., & Garimella, U. (2009). Visualizing a plant defense and insect counterploy: Alkaloid distribution in *Lobelia* leaves trenched by a *plusiine* caterpillar. *Journal of Chemical Ecology, 35*, 625–634.
Pate, J., & Dixon, K. (1978). Mineral nutrition of *Drosera erythrorhiza* Lindl. With special reference to its tuberous habit. *Australian Journal of Botany, 26*, 455–464.
Ramos, M. V., Grangeiro, T. B., Freire, E. A., Sales, M. P., Souza, D. P., Araújo, E. S., et al. (2010). The defensive role of latex in plants: Detrimental effects on insects. *Arthropod-Plant Interactions, 4*, 57–67.
Rasmann, S., Johnson, M. D., & Agrawal, A. A. (2009). Induced responses to herbivory and jasmonate in three milkweed species. *Journal of Chemical Ecology, 35*, 1326.
Rees, S. B., & Harborne, J. B. (1985). The role of sesquiterpene lactones and phenolics in the chemical defence of the chicory plant. *Phytochemistry, 24*, 2225–2231.
Rost, K., & Schauer, R. (1977). Physical and chemical properties of the mucin secreted by *Drosera capensis. Phytochemistry, 16*, 1365–1368.
Samuelsson, G., & Bohlin, L. (2009). *Drugs of natural origin* (6th edition, pp. 470–471). Stockholm, Sweden: Swedish Pharmaceutical Press.
Schaefer, H. M., & Ruxton, G. D. (2008). Fatal attraction: Carnivorous plants roll out the red carpet to lure insects. *Biology Letters, 4*, 153–155.
Sessa, R. A., Bennett, M. H., Lewis, M. J., Mansfield, J. W., & Beale, M. H. (2000). Metabolite profiling of sesquiterpene lactones from *Lactuca* species major latex components are novel oxalate and sulfate conjugates of lactucin and its derivatives. *Journal of Biological Chemistry, 275*, 26877–26884.
Sethi, A., McAuslane, H. J., Rathinasabapathi, B., Nuessly, G. S., & Nagata, R. T. (2009). Enzyme induction as a possible mechanism for latex-mediated insect resistance in romaine lettuce. *Journal of Chemical Ecology, 35*, 190–200.

Takeuchi, Y., Salcher, M. M., Ushio, M., Shimizu-Inatsugi, R., Kobayashi, M. J., Diway, B., et al. (2011). In situ enzyme activity in the dissolved and particulate fraction of the fluid from four pitcher plant species of the genus nepenthes. *PLoS One*, *6*, 1–9.

Tomè, F., & Colombo, M. L. (1995). Distribution of alkaloids in *Chelidonium majus* and factors affecting their accumulation. *Phytochemistry*, *40*, 37–39.

Walker, T. S., Bais, H. P., Grotewold, E., & Vivanco, J. M. (2003). Root exudation and rhizosphere biology. *Plant Physiology*, *132*, 44–51.

Yang, Z., Rogers, L. M., Song, Y., Guo, W., & Kolattukudy, P. E. (2005). Homoserine and asparagine are host signals that trigger in planta expression of a pathogenesis gene in *Nectria haematococca*. *Proceedings of the National Academy of Sciences*, *102*, 4197–4202.

Zhu-Salzman, K., Bi, J. L., & Liu, T. X. (2005). Molecular strategies of plant defense and insect counter-defense. *Insect Sci.*, *12*, 3–15.

CHAPTER ELEVEN

A look on target-specificity of eutectic systems based on natural bioactive compounds

Filipe Silva Nunes de Oliveira and Ana Rita Cruz Duarte*
LAQV@REQUIMTE, Departamento de Química, Faculdade de Ciências e Tecnologia, Universidade Nova de Lisboa, Caparica, Portugal
*Corresponding author: e-mail address: ard08968@fct.unl.pt

Contents

Abstract

Natural occurring molecules often represent an interesting starting point for the development of novel greener solvents. Currently, as a result of centuries compiling information about these molecules, we have a vast library of promising raw materials. Natural deep eutectic systems (NADES) are mostly composed by natural primary metabolites such as sugars, sugar alcohols, organic acids, amino acids, and amines, but can also be formed with secondary metabolites such as terpenes and vanillin, or biopolymers. Moreover, NADES have been described as antibacterial and antibiofilm, antifungal, anticancer, antioxidant and wound healing agents. The tailor-made versatility and its compliance with the green chemistry metrics, have pushed forward the possible therapeutic role of these eutectic systems. In this chapter we aim to challenge the community to join the discussion on what we consider to be the next step in these systems comprehension—target-specificity. We already have evidences of its therapeutic role, so questions such as: does these systems work intra or extra cellularly? How do they interact with cellular membranes? Do they have an impact on gene expression? Are of most relevance to fulfill NADES potential as therapeutic agents. Furthermore, considering the current pursue for sustainability, the greenness and biosafety of NADES is

Advances in Botanical Research, Volume 97
ISSN 0065-2296
https://doi.org/10.1016/bs.abr.2020.09.008

briefly addressed coupling the major advances of eutectics as therapeutic agents with its promising contribution as an alternative to the industrial status quo.

1. Introduction

Natural deep eutectic solvents (NADES) are a mixture of natural occurring molecules which, in a certain molar ratio, experience a significant decrease of its melting point when compared to that of either its individual components (Abbott, Capper, Davies, Rasheed, & Tambyrajah, 2003; Abbott, Boothby, Capper, Davies, & Rasheed, 2004; Guthrie, 1884). As liquid, NADES can work as functional liquid media, both to take advantage of its constituent's bioactive properties, or to dissolve natural or synthetic drugs such as active pharmaceutical ingredients (API) with poor water solubility (Duarte et al., 2017). These eutectic systems have been described as thermally stable, highly tunable, biodegradable, easy to prepare and affordable, placing them in agreement with the green chemistry principles, unlike ionic liquids (IL) and other conventional solvents. These properties have aligned NADES with the worldwide demand for greener alternatives, pushing forward the acceptance of these neoteric systems as a sustainable alternative to conventional solvents in a wide range of applications (Paiva et al., 2014). In an effort to establish the natural eutectic phenomenon, in the past years there have been several reports on NADES being applied to industry, health, food and cosmetics. From the therapeutic perspective, NADES have been revealing astonishing and promising applications namely as antibacterial and antibiofilm (Silva, Silva, Reis, & Duarte, 2019; Zhao et al., 2015), antifungal (Juneidi, Hayyan, & Hashim, 2015; Silva, Silva, et al., 2019), anticancer (Pereira et al., 2019; Rodrigues et al., 2020; Silva, Oliveira, Silva, Reis, & Duarte, 2020), antioxidant agents (Radošević et al., 2018), and also as wound healing enhancers (Silva, Pereira, et al., 2019). The therapeutic potential of these eutectics has been more or less well explored, although, and to the best of our knowledge, little work has been done toward understanding their mechanism of action, namely their site specific interaction with tissues, cells and biomolecules, leading to the lack of a strong cause-effect correlation from which NADES therapeutic effect could be better understood. Furthermore, besides manmade NADES, these most surprising liquid media has been associated with several biochemical processes vital for cells and organisms. Verpoorte hypothesis of a new liquid phase media, separated from water and lipid, could at molecular level explain, for example, how compounds with intermediate polarity present in high concentrations could be synthesized and stored, leading to an enlightenment of how, at a more macroscopic point of view, plants withstand periods of drought (Choi et al., 2011). Comprising this evidence and considering the fundamental role NADES may have in life's pursue for homeostasis, one is tempted to push forward these systems comprehension. In other words, considering NADES importance to cell survival, their presence must be somehow cell mediated and there should be a well-defined interaction between the eutectic system and its specific cell target. Therefore, and considering the previously reported work on NADES therapeutic activity, we consider that unveiling NADES's specific targets and correspondent interaction, and correlating it with its therapeutic effect, is the missing link to the full acceptance of these eutectics as an alternative to the established therapies. Thus, this chapter focuses on two major topics: the most promising examples of NADES's therapeutic activity alongside with the discussion of their current limitations, stemmed from the fact that its therapeutic mechanisms are not yet fully understood (Fig. 1).

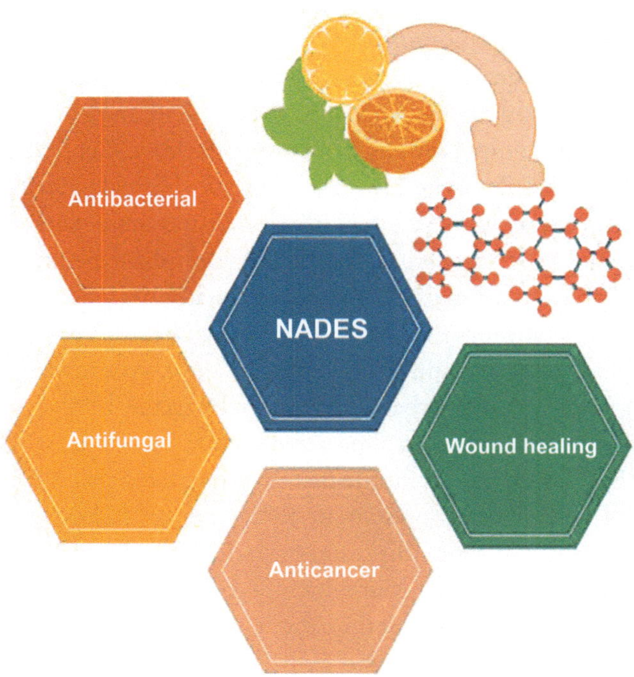

Fig. 1 Graphical abstract.

2. From eutectic to natural deep eutectic systems

Back in the late 19th century when eutectic systems have first been reported, the developments that we now observe in different scientific fields were far from being expected. The classical definition of a eutectic system is a mixture of two substances which at a particular composition possess a lower melting temperature than the initial pure components. Eutectic systems have been described, for example, in metallurgy for the preparation of alloys, where it is crucial that the alloy has a single sharp melting temperature (Guthrie, 1884). The concept is, therefore, not new, so what brings novelty to the field and what makes it a hot topic within the scientific community? Abbot and coworkers have reported in 2003 the preparation of a liquid system from urea and a quaternary ammonium salt with interesting properties for solvent applications (Abbott et al., 2003). The main breakthrough of this work was the demonstration of new sustainable, biodegradable and highly flexible mixtures that could be produced from readily available components and could, therefore, be tailored to particular applications. Following this pioneer work, the same group disclosed the preparation of more systems based on quaternary ammonium salts with carboxylic acids. Choline chloride (ChCl) was the quaternary ammonium salt chosen in combination with molecules such as adipic acid, benzoic acid, citric acid, malonic acid (MA) or oxalic acid (OA) among others (Abbott, Boothby, Capper, Davies, & Rasheed, 2004). While in the first study only the thermal properties are disclosed, on the second study different physical properties of these new liquid systems start to be unraveled. In particular, the phase behavior (thermal properties), surface tension, viscosity and conductivity are herein presented. Furthermore, the authors disclose the solubility of three oxides in the solvents prepared, demonstrating the ability of these systems to replace ILs and overcome the main drawbacks of these, who in the 1990s were thought to be the most promising green alternative solvents. Later in 2011, the study conducted by Verpoorte, has for the first time suggested that DES could be a new type of liquid phase present in living beings (Choi et al., 2011). A steppingstone was set from this manuscript. Herein, the authors have presented different eutectic systems in which both compounds present are naturally occurring molecules, naming them natural deep eutectic solvents (NADES). Major advances in the field came following this manuscript where the authors suggest their potential use in applications as different as solvents for extraction of food flavors, fragrances and dyes, medicines, cosmetics or agrochemicals, as media for synthetic organic chemistry or enzymatic reactions (Choi et al., 2011). In this work, the authors have described the in vitro preparation of several systems starting from molecules, which by nuclear magnetic resonance (NMR)-based metabolomics have been discovered to be present in different living organisms. These include, among others, sugars, amino acids, organic acids and choline derivatives. In 2013, the same authors revealed more than 100 combinations of NADES and have investigated the interactions between the two molecules by NMR spectroscopy (Dai, van Spronsen, Witkamp, Verpoorte, & Choi, 2013). The results obtained lead to the conclusion that the species interact mostly due to hydrogen bonds between the two molecules and that in some cases water present in the system is also strongly bound to the NADES constituents. Since these reports, the number of publications in the field has grown exponentially. The promise of these systems to actually be considered as green solvents, complying with most of the green chemistry principles, from the use of renewable materials, to less hazardous substances, biodegradable and nontoxic, makes them extremely appealing not only for the academia but also for the industry who has long been waiting for alternatives to conventional organic solvents, and this will briefly be discussed in Section 4. Today there are more than 1000 different systems reported in the literature and the scientific community is actively working toward the understanding of the mechanisms which govern the formation of a NADES as well as to develop models to be able to predict their physico-chemical properties. Meanwhile scientists have also been actively focused in the application development. Scientists have been inspired on current drawbacks of industrial processes, limitations in existing technologies, and the design of novel products, and the documented outcomes are quite promising and exciting.

3. Biomedical and pharmaceutical potential/applications of NADES

In biomedical, and in particular pharmaceutical applications, the main interest arises from the current limitations of conventional therapeutics, such as poor solubility, permeability, and bioavailability of drugs/APIs. NADES high tunability and lower toxicity, have been putting them in the spotlight as alternative therapeutics efficiency enhancers. Naturally occurring molecules with reported bioactivity can be combined in a eutectic composition conferring them enhanced properties. Thereafter, either by integrating the drugs/APIs in its composition or by dissolving it in the eutectic system, drawbacks such as lower rates of dissolution and permeation, and also aggregation and consequent precipitation of the bioactive compound can be circumvented. In this sense, most recently, NADES have been described in therapeutic applications such as antibacterial and antibiofilm, antifungal, antiviral, anticancer, and also as wound healing enhancers. Herein, an emphasis on current progress and most promising results of this field are highlighted, together with further discussion on pushing forward these preliminary results by exploring the specific cause effect of NADES as therapeutic agents.

3.1 Antibacterial activity

For thousands of years bacteria have survived relying on their ability to adapt to antimicrobial agents. Through evolutionary mechanisms, such as mutations or horizontal gene transfer, these single cell organisms are capable to evolve and oppose the action of certain antibiotics, rendering them ineffective (Bennett, 2008). With the ubiquitous antibiotic use in modern medicine and consequent selective pressure, we have created a major driving force for bacterial evolution toward multidrug resistance. Thus, there is an urgent need for antimicrobial approaches that are able to inactivate and/or kill these pathogens efficiently with a minor risk of promoting resistance. Considering the wide variety of well described natural occurring molecules available, and the customization advantages of a tailor-made system, many NADES applications have been developed in the past several years, namely working as promising antibacterial and antibiofilm agents. Hayyan, Hashim, Al-Saadi, et al. (2013) reported the cytotoxicity of phosphonium-based DES with three different HBD, such as glycerol, ethylene glycol (EG), and triethylene glycol (TEG). The antibacterial cytotoxic effect was tested toward two Gram positive (Gram$^+$) bacteria *Bacillus subtilis* and *Staphylococcus aureus* (see Fig. 2), and two Gram negative (Gram$^-$) bacteria *Escherichia coli* and *Pseudomonas aeruginosa*. All tested systems revealed antibacterial effect toward Gram$^-$ bacteria, but only methyltriphenylphosphonium bromide (MTPB):EG revealed effective cytotoxicity toward Gram$^+$ bacteria, revealing an HBD-dependent antibacterial activity.

Interestingly, in another study, Hayyan, Hashim, Hayyan, et al. (2013) described the absence of antibacterial for ChCl-based DES with four hydrogen bond donors (HBD) including Gly, EG, TEG and urea (U). Since besides urea the same HBD were used in both studies, from these results can be speculated that not only the HBD takes part on the eutectic cytotoxic action but also the salt, since the hydrogen bonding between the HBD and

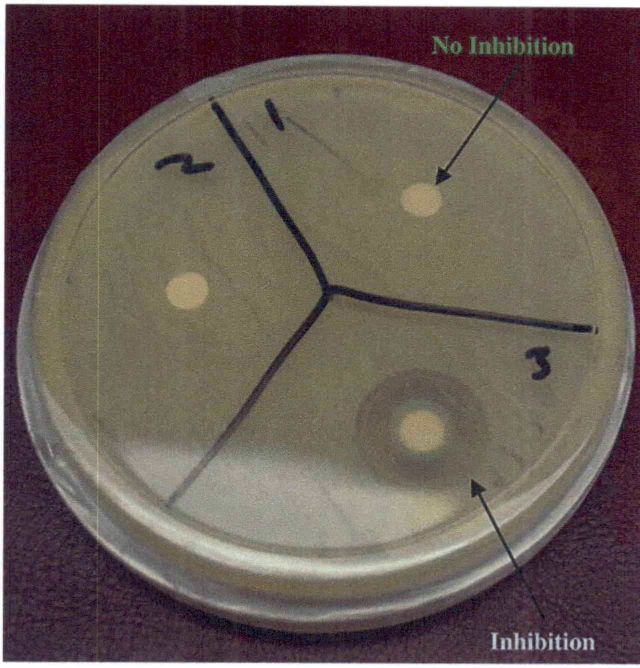

Fig. 2 *Staphylococcus aureus* with DES-dependent growth inhibition. 1—(MTPB):Gly; 2—(MTPB):EG; 3—(MTPB):TEG (Hayyan, Hashim, Al-Saadi, et al., 2013).

the anion of the salt not only affect the physical properties of the mixture but also its chemical structure. Furthermore, since compounds having delocalized charges are more prone to be toxic than chemicals with localized charges, it is excepted that the charge delocalization occurring via hydrogen bonding renders the mixture more toxic in comparison to its individuals components (Clements, Nabholz, & Zeeman, 1994; Modica-Napolitano & Aprille, 2001). Nevertheless, HBD has a preeminent effect on the systems acidity or basicity, which can result in the denaturation of proteins present in the cell outer layer, promoting further disruption of cellular activity and ultimately cell death (Domínguez de María & Maugeri, 2011; Gorke, Srienc, & Kazlauskas, 2010; Kareem, Mjalli, Hashim, & Alnashef, 2010). Zhao et al. (2015) further described the antibacterial effect of ChCl-based DES with amine, alcohol, sugars and organic acids as HBD, toward *E. coli* and *S. enteritidis* ($Gram^-$) and *S. aureus* and *L. monocytogenes* ($Gram^+$). The amine-, alcohol- and sugar-based DES did not inhibit bacterial growth but the organic acid-based DES had a significant inhibitory effect, including ChCl:p-toluenesulfonic acid

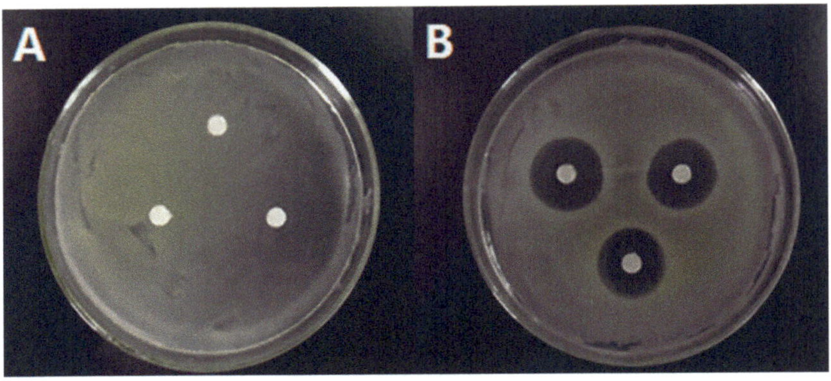

Fig. 3 *Escherichia coli* with NADES-dependent growth inhibition. (A) ChCl:urea; (B) ChCl: citric acid (Zhao et al., 2015).

(PTSA) (1:1), ChCl:oxalic acid (1:1), ChCl:levulinic acid (1:2), ChCl: malonic acid (1:1), ChCl:malic acid (1:1), ChCl:citric acid (1:1) and ChCl: tartaric acid (2:1) (see Fig. 3). These results emphasize once more that depending on system's composition DES can be nontoxic and biocompatible or have targeted toxicity. Similar results were previously reported for the nonorganic acid-based DES (Hayyan, Hashim, Hayyan, et al., 2013).

In Table 1, the minimum inhibitory concentration (MIC) and minimum bactericidal concentration (MBC) values from Gram$^-$ (*E. coli* and *S. enteritidis*) are surprisingly lower than those for Gram$^+$ (*S. aureus* and *L. monocytogenes*). These are somewhat unexpected since higher resistance to toxic compounds is usually attributed to Gram$^-$ bacteria (Russell, 2003). Gram$^-$ bacteria have protective lipopolysaccharide as major protective components of the outer membrane from certain kinds of chemical exposure, this fact foresees a different mechanism of toxicity for ChCl-based DES differing from conventional antibacterial agents.

Since amines, alcohols and sugars can be easily absorbed, via simple or facilitated diffusion, and used as nitrogen or carbon sources, it can be hypothesized that these HBD were simply digested by bacterial cells preventing any associated toxic effect. On the contrary, having an organic acid as HBD inevitably lowers the pH values of the tested DES solutions. In pH under 5.5 the nonionic forms of volatile fatty acids (VFA) are predominant, which are more prone to cross the cell membrane and increase the intracellular H^+ concentration. Since the deportation of H^+ will inevitably increase, the consequent consumption of ATP will decrease its availability for cell metabolism and

Table 1 Antibacterial activity of organic acid-based DESs, expressed as MIC and MBC (Zhao et al., 2015).

DES	*E. coli*		*S. enteritidis*		*S. aureus*		*L. monocytogenes*		pH[a]
	MIC	MBC	MIC	MBC	MIC	MBC	MIC	MBC	
ChCl/p-toluenesulfonic acid (1:1)	18	28	26	40	18	34	30	50	4.60
ChCl/oxalic acid (1:1)	12	18	12	22	12	26	14	30	1.75
ChCl/levulinic acid (1:2)	12	16	12	26	14	22	12	36	4.42
ChCl/malonic acid (1:1)	18	20	20	34	16	30	24	48	2.10
ChCl/malic acid (1:1)	14	20	18	42	14	24	22	48	2.43
ChCl/citric acid (1:1)	12	20	16	38	12	28	20	42	4.08
ChCl/tartaric acid (2:1)	14	18	18	40	12	20	16	44	3.70

[a]pH measured at the 10 mmol/L of DESs aqueous solution. MIC/MBC in mmol/L.

growth. Since for the tested strains the optimal pH is close to neutral (6.5–7.5), this can lead to the inhibition of bacterial growth. By readjusting the pH of the tested DES solutions to neutral before the tests the inhibitory effect was suspended (Zhao et al., 2015). In another report, Radošević et al. (2018) further described the cytotoxic contrast of betaine-, choline-, sugar-, and sugar alcohol-based NADES as having no significant inhibition effect on *Proteus mirabilis*, *Pseudomonas aeruginosa*, *Salmonella typhimurium*, *Staphylococcus aureus* and *E. coli*, while citric acid-based NADES exhibited significant inhibition (higher for citric acid:glucose:glycerol and citric acid:fructose:glycerol systems). Several key aspects can be pointed out to help in further system's design: (i) the benzene ring in p-toluenesulfonic acid can be related to DES toxicity since ChCl:PTSA had less toxic effect; (ii) the elongation of the organic acid carbon chain decreases system's toxicity (ChCl/oxalic acid > ChCl/malonic acid, see Table 1); (iii) in organic acid-based DES the hydrogen bond network is dense and compact, so delocalized charges may have greater impact (see ChCl/citric acid and ChCl/tartaric acid in Table 1); (iv) the presence of hydroxyl groups in the HBD have a slight increasing effect in the system's antibacterial activity (Wikene, Rukke, Bruzell, & Tønnesen, 2017). Furthermore, one of the possible explanations for the high toxicity of ChCl:Oxalic acid could be due to the formation of calcium oxalate crystals in Ca^{2+} containing medium which may result in cell wall and membrane damage (Radošević et al., 2015). In a different approach, Wikene et al. (2017) also reported the antibacterial effect of organic acid-based NADES, as citric acid:sucrose and malic acid:fructose:glucose, toward *E. coli*, *P. aeruginosa*, *K. pneumoniae* and *S. epidermidis*, to be used as solvents for photosensitizers (such as meso-tetra(*p*-hydroxyphenyl)porphine (THPP)) in bacterial photodynamic therapy. By dissolving the THPP in eutectic systems it was possible to obtain a higher degree of phototoxicity in *E. coli* in comparison to phosphate buffer. Mano et al. (2017) described gelatin membranes containing fibers with encapsulated NADES (see Fig. 4), composed by ChCl:mandelic acid, as a fast-dissolving delivery systems (FDDS) in alternative to oral delivery of therapeutic agents.

Interestingly, it was described that the antibacterial properties of mandelic acid were similar for the NADES and fiber encapsulated NADES both for $Gram^-$ (*E. coli*) and $Gram^+$ (*S. aureus*). Even so, there is a decrease in mandelic acid antibacterial activity when compared to it pure form. However, the lack of cytotoxic effect toward an immortalized mouse lung fibroblasts cell line (L929) and retention of a certain degree of antibacterial effect show that both the mandelic acid-based NADES and the fibers with

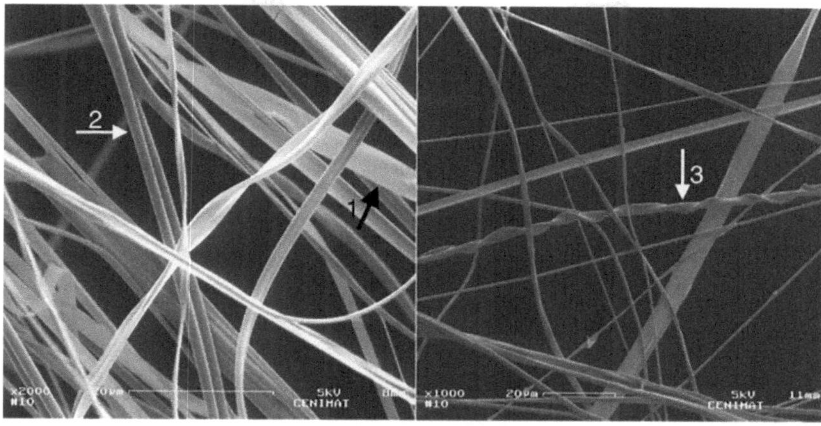

Fig. 4 Scanning electron microscope (SEM) of fiber encapsulated NADES with 2% (v/v) ChCl:mandelic Acid and 30% (w/v) gelatin, showing different conformations. 1—ribbon; 2—ribbon with two tubes; 3—helix (Mano et al., 2017).

NADES retained the antibacterial activity. The potential versatility of NADES as antibacterial have been pinpointed throughout the years in a wide range of systems. Rodrigues et al. (2020) disclosed a terpene-based NADES to be used as sustainable alternative solvents to recover astaxanthin from brown crab shell residues. Besides the promising extraction efficiency, these eutectics also revealed interesting inhibitory effect toward *S. aureus* and *E. coli*, with MICs ranging from 31.25 to 62.50 $\mu L\,mL^{-1}$. The system composed by perillyl alcohol (POH):camphor (CA) has revealed to be the best for inhibiting both *S. aureus* and *E. coli*, with a MIC value of 31 $\mu L\,mL^{-1}$. The same value was also reported using menthol (Me):POH only toward *E. coli*. In another application, namely NADES impregnation in gauzes conferring this medical device antimicrobial properties, Silva, Silva, et al. (2019) reported promising results on saturated fatty acid-based-NADES. Systems combining capric acid (CA):stearic acid (SA), capric acid (CA):myristic acid (MA) and capric acid (CA):lauric acid (LA) were tested toward *S. aureus*, a *S. aureus* Methicillin-resistant strain (MRSA), a *S. epidermis* Methicillin-resistant strain (MRSE). The CA:LA NADES revealed the most promising antimicrobial activity toward the $Gram^{+}$ bacterial strains tested (Table 2), in comparison to the lower values for MA and SA most likely resulting from diffusion limitation in the aqueous growth medium due to their higher hydrophobicity from slightly higher carbon chains.

Table 2 MIC and MBC values, for the system's individual components and NADES, toward the tested bacteria (Silva, Silva, et al., 2019).

	MIC (µg mL^{-1})			MBC (µg mL^{-1})		
DES/individual components	***S. aureus***	**MRSA**	**MRSE**	***S. aureus***	**MRSA**	**MRSE**
CA	625	625	625	1250	1250	1250
LA	312.5	312.5	312.5	625	625	625
MA	625	1250	1250	1250	2500	2500
SA	ND	ND	ND	ND	ND	ND
CA:LA	625	625	625	1250	1250	1250
CA:MA	625	625	625	1250	1250	1250
CA:SA	1250	1250	1250	2500	2500	2500

ND—not dissolved.

Furthermore, Silva, Silva, et al. (2019) reported an antibiofilm activity for the CA:LA, with the promotion of biofilm detachment in all microorganisms tested, including *E. coli* (Gram^{-}) (see Fig. 5).

The antimicrobial properties of natural occurring molecules such as terpenes and fatty acids have been widely described alongside with a synergetic effect within these compounds toward both Gram^{+} and Gram^{-} bacteria (Batovska, Todorova, Tsvetkova, & Najdenski, 2009; Burt, 2004; Kitahara et al., 2004; Knapp & Melly, 1986; Silva, Pereira, et al., 2019). This antimicrobial activity has been mainly attributed to nonspecific action mechanisms, such as membrane destabilization/dissolution. Moreover, within Gram^{+} bacteria, the electrostatic interactions of the NADES/DES with the polysaccharide or peptide chains of the peptidoglycans, the polymers which provides rigidity and strength to bacterial cell walls, may be in the genesis of the antibacterial effect (Wen, Chen, Tang, Wang, & Yang, 2015). On the contrary, by not having an accessible network of peptidoglycan, Gram^{-} bacteria suggest a different cytotoxic mechanism of NADES possibly derived from their potential to solubilize and/or interfere with other bacterial membrane components, such as chelation of membrane-bound divalent cations or transmembrane proteins, as previously discussed.

Considering the current paradigm of antibiotic resistance the nonspecific action of the NADES is able to inactivate the antimicrobial pathogens efficiently without the risk of inducing resistances, alongside with its potential action toward multiple organisms. In contrast, ILs have been reported to

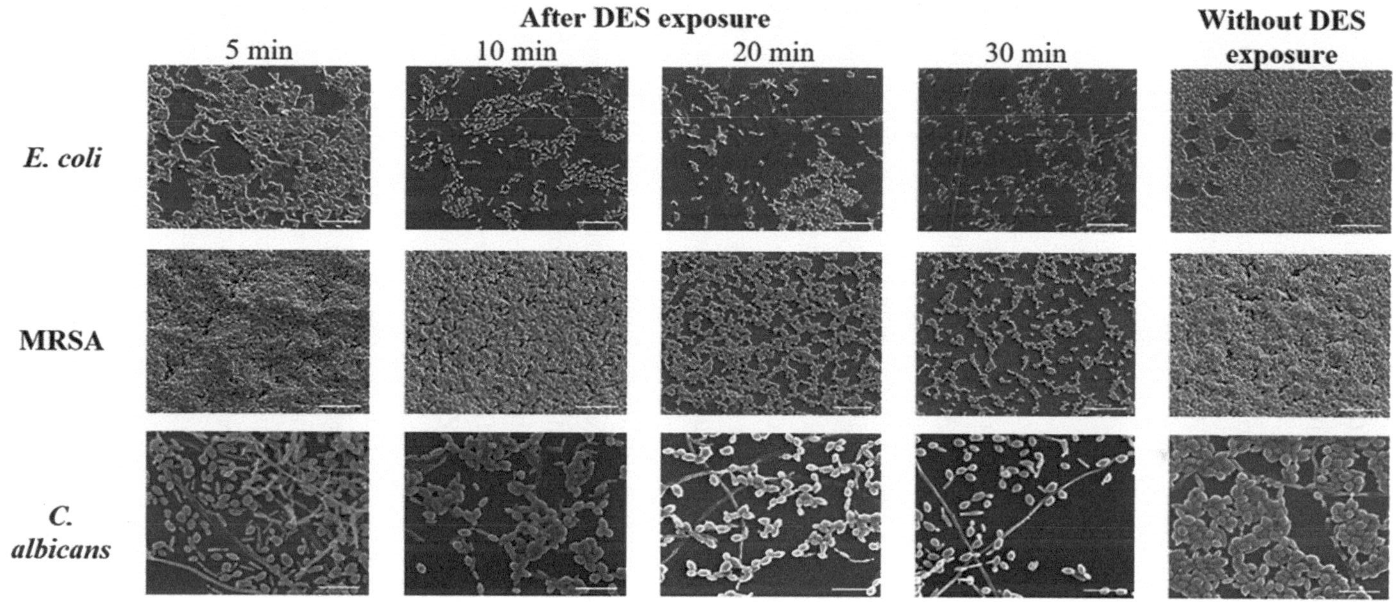

Fig. 5 SEM imaging of the antibiofilm activity of CA:LA eutectic system for *E. coli* and MRSA (Silva, Silva, et al., 2019).

stimulate the expression of antibiotic resistance genes and horizontal gene transfer in freshwater bacteria (Luo, Wang, Lu, Mu, & Mao, 2014). In nowadays quest for sustainable solvents, although NADES/DES present cytotoxicity toward a wide range of microorganisms they can still be referend as greener solvents in comparison to traditional antimicrobial compounds, conventional solvents and ILs. Eutecticity can result in less nonspecific associated toxicity providing tailor-made antimicrobial therapeutic agents with great promise.

3.2 Antifungal activity

In recent years there have also been reports on NADES and DES cytotoxic potential toward fungal cells, such as yeasts and molds, but also dimorphic fungus. Although several fungi and bacteria share the same designation as microorganisms, we believe that it is important to discriminate between the antibacterial and antifungal activity of NADES since the underlying mechanisms of action are most certainly cell type-dependent for either life's domain. However there are cases in which the same system has reported activity toward both bacteria and fungi. Few systems are composed only by naturally occurring molecules, and in some cases their activity is indeed not as pronounced as when metals or other synthetic molecules are included in the DES formulations. Juneidi et al. (Juneidi et al., 2015) describe a comparison between 2-hydroxy-*N,N,N*-trimethylethanaminium chloride-based DES and N,N-Diethyl ethanol ammonium chloride (EAC)-based DES, from which a higher antifungal activity toward *Aspergillus niger* was observed for the EAC-based DES. The highest toxicity was obtained from the EAC: Zinc chloride ($ZnCl_2$) and EAC:ZnN with MICs under $2.2\,mg\,mL^{-1}$. So systems containing metal salts that interact ionically, have higher antifungal activities than systems which contain amides and polyols such as urea, glycerol, and ethylene glycol. Although a modest toxicity was observed for the ChCl-based systems, it was verified for all the described DES a linear relation between the toxicity and the *A. niger* growth inhibition. Interestingly, two systems appear as an exception—ChCl:Urea (U) and ChCl:Glycerol (Gly)—in this cytotoxicity trend, they present even a higher *A. niger* spore growth in comparison to the control. The explanation for this appears to be in the fact that urea can be used as a nitrogen source and fertilizer for fungi and favor molds growth (Abreu, Sanguinetti, Amillis, & Ramon, 2010; Azizullah, Nasir, Richter, Lebert, & Häder, 2011). However, in higher concentrations it was observed an effective inhibition for all systems, even those with urea. From these results, it is once more highlighted the tailor-made potential of DES since the choice of their individual components can dictate its biological activity, as cytotoxic or supplementary nutrients for fungal growth, in a concentration-controlled manner. In another work, it was reported the cytotoxic effect of ChCl-based DES toward four fungi strains—*Phanerochaete chrysosporium*, *A. niger*, *Lentinus tigrinus* and *Candida cylindracea*—here working as a model of eukaryotic microorganisms. The highest antifungal activity was obtained using ChCl:$ZnCl_2$ for all four fungi species, followed by ChCl:malonic acid (MA) and ChCl:PTSA. It was also reported that the DES presented higher antifungal activities than their respective individual components (Juneidi, Hayyan, & Mohd Ali, 2016). Furthermore, there have been several reports on NADES and DES antifungal activity toward *Candida albicans*, a widely studied fungi due to its well-defined presence in human health which possesses a $Gram^+$ like membrane structure (Radošević et al., 2018; Silva et al., 2020). In their study, Radošević et al. (2018) considered several betaine, choline, sugar, and sugar alcohol-based NADES and reported no significant inhibition effect on *Candida albicans* except for Ch:Oxalic acid. In contrast, citric acid-based DES exhibited significant overall antifungal inhibition with maximum inhibition observed for CA: glucose (Glu):Gly and CA: fructose (Fru):Gly. The contribution of the organic acid component in the NADES leads to a decrease of the pH in the growth media (pH <6.0), which is below the optimal pH for the microorganisms to grow (pH 6.5–7.5). Furthermore, the presence of hydroxyl groups in the organic acid most likely contributes to the cell membrane denaturation, and may be the major contributors to these system's antimicrobial activity (Gorke et al., 2010; Wikene et al., 2017). Silva et al. (2020) reported a terpene-based DES combining perillyl alcohol with ibuprofen, in a 4:1, 6:1 and 8:1 M ratio, with antimicrobial activity (see Section 3.1) but their results also suggest an antifungal activity of the NADES similar to POH alone. This is most likely explained due to the absolute mass composition of these formulations where POH takes most part. These are nevertheless interesting results since this antimicrobial effect is combined with IBU antiinflammatory action, which unlocks the range of the therapeutic application action of these systems. Additionally, and as previously discussed, the use of terpene-based antimicrobial systems have the extra advantage of being considered as less prone to microbes development of resistance mechanisms, since its primary actuation method in cell membrane perturbation is nonspecific in nature (Belanger, 1998; Burt, 2004; Isman, 2000). Furthermore, the systems reported to have an effect on biofilm detachment of *C. albicans*, similarly to what reported for *E.coli* and MRSA, which is dependent on the NADES concentration. In summary, the hence reported work once more highlights the difference between the effect of these system's individual components and their effect when combined as an eutectic system. The charge delocalization developed via hydrogen bonding is one of the possible explanations for this cytotoxicity increase (Hayyan et al., 2016; Modica-Napolitano & Aprille, 2001). Another possible explanation is DES toxicity arising as a consequence of dehydrating effects, resembling those induced by dehydrating agents (such as $CaCl_2$) (Cardellini et al., 2014; Cardellini et al., 2015). Either way, there is an evident synergistic effect resulting from DES formation which leads to a notable tailor-made potential of these systems. Either by varying the system concentration or of one of its components, it can go from having a beneficial to a deleterious effect, as it was evidenced from the systems possessing urea. As therapeutic agents, the straightforward deleterious effect of these systems toward pathogenic microorganisms should be explored at its most, but high toxicity against microorganisms used in biotechnology, for example in beverage production, can come as a major inconvenient. Therefore, further work in unraveling the detailed cytotoxic mechanisms toward fungi is necessary so that eutectic systems can be designed accordingly to the desired application.

3.3 Anticancer activity

Despite the fact that the cytotoxicity of several eutectic systems toward certain cancer cell lines has been widely reported, the bioactivity and resulting anticancer potential of these emerging systems is not yet well explored. Hayyan, Looi, Hayyan, Wong, and Hashim (2015) have reported an interesting cytotoxic profile of different ammonium-based eutectic systems on several cancer cell lines. The selectivity index (SI) of a compound is a parameter used to express a compound's in vitro efficacy in the inhibition of a particular target but can also be considered as a preliminary safety indicator of its potential as therapeutic agent. In this sense, it is a measure of the window between anticancer activity and cytotoxicity as a result of the ratio between the cytotoxicity and the antiproliferative effect of the eutectic system. From the resulting SI of the reported systems (showed in Table 3) toward cancer cells, comparing to a normal cell line, it is possible to verify that the IC50 values for the normal cell line (OKF6) was slightly higher than the ones for the cancer cell lines. Such results indicate that these eutectic systems were

Table 3 Selectivity index of ammonium-based eutectic systems for cancer cells (human breast cancer cell line (MCF-7); human prostate cancer cell line (PC3); human malignant melanoma cell line (A375); human liver hepatocellular cell line (HepG2); human colon adenocarcinoma cell line (HT29); carcinoma-derived human oral keratinocyte cells (H413)) compared to a normal cell line of human oral keratinocytes (OKF6).

	Selectivity index					
Eutectic system	**MCF-7**	**PC3**	**A375**	**HepG2**	**HT29**	**H413**
[ChCl]-[Gly]	2.162	1.542	2.615	1.309	1.662	0.864
[ChCl]-[EG]	2.579	2.120	1.979	2.812	2.282	1.232
[ChCl]-[U]	2.789	2.949	1.374	2.172	2.263	1.204
[ChCl]-[TEG]	2.137	1.692	2.797	1.902	1.974	1.782

EG—ethylene glycol; U—urea; TEG—triethylene glycol.
Adapted from Hayyan, M., Looi, C. Y., Hayyan, A., Wong, W. F., & Hashim, M. A. (2015). In vitro and in vivo toxicity profiling of ammonium-based deep eutectic solvents. PLoS One, 10(2), 1–18. https://doi.org/10.1371/journal.pone.0117934.

more toxic toward cancer cells than to normal cells, exhibiting potential as anticancer agents.

Interestingly, it was also demonstrated that two or more components working as an eutectic mixture result in new entity with different and enhanced properties, when compared to its individual components, or its components in a physical mixture—in other words, the aqueous solutions of each component in the DES separately dissolved in distilled water, using the same concentration as in the DES, and/or the two components dissolved in distilled water—thus resulting in a more acute and selective impact on cancer cells (Hayyan et al., 2015; Pereira et al., 2019). It has been reported that the HBD, has a predominant effect on the cytotoxicity of the resulting DES, namely by protein denaturation, which leads to a disruption in cell activity and further death (Hayyan et al., 2016; Hayyan, Hashim, Al-Saadi, et al., 2013). Considering that the hydrogen bonding between HBD and HBA affects not only the systems physical properties, as viscosity, but also its chemical structure, it could explain why not only the concentration of the individual compounds matter when speaking about eutectic systems cytotoxicity. In another example, Badisa et al. (2009) evaluated two different eutectic systems, piperidinyl:diethylstilbestrol and pyrrolidinyl:diethylstilbestrol, on the MCF-7 cancer cell line and rat normal liver cells, and compared them with tamoxifen, a drug used in the treatment of breast hyperplasia by working as an antagonist of the estrogen-related receptors. The resulting selectivity indexes (showed in Table 4) revealed a certain degree of similarity

Table 4 Selectivity index of piperidinyl:diethylstilbestrol (Piperidinyl-DES) and pyrrolidinyl:diethylstilbestrol (Pyrrolidinyl-DES) and 4-Hydroxy tamoxifen (tamoxifen) for MCF-7 cancer cell line compared to rat normal liver cells (CRL-1439) (Badisa et al., 2009).

	Selectivity index
Piperidinyl-DES	1.78
Pyrrolidinyl-DES	>2.84
4-Hydroxy tamoxifen	1.29

between the eutectic systems and tamoxifen, suggesting that they could act in the same cell target site, namely the estrogen receptor. Nevertheless, since the selectivity index can for this matter act as an preliminary safety indicator, namely on the side effects of drugs on normal cells of patients undergoing chemotherapy, Table 4 shows that pyrrolidinyl-DES has an SI value higher than 2 while tamoxifen an SI value much less than 2 indicating higher toxicity of this compound.

The mechanisms underlying eutectics cytotoxicity is still a matter for much debate. It has been suggested that one of DES primary cytotoxic mechanism is the induction of cell death, by increasing cellular membrane permeability leading to its further disruption and consequent cell death (Hayyan et al., 2016). In an example, the MCF-7 cells were exposed to the eutectic system and further stained with a green fluorescent membrane permeability dye and a blue fluorescent DNA biding dye, working as membrane integrity and cell viability indicators, respectively (Fig. 6).

The cancer cells exposed to the four DES revealed higher green fluorescence, indicating an increase of cell membrane permeation possibly due to damage to cell membrane in comparison with the control, where there was less uptake of the permeability dye to the cytosol. A decrease of cell viability is observed by the contrast in blue fluorescence between the control and the DES exposed cells. A higher blue fluorescence, characteristic of cell confluency, was observed in the control cells while the smaller amount of viable cells in the treated samples is emphasized by the lower blue fluorescence (Hayyan et al., 2015). In another study, Mbous, Hayyan, Wong, Looi, & Hashim (2017) also reported damage on the cell membrane as a consequence of exposure to two ChCl-based NADES with glucose and fructose as HBD, and a DES based on *N,N*-diethylethanol ammonium chloride (DAC) and TEG. Similarly to aforementioned, they report an increase in green

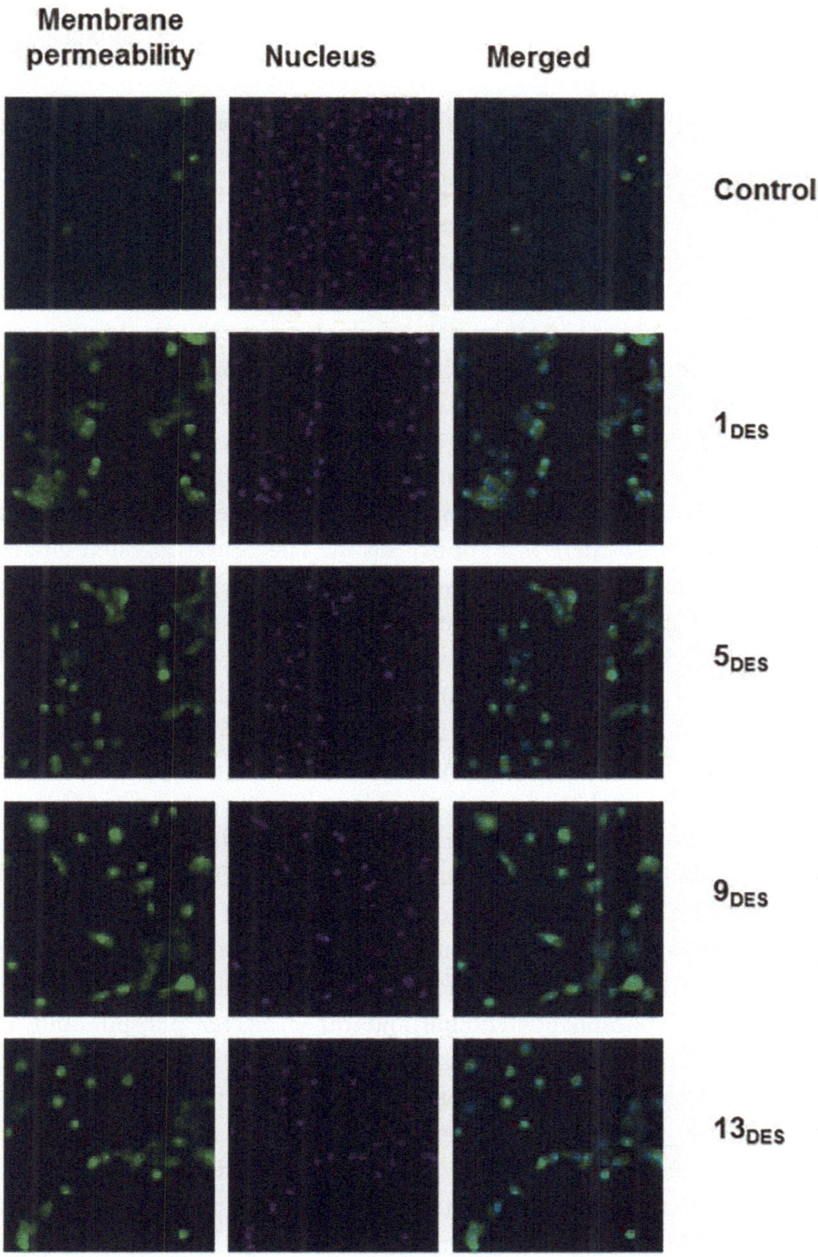

Fig. 6 Membrane permeability and cell viability assessment of MCF-7 cell line after treatment with LC_{50} concentrations of DES. Cells were stained with a green fluorescent membrane permeability dye and a blue fluorescent DNA biding dye (Hoechst 333258 dye). 1DES—ChCl:Gly; 5DES—ChCl:EG; 9DES—ChCl:U; 13DES—ChCl:TEG (Hayyan et al., 2015).

fluorescence in cells exposed to the eutectic systems, implying an increase in the cell membrane porosity, while in control cells almost no fluorescence was observed. A more acute generalized cytotoxicity toward cancer and normal cells was described for the DES rather than for both NADES. There is a trend between the eutectic systems cytotoxicity and several factors such as the physical properties of the system, its chemical structure, cellular requirements of the cancer cells, presence of water in the systems, and nature and interactions of the system's individual components with the cell (Hayyan et al., 2016). From a practical point of view, for example, if one considers a system based on essential carbohydrates such as fructose and glucose, it has also to be considered that their metabolism provides the energy required for the essential cellular functions. At a closer look into the role of fructose and glucose as cell energy sources, two things stand out, both are reducing sugars, but one is the primary energy source for all tissues whereas fructose is metabolized primarily by the liver. Glucose metabolism is regulated by insulin, so there is a control on the fate of the ingested glucose. On the other hand, fructose processing is not controlled by insulin, meaning that fructose can nearly continuously enter the glycolytic pathway and be metabolized. If in energy-deficient conditions, this pathway is an effective beneficial measure for keeping cellular essential functions but, if at homeostasis, this ongoing fructose metabolization can have fatal intracellular effects (Ludwig, 2013). Particularly, when considering cancer cells, that are undergoing abnormal cell growth, there is an innate requirement for more energy than normal cells, to undergo for metastatic phase of migration and tissue invasion (Port, Ruth, & Istfan, 2012; Smith, Abbott, & Ryder, 2014). Therefore, it is expected a higher cellular tolerance of these carbohydrates-based eutectics when applied as anticancer therapeutics. Eutectic systems based on ChCl are among the most common. Choline is essential in the metabolism of phosphatidylcholine and other bilayer phospholipids, so its import is crucial either for normal or cancer cell lines (Michel, Yuan, Ramsubir, & Bakovic, 2006). For that, there are several protein families in the cell phospholipid bilayer which are responsible for the intracellular diffusion of choline. Although, it is generally considered as a safe ingredient which leads to the assumption that the resulting DES possesses insignificant toxicity profiles, there have been reports where its cytotoxicity did not go unnoticed (Hayyan et al., 2015; Radošević et al., 2015; Wen et al., 2015). For the most part this cytotoxic effect has been attributed to the HBD, but there might be another contribution. As the proteins responsible for choline intracellular diffusion reach their maximum transport rate, this extracellular compound might begin to accumulate on cell surface and thereby contribute to the accumulation of ammonium cations, promoting the disruption of the phospholipid bilayer and inducing cell death (Mbous et al., 2017). Upon cell penetration, the eutectic systems contribute to an increase of reactive oxygen species (ROS) which increases intracellular oxidative stress and ultimately result in apoptosis (Hayyan et al., 2015). This was observed in MCF-7 cancer cells for ammonium-based DES (Hayyan et al., 2015), but also in HelaS3, AGS and WRL-68, were a comparison between NADES (ChCl:Glu and ChCl:Fru) and a DES (DAC:TEC) revealed DES as a greater inducer of redox stress possibly due to NADES operating through a disparate mode of action (Hayyan et al., 2015; Mbous et al., 2017). Fructose-based systems may increase the synthesis of advanced glycation end products (AGEs) faster than when glucose-based systems are in use. The AGEs are powerful enhancers of ROS synthesis, leading to impaired metabolic functions related to nucleic acid damage, lipid peroxidation, oxidative damage and membrane damage.

Furthermore, plant secondary metabolites, such as terpenes, have been playing an interesting part in the search for natural occurring molecules with anticancer potential. Limonene (Lim) is a cyclic monoterpene studied at preclinical and clinical stages due to its chemopreventive and chemotherapeutic activities toward lung, gastric and prostate cancers (Bicas, Neri-Numa, Ruiz, De Carvalho, & Pastore, 2011; Miller et al., 2013). Particularly, in colorectal cancer limonene has been reported to induce apoptosis via mitochondrial pathway and affect PI3K/Akt signaling pathway of survival and apoptosis (Jia et al., 2013). Recently, Pereira et al. (Pereira et al., 2019) reported the anticancer activity of Lim-based therapeutic deep eutectic systems (THEDES) toward HT29 cancer cell line. The system combining limonene: ibuprofen (4:1) was able to inhibit selectively the cancer cells without compromising normal cells viability. Moreover, by comparing the antiproliferative profile of this system with a physical mixture of its individual compounds, it is possible to recognize a different mechanism of action of the eutectic system from its individual compounds, emphasizing an eutectic-promoted synergetic effect (Fig. 7).

Considering that inflammation and angiogenesis are major contributors to tumor progression and metastasis, nonsteroidal antiinflammatory drugs such as ibuprofen can play a very interesting therapeutic effect in cancer therapy (Coussens & Werb, 2002). Nevertheless, the limited solubility and consequent restricted bioavailability of these APIs have been a major setback. The Lim:Ibu (4:1 and 8:1 M ratios) systems have revealed that

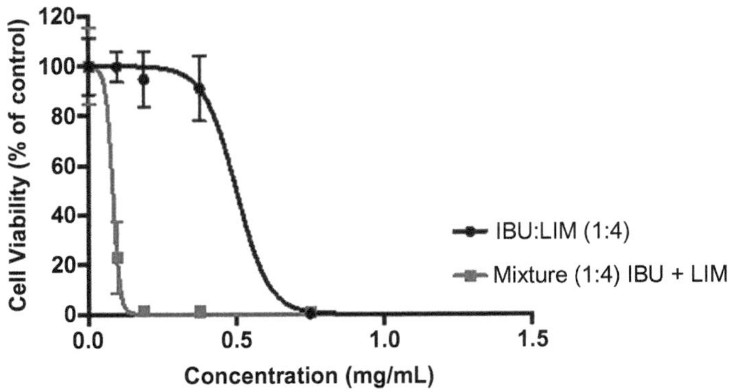

Fig. 7 Comparation of the antiproliferative effect of Lim:Ibu (4:1) with its individual components physical mixture (Pereira et al., 2019).

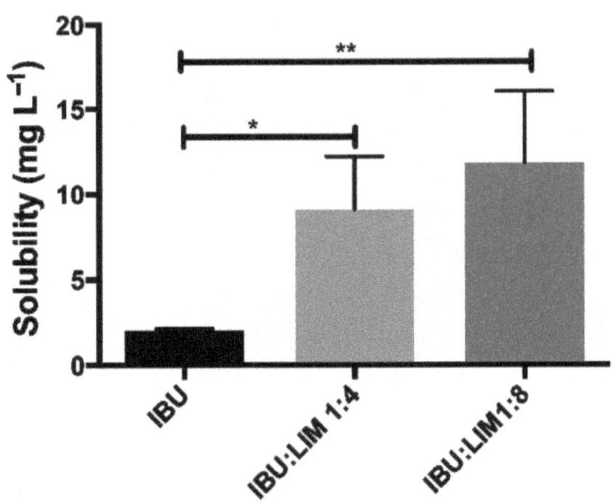

Fig. 8 Comparation of the solubility of Lim:Ibu (4:1 and 8:1) with Ibu as individual component (Pereira et al., 2019).

eutecticity can be a strategy to overcome this limitation (Fig. 8), promoting in this way ibuprofen antiinflammatory activity which may lead to a better control on tumor progression. These eutectic-promoting effect on APIs solubility in accordance to recent publications by Duarte et al. (Duarte et al., 2017).

In another example, POH, a natural occurring monoterpene produced, via the mevalonate pathway, by the hydroxylation of its precursor limonene has shown antitumoral activity, as chemoprotective and chemotherapeutic, toward several cancer cell lines such as colorectal cancer. Due to its amphipathic nature and consequent limitations within its administration form, it has failed to produce satisfactory results in a human clinical trial setting (Belanger, 1998; Gould, Tutsch, Stewart, & Arzoomanian, 1998). POH in combination with ibuprofen as an eutectic system have revealed promising results as a selective system against colorectal cancer cells (HT29) while having low toxicity toward a model of normal intestinal cells (caco-2) (see Table 5) (Silva et al., 2020).

The higher selectivity index reported for the POH:Ibu (3:1) system reveals that although both systems presented similar anticancer activity, since the antiproliferative EC_{50} is almost identical, the 3:1 M ratio was the most able to inhibit the cancer cells without compromising normal cell viability. These results are a preliminary indication of a selective deleterious action toward cancer cells, most likely attributed to POH itself as previously briefly mentioned (see Section 3.2). Considering the amphipathic nature, of both POH and cell membranes, when this monoterpene interacts with the phospholipid bilayer its polar end will bind to the membrane, but its hydrophobic end will prevent its full cellular uptake. This POH-membrane interaction most likely alters cell membrane mechanical fluidity which has a deleterious effect in its stability. In cells experiencing an increased abnormal cell division rate, such as cancer cells, this effect can be of most importance since this mechanical disturbance may cause membrane disruption and eventual cell death. In contrast, in normal cells POH is eventually disengaged by the energy supplied from thermal agitation before the event of cell division (Duelund, Amiot, Fillon, & Mouritsen, 2012).

Perhaps the most important criteria for an efficient therapeutic drug for cancer treatment, is to have as less side effects as possible on normal patients' cells. This can be achieved by using lower doses of the drugs or by designing them to be highly selective toward its target. Based on the abovementioned reports, the existence of eutectic systems with acute toxicity against

Table 5 EC50 values for the cytotoxicity toward Caco-2 cell line (cytotoxicity assay) and cytotoxicity toward HT29 line (antiproliferative assay) and resulting selectivity indexes (Silva et al., 2020).

	EC_{50} values (mM)		
System	**Cytotoxicity assay**	**Antiproliferative assay**	**Selectivity index**
POH:Ibu (3:1)	8.46	1.43	5.89
POH:Ibu (8:1)	4.35	1.68	2.60

cancerous cells while being harmless to normal cells is not implausible. The high tunability of DES properties, via the endless possible combinations between interesting and promising compounds, are a key feature in the context of creating a therapeutic agent adapted to the challenge imposed by each different cancer cell line. Either by membrane damage and disruption, ROS induction, apoptosis induction, herein exemplified, the toxicity mechanisms of DES against both malignant and nonmalignant cells are still poorly understood, thus better understanding of the mechanisms behind NADES/DES cytotoxicity toward cancer cells is a necessary step for generating tailor-made systems for each anticancer application, consolidating these systems as real alternatives to the stablished anticancer treatments.

3.4 Wound healing activity

One of the most remarkable aspects on eutecticity is probably its vast range of applications. There have been attempts to stimulate the wound healing event, with a special emphasis on patients suffering from diabetes which undergo an altered skin healing process, in order to overcome the major setbacks of currently available therapeutic solutions with high cost, prolonged duration of treatment, and side effects (Tan et al., 2019). Diabetic patients have an increased rate of colonization microorganisms such as *Staphylococcus aureus* in skin folds and nares, and also have a greater risk of postoperative wound infections (Tripathi & Srivastava, 2006). For this matter, therapeutic agents based on natural occurring molecules represent an attractive alternative in wound care, avoiding the recurrent use of antibiotics (Beserra et al., 2019; Chen et al., 2012; Mukherjee et al., 2013). In recent work Silva, Pereira, et al. (2019) reported the application of NADES based on menthol and fatty acids (lauric acid, myristic acid and stearic acid) as wound healing enhancers upon superficial epidermis disruption. Menthol has already been widely used for eutectic systems preparation in combination with a wide range of compounds including other terpenes and APIs such as ibuprofen (Duarte et al., 2017; Martins et al., 2019). This is a well-known organic compound found in *Mentha* species, known for its effectiveness as permeation enhancer together with a well-known antiinflammatory and antimicrobial effect (April, 1994; Juergens, Stober, & Vetter, 1998; Tsuk, 1990). Lauric acid and myristic acid are saturated fatty acids commonly present in vegetal and animal fats which have an effect in human skin natural disinfection (Nakatsuji et al., 2009). Both terpene and fatty acids combined as an eutectic system revealed a new therapeutic agent, with a combined

Table 6 MIC and MBC values of NADES and individual components.

Microbial strain	MIC (mM)			MBC (mM)		
	Me	SA	Me:SA	Me	SA	Me:SA
S. *aureus*	4	ND	3.26 + 1.79	8	ND	6.52 + 3.58
MRSA	8	ND	6.52 + 3.58	16	ND	13.03 + 7.16
MRSE	8	ND	6.52 + 3.58	16	ND	13.03 + 7.16

The Me:SA system was tested with 8:1 M ratio.
Me—menthol; SA—stearic acid; ND—not dissolved.

promising therapeutic action of both compounds. Menthol alone tested against S. *aureus*, and *S. aureus* Methicillin-resistant strains (MRSA and MRSE), revealed a MIC value of 4 mM and 8 mM, respectively (see Table 6). In contrast, stearic acid did not present antibacterial activity against any of the tested microorganisms, possibly as a consequence of its low solubility and diffusion rate.

The antibacterial effect of the system appears to be directly corelated to menthol itself since the MIC/MBC values overlap. Nevertheless, and considering that in the 8:1 M ratio eutectic system, 1 mol of Me:SA is composed of about 0.82 mol of menthol and 0.18 mol of stearic acid, this means that the tested systems effectively contains a lower amount of menthol, which strongly suggests a synergistic interaction between menthol and SA when as an eutectic mixture, that potentiates its antibacterial activity. Since keratinocytes have a major role in wound healing (Singer & Clark, 1999), the potential of Me:SA to be used as wound healing enhancer was reported via an assessment of keratinocytes (HaCaT cell line) migration potentiation (Fig. 9).

The two highest NADES concentrations induced significantly keratinocytes migration, which in practical terms, corresponds to a higher degree of wound closure. An increase of nearly 40% when compared to the control, represents a remarkable potential of these NADES to work as wound healing enhancers at least on epidermis superficial wounds. When giving a look on the individual components isolated effect, stearic acid reveals no enhancement effect, but 0.75 mM menthol is significantly different. Despite that, the NADES higher concentration (0.75 menthol +0.09 SA mM) was more effective than isolated menthol, which might suggest that the hydrogen bonding interaction of SA with menthol potentiates indeed its activity.

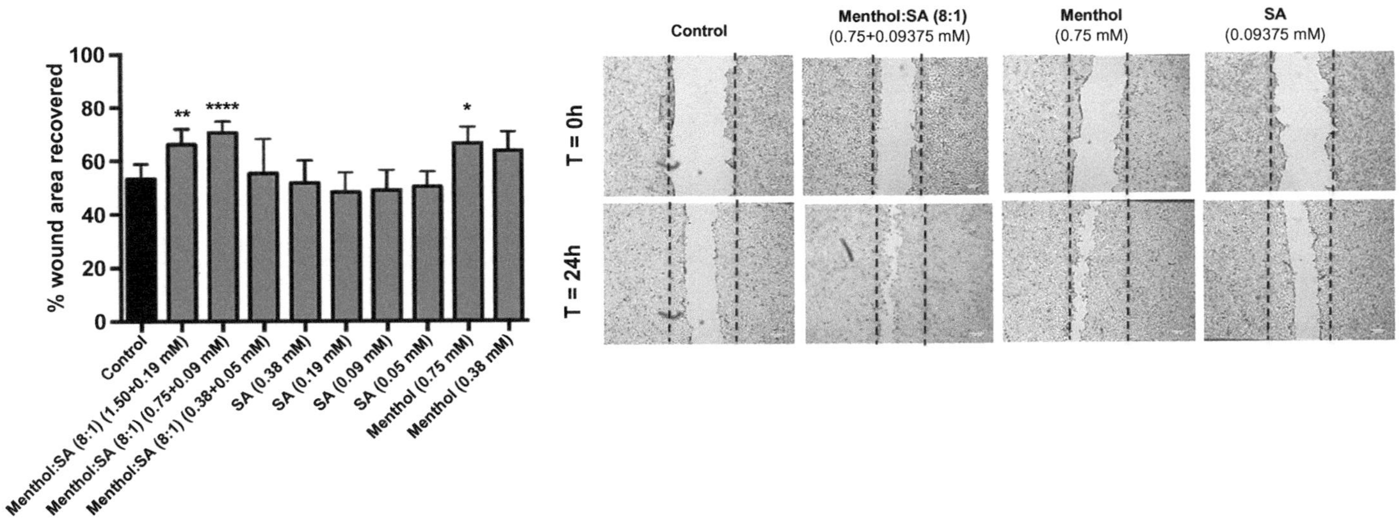

Fig. 9 Reported results from the wound healing assay with different concentrations tested for the Me:SA system along with the corresponding individual components (Silva, Pereira, et al., 2019).

A wound, or in other words, a disruption in epidermis is always a stress inductor for the body. Specially, among individuals with a chronical medical condition such as diabetes, wounds can result in serious challenges due to delayed wound healing mechanisms such as prolonged inflammatory stages and postponed proliferation stages (Landén, Li, & Ståhle, 2016). Several biomolecules such as the nuclear factor kappa B (NF-κB)—which regulates the gene expression of several proinflammatory cytokines such as interleukin-1beta (IL-1β), interleukin-6 (IL-6), tumor necrosis factor-alpha (TNF-α), and interleukin-10 (IL-10)—but also nitric oxide synthases (NOS), chemotactic and matrix proteins, have a key role on the immunological response and cell proliferation (DeClue & Shornick, 2015; Mezzano et al., 2004). Moreover, a chronic inflammatory state has been associated with oxidative stress which further delays the wound healing process. The consequent increase production of free radicals and decrease in the antioxidant enzymes activity, such as superoxide dismutase (SOD), glutathione peroxidase (GPx), heme oxygenase-1, (HO-1) and heme oxygenase-2 (HO-2), are additional delaying factors in the wound healing process (Soneja, Drews, & Malinski, 2005). Furthermore, growth factors such as epidermal growth factor (EGF), fibroblast growth factor (FGF), transforming growth factor-beta1 (TGF-β1), and vascular endothelial growth factor (VEGF), are also within the key factors that contribute to the healing process by stimulating and activating cell proliferation, through several processes such as angiogenesis, reepithelialization, and differentiation and production of the extracellular matrix (Landén et al., 2016; Singer & Clark, 1999; Xie et al., 2013). As mentioned before, there have been several natural occurring molecules, such as terpenes and fatty acids, associated with beneficial effects in wound healing. This effect is most likely due to their ability to modulate the inflammatory mediators, growth factors and oxidative stress response, in combination with the maintenance of a constricted degree of microorganism invasion and proliferation. Based on the aforementioned promising results by Silva and coworkers (Silva, Pereira, et al., 2019), a wound healing effect combined with antibacterial activity could be attributed to NADES in a dose dependent manner, most likely as a result of the system individual compounds bioactivities, but most importantly from the new entity formed by their combination as an eutectic system. Thus, better understanding of NADES wound healing mechanism is required and may generate new ideas for tailor-made systems with application on skin disorders. This discussion emphasizes once more the wide versatility of NADES as a promising approach toward the development of more effective therapies, and highlights the initial stages of scientific developments in this field.

4. "Greenness" and biosafety

In the past two decades several attempts have been made for the design and implementation of more efficient processes with the reduction of waste generated, however the pharmaceutical industry, as well as other fine chemical industries are still major waste producing manufacturers. The truth is that these industries are far from being green. In a realistic perspective, it is still unlikely to aim for completely green processes. In 1992, shortly after the scientific community draw their attention to green chemistry and sustainability, Roger Sheldon defined one of the first green chemistry metrics (Sheldon, 1992, 2017). A measure that has been known since then as the *E*-factor and which relates the mass of waste produced per mass of product formed. Together with his team, they made an inventory of the mass of waste produced in different processes related with fine chemical industry, which revealed that tons of waste are produced every year. The values differ significantly whether it is an oil refining industry, bulk chemicals or pharmaceuticals. The numbers in the pharmaceutical industry are undoubtedly the highest and the E-factor varies within the range of 25 up to 100 (kg waste/kg product) (Roger A. Sheldon, 2017). After raising awareness for process sustainability, it was clear that changes where needed. Following this, Anastas and Warner defined in 1998 the 12 principles of green chemistry (Anastas & Warner, 1998a; Anastas & Warner, 1998b). The principles defined served since then as directives toward the development of greener processes, however fulfilling all the 12 principles seems highly improbable. For this reason, scientists are pointing out to the use of terms like "greener" or "more sustainable" than exactly green processes (Moity et al., 2012). Most efforts were placed in the design of greener chemical reactions, in particular the design of less hazardous chemical synthesis with a higher atom economy and reduction of waste produced, together with the design of more efficient catalysts and the replacement of conventional solvents by safer ones. The best solvent would be no solvent (Sheldon, 2017). Unfortunately this is an utopic perspective and the solvent market has a significant volume, estimated to represent $57.34 billion by 2023 (MarketsandMarkets, 2018). Solvents are required in a wide range of industrial applications, from polymers, adhesives, paints, cosmetics, pharmaceuticals, i.e., nearly in every industry that one might think there is the need for solvent use. Moity et al. (2012) have created a map of sustainable solvents using the COSMO-RS approach and have identified seven classes of "greener" solvents, alternatives to conventional organic solvents. Water, supercritical fluids, bio-sourced solvents, ionic liquids and eutectic mixtures, fluorinated solvents, liquids polymers and eco-friendly solvents have been identified. From the work reported it is clearly unrealistic to assume one solution fitting all, so the options have to be weighted based on the objective for which the solvent is required. This study was conducted when eutectic mixtures were still somewhat paired with the family of ionic liquids. With increasing knowledge on eutecticity the panorama of the green solvents at our disposal should be revisited. While the first green chemistry principle that one might think of when considering NADES is "safer solvents and auxiliaries", they also fall in other categories. In particular, analyzing the greenness of these eutectic systems from the perspective of their use in pharmaceutical industries, which is the aim of this chapter, we can argue that NADES also fulfill the following principles: (i) atom economy—considering that the preparation of these eutectics involves solely the mixing of two compounds and that no chemical reaction is involved in this process the atoms of the reactants are all incorporated into the final product and no side products are formed, therefore there is no atom waste; (ii) less hazardous chemical synthesis—the preparation of NADES does not involve a chemical reaction, hence one cannot use the term synthesis for the physical process which takes place upon NADES production; (iii) design safer chemicals and (iv) use of renewable feedstocks—undoubtedly starting from naturally occurring molecules, NADES are a priori much safer than volatile organic solvents or, for instance ionic liquid the technical challenge in the use of such renewable feedstocks is to develop sustainable process to recover these natural source chemicals in a manner that does not jeopardize their carbon footprint; (v) design for degradation—NADES constituents can be metabolized by living organisms, hence these are naturally biodegradable and break down into innocuous degradation products that do not persist in the environment. The starting molecules present in the new eutectic therapeutic formulations fall mostly on the category of the fine chemicals. Therefore, considering that the E-factor reported for fine chemicals production is within the range of 5–50 kg waste/kg product, it would be possible to hypothesize that by boosting the large scale applications of NADES, compared to conventional APIs, could mean a decrease of 50% in the E-factor of the pharmaceutical industry.

A thorough analysis of the green chemistry metrics coupled with a life cycle assessment would provide insightful information of these new class of systems, nevertheless, these assumptions already provide good indications of the sustainability and "greenness" of NADES. The "output pulled" approach of the manufacture of these systems, in opposition to the conventional "input pushed" widely used in the chemical industry just by taking advantage of the abundance of raw materials, allows the outset of the development of highly application-adapted systems. Furthermore, this analysis is also a good premise which concerns the biosafety of NADES.

The design of new systems, particularly with the aim of designing products for human consumption, has to account for detailed toxicity studies that provide evidence of their efficacy but also their safety. There have been studies on NADES toxicology, but the information is scattered and mostly focused on the specific application envisaged by the authors of each study. Adding to this, there is still no consensus in what are the best techniques to evaluate the NADES toxicity and no directives or guidelines are yet available (Torregrosa-Crespo, Marset, Guillena, Ramón, & María Martínez-Espinosa, 2019). Toxicity studies have been adapted from the literature following the work conducted earlier within ionic liquids. While some cytotoxicity data are available, as described in the previous section, there is still a lack of more systematic toxicological studies of NADES, describing not only cell specific action but also the whole organism's response to these stimuli. There is now the need to go further in life's hierarchy to unravel not only the organism's response but also the environmental impact as an abiotic and biotic stimulus. The greenness effect of NADES cannot be fully ensured unless the environmental impact of these new systems is addressed, which gives a whole story like perspective. Toxicological studies have been carried out by different groups using different models. To the best of our knowledge the first studies were reported in 2013, by Hayyan and coworkers, where different DES were evaluated toward *Artemia salina,* a species of brine shrimp, widely used as standard organisms for testing toxicity (Hayyan, Hashim, Al-Saadi, et al., 2013; Hayyan, Hashim, Hayyan, et al., 2013). In 2015, new studies revealed the effect of DES on a plant, *Allium sativum* (garlic) and an invertebrate, hydra (Wen et al., 2015). At the same time the work of a different research group was reported on phytotoxicity of DESs on common wheat seeds (*T. aestivum*) (Radošević et al., 2015). Moreover, much work has been centered on the effect toward marine bacteria *Aliivibrio fischeri* (De Morais, Gonçalves, Coutinho, & Ventura, 2015; Macário et al., 2018). Recently, Giner and coworkers have developed a mathematical model based on the quantitative structure-activity relationship (QSAR) methodology to predict the ecotoxicity of NADES. Toxicity screening of both organic and inorganic compounds has been carried out using *A. fischeri*, which has demonstrated to be a reliable, reproducible and cost effective methodology (Giner, Lafuente, Lapeña, Errazquin, & Lomba, 2020). With the advances in both reported experimental data and the development of mathematical models able to provide cues on the toxicology of NADES, future development on the design of safer and less hazardous system is foreseen. Both in vivo and in silico tools are essential in the development of greener and more sustainable NADES for a myriad of different applications including biotechnological, pharmaceutical and biomedical sciences.

5. Conclusion and perspectives

In the current scenario of a worldwide quest for sustainable and "greener" alternatives for current applications, in the past two decades the number of reports on NADES/DES applicability in various situations has rapidly grown. The tailor-made versatility of these eutectics, meaning in practical terms a wide range of chemical and physical properties, has resulted in promising studies applied in industry, biotechnology, cosmetics, pharmaceuticals, food, and health. From this book chapter several key aspects can be pinpointed providing the reader basic information on NADES therapeutic effect regarding its individual constituents and cellular target, namely as antibacterial, antifungal, anticancer and/or as wound healing agents. Though to some extent these molecules, their interactions and their biological output are increasingly being understood, the primary mechanisms behind the regulation of this physicochemically induced biological functions remain unknown. The increasing knowledge on metabolic pathways, their genetic control and proteomic blueprints, alongside with life's pursue for parsimony, have showed that the production of each of these natural molecules by an organism is connected to a designated biological function. Considering that it has been proven that these ubiquitous natural molecules are present in high concentrations in all living cells, and certain combinations form NADES (Choi et al., 2011), the question of what is NADES role in certain biological functions still remains unanswered. In other words, if these compounds are present in excess, they must have some sort of biological function, or else they would compromise parsimony and homeostasis. Most likely they have a specific cell target to exert that function and some sort of feedback mechanism associated, resembling already known examples such as quorum sensing or hormonal effects. If so, NADES/DES specific target can be a major key in the path to unravel the eutectic-promoting therapeutic effects. As discussed by Lakhani et al. (2019), it is of most interest to have a well-defined distinction between NADES individual compounds biosynthetic assembly, such as for instance primary and secondary metabolites—as lactic acid or terpenes—and its biological function, since one is not necessarily directly correlated to another. As described in the previous sections, NADES individual compounds can exert different biological functions, along with the different possible combinations of these same compounds, rendering different outcomes. Adding to this fact, the versatility of targets in these systems, the dichotomy of metabolites vs their functions, lead to a number of hypothesis about the understanding of the multiple roles NADES may have as therapeutic agents. The advantages of the use of NADES/DES in pharmaceutical sciences have even more relevance when considering the greenness of these systems, which could greatly lower the large e-factor associated to the pharmaceutical industry. In opposition to ILs, DES are generally considered as safe from the start, due to the well described nontoxic effect of the individual components. However, considering what has been previously described in this chapter, the final eutectic mixture can have different toxic effects from its individual components as a result of synergistic effects of DES individual components in the eutectic form. There is still lack of consolidated information that clearly demonstrates the low toxicity, ecotoxicity and biodegradability of these systems, but the scientific community is actively pursuing these answers. Only through an orchestrated effort of the community will we, as scientists in the field boost the widespread application of NADES/DES and its implementation at an industrial and commercial scale.

Acknowledgments

The research leading to these results has received funding granted through the project Des. solve (ERC consolidator), ERC-2016-COG 725034. This work was supported by the Associate Laboratory for Green Chemistry-LAQV which is financed by national funds from FCT/MCTES (UIDB/50006/2020).

References

Abbott, A. P., Boothby, D., Capper, G., Davies, D. L., & Rasheed, R. K. (2004). Deep eutectic solvents formed between choline chloride and carboxylic acids: versatile alternatives to ionic liquids. *Journal of the American Chemical Society*, *126*(29), 9142–9147. https://doi.org/10.1021/ja048266j.

Abbott, A. P., Capper, G., Davies, D. L., Rasheed, R. K., & Tambyrajah, V. (2003). Novel solvent properties of choline chloride/urea mixtures. *Chemical Communications*, *9*(1), 70–71. https://doi.org/10.1039/b210714g.

Abreu, C., Sanguinetti, M., Amillis, S., & Ramon, A. (2010). UreA, the major urea/H+ symporter in Aspergillus nidulans. *Fungal Genetics and Biology*, *47*(12), 1023–1033. https://doi.org/10.1016/j.fgb.2010.07.004.

Anastas, P. T., & Warner, J. C. (1998a). *Green chemistry: Theory and practice, 12 principles of green chemistry*. Oxford University Press.

Anastas, P. T., & Warner, J. C. (1998b). *Green chemistry: Theory and practice*. Green Chemistry: Theory and Practice, Oxford University Press, New York.

April, A. (1994). Menthol and related cooling compounds. *Journal of Pharmacy and Pharmacology*, *46*, 618–630.

Azizullah, A., Nasir, A., Richter, P., Lebert, M., & Häder, D. P. (2011). Evaluation of the adverse effects of two commonly used fertilizers, DAP and urea, on motility and orientation of the green flagellate Euglena gracilis. *Environmental and Experimental Botany*, *74*(1), 140–150. https://doi.org/10.1016/j.envexpbot.2011.05.011.

Badisa, R. B., Darling-Reed, S. F., Joseph, P., Cooperwood, J. S., Latinwo, L. M., & Goodman, C. B. (2009). Selective cytotoxic activities of two novel synthetic drugs on human breast carcinoma MCF-7 cells. *Anticancer Research*, *29*(8), 2993–2996.

Batovska, D. I., Todorova, I. T., Tsvetkova, I. V., & Najdenski, H. M. (2009). Antibacterial study of the medium chain fatty acids and their 1-monoglycerides: Individual effects and synergistic relationships. *Polish Journal of Microbiology*, *58*(1), 43–47.

Belanger, J. T. (1998). Perillyl alcohol: Applications in oncology. *Alternative Medicine Review*, *3*(6), 448–457.

Bennett, P. M. (2008). Plasmid encoded antibiotic resistance: Acquisition and transfer of antibiotic resistance genes in bacteria. *British Journal of Pharmacology*, *153*(SUPPL. 1), 347–357. https://doi.org/10.1038/sj.bjp.0707607.

Beserra, F. P., Vieira, A. J., Gushiken, L. F. S., de Souza, E. O., Hussni, M. F., Hussni, C. A., et al. (2019). Lupeol, a dietary triterpene, enhances wound healing in streptozotocin-induced hyperglycemic rats with modulatory effects on inflammation, oxidative stress, and angiogenesis. *Oxidative Medicine and Cellular Longevity*, *2019*, 3182627. https://doi.org/10.1155/2019/3182627.

Bicas, J. L., Neri-Numa, I. A., Ruiz, A. L. T. G., De Carvalho, J. E., & Pastore, G. M. (2011). Evaluation of the antioxidant and antiproliferative potential of bioflavors. *Food and Chemical Toxicology*, *49*(7), 1610–1615. https://doi.org/10.1016/j.fct.2011.04.012.

Burt, S. (2004). Essential oils: Their antibacterial properties and potential applications in foods - a review. *International Journal of Food Microbiology*, *94*(3), 223–253. https://doi.org/10.1016/j.ijfoodmicro.2004.03.022.

Cardellini, F., Germani, R., Cardinali, G., Corte, L., Roscini, L., Spreti, N., et al. (2015). Room temperature deep eutectic solvents of (1S)-(+)-10-camphorsulfonic acid and sulfobetaines: Hydrogen bond-based mixtures with low ionicity and structure-dependent toxicity. *RSC Advances*, *5*(40), 31772–31786. https://doi.org/10.1039/c5ra03932k.

Cardellini, F., Tiecco, M., Germani, R., Cardinali, G., Corte, L., Roscini, L., et al. (2014). Novel zwitterionic deep eutectic solvents from trimethylglycine and carboxylic acids: Characterization of their properties and their toxicity. *RSC Advances*, *4*(99), 55990–56002. https://doi.org/10.1039/c4ra10628h.

Chen, X., Peng, L. H., Li, N., Li, Q. M., Li, P., Fung, K. P., et al. (2012). The healing and anti-scar effects of astragaloside IV on the wound repair in vitro and in vivo. *Journal of Ethnopharmacology*, *139*(3), 721–727. https://doi.org/10.1016/j.jep.2011.11.035.

Choi, Y. H., van Spronsen, J., Dai, Y., Verberne, M., Hollmann, F., Arends, I. W. C. E., et al. (2011). Are natural deep eutectic solvents the missing link in understanding cellular metabolism and physiology? *Plant Physiology*, *156*(4), 1701–1705. https://doi.org/10.1104/pp.111.178426.

Clements, R. G., Nabholz, J. V., & Zeeman, M. (1994). *Estimation toxicity of industrial chemicals to aquatic organisms using structure activity relationships*.

Coussens, L. M., & Werb, Z. (2002). Inflammation and cancer. *Nature*, *420*(6917), 860–867. https://doi.org/10.1038/nature01322.Inflammation.

Dai, Y., van Spronsen, J., Witkamp, G. J., Verpoorte, R., & Choi, Y. H. (2013). Natural deep eutectic solvents as new potential media for green technology. *Analytica Chimica Acta*, *766*, 61–68. https://doi.org/10.1016/j.aca.2012.12.019.

De Morais, P., Gonçalves, F., Coutinho, J. A. P., & Ventura, S. P. M. (2015). Ecotoxicity of cholinium-based deep eutectic solvents. *ACS Sustainable Chemistry and Engineering.*, *3*(12), 3398–3404. https://doi.org/10.1021/acssuschemeng.5b01124.

DeClue, C. E., & Shornick, L. P. (2015). The cytokine milieu of diabetic wounds. *Diabetes Management*, *5*(6), 525–537. https://doi.org/10.2217/dmt.15.44.

Domínguez de María, P., & Maugeri, Z. (2011). Ionic liquids in biotransformations: From proof-of-concept to emerging deep-eutectic-solvents. *Current Opinion in Chemical Biology*, *15*(2), 220–225. https://doi.org/10.1016/j.cbpa.2010.11.008.

Duarte, A. R. C., Ferreira, A. S. D., Barreiros, S., Cabrita, E., Reis, R. L., & Paiva, A. (2017). A comparison between pure active pharmaceutical ingredients and therapeutic deep eutectic solvents: Solubility and permeability studies. *European Journal of Pharmaceutics and Biopharmaceutics*, *114*, 296–304. https://doi.org/10.1016/j.ejpb.2017.02.003.

Duelund, L., Amiot, A., Fillon, A., & Mouritsen, O. G. (2012). Influence of the active compounds of Perilla frutescens leaves on lipid membranes. *Journal of Natural Products*, *75*(2), 160–166. https://doi.org/10.1021/np200713q.

Giner, B., Lafuente, C., Lapeña, D., Errazquin, D., & Lomba, L. (2020). QSAR study for predicting the ecotoxicity of NADES towards Aliivibrio fischeri. Exploring the use of mixing rules. *Ecotoxicology and Environmental Safety*, *191*, 110004. https://doi.org/10.1016/j.ecoenv.2019.110004.

Gorke, J., Srienc, F., & Kazlauskas, R. (2010). Toward advanced ionic liquids. Polar, enzyme-friendly solvents for biocatalysis. *Biotechnology and Bioprocess Engineering*, *15*(1), 40–53. https://doi.org/10.1007/s12257-009-3079-z.

Gould, N., Tutsch, D., Stewart, A., & Arzoomanian, Z. (1998). Phrase I clinical of perillyl alcohol administered. *Clinical Cancer Research*, *4*, 1159–1164.

Guthrie, F. (1884). On eutexia. *Communicated by the Physical Society*, 462–482.

Hayyan, M., Hashim, M. A., Al-Saadi, M. A., Hayyan, A., AlNashef, I. M., & Mirghani, M. E. S. (2013). Assessment of cytotoxicity and toxicity for phosphonium-based deep eutectic solvents. *Chemosphere*, *93*(2), 455–459. https://doi.org/10.1016/j.chemosphere.2013.05.013.

Hayyan, M., Hashim, M. A., Hayyan, A., Al-Saadi, M. A., AlNashef, I. M., Mirghani, M. E. S., et al. (2013). Are deep eutectic solvents benign or toxic? *Chemosphere*, *90*(7), 2193–2195. https://doi.org/10.1016/j.chemosphere.2012.11.004.

Hayyan, M., Looi, C. Y., Hayyan, A., Wong, W. F., & Hashim, M. A. (2015). In vitro and in vivo toxicity profiling of ammonium-based deep eutectic solvents. *PLoS One*, *10*(2), 1–18. https://doi.org/10.1371/journal.pone.0117934.

Hayyan, M., Mbous, Y. P., Looi, C. Y., Wong, W. F., Hayyan, A., Salleh, Z., et al. (2016). Natural deep eutectic solvents: Cytotoxic profile. *Springerplus*, *5*(1), 913. https://doi.org/10.1186/s40064-016-2575-9.

Isman, M. B. (2000). Plant essential oils for pest and disease management. *Crop Protection*, *19*(8–10), 603–608. https://doi.org/10.1016/S0261-2194(00)00079-X.

Jia, S. S., Xi, G. P., Zhang, M., Chen, Y. B., Lei, B., Dong, X. S., et al. (2013). Induction of apoptosis by D-limonene is mediated by inactivation of Akt in LS174T human colon cancer cells. *Oncology Reports*, *29*(1), 349–354. https://doi.org/10.3892/or.2012.2093.

Juergens, U. R., Stober, M., & Vetter, H. (1998). The anti-inflammatory activity of L-menthol compared to mint oil in human monocytes in vitro: A novel perspective for its therapeutic use in inflammatory diseases. *European Journal of Medical Research*, *3*(12), 539–545.

Juneidi, I., Hayyan, M., & Hashim, M. A. (2015). Evaluation of toxicity and biodegradability for cholinium-based deep eutectic solvents. *RSC Advances*, *5*(102), 83636–83647. https://doi.org/10.1039/c5ra12425e.

Juneidi, I., Hayyan, M., & Mohd Ali, O. (2016). Toxicity profile of choline chloride-based deep eutectic solvents for fungi and Cyprinus carpio fish. *Environmental Science and Pollution Research, 23*(8), 7648–7659. https://doi.org/10.1007/s11356-015-6003-4.

Kareem, M. A., Mjalli, F. S., Hashim, M. A., & Alnashef, I. M. (2010). Phosphonium-based ionic liquids analogues and their physical properties. *Journal of Chemical and Engineering Data, 55*(11), 4632–4637. https://doi.org/10.1021/je100104v.

Kitahara, T., Koyama, N., Matsuda, J., Aoyama, Y., Hirakata, Y., Kamihira, S., et al. (2004). Antimicrobial activity of saturated fatty acids and fatty amines against methicillin-resistant Staphylococcus aureus. *Biological and Pharmaceutical Bulletin, 27*(9), 1321–1326. https://doi.org/10.1248/bpb.27.1321.

Knapp, H. R., & Melly, M. A. (1986). Bactericidal effects of polyunsaturated fatty acids. *Journal of Infectious Diseases, 154*(1), 84–94. https://doi.org/10.1093/infdis/154.1.84.

Lakhani, C. M., Tierney, B. T., Manrai, A. K., Yang, J., Visscher, P. M., & Patel, C. J. (2019). Natural deep eutectic solvents: Properties, applications, and perspectives. *Physiology & Behavior, 176*(3), 139–148. https://doi.org/10.1016/j.physbeh.2017.03.040.

Landén, N. X., Li, D., & Ståhle, M. (2016). Transition from inflammation to proliferation: A critical step during wound healing. *Cellular and Molecular Life Sciences, 73*(20), 3861–3885. https://doi.org/10.1007/s00018-016-2268-0.

Ludwig, D. S. (2013). Examining the health effects of fructose. *JAMA - Journal of the American Medical Association, 310*(1), 33–34. https://doi.org/10.1001/jama.2013.6562.

Luo, Y., Wang, Q., Lu, Q., Mu, Q., & Mao, D. (2014). An ionic liquid facilitates the proliferation of antibiotic resistance genes mediated by class i Integrons. *Environmental Science and Technology Letters, 1*(5), 266–270. https://doi.org/10.1021/ez500103v.

Macário, I. P. E., Jesus, F., Pereira, J. L., Ventura, S. P. M., Gonçalves, A. M. M., Coutinho, J. A. P., et al. (2018). Unraveling the ecotoxicity of deep eutectic solvents using the mixture toxicity theory. *Chemosphere, 212*, 890–897. https://doi.org/10.1016/j.chemosphere.2018.08.153.

Mano, F., Martins, M., Sá-Nogueira, I., Barreiros, S., Borges, J. P., Reis, R. L., et al. (2017). Production of electrospun fast-dissolving drug delivery systems with therapeutic eutectic systems encapsulated in gelatin. *AAPS PharmSciTech, 18*(7), 2579–2585. https://doi.org/10.1208/s12249-016-0703-z.

MarketsandMarkets. (2018). *Market Reserach Report - Solvents Market by Type (Alcohols, Ketones, Esters, Glycol Ethers, Aromatic, Aliphatic), Application (Paints & Coatings, Polymer Manufacturing, Printing Inks), Source (Petrochemical-based, Bio & Green), and Region - Global Forecast to 2023*.

Martins, M. A. R., Silva, L. P., Schaeffer, N., Abranches, D. O., Maximo, G. J., Pinho, S. P., et al. (2019). Greener terpene—Terpene eutectic mixtures as hydrophobic solvents. *ACS Sustainable Chemistry & Engineering.*, 7, 17414–17423. https://doi.org/10.1021/acssuschemeng.9b04614.

Mbous, Y. P., Hayyan, M., Wong, W. F., Looi, C. Y., & Hashim, M. A. (2017). Unraveling the cytotoxicity and metabolic pathways of binary natural deep eutectic solvent systems. *Scientific Reports*, 7, 1–14. https://doi.org/10.1038/srep41257.

Mezzano, S., Aros, C., Droguett, A., Burgos, M. E., Ardiles, L., Flores, C., et al. (2004). NF-κB activation and overexpression of regulated genes in human diabetic nephropathy. *Nephrology Dialysis Transplantation, 19*(10), 2505–2512. https://doi.org/10.1093/ndt/gfh207.

Michel, V., Yuan, Z., Ramsubir, S., & Bakovic, M. (2006). Choline transport for phospholipid synthesis. *Experimental Biology and Medicine, 231*(5), 490–504. https://doi.org/10.1177/153537020623100503.

Miller, J. A., Lang, J. E., Ley, M., Nagle, R., Hsu, C.-H., Thompson, P. A., et al. (2013). Human breast tissue disposition and bioactivity of limonene in women with early stage

breast cancer. *Cancer Prevention Research.*, *6*(6), 577–584. https://doi.org/10.1158/1940-6207.CAPR-12-0452.
Modica-Napolitano, J. S., & Aprille, J. R. (2001). Delocalized lipophilic cations selectively target the mitochondria of carcinoma cells. *Advanced Drug Delivery Reviews*, *49*(1–2), 63–70. https://doi.org/10.1016/S0169-409X(01)00125-9.
Moity, L., Durand, M., Benazzouz, A., Pierlot, C., Molinier, V., & Aubry, J. M. (2012). Panorama of sustainable solvents using the COSMO-RS approach. *Green Chemistry*, *14*(4), 1132–1145. https://doi.org/10.1039/c2gc16515e.
Mukherjee, H., Ojha, D., Bharitkar, Y. P., Ghosh, S., Mondal, S., Kaity, S., et al. (2013). Evaluation of the wound healing activity of Shorea robusta, an Indian ethnomedicine, and its isolated constituent(s) in topical formulation. *Journal of Ethnopharmacology*, *149*(1), 335–343. https://doi.org/10.1016/j.jep.2013.06.045.
Nakatsuji, T., Kao, M. C., Fang, J., Zouboulis, C. C., Zhang, L., Gallo, R. L., et al. (2009). Antimicrobial property of Lauric acid against Propionibacterium acnes: Its therapeutic potential for inflammatory acne vulgaris. *Journal of Investigative Dermatology.*, *129*(10), 2480–2488. https://doi.org/10.1038/jid.2009.93.
Paiva, A., Craveiro, R., Aroso, I., Martins, M., Reis, R. L., & Duarte, A. R. C. (2014). Natural deep eutectic solvents—Solvents for the 21st century. *ACS Sustainable Chemistry and Engineering*, *2*(5), 1063–1071. https://doi.org/10.1021/sc500096j.
Pereira, C. V., Silva, J. M., Rodrigues, L., Reis, R. L., Paiva, A., Duarte, A. R. C., et al. (2019). Unveil the anticancer potential of Limomene based therapeutic deep eutectic solvents. *Scientific Reports*, *9*(1), 1–11. https://doi.org/10.1038/s41598-019-51472-7.
Port, A. M., Ruth, M. R., & Istfan, N. W. (2012). Fructose consumption and cancer: Is there a connection? *Current Opinion in Endocrinology, Diabetes and Obesity*, *19*(5), 367–374. httpse://doi.org/10.1097/MED.0b013e328357f0cb.
Radošević, K., Čanak, I., Panić, M., Markov, K., Bubalo, M. C., Frece, J., et al. (2018). Antimicrobial, cytotoxic and antioxidative evaluation of natural deep eutectic solvents. *Environmental Science and Pollution Research*, *25*(14), 14188–14196. https://doi.org/10.1007/s11356-018-1669-z.
Radošević, K., Cvjetko Bubalo, M., Gaurina Srček, V., Grgas, D., Landeka Dragičević, T., & Redovniković, R. I. (2015). Evaluation of toxicity and biodegradability of choline chloride based deep eutectic solvents. *Ecotoxicology and Environmental Safety*, *112*, 46–53. https://doi.org/10.1016/j.ecoenv.2014.09.034.
Rodrigues, L. A., Pereira, C. V., Leonardo, I. C., Fernández, N., Gaspar, F. B., Silva, J. M., et al. (2020). Terpene-based natural deep eutectic systems as efficient solvents to recover Astaxanthin from Brown crab Shell residues. *ACS Sustainable Chemistry and Engineering*, *8*(5), 2246–2259. https://doi.org/10.1021/acssuschemeng.9b06283.
Russell, A. D. (2003). Similarities and differences in the responses of microorganisms to biocides. *The Journal of Antimicrobial Chemotherapy*, *52*(5), 750–763. https://doi.org/10.1093/jac/dkg422.
Sheldon, R. A. (1992). Organic synthesis—Past, present, and future. *Chemistry and Industry*, *23*, 903–906.
Sheldon, R. A. (2017). The: E factor 25 years on: The rise of green chemistry and sustainability. *Green Chemistry*, *19*(1), 18–43. https://doi.org/10.1039/c6gc02157c.
Silva, E., Oliveira, F., Silva, J. M., Reis, R. L., & Duarte, A. R. C. (2020). *Therapeutic roles of deep eutectic systems based on perillyl alcohol and ibuprofen*. Pharmaceutics MDPI.
Silva, J. M., Pereira, C. V., Mano, F., Silva, E., Castro, V. I. B., Sá-Nogueira, I., et al. (2019). Therapeutic role of deep eutectic solvents based on menthol and saturated fatty acids on wound healing. *ACS Applied Bio Materials*, *2*(10), 4346–4355. https://doi.org/10.1021/acsabm.9b00598.
Silva, J. M., Silva, E., Reis, R. L., & Duarte, A. R. C. (2019). A closer look in the antimicrobial properties of deep eutectic solvents based on fatty acids. *Sustainable Chemistry and Pharmacy*, *14*, 100192. https://doi.org/10.1016/j.scp.2019.100192.

Singer, A. J., & Clark, R. A. F. (1999). Cutaneous wound healing. *The New England Journal of Medicine.*, *341*, 738–746.
Smith, E. L., Abbott, A. P., & Ryder, K. S. (2014). Deep eutectic solvents (DESs) and their applications. *Chemical Reviews*, *114*(21), 11060–11082. https://doi.org/10.1021/cr300162p.
Soneja, A., Drews, M., & Malinski, T. (2005). Role of nitric oxide, nitroxidative and oxidative stress in wound healing. *Pharmacological Reports*, *57*(Suppl), 108–119.
Tan, W. S., Arulselvan, P., Ng, S., Norma, C., Taib, M., & Sarian, M. N. (2019). Improvement of diabetic wound healing by topical application of Vicenin-2 hydrocolloid film on Sprague Dawley rats. *BMC Complementary and Alternative Medicine*, *19*(1), 20. https://doi.org/10.1186/s12906-018-2427-y.
Torregrosa-Crespo, J., Marset, X., Guillena, G., Ramón, D. J., & María Martínez-Espinosa, R. (2019). New guidelines for testing "deep eutectic solvents" toxicity and their effects on the environment and living beings. *Science of the Total Environment.*, *704*, 135382. https://doi.org/10.1016/j.scitotenv.2019.135382.
Tripathi, B. K., & Srivastava, A. K. (2006). Diabetes mellitus: Complications and therapeutics. *Medical Science Monitor*, *12*(7), 130–147.
Tsuk, A. G. (1990). *Menthol enhancement of transdermal drug delivery*. Patent 4,931,283.
Wen, Q., Chen, J. X., Tang, Y. L., Wang, J., & Yang, Z. (2015). Assessing the toxicity and biodegradability of deep eutectic solvents. *Chemosphere*, *132*, 63–69. https://doi.org/10.1016/j.chemosphere.2015.02.061.
Wikene, K. O., Rukke, H. V., Bruzell, E., & Tønnesen, H. H. (2017). Investigation of the antimicrobial effect of natural deep eutectic solvents (NADES) as solvents in antimicrobial photodynamic therapy. *Journal of Photochemistry and Photobiology B: Biology*, *171*, 27–33. https://doi.org/10.1016/j.jphotobiol.2017.04.030.
Xie, Z., Paras, C. B., Weng, H., Punnakitikashem, P., Su, L. C., Vu, K., et al. (2013). Dual growth factor releasing multi-functional nanofibers for wound healing. *Acta Biomaterialia*, *9*, 9351–9359. https://doi.org/10.1016/j.actbio.2013.07.030.
Zhao, B. Y., Xu, P., Yang, F. X., Wu, H., Zong, M. H., & Lou, W. Y. (2015). Biocompatible deep eutectic solvents based on choline chloride: Characterization and application to the extraction of Rutin from Sophora japonica. *ACS Sustainable Chemistry and Engineering*, *3*(11), 2746–2755. https://doi.org/10.1021/acssuschemeng.5b00619.

CHAPTER TWELVE

The use of NADES to support innovation in the cosmetic industry

Caprin Benoit*, Charton Virginie, and Vogelgesang Boris
GATTEFOSSÉ Research & Development Centre, Saint-Priest Cedex, France
*Corresponding author: e-mail address: bcaprin@gattefosse.com

Contents

Abstract

Since their introduction, natural deep eutectic solvents (NADES) have emerged as a promising, eco-friendly alternative to petrochemicals to dissolve plant metabolites. As the demand for sustainable, green cosmetics is growing, those solvents represent unexplored opportunities to develop innovative extracts with unique phytochemical footprints and biological activities. However, despite the number of NADES described in the literature, only a few can actually be used for cosmetic applications because of safety or regulatory issues.

This chapter will first introduce the regulatory framework and green chemistry requirements that cosmetic active ingredients have to comply with. To answer the challenging question of how NADES can support innovation in the cosmetic industry, the second part will present the strategy chosen by Gattefossé to develop, characterize and industrialize solvents and extraction processes that overcome cosmetic constraints. Finally, the value-added of NADES compared to traditional extraction solvents in terms of their phytochemical composition and biological efficacy on human skin and skin models will be discussed through concrete examples.

Advances in Botanical Research, Volume 97
ISSN 0065-2296
https://doi.org/10.1016/bs.abr.2020.09.009

1. Regulatory framework, market trends and botanical cosmetic bioactives

1.1 A few words on the cosmetic regulatory and market environments

According to the European Regulation (EC) No 1223/2009 a "cosmetic product" is defined as "*any substance or mixture intended to be placed in contact with the external parts of the human body (epidermis, hair system, nails, lips and external genital organs) or with the teeth and the mucous membranes of the oral cavity with a view exclusively or mainly to cleaning them, perfuming them, changing their appearance, protecting them, keeping them in good condition or correcting body odours*" (Regulation (EC) No 1223/2009, n.d.). The European personal care association, namely Cosmetics Europe, describes seven main categories of cosmetics and personal care products: oral care, skin care, sun care, hair care, decorative cosmetics, body care and perfumes (Cosmetics Europe, n.d.).

The Federal Food, Drug, And Cosmetic Act codified into Title 21 Chapter 9 of the United States Code (sometimes referred to as 21CFR) offers a slightly different definition, as the term "cosmetic" identifies *"articles intended to be rubbed, poured, sprinkled, or sprayed on, introduced into, or otherwise applied to the human body or any part thereof for cleansing, beautifying, promoting attractiveness, or altering the appearance, and articles intended for use as a component of any such articles; except that such term shall not include soap."* For regulatory purposes, the term "cosmetics" includes products for the eyes, face, nails, hair, skin, and mouth, which may be in the form of products such as makeup, polish, hair dyes and coloring, sunscreens, fragrances, shave gel, oral care and bath products, and products for infants and children (United States Code, n.d.).

Cosmetic formulation is primarily a matter of chemical substances, even when substances are from natural origin. REACH—which stands for Registration, Evaluation, Authorization and Restriction of Chemicals—has entered into force in 2007 and has set the regulatory framework of chemical companies operating in Europe. In short, chemical manufacturers must identify, declare, register, and provide safety data and information for the substances they manufacture (a substance is defined as a chemical element and its compounds in the natural state or the result of a manufacturing process). Requirements increase with increasing production volumes.

Over the past decade, with growing environmental awareness and consumer education thanks notably to digital technologies and social networks, the demand for natural cosmetics has been on the rise. As a result, the pressure on the industry for providing more sustainable raw materials of natural origin has increased likewise. For instance, even though preservatives are strictly regulated in most countries (with a restricted list of ingredients and appropriate concentrations of use, based on available safety data), concerns about preservatives of petrochemical origin such as phenoxyethanol have been rising at the same time.

Given the wide range of ingredients required for cosmetic formulation (oils, emollients, surfactants, preservatives, texturing agents, actives and auxiliaries, pigments, fragrance), the challenge for a green transition is immense. The recently released ISO16128 standard is an interesting example of how the industry is moving to help formulators and consumer make more sustainable decisions.

Bioactive ingredients take a very special part in that landscape. If some chemicals such as retinol, hyaluronic acid and collagen have long acquired a cosmetic fame, plants have always been a source of inspiration for the personal care industry. Most traditional medicines are based on herbal remedies and beauty rituals have always been close to folk medicine. Licorice, calendula, mallow, rose, soy, ginseng, *aloe vera*, among others, are widely associated with cosmetic benefits. Since the late 1990s and the scandal of Creutzfeld-Jacob disease, the interest of the industry in plant remedies has renewed. But the green challenge is not as easy as it may appear. Indeed, until recently, a natural origin of the feedstock was enough for a botanical extract to claim "natural." Nowadays, expectations have become much higher and the naturalness of a commercially available plant extract must be considered as a whole: feedstock, farming type, eco-design, bio-based solvents and preservatives. This seems to be a smart direction when considering the current environmental challenges, but it also requires companies to be more creative and innovative. And one way to innovate in plant extraction is undoubtedly working on the extraction process itself.

1.2 Plant extracts, green chemistry and naturalness

Green chemistry is often cited as a master guidance to make a mindful use of biodiversity. This concept is based on 12 principles that aim to reduce manufacturing wastes of existing processes or make use of renewable feedstocks but also to define less polluting processes (Anastas & Warner, 1998; Anastas & Zimmerman, 2003). Naturally, solvents occupy a strategic place in the development of the concept of green chemistry. To be called "green," a solvent must meet various criteria such as safety, biodegradability and recyclability. Most of these principles can be applied to plant extraction processes. Indeed, the first step of all plant extraction processes is a solid-liquid mass transfer, meaning the transfer of phytochemicals contained in the solid plant raw material to an extraction solvent, most of the time a liquid. The higher the affinity of the phytochemicals for the solvent, the better the transfer is expected to be. However, polarity and chemical affinity are not the only requirements for efficient plant extraction. Solid processing prior to extraction (drying, grinding, sieving, etc.), kinetics, temperature, physicochemical characteristics of the solvent, stirring method, are some of the many variables that have to be taken into account to design an extraction process.

A huge number of industrial processes use non-renewable solvents. Efficient extraction solvents are often of petrochemical origin and thus, more and more disregarded. For instance, hexane is the reference extraction solvent when it comes to non-polar compounds such as curcumin. Many resources are invested nowadays to find an efficient, eco-friendly alternative to hexane. Supercritical CO_2 is a promising alternative to hexane when searching for lipophilic compounds, but extraction yields are sometimes low.

The number of extraction solvents suitable for manufacturing plant-derived bioactive ingredients has shrunk with increasing sustainable requirements. For the extraction of polar or semi-polar compounds, solvents such as water, glycols and ethanol (alone or in mixtures in water) are perfectly appropriate but certain drawbacks remain. Because of volatility and flammability, the transportation and handling of ethanol require specific safety measures and labeling. Besides, due to its drying nature, it can disturb the hydrolipidic film of the skin and cause discomfort sensations for people with sensitive skin. It can be removed by distillation but from an energy saving point of view, this downstream step is not ideal. Glycols such as butylene glycol, propylene glycol (1,2-propanediol) and propanediol (1,3-propanediol) can make up interesting extraction solvents when used alone or in combination with water. However, consumers tend to associate glycols with petrochemicals and often try to avoid glycol-containing cosmetic formulations. Nonetheless, manufacturers offer nowadays both butylene glycol and 1,3-propanediol of natural origin. When it comes to aqueous extracts, both liquid and powdered, the microbiological safety is a major limitation. Generally speaking, cosmetic products should not contain more than 100–1000 CFU/mL of total viable aerobic mesophilic microorganisms, and pathogens such as *Staphylococcus aureus* or *Escherichia coli* should be absent. Sterile filtration and sterile packaging can be an alternative but remain relatively expensive and create handling constraints at an industrial scale. Gamma irradiation doesn't obviously comply with natural standards. As a result, aqueous extracts often require the addition of preservatives. And unfortunately, only a few green preservatives can be used without downgrading product naturalness.

2. NADES, an opportunity for sustainable plant extraction

2.1 Low transition temperature mixtures for plant extraction

The discovery of ionic liquids (ILs), sometimes referred to as deep eutectic solvents (DES), natural deep eutectic solvents (NADES) or low transition temperature mixtures (LTTM) has opened new horizons for sustainable, green plant extraction. These names are sometimes used without robust physicochemical characterization and different names can refer to the same solvent. Moreover, in some cases, the identification of the class a mixture belongs to is unclear (Durand, Lecomte, & Villeneuve, 2016). But leaving the naming controversy aside, thanks to their physical properties, which facilitate the recycling (low vapor pressure and high melting point) ionic liquids were quickly classified as "green solvents". However, this affiliation is debated since some papers have reported a poor biodegradability and potential toxicity of some of these mixtures, especially halogenated ionic liquids (Zhao, Liao, & Zhang, 2007). And despite a growing interest of the scientific community in ionic liquids for industrial applications, some of the most widely used salts are incompatible with cosmetic applications.

Deep eutectic solvents are a more sustainable alternative than ionic liquids. They consist of a mixture of salts and organic compounds, one being a hydrogen bond donor (HBD), the other one a hydrogen bond acceptor (HBA). Charge delocalization is achieved through hydrogen bonding between compounds. As a result, those mixtures possess a melting point significantly lower than that of individual components, leading to a stable liquid phase at room temperature (Abbott, Boothby, Capper, Davies, & Rasheed, 2004).

Based on their NMR-based metabolomics knowledge, Choi and coworkers hypothesized that the metabolites that occur in large amounts in plant cells may form a third type of liquid, different from water and lipids. To support this hypothesis, they assessed the solubility of some common natural products in their new viscous liquids and discovered that those mixtures had an important role to play in cellular metabolism. Thus, they coined the term "Natural Deep Eutectic Solvents" for those mixtures, abbreviated NADES (Choi et al., 2011).

Dai has reported several stable NADES, particularly mixtures of primary metabolites, such as organic acids, amino acids, and sugars. The properties of those mixtures can be tailored by the addition of water. Moreover, they can be used as extraction solvents and can thus replace conventional organic solvents thanks to high extraction potential and good stability (Dai, 2013). Considering the diversity of constituents which can enter the preparation of DES and NADES, perspectives offered by these mixtures for the development of new industrial extraction processes are impressive. For instance, Jablonsky and coworkers published in 2018 a review on the up-cycling of agro-forestry biowastes (Jablonský, Škulcová, Malvis, & Šima, 2018). Hundreds of mixtures used to successfully extract several types of compounds of different polarity such as flavonoids, phenolic acids, stilbenes, tannins, protein, carbohydrates, nucleic acids, flavones, lignans or tocopherols, are listed. Choline chloride here appears as the most widely used compound to make up such solvents. However, choline chloride and its derivatives are prohibited in cosmetics in Europe (European Cosmetic Regulation EC No. 1223/2009, annex II). Thus, no solvent mixture containing choline chloride can be used as an extraction solvent for cosmetic plant extracts. In this context, there is a need to design new NADES mixtures compatible with cosmetic applications, and to assess their plant extraction performance in comparison with conventional solvents.

2.2 Designing NADES solvents for cosmetic applications

Gattefossé started working on NADES back in 2012, at a time when the available scientific literature was much more limited than today. We started looking for innovative, effective, bio-based solvents for plant extraction that would widely comply with cosmetic regulations. Those solvents had to meet both the green chemistry principles on one side and comply with industrial constraints on the other side. Keywords were unique phytochemical profile and extraction performance at least equivalent to that of conventional solvents.

First of all, substances should be allowed for cosmetic use and should thus primarily not appear in Annex II of the European Cosmetic Regulation EC No. 1223/2009. Second, those substances should be bio-based and available at a reasonable cost at industrial scale. Third, the solvents should demonstrate characteristic features of NADES mixtures, meaning a supramolecular structure as that described by Choi and coworkers. Fourth, mixtures should have a low or tunable viscosity at room temperature to allow homogeneous dispersion of the plant material and easy solid/liquid separation. Fifth, those mixtures should demonstrate long-term stability, including microbial stability, alone or in combination with green preservative systems. Sixth, the solvents should be compatible with classical cosmetic formulations such as lotions, creams or gels. Last but not least, they must demonstrate an excellent safety profile.

2.2.1 Selection of compounds and mixture preparation

Taking into account the above mentioned criteria, we selected the following ingredients:

- Trimethyl glycine (an amino acid also called betaine) as alternative quaternary ammonium to choline chloride, which as mentioned earlier is forbidden in cosmetic products in Europe. Betaine is non-toxic, already used in food supplements and cosmetics, available in large quantities at a very affordable price and can be approved by the Cosmos standard for natural cosmetics, when manufactured from sugar beet. Two other amino acids were selected according to the same criteria: glycine and arginine.
- Although organic acids are reported to be suitable for NADES preparation, we did not select these compounds because of their acidic nature that can create instability in cosmetic formulations.
- Fructose and glucose were selected among monosaccharides and sorbitol and xylitol among sugar alcohols (or polyols).
- Three short-chain alcohols were selected: two dihydroxy alcohols (propanediol and butylene glycol) and one trihydroxy alcohol (propane-1,2,3-triol or glycerol or glycerine). Even if these short-chain alcohols are rarely present in the plant kingdom, their use is well described in the literature for the preparation of NADES and they offer two strong advantages for cosmetic and industrial applications: they are liquid at room temperature, can be bio-based and are available at large scale for a reasonable price. The chemical structures of those compounds are presented in Fig. 1.

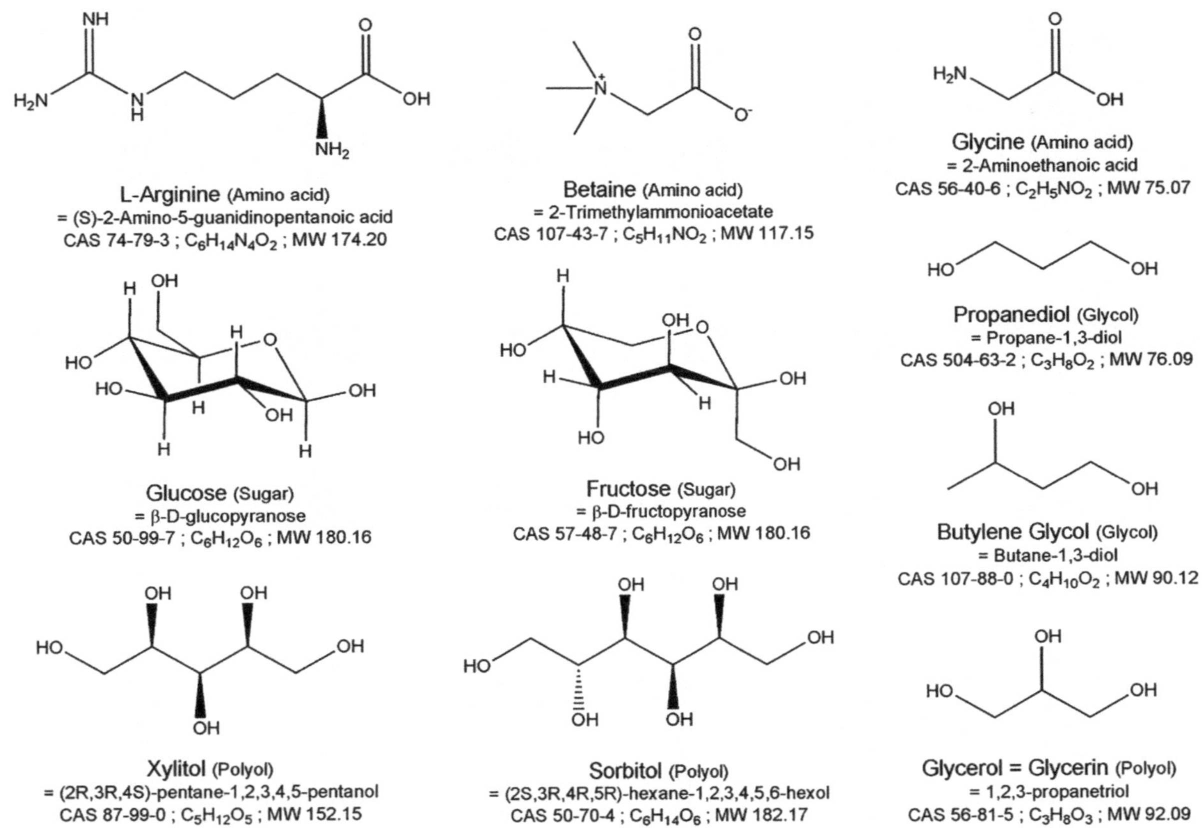

Fig. 1 Chemical structures of compounds selected for designing cosmetic NADES.

We next studied the feasibility of producing NADES using these 10 organic molecules. Different preparation methods can be used depending on the mixture, water content and intended application (Gomez, Espino, Fernandez, & Silva, 2018). Most of the time, compounds should be heated under stirring. Water can be added before being later evaporated using classical rotary evaporator, desiccator, or freeze-drying. Mixing and heating can be intensified using microwaves. From an energy-saving point of view, water evaporation, freeze-drying and process intensification are to be used with caution. We thus chose to use primarily a low-energy demanding process, which is conventional heating coupled with stirring.

Using this method, only a few combinations lead to a clear, stable, homogenous liquid at room temperature, which is an important selection criterion. Water is often added as a third compound for NADES, DES, LTTM preparation in order to decrease viscosity and promote liquefaction of solid organic compounds. We thus studied the effect of adding water to the mixture. The most successful trials were obtained with ternary mixtures based on the equimolar association of two organic compounds (1:1) added with water. The six most promising examples are integrated in the following discussion and reported in Table 1. To the best of our knowledge, it was the first time that those six ternary mixtures were described for potential use as plant extraction solvents (Charton & Clairet, 2018).

Once the composition of easy-to-use NADES mixtures was defined, we next proceeded to their physicochemical characterization.

Table 1 Example of homogeneous, limpid and stable liquid phases obtained.

Acronyms	Studied mixtures			Mass fraction (g/100g)			Equivalent molar amount		
BFW115	Betaine	Fructose	Water	30.2	46.5	23.3	1	1	5
FGW115	Fructose	Glycerol	Water	25.4	49.7	24.9	1	1	5
FBgW113	Fructose	Butylene glycol	Water	55.5	27.8	16.7	1	1	3
BSW115	Betaine	Sorbitol	Water	30.1	46.8	23.1	1	1	5
BGluW115	Betaine	Glucose	Water	30.2	46.5	23.3	1	1	5
FSW115	Fructose	Sorbitol	Water	39.8	40.3	19.9	1	1	5

2.2.2 NADES characterization and physicochemical properties

The characterization method of these six systems was inspired by the literature and different techniques have been selected considering our goal that is to evaluate their potential use as plant extraction solvents.

Differential scanning calorimetry was first used in order to evaluate the melting point of each mixture. Samples were analyzed at a temperature ranging from −50 °C to 30 °C. In these experimental conditions, no melting point could be detected. Moreover, we observed the absence of thermal phenomena in the studied range of temperature. This result indicates that the ternary mixtures included in this study present a liquid phase stability over a large range of temperatures. The absence of melting point for NADES ternary mixtures that contain water has already been reported (Dai, 2013). Instead of Deep Eutectic Solvents, Francisco and coworkers named those mixtures "Low Transition Temperature Mixtures" because they are characterized by glass transitions instead of melting points (Francisco, van den Bruinhorst, & Kroon, 2012). The most documented structural hypothesis indicates that the components of a NADES mixture interact via strong intermolecular hydrogen bonds when the molar ratio and the water content allow it. The existence of these non-covalent bonds can be demonstrated by nuclear magnetic resonance (NMR) spectroscopy. NMR is a non-destructive spectroscopic technique which gives access to the structure of molecules in a liquid or solid medium. The two-dimensional NOESY (Nuclear Overhauser Effect SpectroscopY) experiment offers a way to obtain all NOE information from a molecule or a mixture. Applied to NADES, this single experiment is used to study the spatial proximity between protons present in the mixture. This analysis provides relevant data to discuss the existence of hydrogen bonds and to identify the protons that contribute to these interactions. $^{1}H—^{1}H$ NOESY was thus used to characterize the mixtures from a physicochemical point of view with the help of Leiden University and further analyzed with Topspin software.

For the sake of this experiment, two mixtures were compared: fructose-glycerol-water mixture, in a 1:1:5 ratio (abbreviated FGW115) and fructose-glycerol-water mixture, in a 1:1:10 ratio (abbreviated FGW1110). The spectrum of the FGW115 mixture shows strong positive correlations between all protons, thus indicating a spatial proximity between molecules (Fig. 2A). Those correlations appear as red signals observed outside the diagonal. The presence of those correlations indicates a low correlation time of molecules in the system. The NOESY spectrum obtained with the FGW1110 mixture is clearly different (Fig. 2A). Apart from the typical

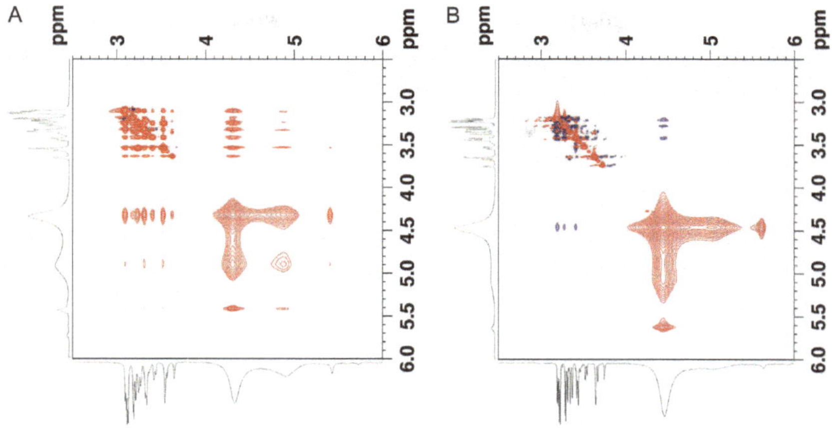

Fig. 2 Phase sensitive $^1H—^1H$ NOESY spectra recorded at 298 K for (A) FGW115 (B) FGW1110.

exchange of the hydroxyl protons in a water solution, correlations between other protons are negative or have disappeared. This shows that the amount of water and resulting dissolution is too high, and that in the system, interactions with water overcome other interactions. Consequently, water becomes the principal platform of hydrogen bonds and interactions between glycerol and fructose disappear. FGW1110 mixture should thus be considered a water solution of glycerol and fructose, but not a NADES.

NMR analysis is crucial to demonstrate the strong network of hydrogen bonds representative of a NADES mixture and to differentiate it from an aqueous solution. These results confirm that NMR analysis appears as a very effective guide to determine the maximum quantity of water that can be added to reduce the viscosity of mixtures, while preserving the supramolecular structure of a NADES. This demonstration first made by Dai and coworkers (Leiden University) is nowadays widespread in the literature (Delso, Lafuente, Muñoz-Embid, & Artal, 2019; López, Delso, Matute, Lafuente, & Artal, 2020).

Due to the network of intermolecular hydrogen bonds, liquids obtained display a relatively high viscosity. A simple way to reduce viscosity, besides the addition of water, is heating. Indeed, during heating, hydrogen bonds are broken or weakened. The study of viscosity as a function of temperature is helpful to define the ideal temperature at which plant extraction can be performed. Viscosity measurements were performed at a temperature ranging from 20 °C to 70 °C. The viscosity of NADES systems was compared to that of viscous conventional solvents: butylene glycol, propanediol and glycerol. Results are presented in Fig. 3. All studied solvents demonstrate a drastic drop in viscosity with increasing temperature. At temperatures above 50 °C, their viscosities fall down below 0.2 Pa.s, which is acceptable to perform plant extraction and solid-liquid separation.

On the basis of "like dissolves like," an extraction process, and more particularly the extraction solvent selection is often based on polarity. We used Reichardt's solvatochromic dye to determine the polarity of each solvent (Reichardt, 2004). The term solvatochromism refers to a change in the absorption spectrum of a chromophore in solvents of different polarity. Using such dye, the polarity of each sample solvent can thus be characterized by a maximum absorption wavelength. According to this method, polarity is scaled from 0 to 1. The polarity values of NADES and conventional solvents are presented in Table 2. Neither fructose-butylene glycol-water 1:1:3, nor butylene glycol-water, has any characteristic maximum wavelength so their polarity could not be determined using this method. The polarity of the five other NADES is comprised between those of glycerol and water. According to those results, those solvents should be efficient to extract polar and medium polar phytochemicals from plants.

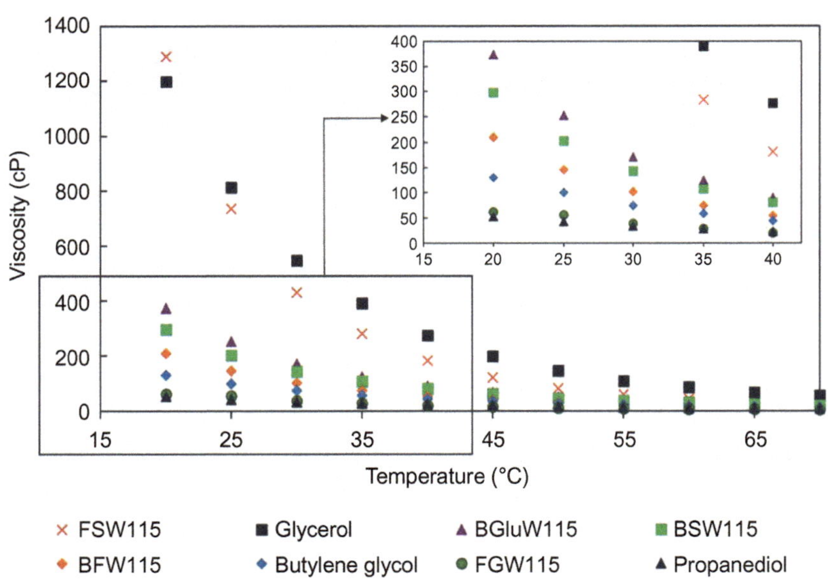

Fig. 3 Evolution of viscosity as a function of temperature for NADES and conventional solvents.

Table 2 Solvent polarity of NADES and conventional solvents obtained with Reichardt's method.

Solvents	λ_{max}(nm)	ET(30)	E^{N}_{T}
Ethanol	473.2	51.7	0.648
Propanediol (PDO)	520.3	55	0.75
Butylene Glycol (BG)	497.6	57.5	0.827
PDO/Water (60/40)	496	57.6	0.83
Glycerol	429	57.8	0.836
BFW115	428.2	57.9	0.84
BSW115	424	58.6	0.86
BGluW15	423	58.7	0.865
FGW115	415.6	59.9	0.901
FSW115	414.4	60.1	0.907
FBgW113		No characteristic λ_{max}	
BG/Water (60/40)			

2.2.3 *Study of the microbial safety profile, formulability and skin cell tolerance*

Microbiological safety is essential to cosmetic ingredients. Water and nutrients such as sugars favor bacterial growth. Challenge tests are commonly used to assess the efficacy of microbial preservation. Challenge-tests were thus carried out following the European Pharmacopeia monograph (9th edition, 2019–5.1.3: effectiveness of the antimicrobial conservation, preparations for local applications). The test consists of challenging a preparation with a prescribed inoculum of suitable microorganisms (two bacteria, one mold and one yeast: *Pseudomonas aeruginosa*, *Staphylococcus aureus*, *Candida albicans* and *Aspergillus brasiliensis*, respectively), storing the inoculated preparation at ambient temperature in the dark, withdrawing samples from the container at specified intervals of time and counting the microorganisms in the samples so removed. The preservative properties of the preparation are adequate if, in the conditions of the test, there is a significant fall or no increase, in the number of microorganisms in the inoculated preparation after 2, 7, 14 and 28 days. The acceptance criteria, in terms of decrease in the number of microorganisms with time, vary for different types of preparations according to the degree of protection intended. Fig. 4 shows that the

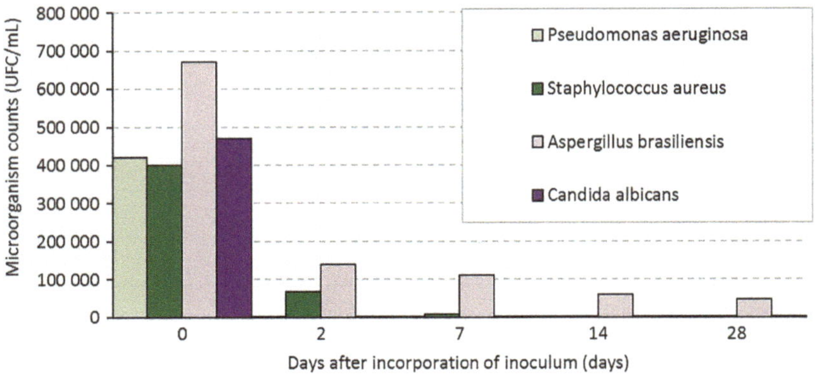

Fig. 4 Microbial stability study of FGW115 according to European Pharmacopeia challenge test method.

microorganism count of NADES FGW115. Microorganisms decrease as early as the second day, which is however not quick enough to meet the so-called A criteria. As a result, in order to ensure the microbiological safety of future FGW115 extracts, a final sterile filtration step with a 0.2 μm cut-off will be necessary.

The compatibility of the selected NADES in formulation was also checked by adding 5% of solvent in a cream and a gel. Formulations are considered stable when no significant changes in organoleptic properties are observed after 1 month of storage at different temperatures (4 °C, RT, 45 °C). In addition, the impact of 5% NADES on the sensory profiles of emulsions was assessed. Several criteria during and after application on the skin were evaluated by a panel of experts. Very good profiles were obtained for the NADES. In particular, despite an important amount of sugar in some NADES, the tacky after-feel was deemed lower than that of glycerine.

Finally, the viability of skin cells cultured in the presence of those NADES was evaluated using MTT assay (Riss et al., 2016). Normal human epidermal keratinocytes (NHEK) and normal human dermal fibroblasts (NHDF) were cultured for 48 h in the absence (untreated control) or presence of solvents at 1%. Results are presented in Fig. 5A and B for keratinocyte and fibroblast viability, respectively. No cytotoxicity was observed either with NHEK or NHDF further to NADES treatment. Consequently, those NADES can be considered harmless for cultured skin cells.

As a conclusion, among the many liquids obtained by combining compounds compatible with cosmetic requirements, six have been selected for

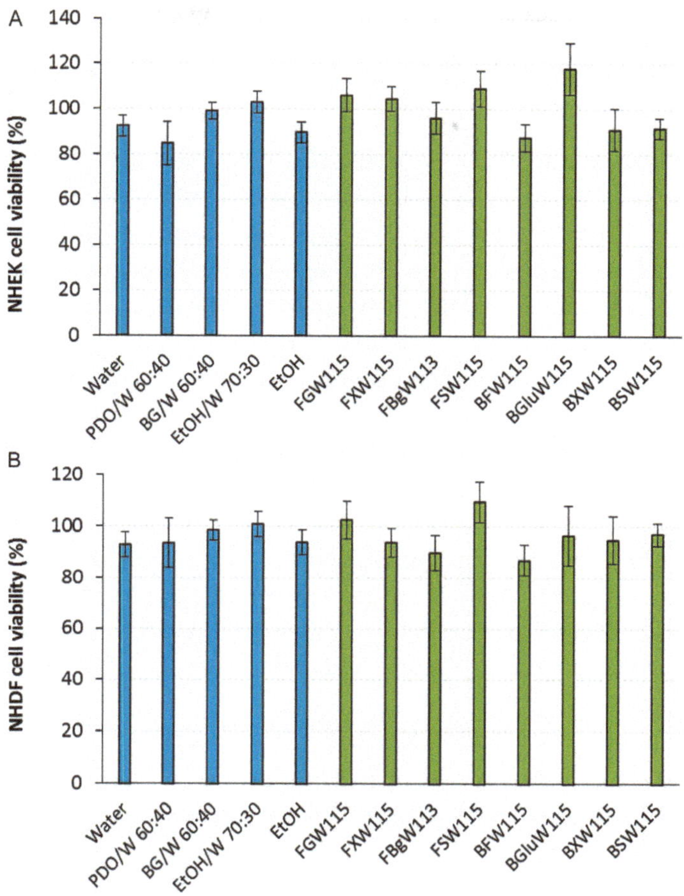

Fig. 5 Cell viability (MTT assay) of normal human epidermal keratinocytes (NHEK) (A) and normal human dermal fibroblasts (NHDF) (B) culture containing 1% of NADES or conventional solvents.

further evaluation as plant extraction solvent. A network of hydrogen bonds has been demonstrated by NMR for each of them, which results in a high viscosity that can be reduced by increasing temperature. These new solvents exhibit an intermediate polarity between those of glycerol and water. They show good microbial safety, are perfectly compatible with standard cosmetic formulations and well tolerated by human skin cells in culture. Those new solvents thus meet all the main characteristic features required for plant extraction.

3. Benefits of NADES extracts compared to traditional extracts

The extraction efficiency of the six selected NADES toward different plant metabolites was evaluated. The plant material chosen herein as an example is a flower long used in cosmetics: *Calendula officinalis* L., also referred to as pot marigold. Flowers were used dried and ground to increase the contact surface between solid and liquid and thus promote mass transfer. Extracts have been prepared according to a standardized protocol regardless of the solvent used. The mass ratio of plant to solvent was fixed at 5% and extraction was performed at 70 °C during 3 h under mechanical stirring. At the end of the extraction step, the solid/liquid separation was carried out after extraction using a coarse filter (100 μm). The extract was then centrifuged before sterile filtration (0.2 μm). Many polyphenols display biological properties and are thus of particular interest for cosmetic science. Phenolic acids and flavonoids are two classes of polyphenols described in *C. officinalis* flower. The contents of those two phytochemical classes were thus quantified using high pressure liquid chromatography with UV detection (HPLC-UV).

3.1 Evaluation of extraction performance of NADES in comparison with conventional solvents

Calendula officinalis flower extracts were produced from 11 different solvents: six NADES mentioned earlier (BSW115, FGW115, BGluW115, BFW115, FSW115 and FBgW113) and five conventional solvents (water, ethanol, ethanol/water 50/50 (w/w), propanediol/water 60/40 (w/w) and butylene glycol/water 60/40 (w/w)). The extraction efficiency of each solvent was expressed as % vs. Ethanol/water 50/50 used as reference. Total phenolic acid content was expressed in mg chlorogenic acid equivalent/100 g (UV detection at 327 nm) and total flavonoid was expressed in mg isorhamnetine-3-O-rutinoside equivalent/100 g (UV detection at 355 nm). Results obtained are presented in Fig. 6.

As expected, considering the targeted phytochemicals and individual solvent polarity, water and ethanol alone delivered poor extraction performance. In contrast, large amounts of both phenolic acids and flavonoids from *Calendula officinalis* flowers were extracted with the studied NADES. Each of them exhibited an extraction efficiency toward phenolic acids equal or higher than that of the best conventional solvents. On the contrary, the

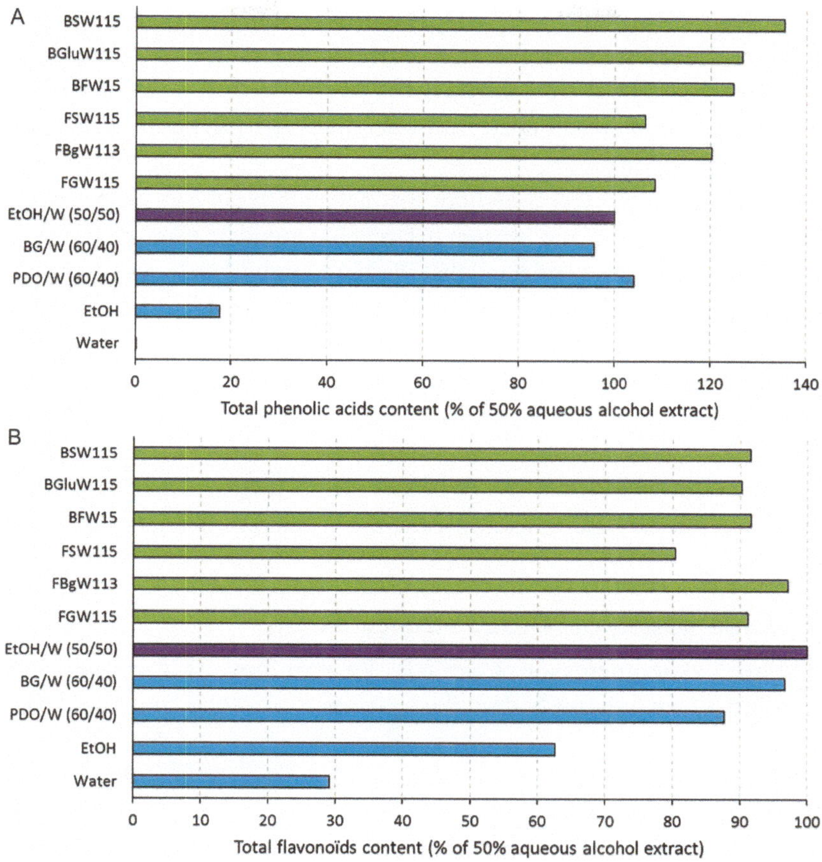

Fig. 6 Relative extraction efficiency of both NADES and conventional solvents for (A) phenolic acids and (B) flavonoids of *Calendula officinalis* compared to ethanol/water standardized at 100%.

extraction performance was similar or lower for flavonoids. The extraction efficiency of phenolic acids with NADES BSW115 was 35% higher than that of ethanol-water and that of flavonoids 10% lower. NADES FGW115 slightly outperformed ethanol-water with a total phenolic acid content close to that obtained with propanediol/water 60:40. The best NADES for flavonoid extraction was NADES FBgW113, with a performance close to that of ethanol-water.

As mentioned earlier, the choice of a solvent is often based on the similarity between its polarity and that of the targeted molecules. However, this study also showed that polarity alone is insufficient. Indeed, BFW115 has a similar polarity to propanediol / water 60/40 (respectively 0.84 and 0.83) but the extraction performance was generally better, especially for phenolic acids. Likewise, FGW115 extracts had a phytochemical content similar to that of propanediol/water 60/40 although its polarity is significantly higher (0.90).

Others have reported elsewhere that parameters such as dielectric constant, viscosity, density and Hansen parameters should be taken into account. Other internal studies confirmed that the high solubility of a given molecule in a specific solvent does not necessarily predict the extraction performance of the same solvent toward the same molecule contained in a plant raw material. This is true for both NADES and conventional solvents as reported by others (González, Mustafa, Wilson, Verpoorte, & Choi, 2018). We attempted to better understand the mechanisms underlying extraction performance and asked ourselves whether the supramolecular structure of NADES could explain, at least in part, the extraction efficiency.

3.2 Value-added of NADES structures

Additional experiments on *Calendula officinalis* flowers were performed according to the previously described extraction protocol and sample analysis procedure.

Here, NADES FGW115 was taken as reference regarding the extraction efficiency of phenolic acids. The aim was here to compare extraction efficiency of NADES FGW115 to other ternary or binary mixtures of fructose, glycerine and water or to single compounds whenever possible. Two ternary mixtures were selected: FGW113 (55:28:17, w/w/w), which contains less water than FGW115 and whose specific NADES supramolecular structure was confirmed based on NMR data; FGW117 (45:23:32, w/w/w), which contains more water than FGW115 and whose typical NADES supramolecular structure could not be observed. The selected binary mixtures were fructose/water 1:10 (50:50, w/w) and glycerol/water GW1:12(30:70, w/w). The extraction with pure fructose is of course impossible due to its solid nature and the extraction with glycerol is not feasible at industrial scale due to its high viscosity. Consequently, only water was used for this comparison. Due to its high viscosity, the FGW113 *C. officinalis* extract was very difficult to filtrate at 0.2 μm and was thus not relevant from an industrial perspective. Therefore, this sample was not analyzed. The phenolic acid content of other extracts is shown in Fig. 7.

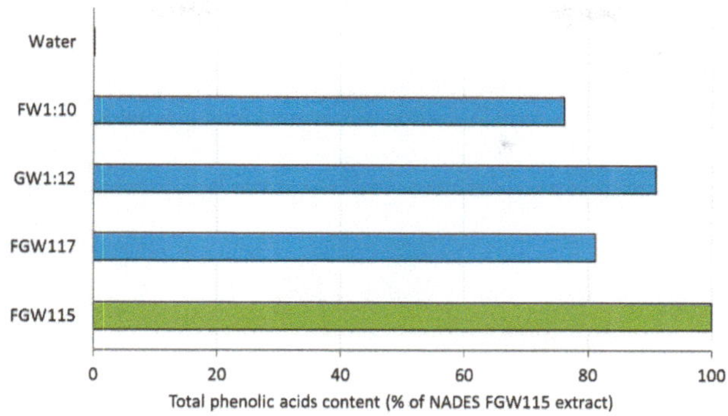

Fig. 7 Relative extraction efficiency of solvents composed of fructose, glycerol and water regarding phenolic acids extracted from *Calendula officinalis* flowers.

Concerning the ternary mixtures, the total phenolic acid content of FGW117 was lower than that of FGW115. The addition of water in this NADES reduced the viscosity but also significantly reduced the extraction efficiency. A similar observation was made with the extraction of phenolic metabolites from *Carthamus tinctorius* (Dai, Witkamp, Verpoorte, & Choi, 2013) and more recently with the extraction of phytochemicals from *Ginkgo biloba* and *Panax ginseng* leaves (Liu et al., 2018). Extracts obtained with binary mixtures had lower phytochemical contents than FGW115. Barely any phenolic acid could be extracted using water.

If these results are insufficient to establish a direct correlation between the NADES supramolecular structure and the extractive performance, this study shows however that NADES mixtures can offer superior extraction ability compared to aqueous solutions made of the same components. This is of particular interest for cosmetic applications. Indeed, NADES give access to plant extracts with unique phytochemical signatures and potentially unique biological activities. The biological activity of different NADES extracts was thus further assessed to identify *in vitro* potential skin benefits.

3.3 Evaluation of biological activity of NADES extracts

Most polyphenols such as phenolic acids and flavonoids exhibit antioxidant properties. The previously selected NADES BSW115 extract from *Calendula officinalis* flowers demonstrated an interesting phytochemical composition with an interesting balance between phenolic acids and flavonoids. Its antioxidant activity was thus assessed using DPPH assay and compared to

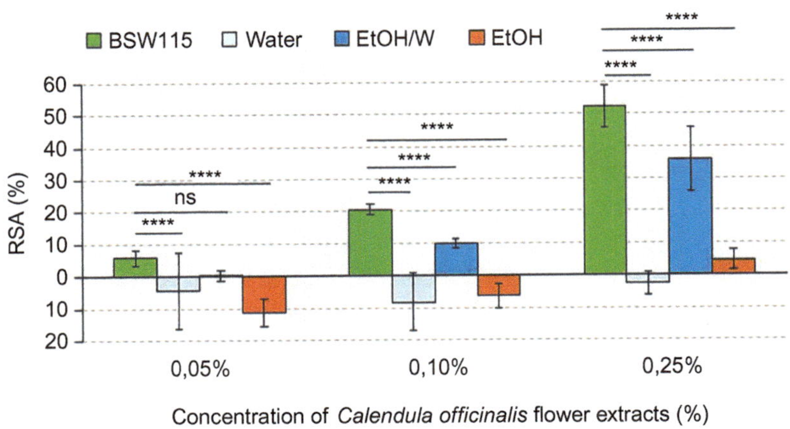

Fig. 8 Radical scavenging activity of a BSW115 extract compared to conventional solvents (ns, non-significant; ****, statistically significant vs untreated control $P < 0.0001$).

that of three conventional extracts obtained with water, ethanol/water 50:50, w/w, and ethanol. The DPPH *in tubo* assay allows measuring the radical scavenging activity. At room temperature, the DPPH reagent (1,1-diphenyl-2-picrylhydrazyl) is a stable radical with a strong absorption band centered at about 518 nm. It has a deep violet color in solution and becomes colorless or pale yellow when neutralized by radical scavengers. It thus offers a quick and easy method to measure the radical scavenging activity. Increasing concentrations of plant extract (0.05%, 0.1% and 0.25%) were tested for their radical scavenger activity (RSA). Three independent experiments were performed in sextuplicate and the statistical analysis was carried out using One-way ANOVA followed by Tukey's multiple comparison test. Results are expressed as the percentage of radical scavenger activity vs untreated control. Results are presented in Fig. 8. The four solvents were tested alone and showed no significant antioxidant activity in this model (data not shown).

The NADES extract showed a dose-related radical scavenging activity with significantly greater activity than that of the conventional ethanol/water solvent used as positive reference.

4. From the laboratory proof of concept to the market

Numerous plants extracts were developed at lab scale based upon the strategy presented above. Among these, the best ones were selected and scaled-up to our industrial production plant. For instance, a cosmetic ingredient based on a FGW115 NADES extract of horse chestnut (*Aesculus hippocastanum*) flowers was recently launched on the market. Same as previously, NADES FGW115 was selected for its unique phytochemical content. More specifically, this extract contains very interesting plant metabolites, such as flavonol glycosides among which astragalin attracted great attention. Several biological assays were performed using monolayer cultures of normal human skin cells to identify possible skin benefits of this extract. Among the different biological pathways evaluated, the extract exhibited a strong collagen IV boosting effect in normal human dermal fibroblast cultures. Noteworthy is that this extract outperformed the ethanol/water 70:30 (w/w) extract that was used as reference in the test. Indeed, the NADES extract showed a fourfold increase in collagen IV synthesis, while the hydroethanolic extract and the positive control vitamin C both delivered a twofold increase. FGW115 and ethanol/water tested alone showed no effect. Collagen IV is a major component of the dermal-epidermal junction that is both a structural and biological membrane at the interface between the dermis and epidermis. Collagen IV is known to decrease with skin aging and is thus a target of interest for anti-aging purpose. This result therefore allowed us to initiate and further assess the anti-aging properties of the FGW115 NADES extract of horse chestnut flowers. Results are reported in Fig. 9.

Interestingly, the flavonol content in both the NADES and the ethanol/water extracts was similar. Consequently, is clear that other molecules identified in this NADES such as procyanidins, amino acids or even minerals contribute to the biological activity. No single class of molecules can probably explain the difference in biological activity. We can also hypothesize

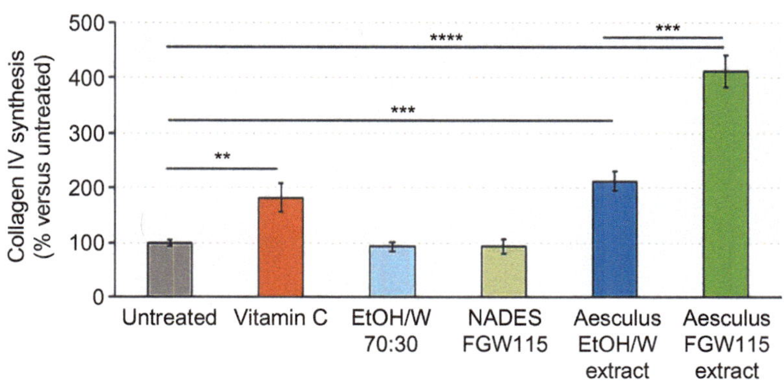

Fig. 9 Effect of two *Aesculus hippocastanum* extracts on collagen IV synthesis.

that the NADES solvent itself may not only demonstrate unique extraction properties but also modulate the bioavailability of phytochemicals for skin cells.

5. Conclusion and perspectives

Over the past decade, design and development of cosmetic ingredients has dramatically evolved. The industry has operated a shift toward green chemistry to meet market trends but also deliver more sustainable and responsible ingredients. Basic research on NADES has laid the ground for cosmetic science to develop eco-friendly bioactive ingredients without compromising on biological efficacy. Several years of research were needed before coming up with a commercially bioactive ingredient. The collaboration between Gattefossé and Leiden University has been essential to the successful development of innovative, economically viable, yet efficient bioactive ingredients.

These easy-to-use, stable liquids composed of safe molecules form a unique network of hydrogen bonds and offer terrific opportunities of innovation. Our work has shown that each solvent has its own extraction capacity and that the resulting phytochemical signature is unique. Biological assays have demonstrated that different extracts obtained with solvents of different polarities, including NADES, could deliver significantly different biological activities in cultured skin cells. It is not always clear exactly how different plant extracts obtained using different solvents can display significantly different biological activities. But one thing is certain: by giving access to unique phytochemicals, NADES offer a new opportunity to explore biodiversity and identify new plant extracts with new benefits for skin health. Important questions remain regarding the mechanisms underlying the original extraction performance of NADES. What is the relation between these particular assemblies and their particular extraction capacities? Can they affect the bioavailability of phytochemicals for skin cells? Do they actually play a role in nature for the transport or preservation of plant metabolites? Further basic and applied research is needed to help answer those questions.

References

Abbott, A. P., Boothby, D., Capper, G., Davies, D. L., & Rasheed, R. K. (2004). Deep eutectic solvents formed between choline chloride and carboxylic acids: A versatile alternatives to ionic liquids. *Journal of the American Chemical Society*, *126*(29), 9142–9147. https://doi.org/10.1021/ja048266j.

Anastas, P. T., & Warner, J. C. (1998). *Green chemistry: Theory and practice*. Oxford England, New York: Oxford University Press.

Anastas, P. T., & Zimmerman, J. B. (2003). Design through the 12 principles of green engineering. *Environmental Science and Technology*, *37*(5), 95A–101A.

Charton, V., & Clairet, A. (2018). Using nades extraction to create natural, sustainable, high-performance actives. *Household and Personal Care Today*, *13*(2), 37–40.

Choi, Y. H., van Spronsen, J., Dai, Y., Verberne, M., Hollmann, F., Arends, I. W. C. E., et al. (2011). Are natural deep eutectic solvents the missing link in understanding cellular metabolism and physiology? *Plant Physiology*, *156*(4), 1701–1705. https://doi.org/10.1104/pp.111.178426.

Cosmetics Europe. n.d. Cosmetics Directive 76/768/EEC Explanatory Brochure. Retrieved from https://cosmeticseurope.eu/files/4114/6400/4203/The_European_Union_Cosmetics_Directive_Explanatory_Brochure_2004.pdf (assessed on 13 August 2020).

Dai, Y. (2013). *Natural deep eutectic solvents and their application in natural product research and development* (Ph.D. Thesis). Faculty of Science, Leiden University (The Netherlands).

Dai, Y., Witkamp, G. J., Verpoorte, R., & Choi, Y. H. (2013). Natural deep eutectic solvents as a new extraction media for phenolic metabolites in *Carthamus tinctorius L. Analytical Chemistry*, *85*(13), 6272–6278. https://doi.org/10.1021/ac400432p.

Delso, I., Lafuente, C., Muñoz-Embid, J., & Artal, M. (2019). NMR study of choline chloride-based deep eutectic solvents. *Journal of Molecular Liquids*, *290*, 111236. https://doi.org/10.1016/j.molliq.2019.111236.

Durand, E., Lecomte, J., & Villeneuve, P. (2016). From green chemistry to nature: The versatile role of low transition temperature mixtures. *Biochimie*, *120*, 119–123. https://doi.org/10.1016/j.biochi.2015.09.019.

Francisco, M., van den Bruinhorst, A., & Kroon, M. C. (2012). New natural and renewable low transition temperature mixtures (LTTMs): Screening as solvents for lignocellulosic biomass processing. *Green Chemistry*, *14*(8), 2153. https://doi.org/10.1039/C2GC35660K.

Gomez, F. J. V., Espino, M., Fernandez, M. A., & Silva, R. (2018). A greener approach to prepare natural deep eutectic solvents. *ChemistrySelect*, *3*(22), 6122–6125. https://doi.org/10.1002/slct.201800713.

González, C. G., Mustafa, N. R., Wilson, E. G., Verpoorte, R., & Choi, Y. H. (2018). Application of natural deep eutectic solvents for the "green" extraction of vanillin from vanilla pods. *Flavour and Fragrance Journal*, *33*(1), 91–96. https://doi.org/10.1002/ffj.3425.

Jablonský, M., Škulcová, A., Malvis, A., & Šima, J. (2018). Extraction of value-added components from food industry based and agro-forest biowastes by deep eutectic solvents. *Journal of Biotechnology*, *282*, 46–66. https://doi.org/10.1016/j.jbiotec.2018.06.349.

Liu, X., Ahlgren, S., Korthout, H. A. A. J., Salomé-Abarca, L. F., Bayona, L. M., Verpoorte, R., et al. (2018). Broad range chemical profiling of natural deep eutectic solvent extracts using a high performance thin layer chromatography-based method. *Journal of Chromatography. A*, *1532*, 198–207. https://doi.org/10.1016/j.chroma.2017.12.009.

López, N., Delso, I., Matute, D., Lafuente, C., & Artal, M. (2020). Characterization of xylitol or citric acid: Choline chloride:water mixtures: Structure, thermophysical properties, and quercetin solubility. *Food Chemistry*, *306*, 125610. https://doi.org/10.1016/j.foodchem.2019.125610.

Regulation (EC) No 1223/2009. n.d. Regulation (EC) No 1223/2009 of the European Parliament and of the Council of 30 November 2009 on Cosmetic Products. Retrieved from https://ec.europa.eu/health/sites/health/files/endocrine_disruptors/docs/cosmetic_1223_2009_regulation_en.pdf (assessed on 13 August 2020).

Reichardt, C. (2004). Pyridinium N-phenolate betaine dyes as empirical indicators of solvent polarity: Some new findings. *Pure and Applied Chemistry*, *76*(10), 1903–1919. https://doi.org/10.1351/pac200476101903.
Riss, M. R. A., Niles, A. L., Duellman, S., Benink, H. A., Worzella, T. J., & Minor, L. (2016). *Cell viability assays*. Retrieved from https://www.ncbi.nlm.nih.gov/books/NBK144065/.
United States Code. n.d. Federal Food, Drug, and Cosmetic Act, 21 U.S.C. §§ 301-392 Suppl. Retrieved from https://tile.loc.gov/storage-services/service/ll/uscode/uscode1934-00602/uscode1934-006021009/uscode1934-006021009.pdf (assessed on 13 August 2020).
Zhao, D., Liao, Y., & Zhang, Z. (2007). Toxicity of ionic liquids. *CLEAN– Soil, Air, Water*, *35*(1), 42–48. https://doi.org/10.1002/clen.200600015.

CHAPTER THIRTEEN

Applications of NADES in stabilizing food and protecting food compounds against oxidation

Aleksandra Mišan* and Milica Pojić
Institute of Food Technology, University of Novi Sad, Novi Sad, Serbia
*Corresponding author: e-mail address: aleksandra.misan@fins.uns.ac.rs

Contents

Abstract

In new food product development, one of the major challenges is to enable the preservation of the designed food properties throughout its shelf life. During the shelf life, food ingredients undergo different deteriorative changes thus affecting their chemical, sensory and nutritional properties (e.g. texture, appearance, flavor, nutritional value, beneficial health effects). Different strategies have been applied to food products to stabilize food ingredients and improve the oxidative stability, such as: the use of

Advances in Botanical Research, Volume 97
ISSN 0065-2296
https://doi.org/10.1016/bs.abr.2020.09.010

chemical additives, antioxidants, and utilization of encapsulation/microencapsulation techniques. However, in recent years the change of consumer needs and expectations from food industry have dramatically changed, requiring more natural, less processed and "free from" ingredients food. Therefore, the importance of natural deep eutectic solvents (NADES) made of natural primary metabolites is rising especially in the food sector. However, studies presenting the stabilization ability of NADES are still scarce. This chapter overviews the unique physicochemical properties of NADES accountable for stabilization of solutes and their application related to stabilization of food and its components.

1. Introduction: The major deteriorating changes in foods

Foods are often classified based on their stability as: non-perishable, semi-perishable and perishable. Most foods (e.g. most fresh fruits and vegetables, meat and fish, milk and eggs) are classified as perishable, due to their high-water content. Low moisture content foods (e.g. cereal grains, dried fruit and vegetables) and frozen foods are classified as semi-perishable, while canned foods can sometimes be stored for years. Food deterioration during storage is a result of enzymatic, chemical, biological and physical changes (Rahman, 2007).

Enzymatic reactions in foods lead for example to: (a) the post-harvest senescence and spoilage of fruits and vegetables (Toivonen & Brummell, 2008); (b) oxidation of phenolic substances in plant tissues by polyphenol oxidase leading to browning due to the accumulation of melanin (Toivonen, 2004); (c) sugar-starch conversion in plant tissues by amylases (Gao, Huang, Dong, Yang, & Yi, 2016); (d) post-harvest demethylation of pectic substances in plant tissues resulting in their softening (González-Cabrera, Domínguez-Vidal, & Ayora-Cañada, 2018).

The two most important chemical reactions which lead to sensory quality deterioration are lipid oxidation in meat (Xiong et al., 2020) and fish tissues (Pérez-Andrés et al., 2020) and nonenzymatic browning, which can be categorized into Maillard reaction, sugar pyrolysis (caramelization), ascorbic acid degradation, oxidation and condensation of phenolic compounds (He et al., 2019). Oxidative changes are a cause of undesirable changes in the color and flavor of foods during processing and storage. For example, food natural pigments, anthocyanins are highly unstable and easily degradable compounds under various conditions such as alkaline pH, prolonged light durations, increased temperature and oxygen concentrations, presence of various enzymes and ingredients, which lead to color fading and loss of bioactivity (Bordenave, Hamaker, & Ferruzzi, 2014; Gérard et al., 2019; He, Xu, Zeng, Qin, & Chen, 2016a, 2016b).

Contrary to them, biological changes induced by microorganisms can induce either desirable or undesirable changes depending on whether they are introduced as an essential part of the food preservation process (Cvetković et al., 2015) or grow spontaneously to cause food spoilage (Gram & Dalgaard, 2002).

Physical changes in food are associated with moisture absorption in dried products, for example, leading to undesirable increase in water activity (Barreiro & Sandoval, 2020) or freezing damage of water leading to irreversible changes in food texture after thawing (Tian, Zhu, & Sun, 2020).

2. Food stability and preservation

Food preservation applies a number of different actions and processing techniques to preserve the desired properties of foods during the defined shelf life, to keep it safe and pleasant to consume (Rahman, 2007). Although new concepts and hypotheses are being developed and proposed, shelf-life prediction from a scientific basis rather than empiricism still presents a challenge to food scientists and engineers.

2.1 Water activity (a_w) concept

Water presents a universal solvent in living organisms and food products. Serving as a medium for many chemical and biochemical reactions it is essential for growth and metabolism in living organisms, as well as for deterioration reactions in food products. Free water in foods is available for chemical reactions, microbes and for transporting different compounds. On the other hand, bound water is amount of water which is bound to molecules such as by the hydroxyl groups of polysaccharides, proteins, and others to which water can be held by hydrogen bonding, by ion-dipole bonds, or by other strong interactions. As it was observed that the free water is much more important to the stability of a food than the total amount of water present, general rules or limits for the stability of foods using water activity were made. Water activity (a_w) is the partial vapor pressure of water in a substance divided by the partial vapor pressure of pure water at the same temperature or the equilibrium relative humidity of the air surrounding the system at the same temperature (Barbosa-Cánovas, Fontana, Schmidt, &

Labuza, 2007). In other words, water activity is the measure of availability of water to enter microbial, enzymatic, physical and chemical reactions causing the deterioration of foods during storage.

Water activity concept has been mostly applied in predicting the growth of bacteria, yeasts and molds as it is known that each microorganism has a water activity below which it cannot grow. For instance, pathogenic bacteria cannot grow below $a_w < 0.85$, whereas yeasts and molds are more tolerant to reduced water activity, but usually no growth occurs below a water activity of about 0.6 (Rahman, 2010). The so-called intermediate moisture foods (IMF) have a_w values in the range of 0.65–0.9 (Barbosa-Cánovas et al., 2007).

Water activity concept is used in process design and control, ingredient and packaging selection. For example, the endpoint of drying presents the equilibrium between foods and relative air humidity, while in an osmotic or salting process, foods equilibrate with the water activity of osmotic solution (Rahman & Labuza, 2007).

Referring to the water activity concept, a food product is the most stable at its monolayer moisture content (BET monolayer) or monolayer water activity and unstable above or below the monolayer (Rahman, 2009). The food stability map indicates the trends in microbial growth, biochemical reactions and mechanical characteristics in the three zones of water activity (Fig. 1). Zone I correspond to a_w value below monolayer in which: (i) stability can be decreased with decreasing water activity, for example lipid oxidation (line ob); (ii) stability can remain constant (line ab); and (iii) stability can increase (line nb). Zone II presents adsorbed layer in which: (i) stability can increase (line bp) and (ii) stability can decrease (line bc). In multi-layer (zone III): stability can increase, decrease, or remain constant.

Referring to Rahman and Labuza (2007), the water activity concept has a lot of limitations. At first place, water activity is defined at equilibrium, whereas foods may not be in a state of equilibrium. The critical limits of water activity may also be shifted to higher or lower levels by other factors, such as pH, salt, antimicrobial agents, heat treatment, and temperature and the nature of the solute used also plays an important role. Above all, the water activity concept does not indicate the state of the water present, how it is bound to the substrate and the information about the mobility or reactivity of water, it only relates to the amount of strongly bound and free water without their precise reactivity. As physical changes, such as, caking, stickiness, gelatinization, diffusivity cannot be explained based on the water activity alone, the glass transition concept was proposed.

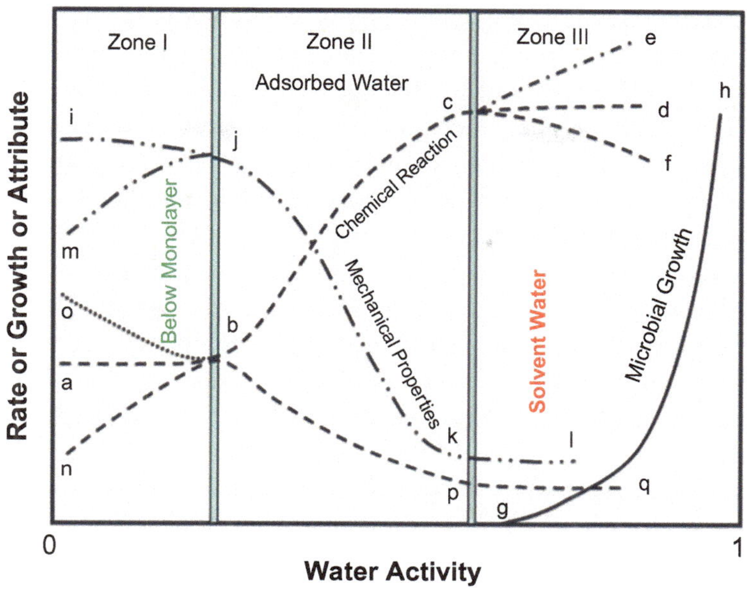

Fig. 1 The food stability map based on the water activity concepts. *Taken from Rahman, M. S. (2019).* Water activity and glass transition of foods. Reference module in food science, *Elsevier.*

2.2 Glass transition concept

As explained by Rahman (2010) glass is a liquid that has lost its ability to flow. As its structure is not crystallin but amorphous, glass is physically a solid, but resembles closely a thermodynamic liquid as it has the disorder of the liquid state. Amorphous solids are commonly formed by rapid cooling of a liquid melt when the molecules in the melt do not have enough time to rearrange but are frozen in their original position. Glass transition refers to the phase transition when a glass is changed into a supercooled melt or the reverse and the glass transition temperature (Tg) is defined as the temperature range corresponding to the glass-liquid transition (Liu, Bhandari, & Zhou, 2006).

Referring to the glass transition concept, food materials are considered to be very stable at the glassy state, since below glass temperature compounds involved in the deterioration reactions diffuse very slowly over molecular distances and approach each other to react and the water becomes kinetically immobilized to support or participate in reactions, the higher $T-Tg$ or T/Tg (i.e. above glass transition), the higher the deterioration or reaction rates (Slade & Levine, 1991).

Cases supporting the glass transition concepts can be found in nature. For example, in desiccation-tolerant plants (anhydrobiotes) seeds can survive the removal of their cellular water, so that the life span of seeds can be extremely long. To explain this phenomenon, Burke (1986) postulated the hypothesis that the cytoplasm of seeds could enter a glassy state as it was known that during drying, cytoplasmic viscosity increased dramatically. He suggested that in dry anhydrous seeds, glasses could be formed from cell solutes like sugars which were known to provide protection from denaturation of large molecules and formation of molecular aggregates, and high viscosity of glass state may stop all chemical reactions that require molecular diffusion. Plant anhydrobiotes possess large amounts of soluble non-reducing sugars and their state diagrams resemble those of simple sugar mixtures (Buitink & Leprince, 2004).

As far as food systems are concerned, glass transition concept was tested in a number of papers as a framework for controlling diffusion, structure, crystallization, stickiness, grain damage, pore formation, microbial stability, seed stability, oxidation, nonenzymatic browning, enzymatic reaction, protein denaturation, hydrolysis, and enzyme inactivation. These applications were reviewed by Rahman (2006, 2010), Liu et al. (2006), and Le Meste, Champion, Roudaut, Blond, and Simatos (2002).

As concluded by Rahman (2010) all the experimental results could not be explained by the above rules, so it is important to identify when the concept fails and why. A combination of different concepts is offered (Rahman, 2006, 2009, 2010).

2.3 Macro-micro region concept

Rahman (2006) combined both the water activity and glass transition concepts in the state diagram, by plotting the BET-monolayer value as the LNO line in the state diagram, proposing 13 micro-regions for determining food stability (Fig.2). According to Rahman (2006), a state diagram presents a stability map of different states of a food as a function of solute mass fraction and temperature which can be used to identify the suitable processing (freezing and drying) to decrease deterioration reactions. As can be seen in Fig. 2, state diagram consists of the freezing line (ABC), solubility line (BDL) and the glass transition line (EQFS).

Regarding food processing and preservation, crystallization, stickiness, and collapse are phenomena that are observed in the region BQEL, while the melting line BDL is important for high temperature processing, for example frying, baking, roasting, and extrusion cooking.

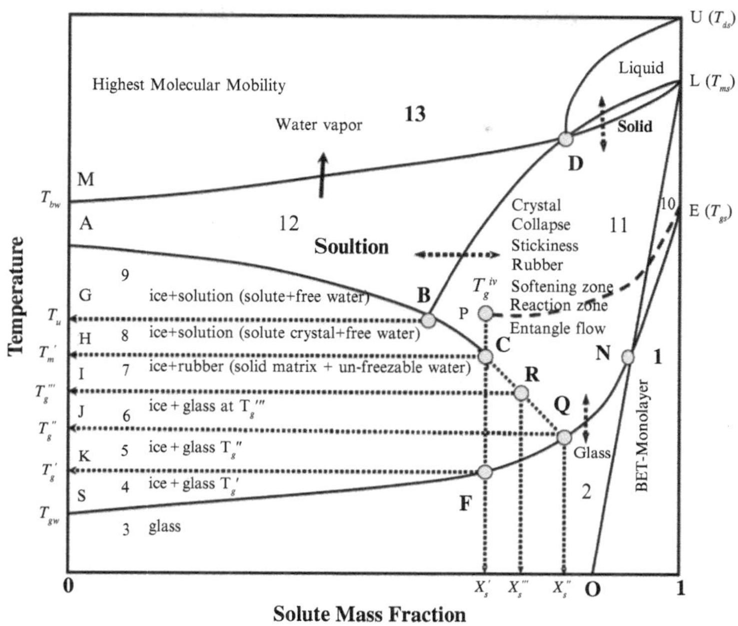

Fig. 2 State diagram presenting different regions and states of foods (*Tbw* boiling point; *Tu* eutectic point; *Tm'* end point of freezing; *Tg'* glass transition at end point of freezing; *Tgw* glass transition of water; *Tms* melting point of dry solids; *Tgs* glass transition of dry solids). *Taken from Rahman, M. S. (2006). State diagram of foods: Its potential use in food processing and product stability.* Trends in Food Science & Technology, 17 *(3), 129–141.*

Rahman (2006) proposed a hypothesis which recognizes 13 micro-regions of food stability having highest to lowest stability based on the locations from glass and BET-monolayer lines, the region-1 is below the BET-monolayer line and glass line and presents the most stable while region-13 presents least stable, highly reacting zone as it is far from BET-monolayer line and glass line. The stability decreases as the zone number increases.

According to Rahman (2010) stability estimation of a multi-component mixture such as food, based on macro-regions in the state diagram can be deficient as there are a lot of examples of food systems in the literature which were stable above glass transition and unstable below glass transition. However, although both glass transition and water activity concepts have some limitations, they provide the same stability criteria at one location, point N. It is still not known how pH, and preservatives could be linked with these concepts. At present, the scientific community far from developing an unified theory (Rahman, 2010).

3. NADES and food stability

Natural deep eutectic solvents (NADES) were firstly described by Choi et al. (2011) as a third class of liquids present in living cells, different from water and lipids, which play an important role as an alternative medium for biosynthesis, transport and storage of compounds with intermediate polarity. Referring to the same authors, NADES have a pivotal role in maintaining metabolism in plants in the absence of water during drought or cold conditions.

If NADES exist in the plant world, it is logical to assume that NADES either exist or are formed in food systems during processing (e.g. drying and cooling) and influence its stability. Regarding the water activity concept (Fig. 1), the increase in the rate of some reactions, such as lipid oxidation in the Zone 1, which refers to a_w activities below BET monolayer, might be explained by the formation of NADES.

Referring to the macro-micro region concept (Fig. 2), below the freezing line (ABC), in the regions 4–9 there is a presence of unfrozen water. For example, the water content at the point F (1-Xs') or C is the unfrozen water, which is defined by Rahman (2010) as a mass fraction of water remaining unfrozen even at very low temperature, which includes both un-crystallized free water and bound water attached to the solids matrix. It has been observed that when water is present as part of NADES, it is strongly retained in the liquid and cannot be easily evaporated (Choi et al., 2011).

The nature of interactions taking place in food systems, enabling the existence of unfrozen water is not clear. However, there is experimental evidence that mobility of water exists even below glass transition temperature (Rahman, 2009).

Although there is no scientific evidence yet supporting the hypothesis of NADES formation in foods and the way they might contribute to the food stability, some unique physicochemical properties of synthesized NADES may play a role in food preservation (Mišan et al., 2020).

3.1 Prolonging the stability of high moisture foods

High moisture foods (a_w 0.9–0.999) usually contain more than 50% (w/w) of water. These foods include fresh fruits and vegetables, meat and dairy products and their shelf life is largely controlled by the growth of microorganisms. The key factors controlling the deterioration changes are storage temperature, pH and protective package. These factors influence not only the growth of microorganism, but also other components and processes, such as enzyme stability, gel formation, and stability of proteins and vitamins.

3.2 Antimicrobial activity of NADES

The shelf life of high moisture foods can be prolonged by inactivating the microbes present in them. Referring to the Gómez et al. (2019) who measured a_w value of a range of NADES prepared with citric acid (CA), malic acid (MA), glucose (Glu), fructose (Fru), urea (U), β-alanine (BA) and choline chloride (CC) and found that it ranges between 0.653 and 0.177, so in line with the water activity concept, their antimicrobial activity can be expected. The same authors demonstrated that the incorporation of water to NADES causes a gradual increase in the water activity.

Radošević et al. (2018) proved that NADES can exhibit antimicrobial activity to most common Gram-positive (i.e. *Staphylococcus aureus*) and Gram-negative bacteria (*Escherichia coli, Proteus mirabilis, Salmonella typhimurium, Pseudomonas aeruginosa*) and yeast (*Candida albicans*). It was shown that NADES containing choline, betaine, sugar and sugar alcohols influenced no antimicrobial activity of NADES, such as examined choline chloride:xylitol (5:2), choline chloride:sorbitol (2:3) and betaine:glucose (5:2). However, other examined NADES especially those containing citric acid exhibited microbial growth inhibition, whereby citric acid:glucose: glycerol (1:1:1) and citric acid:fructose:glycerol (1:1:1) NADES indicated the potential for their use as antimicrobial agents. There are several mechanisms that explain antimicrobial activity of NADES depending on the type of microorganism, differing in the structure and composition of cell walls (Juneidi, Hayyan, & Hashim, 2018). One mechanism can be associated with low pH value of the system, which damages the cell membrane and its proteins causing the death of microorganism (Radošević et al., 2018). This mechanism is explained by the fact that the optimal pH for bacterial growth is between 6.5 and 7.5, whereby the pH values of common NADES are pretty low, even lower than 3 (Juneidi et al., 2018). The organic acid-based NADES are the most polar, while sugar- and polyalcohol-based NADES are the least polar NADES. NADES in water are weakly acidic (sugar- and sugar alcohol-based NADES with pH ca. 4.5), while carboxylic acid-containing NADES are strongly acidic ($pH < 3$). Moreover, it was shown that the antimicrobial activity of NADES-forming compounds and NADES is different, where NADES having higher antimicrobial activity in comparison to their individual components due to delocalized charges as a result of the hydrogen bond, especially in acid-based NADES.

Another mechanism is associated with the structure of the bacterial cell wall, mainly composed of peptidoglycan, which due to its high porosity does not represent a physical barrier, so that NADES diffusion occurs within the cell to cytoplasm causing enzyme denaturation, oxidative stress, and consequently cell death (de Morais, Gonçalves, Coutinho, & Ventura, 2015). Moreover, the antimicrobial effect of NADES toward yeasts and fungi is associated with the dehydrating effect similar to that produced by $CaCl_2$, whereby a very rapid exit of the cell water occurs causing the cell death (Cardellini et al., 2015, 2014).

Antimicrobial photodynamic therapy (aPDT), also known as photodynamic antimicrobial chemotherapy (PACT), combines a photosensitiser (PS), oxygen and visible light to produce reactive oxygen species (ROS) and other reactive photoproducts and has demonstrated antimicrobial efficacy against several antibiotic-sensitive and antibiotic resistant bacteria (Wikene, Rukke, Bruzell, & Tønnesen, 2016). PS is a "light-activatable dye" which is capable of absorption of a photon (A), which transfers PS from its ground singlet state (S_0) to an excited singlet state (Sn). Subsequently, the PS can return to S_0 by generating reactive oxygen species (ROS). After such a cycle of absorption and generation of ROS, the PS is in its ground state (S_0) again and therefore ready for absorption of a new photon (Cieplik et al., 2018). It was demonstrated that NADES may increase the phototoxic effect of an anionic *meso*-tetra-(4-carboxyphenyl)-porphine photosensitiser (Wikene et al., 2016) and the solubility, physical stability and bacterial phototoxicity of a neutral porphyrin, *meso*-tetraphenylporphyrin (Wikene, Bruzell, & Tønnesen, 2015). These findings point out that a proper selection of NADES as a solvent for PS can highly increase the efficiency of photodynamic therapy. This may lead to a method to decrease the initial level of antimicrobial load in food and to prolong the shelf life of high moisture foods.

3.3 Cryoprotective ability of NADES

Freezing technology provides a series of the beneficial effects of low temperatures for the suppressing of the growth of microorganisms, the reduction of detrimental chemical and cellular metabolic reactions during long storage periods. Although foods have been commercially frozen for more than 140 years, the significance of freezing technology is still actual due to the increasing demand for minimally processed foods required by the consumers. Despite the fact that it generally allows the preservation of taste and nutritional value of foods better than any other method, freezing very often compromise the texture of a frozen product (http://www.fao.org/3/y5979e/y5979e03.htm). The frozen damage is induced by ice crystal growth, which affects the microstructure of the frozen food. These changes are reflected in excessive drip, texture breakdown and color and flavor changes of a product, negatively perceived by the consumers (Charoenrein, 2018; Tian et al., 2020). This problem can be overcome by regulating the freezing point that promotes and inhibits ice nucleation, and controls ice crystal growth and recrystallization, and/or utilization of freezing systems that produce small ice crystals and maintain them during storage. To assist the regulation of freezing point several naturally sourced, non-toxic, eco-friendly bio-substances were suggested, including NADES in addition to ice-nucleating proteins and anti-freezing proteins (Tian et al., 2020). This approach is especially important in freeze-thaw treatment of berry fruits (e.g. strawberries, raspberries, blueberries) freezing of which often does not meet commercial quality requirements (Tian et al., 2020). The cryoprotective mechanism of NADES is based on the supramolecular network structure of NADES affecting the high viscosity of NADES. In such a system, the molecular motion and reorientation of free water is limited, so that its nucleation is difficult. Another cryoprotective mechanism of NADES is attributed to increase of the NADES viscosity caused by the decrease of the temperature during freezing, whereby NADES is in supercooled melt state and converted into a nonequilibrium, metastable and amorphous glass-like state in which nucleation does not occur. Castro et al. (2018) demonstrated the potential of NADES based on trehalose and glycerol (1:30) to act as a cryoprotective agent. This is associated with the NADES effect on the water crystallization/freezing and melting process, affecting the reduction of the number of ice crystals and consequently ice crystal damage in cells, which can be utilized in cryopreservation methods in industrial-scale food freezing. NADES influenced the thermophysical properties of water depending on its concentration in the system in a way that higher wt% of NADES decreased the melting temperature of the mixture, and consequently increased the melting enthalpy. It was shown that the NADES is able to prevent crystallization at lower temperatures, and that it can have a role in the vitrification process of water. To provide cryoprotective action of biological systems such as fresh fruit and vegetables, NADES viscosity is the property of interest. High values of viscosity are reported to have a negative effect to the cell viability, while systems with lower viscosity provide a higher stabilization capacity of the cells integrity. Although this study was predominantly aimed at cryopreservation of human cells and tissues, and since the application of NADES has been scarcely reported in food-related areas, its outcomes can be applied in the food freezing applied to extend the shelf life of food products. The reduction of freezing point and inhibition of ice formation by NADES is explained by its supramolecular network structure and the viscosity of NADES (Tian et al., 2020). The molecular motion and reorientation of free water available in the NADES is difficult, which also makes water difficult to nucleate. With the temperature decrease during freezing, the viscosity of NADES is further increased but without nucleation, whereby it turns into a supercooled melt state and finally into a nonequilibrium, metastable and amorphous glass-like state–vitrification state. The mechanism of cryoprotective ability of NADES is ascribed to the existence of a concentration gradient between the inside and outside of the cell and osmotic stress causing the migration of bio-substances through the cell membrane. Thus, the penetrating molecules (e.g. glycerol) enter the interior of the cell, while water diffuses to the exterior of the cell, creating a viscosity increment within cells. After the equilibrium is reached further dehydration of the cell is constrained, so that with the severe temperature decrease cells will be protected from severe dehydration, shrinkage or damage. Qiao, Cai, Yang, Zang, and Chen (2018) demonstrated the cryoprotective effects of NADES based on choline chloride, glycerol, and ethylene glycol (choline chloride:xylitol:water (1:2:3); ethylene glycol:L-proline (3:1); choline chloride:glucose:water (2:5:5); glycerol:L-proline (3:1); choline chloride:D-sorbitol:water (2:5:6)) on lactic acid bacteria (*Streptococcus thermophilus*) viability during frozen storage at −20 °C for 1 day and for 6 months. They observed an integrated cell membrane structure and enhanced intracellular enzyme activity, confirming the lactic acid bacteria viability, being of high importance for food production and probiotics industry.

3.4 Shelf life of low moisture foods and NADES

Low moisture foods like pasta (Jambrec, Sakač, Mišan, Mandić, & Pestorić, 2015), cookies (Mišan et al., 2011; Šarić et al., 2016, 2019) and crackers

(Sedej et al., 2011) have relatively long shelf life, which is determined by the stability of its constituents, content and composition of lipids and other color and aroma compounds prone to oxidation. It is known that the storage temperature, a_w value and the type of packaging can influence the stability of these products, but prediction of shelf life without a real time experiment presents a challenge due to the complexity of these systems (Sakač et al., 2016).

Referring to Vu, He, McClements, and Decker (2020) water activity (a_w) presents one of the factors that can have an impact on lipid oxidation in low moisture foods. In their experiment the influence of water activity (a_w), sugars (glucose, maltose, maltodextrin, and cyclodextrin), and proteins (casein and gluten) on the lipid hydroperoxide and hexanal lag phases of model crackers was examined. They found that oxidative stability of crackers was in an order: a_w 0.7 > a_w 0.4 > a_w 0.2 > a_w 0.05 and that reducing sugars including glucose, maltose, and maltodextrin at the same dextrose equivalence increased both hydroperoxide and hexanal lag phases. In their experiments, gluten was able to inhibit lipid oxidation with activity increasing with increasing a_w while casein showed minimal antioxidant impact. Yamauchi, Aoki, Sugiura, Kato, and Ueno (1982) found as well that reducing sugars can increase the oxidative stability of low moisture model systems, but not in oil-in-water emulsion.

From these and many other experiments it is evident that the state and composition of food system and water availability and mobility largely influence the rate of the oxidation, but the exact mechanism is still not clear.

3.5 NADES application in the production of biopolymers and edible coatings

Suitable packaging material can prolong the shelf life of foods. However, synthetic polymers traditionally used for their production have caused a serious ecological problem producing large amounts of waste, which can be toxic to biological life due to their total non-biodegradability. These environmental issues have put a demand for the production of eco-friendly packaging materials. In this sense, research on packaging materials based on biopolymers has been growing. Gouveia, Biernacki, Castro, Gonçalves, and Souza (2019) demonstrated that NADES–ChCl:Gly (1:2) or its individual components glycerol (Gly) and choline chloride (ChCl) can be used as plasticizers in the production of pectin-based films.

4. NADES and the protection of food components

4.1 Stabilization of enzymes

Enzymes are used in a variety of food processing operations due to their ability to modify all the major biomolecules—proteins, carbohydrates, lipids, and nucleic acids as well as smaller molecules such as amino acids, sugars, and vitamins. By modifying biomolecules, their utilization enables the reduction of viscosity, improvement of extractions and separations in processing, as well as the development of functionality, creation of flavor and enhancement of texture and structure of food products (Ashie, Simpson, & Smith, 1996). Due to the fact that enzyme-catalyzed reactions are specific and require exact reaction conditions, novel protein engineering has been striving to improve the specific activity of enzymes, modify their pH-activity profile, increase their specificity, but also to improve their thermal and operational stability (Fernandes, 2010). It was shown that deep eutectic solvents are of great importance in enzymatic reactions since they facilitate the effective contact between reactants and enzymes and thus affect the activity, structure and stability of enzymes (Daneshjou et al., 2017). Therefore, the significance of utilization of NADES in biotransformation processes is getting more and more important, since they are able to improve the activity and stability of the enzyme, in contrast to organic solvents which are toxic, flammable and cause enzyme (Juneidi et al., 2018). By the selection of different types of DESs as cosolvents, the regioselectivity of the enzyme can be modified (Juneidi et al., 2018).

Wu, Wen, Xu, and Yang (2014) demonstrated high stabilization ability of 24 NADES prepared from choline chloride and choline acetate, and urea, glycerol, acetamide, ethylene glycol as H-bond donors (mixed in three molar ratios of 1:2, 1:1 or 2:1) on horseradish peroxidase. They suggested that the enzyme in aqueous solution was affected by the DES itself rather than by the synergistic action of the individual NADES components. Khodaverdian et al. (2018) demonstrated the high stabilization ability of sorbitol:betaine:water (1:1:1) and glycerol:betaine (2:1) NADES for laccase from *Bacillus HR03* in comparison to the aqueous buffer at 80 and 90 °C. It was shown that the enzyme was more stable in the glycerol:betaine (2:1) NADES itself, in comparison to the solution of individual NADES components used in the same ratio in a buffer, due to the strong hydrogen-bonding network. These findings are of high industrial relevance, implying the possibility of performing reactions at higher temperatures. Huang, Wu, Wen, Yang, and Yang (2014) indicated that the stability of fungal lipase from *Penicillium expansum* in NADES is dependent on the choice of the choline salt (choline chloride and choline acetate) and the H-bond donors and their molar ratio, affecting the solvent properties—polarity, viscosity, and surface tension. Choline acetate exhibited superior stabilizing ability in comparison to choline chloride, while glycerol and ethylene glycol triggered the enzyme to exhibit a relatively higher hydrolytic reaction rate and a markedly improved half-life than acetamide and urea. Urea itself can denature the enzyme, while glycerol and ethylene glycol can actively participate in the enzymatic reaction, it was demonstrated that when they are incorporated in the DES, their reactivity is decreased due to the extensive hydrogen-bond network formed. Higher stabilization ability of NADES containing glycerol and ethylene glycol is due to 3 and 2 hydroxyl groups in the molecule, respectively, having more powerful strength than urea and acetamide in forming H-bonds with the cholinium salt anions, chloride and acetate. Elgharbawy et al. (2018) demonstrated the stabilization ability of NADES prepared from choline chloride and sucrose for lipases, as one the most widely used enzymes in industry for hydrolysis, alcoholysis and esterifications (Juneidi et al., 2018). It was shown that the initial activity of lipase in NADES was maintained at elevated temperatures (70–80 °C) during 1 h, while their activity in phosphate buffer started to decrease at 60 °C. The better protective properties of lipases showed non-aqueous NADES due to its high viscosity and the low solubility of lipases in the absence of water. Kim et al. (2016) reported enhanced thermal, pH and storage stability of lipase in NADES. It was shown that glycerol and ethylene glycol as H-bond donors for DES mixtures significantly enhanced the thermal stability of lipase in aqueous DES solutions they were used as cosolvents in buffer (pH 7.0), unlike formamide and thiourea containing mixtures. The half-life of lipase in aqueous solutions of glycerol-containing DES mixtures was increased by 2.9–9.2 times. The best lipase stability under acidic conditions provided aqueous solutions of glycerol-containing DES mixtures, unlike the aqueous solutions of formamide-containing DES mixtures, although the residual activity of lipase was higher than that in buffer without DES mixtures. The aqueous solutions of glycerol-containing DES mixtures also provided high lipase stability, but the best results were obtained with choline chloride:ethylene glycol:formamide (1:1:1), exhibiting 117% higher lipase activity at pH 7.0 during incubation at 40 °C and 610% higher half-time than those in buffer without DES mixtures. Glycerol-containing DES mixtures enhanced the storage stability for a long time, where choline chloride:glycerol:thiourea (1:1:1) exhibited the highest residual activity of lipase after incubation for 48h. On the contrary, urea- or formamide-containing DES mixtures were less efficient, while choline chloride:formamide:thiourea (1:1:1) could not enhance the stability of lipase.

4.2 Stabilization of natural food colorants

As being an indicator of food quality, defects and grade, color of food greatly influences the food choices (Cömert, Mogol, & Gökmen, 2020). While the majority of natural products tend to change both appearance and color with the ripening process, many processed foods are formulated to maintain a consistent appearance and color during the product shelf life that is achieved by the use of food additives (Spence, 2019). Since consumers associate the color of food with its nutritional composition and health benefits, it is of utmost importance to preserve the color of the product and avoid the usage of artificial color nowadays associated with food safety concerns (Cömert et al., 2020; Oplatowska-Stachowiak & Elliott, 2017).

Anthocyanins can be used as natural food colorants, but their relative low stability during storage can be an obstacle for this application. Low stability of anthocyanins is attributed to the unstable structure of the anthocyanin core flavylium cation (Guo et al., 2019). Bosiljkov et al. (2017) reported the superiority of NADES prepared with organic acids for the extraction of anthocyanins from wine lees over a variety of NADES based on choline chloride and xylose, fructose, glucose and glycerol. This NADES provides higher stability of extracted anthocyanins due to acidic pH conditions and polarity. The flavylium ion is stable under acidic conditions, while with the increase in pH the structures of the bonds around the flavylium ion change to form the blue quinoidal species (at pH 2–4). At pH between 5 and 6, the colorless carbinol pseudobase and chalcone are predominant, while at pH > 7 the anthocyanins are degraded rapidly. Moreover as the organic acid-based NADES are more polar than the sugar- and polyalcohol-based NADES they favor the stability of the highly polar anthocyanins. Guo et al. (2019) confirmed the suitability of acid-based NADES for thermal stabilization of mulberry anthocyanins at 60 and 80 °C during 8h. It was shown that the degradation of anthocyanin content is faster than the color fading due to higher thermal resistance of anthocyanin self-stacking than co-pigmentation with phenolic acids via p–p overlap.

By determining the stability of different NADES based on choline chloride, proline, betaine and malic acid on grape-pomace anthocyanins, Panić, Gunjević, Cravotto, and Redovniković (2019) reported no differences in anthocyanin stability in the various NADES at room temperature, while choline chloride:citric acid (2:1) exhibited the highest stabilizing capacity at 4 °C and −18 °C for 60 days, being 14% and 10% degraded, respectively. For comparison, the degradation of anthocyanins in choline chloride: proline:malic acid NADES at 4 °C was 70%.

Liu et al. (2019) examined the thermal and storage stability of natural pigments from *Curcuma longa*—curcuminoids in citric acid:glucose NADES (1:1) and 15% water and compared it with the conventional solvents (ethanol and methanol). At 50 °C the degradation of curcuminoids extracted by NADES ranged between 1% and 3%, being similar to that extracted by ethanol and methanol. However, with the increase of the temperature, the degradation of curcuminoids increased being in the range 2–6.5% at 80 °C, whereby the stability of curcuminoids in NADES was higher than that in ethanol and methanol. During 30 days of storage at 4 and 25 °C, curcuminoids in citric acid:glucose NADES were more stable than that in ethanol and methanol. The stabilization ability of citric acid:glucose NADES for curcuminoids can be associated with the formation of hydrogen bonds between hydroxyl (OH) and methoxy groups, as well as between oxygen atoms of the OH groups and components of citric acid:glucose NADES which affected the decreased diffusion of curcuminoids and protected them from oxygen. Moreover, acidic pH conditions of citric acid: glucose NADES contributed to the suppression of oxidative changes of curcuminoids, since bis-keto tautomer of curcuminoids exhibited higher stability in acidic solutions.

Curcumin dissolved in choline chloride and glycerol NADES in unimolar proportions, remained stable when exposed to artificial sunlight without observing any degradation over 120 min, contrary to the degradation of curcumin in the powder form and in methanol extract (Jeliński, Przybyłek, & Cysewski, 2019). The concentration of curcumin in methanol solution dropped to 5% of its initial concentration, while the decrease of curcumin concentration in powder form was less pronounced, with 19% of residual curcumin after 120 min. The NADES stabilization ability for curcumin was explained by hydrogen bonding of vicinal hydroxyl group with one of the central keto-groups. It was found that variation of the choline chloride amount in the system had a very strong influence on self-association of curcumin and the rise of choline chloride share significantly reduces the stability of curcumin dimers in the solution. The overall concentration of bound curcumin decreases with the increase of choline chloride in the system.

Dai, Verpoorte, and Choi (2014) demonstrated how the heating, light, storage time and water content in NADES differing in pH, viscosity and polarity affected the stability of carthamin (red pigment), hydroxysafflor yellow A and cartormin (major components of safflor yellow) or safflower extracts. Xylitol:choline chloride NADES (1:4) provided a protective effect against thermal degradation of carthamin at 40 and 60 °C, being longer stable in xylitol:choline chloride NADES than in aqueous solution expressed as a longer half-time. Glucose:choline chloride (2:5) and sucrose:choline chloride (1:4) provided a higher stability of carthamin under 24h light exposure over a 15-day period in comparison to 40% ethanol and water solution, exhibiting 90% and 75% degradation, respectively. Carthamin in all NADES at 20 °C was not affected during 7 days of storage, but with the extension of the storage time to 15 days carthamin remained unaffected only in sucrose:choline chloride NADES, while in other NADES carthamin was degraded less than 10%. Simultaneously, the substantial degradation of carthamin was observed in water and 40% ethanol. Sucrose:choline chloride NADES appeared to be a good stabilizing agent for carthamin at lower storage temperatures (4 °C) during 1 month period, unlike other NADES and conventional solvents. It was shown that the stability of carthamin in sucrose:choline chloride and glucose:choline chloride (2:5) NADES was better than that in proline:malic acid (1:1) and lactic acid:glucose (5:1) due to higher viscosity of sugar-based NADES in which the movement of molecules is limited and stable molecular interactions between solvents and carthamin occur which reduce the degradation of carthamin.

4.3 Stabilization of food aromas

The manufacturing of flavor ingredients from natural resources to be used in the food or other types of products is increasingly gaining importance due to the emerging trends in new food product development tending to avoid artificial ingredients and satisfy the requirements for clean-labeled food products. Food flavors comprising short-chain alkanes and alkenes (with or without oxygen, nitrogen, or sulfur), alcohols, esters, ketones, organic acids, with terpenes as the largest and the most diverse group of organic compounds responsible for nature's aromas. All of these organic compounds can easily be degraded by hydrolysis, oxidation and radical reaction, as well as thermal degradation, which limits the utilization of certain conventional techniques for their extraction (Castro-Muñoz, 2019). In order to preserve the functional properties of the flavor and aroma compounds and reduce the risks of product contamination, the utilization of non-chemical agents or solvents is advisable.

Kranz and Hofmann (2018) used NADES as a reaction media, and demonstrated the stabilizing ability of NADES for Amadori product, being the main precursor of Maillard aromas and colors. They investigated the Amadori product formation in NADES and in aqueous buffer by dissolving glucose and carnosine and subjecting it to a temperature between 60 and 100 °C for 2 h. It was found that yield of Amadori product up to 80 °C was higher in all examined NADES in comparison to the buffer solution. Namely, betaine/glycerol NADES yielded 120.3 (60 °C), 340.4 (80 °C), and 312.7 μmol/mmol (100 °C) of Amadori product, in comparison to the maximal 12.2 μmol/mmol at 80 °C in aqueous solution. Further temperature increase up to 100 °C, accelerated degradation in a greater extent than conversion was increased, whereby betaine/sucrose and malic acid/sucrose NADES yielded the highest amount of target compound at 100 °C, being 306.4 and 474.9 μmol/mmol, respectively, indicating high stabilization ability of betaine/sucrose and malic acid/sucrose NADES for Amadori product.

4.4 Stabilization of bioactives

The thermal and storage stabilization ability of NADES composed of lactic acid, glucose and 15% water for different phenolic compounds from agri-food by products (e.g. by products of olive oil industry, onion seed production, tomato and pear canning industry) was determined and compared with water and methanol. All examined phenolic compounds (oleuropein, 3-hydroxytyrosol, coumaric, cinnamic, caffeic and ferulic acids, quercetin, and apigenin) were stable at −18 and 4 °C over 2 months in NADES and methanol, whereas degradation of quercetin and apigenin in water was up to 90%. The degradation rate of oleuropein, 3-hydroxytyrosol, quercetin, and apigenin in NADES at 25 °C was in the range 30–50% over 2 months, while coumaric, cinnamic, caffeic and ferulic acids were stable in the same conditions. Sugar-based NADES appeared to have high stabilizing ability for phenolic compounds, explained by the formation of hydrogen bonds between solute and NADES (de los Ángeles Fernández, Espino, Gomez, & Silva, 2018).

Chemical instability is one of the major hurdles that complicates the utilization of catechins in the different brunches of industry—food, cosmetic and pharmaceutical. Unfavorable changes of catechins occur during the storage and processing, and they are mainly affected by the temperature and the storage duration. Jeong et al. (2017) demonstrated the stabilization ability of betain:glycerol:glucose (4:20:1) NADES for epigallocatechin-3-gallate from green tea at 60 °C over 3 weeks of storage. Over a 3-week period, 27% of epigallocatechin-3-gallate remained preserved from 70% MeOH extract, compared to 60% of EGCG in the NADES extract (Jeong et al., 2017).

Funari et al. (2019) demonstrated the stability of propolis over 17 months in choline chlorine:propylene glycol (1:2) and choline chlorine:lactic acid: water (1:1:1) remaining the transparent liquid, which demonstrated the potential of NADES to replace alcoholic extracts of propolis, not always suitable for certain groups of consumers.

NADES can be used as an aid in rice bran oil processing to remove free fatty acid from crude rice bran oil, and at the same time preserve γ-oryzanol, which is normally lost during chemical and/or physical deacidification. Chemical deacidification comprises the adding alkaline solutions to the oil, while physical process comprises the utilization of high temperatures all of which negatively affect the preservation of γ-oryzanol. The removal of free fatty acids (19.03 ± 2.33%) and the preservation of γ-oryzanol (51.30 ± 1.77%) was achieved by choline chloride:ethylene glycol NADES (1:2) at the temperature of 50 °C over 240 min (Zullaikah, Wibowo, Wahyudi, & Rachimoellah, 2019).

Similarly, the preservation of α-tocopherol and β-carotene in palm oil with the efficacy of up to 99% was achieved in deacidification process of with betaine monohydrate:glycerol NADES (1:8) (Zahrina, Nasikin, Krisanti, & Mulia, 2018).

5. Conclusions and future perspectives

Despite the number of experiments, concepts and hypotheses dealing with food stability, shelf-life prediction from a scientific basis rather than empiricism still presents a challenge to food scientists and engineers. Food deterioration reactions depend on the state and composition of a food product, and it is evident that water availability and mobility largely influence their rates.

If NADES exist in living cells as a third class of liquids, which is different from water and lipids, it is logical to assume that NADES either exist or are formed in food systems during processing (e.g. drying and cooling) and influence its stability. Although there is no scientific evidence yet which supports the hypothesis of NADES formation in food and the way they contribute to the food stability, some of the findings related to the unique physicochemical properties of synthesized NADES that might play a role in food preservation as summarized in Table 1.

Table 1 NADES application in the stabilization of foods and protection of food components.

Property	Potential application	NADES	Reference
Antimicrobial activity	Gram-positive bacteria (*Staphylococcus aureus*) Gram-negative bacteria (*Escherichia coli*, *Proteus mirabilis*, *Salmonella typhimurium*, *Pseudomonas aeruginosa*) Yeast (*Candida albicans*)	Citric acid:glucose: glycerol (1:1:1) Citric acid:fructose: glycerol (1:1:1) NADES	Radošević et al. (2018)
Antimicrobial photodynamic therapy	Increase the phototoxic effect of the photosensitiser	Citric acid:sucrose (1:1) DL-malic acid:D-(−)-fructose:D-(+)-glucose (1:1:1) Citric acid:D-(−)-fructose (1:1)	Wikene et al. (2016)
	Increase the solubility, physical stability and bacterial phototoxicity of a neutral porphyrin	Citric acid:sucrose (1:1) D-(+)-glucose: DL-malic acid (1:1)	Wikene et al. (2015)
Cryoprotective activity	Preservation by freezing	Trehalose:glycerol (1:30)	Castro et al. (2018)
	Food production and probiotics industry	Choline chloride: xylitol:water (1:2:3); ethylene glycol: L-proline (3:1); choline chloride:glucose:water (2:5:5); glycerol: L-proline (3:1); choline chloride:D-sorbitol: water (2:5:6)	Qiao et al. (2018)
Plasticizers	Production of pectin-based films	Choline chloride: glycerol (1:2)	Gouveia et al. (2019)

Continued

Table 1 NADES application in the stabilization of foods and protection of food components.—cont'd

Property	Potential application	NADES	Reference
Stabilization of enzymes	Stabilization of horseradish peroxidase	24 NADES prepared from choline chloride and choline acetate, and urea, glycerol, acetamide, ethylene glycol	Wu et al. (2014)
	Stabilization of laccase from *Bacillus HR03*	Osorbitol:betaine: water (1:1:1) Glycerol:betaine (2:1)	Khodaverdian et al. (2018)
	Stabilization of fungal lipase from *Penicillium expansum*	NADES composed of choline acetate and glycerol or ethylene glycol	Huang et al. (2014)
	Stabilization of lipases	NADES prepared from choline chloride and sucrose	Elgharbawy et al. (2018)
		Choline chloride: ethylene glycol: formamide (1:1:1) Chloride:glycerol: thiourea (1:1:1)	Kim et al. (2016)
Stabilization of natural food colorants	Stabilization of anthocyanins	NADES prepared with organic acids	Bosiljkov et al. (2017) and Guo et al. (2019)
		Choline chloride:citric acid (2:1)	Panić et al. (2019)
	Stabilization of curcuminoids	Citric acid:glucose (1:1) and 15% water	Liu et al. (2019)
	Stabilization of curcumin	Choline chloride: glycerol (1:1)	Jeliński et al. (2019)
	Stabilization of carthamin, hydroxysafflor yellow A and cartormin	Proline-malic acid (1:1), lactic acid–glucose (5:1), glucose-choline chloride (2:5), sucrose-choline chloride (1:4), and xylitol-choline chloride (1:4)	Dai et al. (2014)

Table 1 NADES application in the stabilization of foods and protection of food components.—cont'd

Property	Potential application	NADES	Reference
Stabilization of food aromas	Stabilization of Amadori product in the production of Maillard aromas and colors	Betaine/glycerol and malic acid/sucrose NADES	Kranz and Hofmann (2018)
Stabilization of bioactives	Stabilization of different phenolic compounds from agri-food by products	NADES composed of lactic acid, glucose and 15% water	de los Ángeles Fernández et al. (2018)
	Stabilization of epigallocatechin-3-gallate from green tea	Betain:glycerol:glucose (4:20:1)	Jeong et al. (2017)
	Stabilization of propolis	Choline chlorine: propylene glycol (1:2) Choline chlorine:lactic acid:water (1:1:1)	Funari et al. (2019)
	Preservation of γ-oryzanol	Choline chloride: ethylene glycol NADES (1:2)	Zullaikah et al. (2019)
	Preservation of α-tocopherol and β-carotene in palm oil	Betaine monohydrate: glycerol NADES (1:8)	Zahrina et al. (2018)

Acknowledgment

The work is a part of the project: Natural deep eutectic solvents for green agri-food solutions (No 6060592) that has received funding from the Science Fund of the Republic of Serbia.

References

Ashie, N. A., Simpson, B. K., & Smith, J. P. (1996). Mechanisms for controlling enzymatic reactions in foods. *Critical Reviews in Food Science and Nutrition, 36*(1–2), 1–30.

Barbosa-Cánovas, G. V., Fontana, A. J., Schmidt, S. J., & Labuza, T. P. (2007). *Water activity in foods: Fundamentals and applications.* USA: Blackwell Publishing and Institute of Food Technologists.

Barreiro, J. A., & Sandoval, A. J. (2020). Kinetics of moisture adsorption during simulated storage of whole dry cocoa beans at various relative humidities. *Journal of Food Engineering, 273*, 109869.

Bordenave, N., Hamaker, B. R., & Ferruzzi, M. G. (2014). Nature and consequences of non-covalent interactions between flavonoids and macronutrients in foods. *Food & Function, 5*(1), 18–34.

Bosiljkov, T., Dujmić, F., Cvjetko Bubalo, M., Hribar, J., Vidrih, R., Brnčić, M., et al. (2017). Natural deep eutectic solvents and ultrasound-assisted extraction: Green approaches for extraction of wine lees anthocyanins. *Food and Bioproducts Processing, 102*, 195–203.

Buitink, J., & Leprince, O. (2004). Glass formation in plant anhydrobiotes: Survival in the dry state. *Cryobiology, 48*, 215–228.

Burke, M. J. (1986). The glassy state and survival of anhydrous biological systems. In A. C. Leopold (Ed.), *Membranes, metabolism and dry organisms* (pp. 358–363). Ithaca, NY: Cornell University Press.

Cardellini, F., Germani, R., Cardinali, G., Corte, L., Roscini, L., Spreti, N., et al. (2015). Room temperature deep eutectic solvents of (1 S)-(+)-10-camphorsulfonic acid and sulfobetaines: Hydrogen bond-based mixtures with low ionicity and structure-dependent toxicity. *RSC Advances, 5*(40), 31772–31786.

Cardellini, F., Tiecco, M., Germani, R., Cardinali, G., Corte, L., Roscini, L., et al. (2014). Novel zwitterionic deep eutectic solvents from trimethylglycine and carboxylic acids: Characterization of their properties and their toxicity. *RSC Advances, 4*(99), 55990–56002.

Castro, V. I. B., Craveiro, R., Silva, J. M., Reis, R. L., Paiva, A., & Duarte, A. R. C. (2018). Natural deep eutectic systems as alternative nontoxic cryoprotective agents. *Cryobiology, 83*, 15–26.

Castro-Muñoz, R. (2019). Pervaporation: The emerging technique for extracting aroma compounds from food systems. *Journal of Food Engineering, 253*, 27–39.

Charoenrein, S. (2018). Microstructural changes and their relationship with quality and stability of frozen foods. In S. Devahastin (Ed.), *Food microstructure and its relationship with quality and stability* (pp. 123–138). Woodhead Publishing.

Choi, Y. H., van Spronsen, J., Dai, Y., Verberne, M., Hollmann, F., Arends, I. W. C. E., et al. (2011). Are natural deep eutectic solvents the missing link in understanding cellular metabolism and physiology? *Plant Physiology, 156*, 1701–1705.

Cieplik, F., Deng, D., Crielaard, W., Buchalla, W., Hellwig, E., Al-Ahmad, A., et al. (2018). Antimicrobial photodynamic therapy—What we know and what we don't. *Critical Reviews in Microbiology, 44*, 571–589.

Cömert, E. D., Mogol, B. A., & Gökmen, V. (2020). Relationship between color and antioxidant capacity of fruits and vegetables. *Current Research in Food Science, 2*, 1–10.

Cvetković, B. R., Pezo, L. L., Tasić, T., Šarić, L., Kevrešan, Ž., & Mastilović, J. (2015). The optimisation of traditional fermentation process of white cabbage (in relation to biogenic amines and polyamines content and microbiological profile). *Food Chemistry, 168*, 471–477.

Dai, Y., Verpoorte, R., & Choi, Y. H. (2014). Natural deep eutectic solvents providing enhanced stability of natural colorants from safflower (Carthamus tinctorius). *Food Chemistry, 159*, 116–121.

Daneshjou, S., Khodaverdian, S., Dabirmanesh, B., Rahimi, F., Daneshjoo, S., Ghazi, F., et al. (2017). Improvement of chondroitinases ABCI stability in natural deep eutectic solvents. *Journal of Molecular Liquids, 227*, 21–25.

de los Ángeles Fernández, M., Espino, M., Gomez, F. J. V., & Silva, M. F. (2018). Novel approaches mediated by tailor-made green solvents for the extraction of phenolic compounds from agro-food industrial by-products. *Food Chemistry, 239*, 671–678.

de Morais, P., Gonçalves, F., Coutinho, J. A. P., & Ventura, S. P. M. (2015). Ecotoxicity of cho-linium-based deep eutectic solvents. *ACS Sustainable Chemistry & Engineering, 3*(12), 3398–3404.

Elgharbawy, A. A., Hayyan, A., Hayyan, M., Rashid, S. N., Nor, M. R. M., Zulkifli, M. Y., et al. (2018). Shedding light on lipase stability in natural deep eutectic solvents. *Chemical and Biochemical Engineering Quarterly, 32*, 359–370.

Fernandes, P. (2010). Enzymes in food processing: A condensed overview on strategies for better biocatalysts. *Enzyme Research*, *2010*, 862537.

Funari, C. S., Sutton, A. T., Carneiro, R. L., Fraige, K., Cavalheiro, A. J., da Silva Bolzani, V., et al. (2019). Natural deep eutectic solvents and aqueous solutions as an alternative extraction media for propolis. *Food Research International*, *125*, 108559.

Gao, H., Huang, S., Dong, T., Yang, Q., & Yi, G. (2016). Analysis of resistant starch degradation in postharvest ripening of two banana cultivars: Focus on starch structure and amylases. *Postharvest Biology and Technology*, *119*, 1–8.

Gérard, V., Ay, E., Morlet-Savary, F., Graff, B., Galopin, C., Ogren, T., et al. (2019). Thermal and photochemical stability of anthocyanins from black carrot, grape juice, and purple sweet potato in model beverages in the presence of ascorbic acid. *Journal of Agricultural and Food Chemistry*, *67*(19), 5647–5660.

Gómez, A. V., Tadini, C. C., Biswas, A., Buttrum, M., Kim, S., Boddu, V. M., et al. (2019). Microwave-assisted extraction of soluble sugars from banana puree with natural deep eutectic solvents (NADES). *LWT*, *107*, 79–88.

González-Cabrera, M., Domínguez-Vidal, A., & Ayora-Cañada, M. J. (2018). Hyperspectral FTIR imaging of olive fruit for understanding ripening processes. *Postharvest Biology and Technology*, *145*, 74–82.

Gouveia, T. I. A., Biernacki, K., Castro, M. C. R., Gonçalves, M. P., & Souza, H. K. S. (2019). A new approach to develop biodegradable films based on thermoplastic pectin. *Food Hydrocolloids*, *97*, 105175.

Gram, L., & Dalgaard, P. (2002). Fish spoilage bacteria—Problems and solutions. *Current Opinion in Biotechnology*, *13*(3), 262–266.

Guo, N., Jiang, Y. W., Wang, L. T., Niu, L. J., Liu, Z. M., & Fu, Y. J. (2019). Natural deep eutectic solvents couple with integrative extraction technique as an effective approach for mulberry anthocyanin extraction. *Food Chemistry*, *296*, 78–85.

He, Z., Xu, M., Zeng, M., Qin, F., & Chen, J. (2016a). Interactions of milk alpha- and beta-casein with malvidin-3-O-glucoside and their effects on the stability of grape skin anthocyanin extracts. *Food Chemistry*, *199*, 314–322.

He, Z., Xu, M., Zeng, M., Qin, F., & Chen, J. (2016b). Preheated milk proteins improve the stability of grape skin anthocyanins extracts. *Food Chemistry*, *210*, 221–227.

He, S., Zhang, Z., Sun, H., Zhu, Y., Zhao, J., Tang, M., et al. (2019). Contributions of temperature and L-cysteine on the physicochemical properties and sensory characteristics of rapeseed flavor enhancer obtained from the rapeseed peptide and D-xylose Maillard reaction system. *Industrial Crops and Products*, *128*, 455–463.

Huang, Z. L., Wu, B. O., Wen, Q., Yang, T. X., & Yang, Z. (2014). Deep eutectic solvents can be viable enzyme activators and stabilizers. *Journal of Chemical Technology & Biotechnology*, *89*, 1975–1981.

Jambrec, D. J., Sakač, M. B., Mišan, A.Č., Mandić, A. I., & Pestorić, M. V. (2015). Effect of autoclaving and cooking on phenolic compounds in buckwheat-enriched whole wheat tagliatelle. *Journal of Cereal Science*, *66*, 1–9.

Jeliński, T., Przybyłek, M., & Cysewski, P. (2019). Natural deep eutectic solvents as agents for improving solubility, stability and delivery of curcumin. *Pharmaceutical Research*, *36*, 116.

Jeong, K. M., Ko, J., Zhao, J., Jin, Y., Han, S. Y., & Lee, J. (2017). Multi-functioning deep eutectic solvents as extraction and storage media for bioactive natural products that are readily applicable to cosmetic products. *Journal of Cleaner Production*, *151*, 87–95.

Juneidi, I., Hayyan, M., & Hashim, M. A. (2018). Intensification of biotransformations using deep eutectic solvents: Overview and outlook. *Process Biochemistry*, *66*, 33–60.

Khodaverdian, S., Dabirmanesh, B., Heydari, A., Dashtban-moghadam, E., Khajeh, K., & Ghazi, F. (2018). Activity, stability and structure of laccase in betaine based natural deep eutectic solvents. *International Journal of Biological Macromolecules*, *107*(Pt. B), 2574–2579.

Kim, S. H., Park, S., Yu, H., Kim, J. H., Kim, H. J., Yang, Y. H., et al. (2016). Effect of deep eutectic solvent mixtures on lipase activity and stability. *Journal of Molecular Catalysis B: Enzymatic, 128*, 65–72.

Kranz, M., & Hofmann, T. (2018). Food-grade synthesis of Maillard-type taste enhancers using natural deep eutectic solvents (NADES). *Molecules, 23*, 261.

Le Meste, M., Champion, D., Roudaut, G., Blond, G., & Simatos, D. (2002). Glass transition and food technology: A critical appraisal. *Journal of Food Science, 67*(7), 2444–2458.

Liu, Y., Bhandari, B., & Zhou, W. (2006). Glass transition and enthalpy relaxation of amorphous food saccharides: A review. *Journal of Agricultural Food Chemistry, 54*, 5701–5717.

Liu, Y., Li, J., Fu, R., Zhang, L., Wang, D., & Wang, S. (2019). Enhanced extraction of natural pigments from *Curcuma longa* L. using natural deep eutectic solvents. *Industrial Crops and Products, 140*, 111620.

Mišan, A., Mimica-Dukić, N., Sakač, M., Mandić, A., Sedej, I., Šimurina, O., et al. (2011). Antioxidant activity of medicinal plant extracts in cookies. *Journal of Food Science, 76*(9), C1239–C1244.

Mišan, A., Nađpal, J., Tomšik, A., Pojić, M., Mandić, A., & Verpoort, R. (2020). Choi Young Hae: The perspectives of natural deep eutectic solvents in agri-food sector. *Critical Reviews in Food Science and Nutrition, 60*(15), 2564–2592.

Oplatowska-Stachowiak, M., & Elliott, C. T. (2017). Food colors: Existing and emerging food safety concerns. *Critical Reviews in Food Science and Nutrition, 57*(3), 524–548.

Panić, M., Gunjević, V., Cravotto, G., & Redovniković, I. R. (2019). Enabling technologies for the extraction of grape-pomace anthocyanins using natural deep eutectic solvents in up-to-half-litre batches extraction of grape-pomace anthocyanins using NADES. *Food Chemistry, 300*, 125185.

Pérez-Andrés, J. M., de Alba, M., Harrison, S. M., Brunton, N. P., Cullen, P. J., & Tiwari, B. K. (2020). Effects of cold atmospheric plasma on mackerel lipid and protein oxidation during storage. *LWT, 118*, 108697.

Qiao, Y., Cai, H. L., Yang, X., Zang, Y. Y., & Chen, Z. G. (2018). Effects of natural deep eutectic solvents on lactic acid bacteria viability during cryopreservation. *Applied Microbiology and Biotechnology, 102*(13), 5695–5705.

Radošević, K., Čanak, I., Panić, M., Markov, K., Cvjetko Bubalo, M., Frece, J., et al. (2018). Antimicrobial, cytotoxic and antioxidative evaluation of natural deep eutectic solvents. *Environmental Science and Pollution Research, 25*, 14188–14196.

Rahman, M. S. (2006). State diagram of foods: Its potential use in food processing and product stability. *Trends in Food Science & Technology, 17*(3), 129–141.

Rahman, M. S. (2007). *Handbook of food preservation* (2nd ed.). Boca Raton, FL: CRC Press.

Rahman, M. S. (2009). Food stability beyond water activity and glass transition: Macro–micro region concept in the state diagram. *International Journal of Food Properties, 12*(4), 726–740.

Rahman, M. S. (2010). Food stability determination by macro–micro region concept in the state diagram and by defining a critical temperature. *Journal of Food Engineering, 99*, 402–416.

Rahman, M. S., & Labuza, T. P. (2007). Water activity and food preservation. In M. S. Rahman (Ed.), *Handbook of food preservation* (2nd ed., pp. 447–477). New York, USA: CRC Press.

Sakač, M., Pestorić, M., Mandić, A., Mišan, A., Nedeljković, N., Jambrec, D., et al. (2016). Shelf-life prediction of gluten-free rice-buckwheat cookies. *Journal of Cereal Science, 69*, 336–343.

Šarić, B., Dapčević-Hadnađev, T., Hadnađev, M., Sakač, M., Mandić, A., Mišan, A., et al. (2019). Fiber concentrates from raspberry and blueberry pomace in gluten-free cookie formulation: Effect on dough rheology and cookie baking properties. *Journal of Texture Studies, 50*, 124–130.

Šarić, B., Mišan, A., Mandić, A., Nedeljković, N., Pojić, M., Pestorić, M., et al. (2016). Valorisation of raspberry and blueberry pomacethrough the formulation of value-added gluten-free cookies. *Journal of Food Science and Technology*, *53*, 1140–1150.

Sedej, I., Sakač, M., Mandić, A., Mišan, A., Pestorić, M., Šimurina, O., et al. (2011). Quality assessment of gluten-free crackers based on buckwheat flour. *LWT-Food Science and Technology*, *44*(3), 694–699.

Slade, L., & Levine, L. (1991). A food polymer science approach to structure property relationships in aqueous food systems: Non-equilibrium behavior of carbohydrate-water systems. In H. Levine, & L. Slade (Eds.), *Water relationships in food* (pp. 29–101). New York: Plenum Press.

Spence, C. (2019). On the changing colour of food & drink. *International Journal of Gastronomy and Food Science*, *17*, 100161.

Tian, Y., Zhu, Z., & Sun, D. W. (2020). Naturally sourced biosubstances for regulating freezing points in food researches: Fundamentals, current applications and future trends. *Trends in Food Science & Technology*, *95*, 131–140.

Toivonen, P. M. A. (2004). Postharvest storage procedures and oxidative stress. *Hort Science*, *39*, 938–942.

Toivonen, P. M. A., & Brummell, D. A. (2008). Biochemical bases of appearance and texture changes in fresh-cut fruit and vegetables. *Postharvest Biology and Technology*, *48*, 1–14.

Vu, T. P., He, L., McClements, D. J., & Decker, E. A. (2020). Effects of water activity, sugars, and proteins on lipid oxidative stability of low moisture model crackers. *Food Research International*, *130*, 108844.

Wikene, K. O., Bruzell, E., & Tønnesen, H. H. (2015). Improved antibacterial phototoxicity of a neutral porphyrin in natural deep eutectic solvents. *J. Photochem. Photobiol. B: Biol.*, *148*, 188–196.

Wikene, K. O., Rukke, H. V., Bruzell, E., & Tønnesen, H. H. (2016). Physicochemical characterisation and antimicrobial phototoxicity of an anionic porphyrin in natural deep eutectic solvents. *European Journal of Pharmaceutics and Biopharmaceutics*, *105*, 75–84.

Wu, B.-P., Wen, Q., Xu, H., & Yang, Z. (2014). Insights into the impact of deep eutectic solvents on horseradish peroxidase: Activity, stability and structure. *Journal of Molecular Catalysis B: Enzymatic*, *101*, 101–107.

Xiong, Q., Zhang, M., Wang, T., Wang, D., Sun, C., Bian, H., et al. (2020). Lipid oxidation induced by heating in chicken meat and the relationship with oxidants and antioxidant enzymes activities. *Poultry Science*, *99*(3), 1761–1767.

Yamauchi, R., Aoki, Y., Sugiura, T., Kato, K., & Ueno, Y. (1982). Effect of sugars and sugar analogs on autoxidation of methyl linoleate and safflower oil. *Agricultural and Biological Chemistry*, *46*(12), 2997–3002.

Zahrina, I., Nasikin, M., Krisanti, E., & Mulia, K. (2018). Deacidification of palm oil using betaine monohydrate-based natural deep eutectic solvents. *Food Chemistry*, *240*, 490–495.

Zullaikah, S., Wibowo, N. D., Wahyudi, I. M. G. E. D., & Rachimoellah, M. (2019). Deacidification of crude rice bran oil (CRBO) to remove FFA and preserve γ-oryzanol using deep eutectic solvents (DES). *Materials Science Forum*, *964*, 109–114.

CHAPTER FOURTEEN

Honey in traditional Chinese medicine: A guide to future applications of NADES to medicines

Yuntao Dai[a,*], Young Hae Choi[b], and Robert Verpoorte[b]
[a]Institute of Chinese Materia Medica, China Academy of Chinese Medical Sciences, Beijing, China
[b]Natural Products Laboratory, Institute of Biology, Leiden University, Leiden, The Netherlands
*Corresponding author: e-mail address: dai_yuntao@live.cn

Contents

Abstract

Honey has played an important role in traditional medicine for thousands of years. It has been used in different aspects of traditional medicine, as a single medicine, as an additive for *paozhi* processing, and as an adjuvant for different types of pharmaceutical formulations, such as electuary, honeyed pills, etc. Hundreds of compounds have already been found in honeys from different sources. However, the major compounds in honey are glucose and fructose in about equal molar ratios. This means that the basis of honey is a natural deep eutectic solvent (NADES). The physico-chemical properties of NADES may play an important role in the use of honey in traditional medicine. Clarifying the role of honey in traditional medicines helps to rationalize traditional practices, as shown in the example honey-frying of *Astragalus propinquus* (Astragali Radix) root material.

Advances in Botanical Research, Volume 97
ISSN 0065-2296
https://doi.org/10.1016/bs.abr.2020.09.011

Moreover, it may lead to the application of NADES in drug delivering systems. The different applications of honey in traditional medicine will be described, and its function and mechanism will be discussed. With the NADES concept we can now explain the various applications of honey as additive or adjuvant. As shown in numerous studies NADES are excellent solvents for medium polar bioactive compound. Besides improving the efficacy of herbal medicine, NADES also stabilize the active compounds. Honey has been used since ancient times for this purpose. The same effect can be achieved by a pure NADES, and the principle of improving solubility and stability can be applied to any poorly water-soluble medicine.

1. Honeybee products, their ingredients and traditional applications

1.1 Traditional applications of honey

Traditional Chinese medicine (TCM) utilizes numerous combinations of natural products. Most TCM products are prepared through extraction with water (e.g. making decoction). The limited number of extraction solvents raises a problem: not all compounds present in a single preparation will be fully extracted. To solve this problem, several natural materials, such as honey, wax, vegetable oil, and milk, are used as such or as adjuvant for preparing the final formulations. Of these materials, particularly honey has played an important role in traditional medicine for thousands of years. Traditionally it has been used as a medicine in itself, and as an additive, in the so-called *Paozhi* processing in which raw herbal drugs are cut into "decoction pieces" and treated with honey to improve their medical applications. Honey is also used as adjuvant for different types of pharmaceutical formulations, such as electuaries and honeyed pills. Hundreds of compounds have already been found in honeys from different sources. But the constituents present are greatly influenced by many factors, such as species of bees, species of plant sources, seasonal and geographical variation. Despite the large variation, the main components in all honeys are sugars such as glucose, fructose and sucrose. The normally solid sugar ingredients of honey can, in certain molar ratios, form natural deep eutectic solvent (NADES), explaining the liquid state of honey at room temperature. The physicochemical properties of such NADES may play an important role in preparing traditional medicines. Clarifying the role of honey in traditional medicines might help to rationalize traditional practices. Moreover it may lead to the application of NADES in drug delivering systems. Based on the applications of honey in traditional medicines, the potential further applications of honey and NADES will be discussed in this chapter. First of all, the different applications of honey in traditional medicine will be described, and its function will be discussed, including the perspectives of applications of NADES in drug delivering systems.

Honey is made by honeybees and some related insects from sugar containing plant secretions, e.g. flower nectar or honeydew from insects. Flower nectars are aqueous solutions mainly containing various sugars, such as sucrose, glucose and fructose. Honeybees make honey out of this, with glucose and fructose as the main ingredients along with about 18% of water. The brewing of honey involves two processes: the physical process reducing the water content from 40% to about 18% and the chemical process transforming di- and trisaccharides as well as polysaccharides (e.g. starch) into monosaccharides through invertases and amylases present in a hypopharyngeal gland secretion of honeybees (Yuan & Feng, 2000). During these processes, honey matures to the point that the water content gets reduced to about 18%. The honeybees store honey in the honeycombs sealed with beeswax for future consumption (Guo, 2004; Yu, 2017).

In addition to the sugars as main ingredients, there are other minor components in honey. Fresh honey is usually a transparent or translucent viscous liquid. During long-term storage at low-temperature and sometimes even at room temperature, part of honey sugars may crystallize. The color of honey can be nearly colorless to dark amber depending on the species of the source plants. The color generally comes from small amounts of pigments and minerals. Concerning fragrance, honey aroma is usually similar to that of the flowers, from which the nectar was collected. The aroma constituents, like esters, alcohols, phenols and acids are derived from the nectar but they are in part produced during the bees brewing process. The maturation of honey results in a unique aroma of honey. As the honey aroma compounds are volatile and sometimes labile substances, the composition of honey may undergo various changes during acid-base treatments, heating, or long-term storage, and the aroma may thus be lost in part (Guo, 2004; Yu, 2017).

Due to its unique sweetness, honey has been widely used in food since ancient times. It was used either as an additive or as a supplement for food processing. For example, honey is extensively used in various fields: sweeteners in confectioneries, brewing, pastries, and cooking. Also it is used as food additive such as: thickener, clarifier, stabilizer; humectant for beverages; and tobacco manufacturing (Guo, 2004).

Honey is also used in traditional medicines. In fact, recent studies reported a number of pharmacological effects for honey such as wound healing, pain relief, antibacterial activity, tissue regeneration, improving immunity, sedation, promoting digestion, hepatoprotection, protection of the cardiovascular system, and moistening lungs for relieving cough. Various biologically active compounds have been identified in honey. The antibacterial effect of honey might come from some of its constituents, from its high osmolarity in connection with the high concentration of sugars as well as the acidic character of honey (Guo, 2004; Hixon et al., 2019; Kurek-Górecka, Górecki, Rzepecka-Stojko, Balwierz, & Stojko, 2020; Martinotti, Bucekova, Majtan, & Ranzato, 2019; Nolan, Harrison, & Cox, 2019).

Honeybeesmilk, so-called royal jelly, is another product made by the bees to feed the queen and infant worker bees. It is a milky or light yellow mixture secreted by the lingual gland and the mandibular gland of the head of young worker bees. It is based on the nutrients taken by the young worker bees after consumption of the brewed honey and naturally fermented pollen. It is slightly sweet, slightly sour, spicy and fragrant. Fresh royal jelly is reported to contain generally 64.5–68.6% (w/w) of water and 12–14% (w/w) of protein. Dried royal jelly contains about 50% (w/w) of protein, 15.6% (w/w) of total lipids, and 0.8% of amino acids (Tang, Wang, & Ren, 2005).

Propolis is a natural substance made by honeybees and used to protect the honeycomb. Propolis is an opaque sticky jelly which consists of a mixture of plant resins collected by bees and substances secreted by the bees' mandibular glands and wax glands (Commission of the Ministry of Health of the People's Republic of China, 2015). The color of propolis is brown, reddish brown, or grayish brown. Before spawning, it is applied to the honeycomb to seal small openings in the inner parts of the honeycombs, and to keep the honeycombs sterile. Thus, propolis keeps bees healthy as it is a natural broad-spectrum antibiotic (Al-Waili, Al-Ghamdi, Ansari, Al-Attal, & Salom, 2012). Propolis contains 50–60% (w/w) of resin and balm, 30% of beeswax, 10% (w/w) of a volatile oil and pollen inclusions (Tang et al., 2005).

Beeswax (Cera Flava) is a substance secreted by four pairs of wax glands under the abdomen of two honeybees, Asian honeybee (*Apis cerana* Fabricius) or European one (*A. mellifera* Linnaeus) (Commission of the Ministry of Health of the People's Republic of China, 2015). Main components of beeswax are fatty acids and their esters with fatty alcohols. This mixture has good antiseptic and shaping properties (Tang et al., 2005). Beeswax is widely used in medicine, industry and agriculture. It can be used as an excipient for irritating drugs to reduce their degree of irritation.

The character of low water-solubility of beeswax is being employed in drug delivery systems. A drug formulated in beeswax is slowly released via micropores or by slow dissolution of the beeswax. In TCM, beeswax is commonly used to make the shell of pills (Commission of the Ministry of Health of the People's Republic of China, 2015; Yang, 2012).

1.2 Chemical ingredients of honey

The many applications of honey fostered investigations on the ingredients of honey. These studies aimed at the identification of active ingredients and how this information could be used in quality control. The main components of honey are glucose and fructose, accounting for about 60–80% (w/w), followed by sucrose (about 5%, w/w), and organic acids (0.1%, w/w). The descriptions of international food standards (Codex Alimentarius Commission, 2011) and Chinese drugs standards for honey (Commission of the Ministry of Health of the People's Republic of China, 2015) are shown in Table 1. Officially honey should contain no less than 60% (w/w) of fructose and glucose and no more than 5% (w/w) sucrose. The water content should be less than 24% (w/w). Minor sugars, like maltose, raffinose, kojibiose, melezitose, and lactose, have been also identified in honey, but their levels differ depending on source plants. The difference in chemical ingredients means variation of the physico-chemical properties of honey,

Table 1 International food standards, and Chinese drugs standards of honey.

	International food standards	Chinese drug standards
Fructose/glucose/ (g/100 g)	Common honey ≥ 60% (w/w)	≥60% (w/w), fructose/glucose ≥ 1.0
Sucrose	Common honey ≤ 5%	Sucrose ≤ 5.0% (w/w), maltose ≤ 5.0% (w/w)
Water content (%)	Common honey ≤ 20%	≤24.0%
Acidity (1 mol/L NaOH)/ (mL/kg)	–	≤40
Conductivity	Common honey ≤ 0.8 ms/cm	–
Relative density	–	≥1.349
Hydroxymethyl furfural (mg/kg)	–	5-Hydroxymethyl furfural ≤ 40 mg/kg

like acidity, conductivity and relative density. Recent studies on the role of sugars in nature raise some interesting questions about the physicochemistry of honey (see below).

Honey is basically the mixture of components from plants, and bees as well as their fermented products. Therefore, besides the sugar ingredients, hundreds of other components have been identified in honeys. The chemical composition of honey is closely related to its source plants (Chen & He, 2010; Guo, 2004; Hixon et al., 2019; Kurek-Górecka et al., 2020; Kuś, Jerković, Tuberoso, & Šarolić, 2013; Mao, 2011; Martinotti et al., 2019; Nolan et al., 2019; Xia, Zhang, Wang, & Ma, 2010). For example, the content of total flavonoids and phenolic acids was determined in five common honeys (jujube honey, prunella honey, acacia honey, vitex honey, rape flower honey) and the results showed that only jujube and prunella honey contained flavonoids. The highest content of total flavonoids found was 0.0015% (w/w) while the total phenolic acids were found to be 0.050–0.063% (w/w) in these five common honeys. The chemical variation in honeys may well cause a difference in biological activities. The characteristic chemicals of thyme- and chestnut honey are vitamins and polyphenolic acids, respectively, which reflect their antioxidant activity. In fact, compared with high antioxidant activity of chestnut honey, thyme honey does not have such high activity (León-Ruiz et al., 2013). The chemicals originating from source plants may cause not only beneficial effects, but also toxicity problems. For example, honey from Ericaceae family may contain toxic substances like grayanotoxin. The presence of pyrrolizidine alkaloids in honey became a social issue because of their liver toxicity, particularly for infants (Cramer & Beuerle, 2012).

A broad range of other minor ingredients has been identified in honey: vitamins, amino acids, acids, choline, folate, volatiles, hormones, minerals, and proteins including enzymes (Kurek-Górecka et al., 2020; Martinotti et al., 2019; Nolan et al., 2019; Xia et al., 2010). Additionally, there are quite some organic acids in honey, and thus adding water results in weak acidity with pH of 3.2–4.5 (about 10% of NADES in water).

1.3 Honey as a NADES?

Through our ^{1}H NMR-based metabolomics studies we found that many organisms and cells do contain large amounts of a few primary metabolites, often in similar molar concentrations, without any clear reason. We showed that certain combinations of solid primary metabolites, such as sugars, polyalcohols, amino acids, organic acids and bases, in certain molar ratios become a liquid at room temperature. These mixtures have a melting point usually far below that of the individual components, a phenomenon known as deep eutectic mixtures. The individual components in these mixtures have strong hydrogen bond acceptor or hydrogen bond donor characteristics and form a liquid with a supramolecular structure (Choi et al., 2011; Dai, Van Spronsen, Witkamp, Verpoorte, & Choi, 2013). These natural deep eutectic solvents (NADES) can be divided into two groups: ionic liquids (ILs) in which a solid acid is mixed with a solid base, and deep eutectic solvents (DES) in which only neutral compounds (sugars, polyalcohols) are mixed, or a neutral compound is mixed with an amino acid, organic acid or base.

In our view the NADES represent a third liquid phase of intermediate polarity in living organisms (Choi et al., 2011). The occurrence of NADES in cells may explain the biosynthesis of poorly water- or fat-soluble compounds, like many secondary metabolites and macromolecules in plants. Also, in the survival of plants and other organism under cold, and drought conditions NADES might play a role, and in fact metabolomics studies show the presence of NADES ingredients in all kind of organisms under such stress conditions.

The occurrence of NADES in plants was proven by the presence of about equal molar concentrations of some sugars in the flower nectar, maple syrup (Choi et al., 2011). Bees feed on nectars that consist of NADES with different concentrations of water. Bees transform the nectar in to honey. Common honey consists of about 60–80% (w/w) of glucose and fructose in about equal molar ratio with a small amount of sucrose (ca. 5%, w/w) and about 20% (w/w) water. According to the Chinese medicine standards the ratio of fructose/glucose in honey should be $\geq$1.0. This ratio is consistent with the optimal NADES composition ratio for the mixture of fructose and glucose (Choi et al., 2011; Dai, Van Spronsen, et al., 2013). Honey can thus be considered to be a NADES containing up to about 20% of water, in which a plethora of plant compounds is dissolved. The role of nectar concerns three aspects. First of all as NADES are non-volatile, the nectar will remain as a liquid in the flower even under drought conditions. Second, poorly water-soluble volatile compounds that attract pollinators can be dissolved in a NADES like honey. Third, nectar is a major food source for pollinators like bees (Fig. 1).

In hunter-gathering period mankind fully relied on nature for gathering their food, and honey was probably the first sweetener. But they discovered

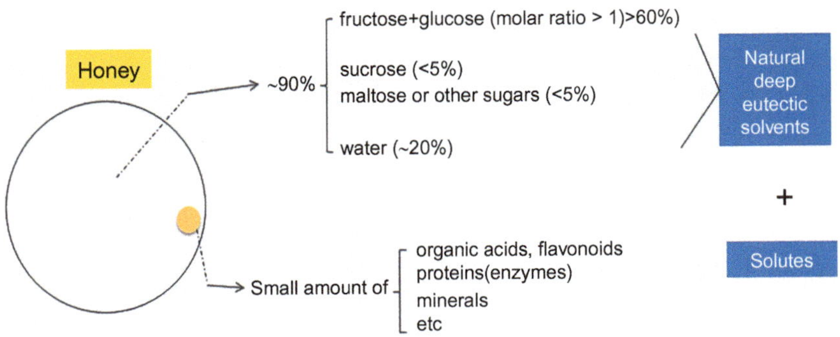

Fig. 1 Schematic diagram of honey composition.

also medical applications. A well-known and proven application is wound healing, which is related to its antibacterial activity (Hixon et al., 2019; Kurek-Górecka et al., 2020; Martinotti et al., 2019; Nolan et al., 2019). Along with its direct use as such, honey has been used as adjuvant in the preparation of various traditional medicines (Chen et al., 2018; Kuropatnicki, Klosek, & Kucharzewski, 2018). To further elaborate the hypothesis that the uses of honey are associated with its NADES characteristics, we will discuss the role of honey in TCM in detail.

2. Honey in traditional Chinese medicine

Most preparations and formulations of TCM products include an extraction step. Though numerous bioactive metabolites with highly different physico-chemical properties are present in plant material and their formulations, extraction solvents are only a few: water, and water mixed with small amounts of ethanol. The most common TCM processing method is water-decoction, but to fully exploit the potential of active compounds, decoctions are not always the most suited method because of solubility issues. To deal with medium and low polar compounds diverse processing methods and formulations have been developed including pills, electuaries, syrups, various oils, tinctures, plasters, suppositories, glues, powders, granules, aerosols, etc. Honey has been applied in processing of various TCM since ancient times. The use of honey was already recorded in a Chinese document more than 2500 years ago. During the Tang Dynasty (AD 618–907) honey was reported to be a potential adjuvant for processing of medicinal plant materials, such as coating licorice with honey before processing. In the Classified Materia Medica written during the Song Dynasty (AD 960–1279) it was mentioned that "raw honey is good for treating eye diseases. Put honey in the eyes, and lie supine for half day." The Compendium of Materia Medica was written during the Ming Dynasty (AD 1368–1644) and described the biological functions of honey as clearing heat, nourishing the stomach and spleen, detoxifying, moistening dryness and relieving pain. In ancient times, it was recorded that honey should be kept beside well-water to keep it fresh for medical use.

The main functions of honey are described in the modern Chinese Pharmacopeia and are as follows: tonifying the stomach and spleen, moistening dryness, relieving pain, and detoxifying. For external use honey helps to regenerate tissues and heal wounds. It is also used for abdominal pain, dry cough caused by lung dryness, constipation caused by intestine dryness and detoxification of Aconite's toxins (e.g. aconitine analogs) in the bulbs of *Aconitum* species (Commission of the Ministry of Health of the People's Republic of China, 2015).

Honey, is also applied to the storage of various foods and medicines (Xie, 1987; Xiong, 2005). These applications of honey in TCM employ two types of honey: crude honey and refined honey. The crude naturally matured honey is used in applications for materials which are rich in nutrients and substances with therapeutic effects. Refined honey is mainly used as additive, for example, for honey stir-frying and as pharmaceutical adjuvant for making pills as well as electuaries. In the following sections, examples of applications of honey as adjuvant are discussed, based on the hypothesis that honey is a sugar-based NADES.

2.1 Honey stir-fried preparations

Paozhi is a treatment to improve the medical applications of herbal medicines in TCM. The functions of *Paozhi* treatment are altering the flavor, moderating the medicinal properties, and eliminating the side effects and detoxification of the crude drugs. The treatment encompasses such techniques as cutting, crushing, steaming, calcining, stir-frying, with or without adjuvants. Especially, honey is one of the most common adjuvants in the stir-frying of *Paozhi*.

Raw honey is usually refined by stir-frying under a gentle heat to reduce its water content down to 10–13%, as well as to remove impurities, deactivate enzymes, kill microorganisms, and enhance adhesion (Yang, 2012). The refined honey, even with a low water content, could keep its liquidity, which is clear evidence for honey behaving like a NADES. The stir-frying with honey requires several steps: first the refined honey is diluted with boiling water (in a ratio of about 1/3 of the weight of honey), this hot solution is added to the cleaned and cut plant materials, mixed thoroughly and the mixture is kept for a while to let the refined honey penetrate into the inner tissues of the plant material. After this preprocessing, the mixture of plant material and honey is stir-fried with gentle heat until the materials gain a specific color (Chinese Pharmacopeia), then the processed material is collected and cooled down. Generally, 25 kg refined honey is used for 100 kg plant material. The honey stir-fried pieces or slices are used to make a decoction with boiling water, the final formulation for the patient. As an example, morphological and metabolic change of the roots of *Astragalus propinquus* Schischkin caused by honey stir-fried process is shown in Fig. 2 (Dai et al., 2020).

According to TCM theory, the honey stir-frying of material of some medicinal plants have the following effects on the pharmacology:

1) moisten lungs for relieving cough, such as Stemona Radix (root of *Stemona sessilifolia* (Miq.) Miq.), Farfara flos (bud of *Tussilago farfara* L.), Asteris radix et rhizoma (root and rhizome of *Aster tataricus* L.f.);

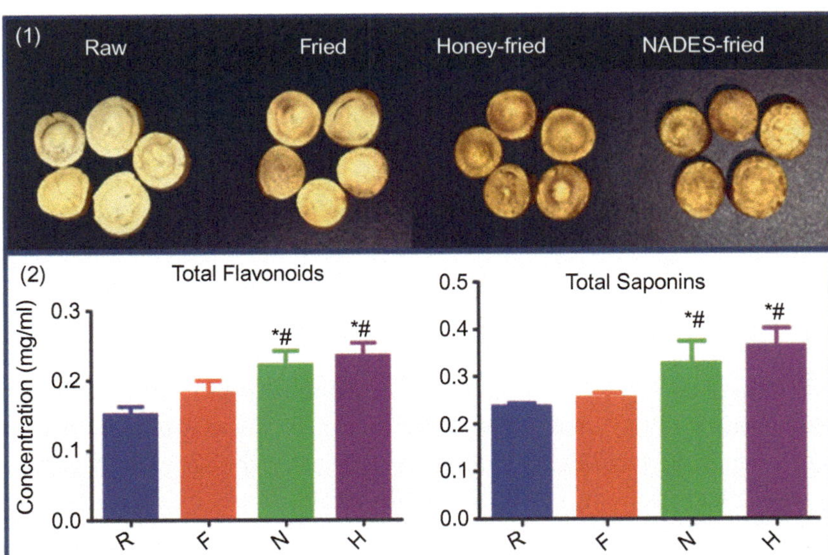

Fig. 2 Differently processed slices of *Astragalus propinquus* (Astragali Radix) and the level of total flavonoids and saponins in the decoctions estimated by UV spectroscopy. R, raw material; F, fried material; N, NADES fried material; H, honey fried material ($n=6$, $*P<0.05$ vs R group; $^{\#}P<0.05$ vs F group) (Dai et al., 2020).

2) enhance the tonifying effects on spleen and nourishing *qi*, such as Astragali radix (root of *Astragalus propinquus*), Glycyrrhizae radix et rhizoma (root and rhizome of *Glycyrrhiza uralensis* Fisch.), Codonopsis radix (root of *Codonopsis pilosula* (Franch.) Nannf.);
3) moderate the properties of crude drugs. For example, Ephedra herba (aerial part of *Ephedra sinica* Stapf.) has a strong sweat driving force, which can be relieved by stir-frying with honey;
4) correct flavor and eliminate the side effects of drugs. For example, Aristolochia fructus (fruit of *Aristolochia contorta* Bunge) has a bitter taste, and is irritating to the stomach. Stir-frying with honey can enhance its antitussive effect, and it can also correct the bitter flavor to avoid vomiting (Jin, 1988).

What role does honey play in the stir-fry processing of plant materials? What components in honey play a major role? We hypothesized that the added honey penetrates the plant material and dissolves compounds present in the plant material during the processing. This may improve the solubilization of compounds in the final decoction step when the treated material is boiled with water. Moreover, the treatment may result in a more stable material to store, and may alter the chemical profile and enhance the biological activity of the final decoction products. This effect might be due to the major constituents of honey, the about equal molar amounts of glucose and fructose which form a NADES, known to readily solubilize medium polar compounds that hardly dissolve in water (Dai, Rozema, Verpoorte, & Choi, 2016; Dai, Van Spronsen, et al., 2013; Dai, Verpoorte, & Choi, 2014; Dai, Witkamp, Verpoorte, & Choi, 2013). To prove this hypothetical mechanism, we studied the processing of Astragali Radix (AR) preparations as an example (Dai et al., 2020). Astragali Radix is a traditional medicine, which has been widely used since ancient times for relieving general weakness (Liu, Wang, & Zhang, 2015). To prove a possible effect of a NADES the AR material was processed in different ways: Stir-frying without additives (samples F); stir-frying with honey (samples H); and stir-frying with artificial honey made by mixing equal molar amounts of glucose and fructose (samples N). Non-processed samples were used as control (Fig. 2). Decoctions were made of the processed AR samples following the method applied in the clinic. The decoction products were evaluated for their chemical ingredients and their bioactivity in some bioassays (Dai et al., 2020).

Total flavonoids and total saponins in the four AR decoctions were to some degree increased after all the processing when compared with the control samples. However, the level of isoflavonoids and saponins were far more increased by honey- and NADES-treated samples (the H and N decoctions) than in the decoction of the raw plant material (R) (Fig. 2). Apparently, the artificial honey using a sugar-based NADES, has the same effect as real honey, which supports the hypothesis that the effect of honey is similar as with a pure NADES in that it improves the solubility of medium polar plant secondary metabolites present in the plant material (Dai, Witkamp, et al., 2013).

The biological effects of these AR decoctions were tested on the endocrine-related estrogen receptor (ER)-α, androgen receptor (AnRec), and antioxidant-related nuclear factor erythroid 2-related factor 2 (Nrf2) signaling pathways. No differences were observed between the H and N decoctions, both of which had significantly higher activities than the R and F decoctions (Fig. 3). Thus, it indicates that honey improved the biological effects of honey-stir-fried AR, which can be attributed to its NADES characteristics. The NADES increases the levels of total flavonoids and saponins in the H and N decoctions as measured by general methods for these types of compounds (Fig. 2) (Dai et al., 2020).

The chemical composition of the four decoctions was determined in detail using UPLC-Qtof-MS based metabolomics to explore the effects of honey. The levels of the aglycones calycosin (Ca) and formononetin (Fo) are relatively the highest in the non-processed samples (R), whereas the levels of their glucosides calycosin-7-*O*-β-D-glucoside (CaG) and formononetin-7-*O*-β-D-glucoside (FoG) were highest in the processed samples, and particularly in the H and the N group. The second notable difference was found in

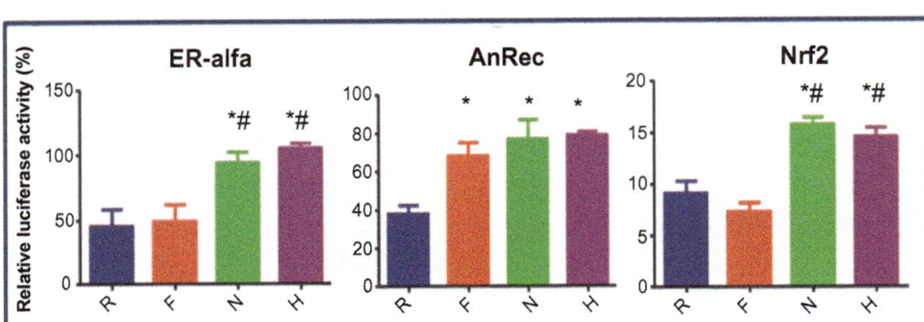

Fig. 3 The bioactivities of the AR decoctions (ER-α, estrogen receptor-α; AnRec, androgen receptor; Nrf2, nuclear factor erythroid-2-related factor; $n=6$, $^{*}P<0.05$ vs R group; $^{\#}P<0.05$ vs F group). R, raw material; F, fried material; N, NADES fried material; H, honey fried material (Dai et al., 2020).

the fried samples (F) group, where the calycosin-7-*O*-β-D-glucoside-6″-*O*-acetate (CaGA) and formononetin-7-*O*-β-D-glucoside-6″-*O*-acetate (FoGA) levels were approximately twofold higher than those in the control (R) samples, whereas the calycosin-7-*O*-β-D-glucoside-6″-*O*-malonate (CaGM) and formononetin-7-*O*-β-D-glucoside-6″-*O*-malonate (FoGM) levels were slightly lower (not significant) in the F group than in the other three decoctions (Fig. 4). A significant increase in acetylastragaloside I (AgA I) relative level occurred in the fried samples (F) (Fig. 4), while the levels of astragaloside I (Ag I), isoastragaloside I (iAg I), and astragaloside II Ag II were lower in the F decoction than that in the control (R) (results not shown). The H and N samples also had slightly higher AGA I levels than the control, but not as much as in the only fried samples (F). These results

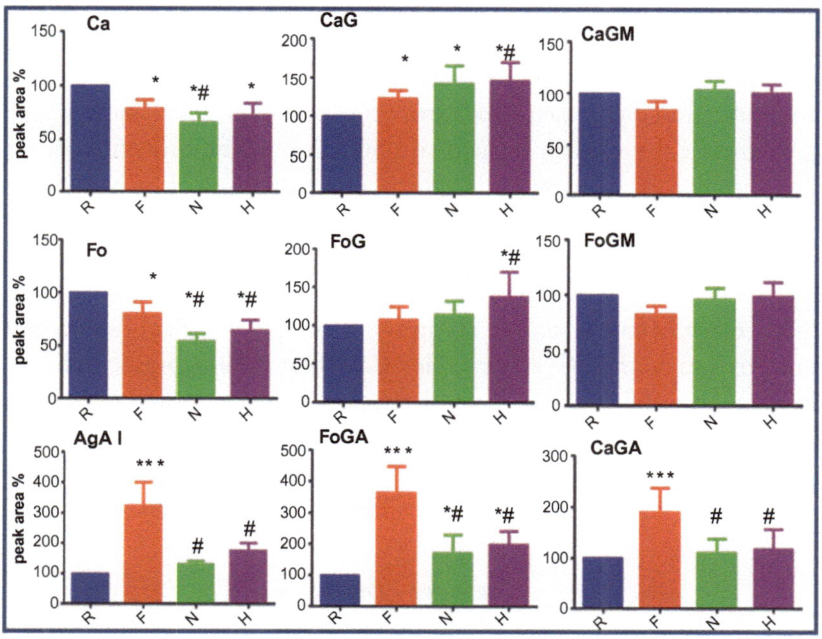

Fig. 4 Relative levels of some chemical constituents in the AR decoctions. ($n = 12$, $^{*}P < 0.05$ vs R group; $^{\#}P < 0.05$ vs F group). Relative contents of components within fried samples were calculated with raw AR decoctions as 100% (Ca, calycosin; Fo, formononetin; CaG, calycosin-7-*O*-β-D-glucoside; FoG, formononetin-7-*O*-β-D-glucoside; CaGM, calycosin-7-*O*-β-D-glucoside-6″-*O*-malonate; FoGM, formononetin-7-*O*-β-D-glucoside-6″-*O*-malonate; CaGA, calycosin-7-*O*-β-D-glucoside-6″-*O*-acetate; FoGA, formononetin-7-*O*-*β*-D-glucoside-6″-*O*-acetate; AgA I, acetylastragaloside I) (R, raw material; F, fried material; N, NADES fried material; H, honey fried material) (Dai et al., 2020).

show that the stir-frying does affect the metabolic profile of the formulations. This is in line with a change in the biological activities.

Several explanations might explain the changes in the metabolic profiles.

The NADES in the honey stir-frying might (Ma, Wei, Wang, & Hattori, 2009; Zheng, Shi, Jie, & Lu, 2014):

- promote the conversion of aglucones to glucosides;
- inhibit the hydrolysis of the glucosides;
- inhibit acetylation of isoflavonoids and astragalosides;
- increase deacetylation of acetyl-containing isoflavonoids and astragalosides;
- inhibit the hydrolysis of the malonylated and/or acetylated glucosides;
- improve the extraction of the isoflavonoids glucosides if compared to the aglycones.

It thus seems that the stir-frying is an important step for the quality of the traditional medicine, so further research is required to study this effect in more detail.

A protective effect of honey might come from the presence of a large amount of reducing sugars, which can prevent the transformations of labile ingredients of the medicinal materials. Such a NADES protective effect is supported by a previous study in which a stabilizing effect of NADES was observed for flavonoids during extraction and storage (Dai et al., 2016). The CaGM might be transformed during frying into an acetyl-containing molecule (CaGA), i.e. a transesterification (Lin et al., 2000; Ma et al., 2009).

The effect of honey on the absorption of active compounds was determined in rats with CG as example. Calycosin-7-*O*-β-D-glucoside is an important flavonoid in AR with the ER-α, AnRec and Nrf2 activities. The bioavailability of CaG was markedly increased with honey or NADES-treated samples, based on the plasma concentration and permanence period in rats (Fig. 5). A similar bioavailability improving effect of NADES

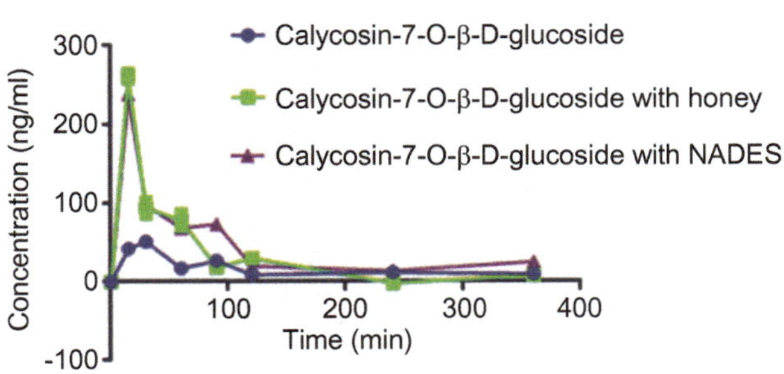

Fig. 5 Time course (mean ± SD) of plasma concentration of calycosin-7-*O*-β-D-glucoside in SD rats following oral administration of calycosin-7-*O*-β-D-glucoside (20 mg/kg) with or without honey, or honey analog (NADES) (Dai et al., 2020).

was reported for rutin, a flavonoid diglycoside (Faggian et al., 2016). The present study provides evidence for a synergistic effect of honey with active compounds in improving activity of AR. This result implies that the NADES structure of honey plays a major role in improving the bioavailability of CG. Apparently the honey- and NADES stir-frying process modify and increase the pharmacological activities of the drug in a favorable way.

2.2 Traditional formulations: Honeyed pills and other dosage forms

In addition to the use as an adjuvant in processing medicinal plants honey is also often used for pharmaceutical purposes in TCM. Especially, honeyed pills and electuaries are quite common formulations. In the production of honeyed pills, the quality of honey is strictly controlled; not only for the chemical ingredients (like water, enzymes, microorganism, and minor impurities) but also for its physical properties. According to the degree of refining, refined honey is divided into three categories: tender honey, medium honey, and old/hard honey. Each of them has a specific viscosity and water content, which eventually decides their specific applications. For example, tender honey with a water content of 17–20%, and a relative density of about 1.35 $g\,cm^{-3}$, is suitable for binding fine sticky powdered crude plant materials; the more viscous medium honey with 14–16% water and about 1.37 $g\,cm^{-3}$ relative density is used to bind fine powders of loose crude plant materials; the most viscous old honey with 10% water content and about 1.40 $g\,cm^{-3}$ relative density is applied for minerals and fibrous fine powders of plant materials.

The preparation of honeyed pills can be divided into several steps: cutting and grinding of the plant material into a fine powder; mixing the powder and honey (about 1:1 to 1:1.5) into a roll with proper hardness and plasticity, which then is cut and made into pills (Fig. 6) (Yang, 2012).

There are also other types of pills derived from honeyed pills (Table 2). In order to reduce the size of honey pills, the powdered plant materials used in honey pills is (partially) replaced by their extracts in so-called concentrated pills. According to the used adjuvants, the concentrated pills can be divided into concentrated water pill, concentrated honeyed pill, and water-honeyed pill. Although concentrated water pills are currently the most common ones, still the other types of honey added pills have some specific applications. Water-honeyed pills are made from a fine powder of the plant material and diluted refined honey as adhesive. Compared with honeyed pills, they save honey, reduce costs and are easier to store. Wax pills are made from a fine powder of the plant material mixed with beeswax as adhesive. Beeswax

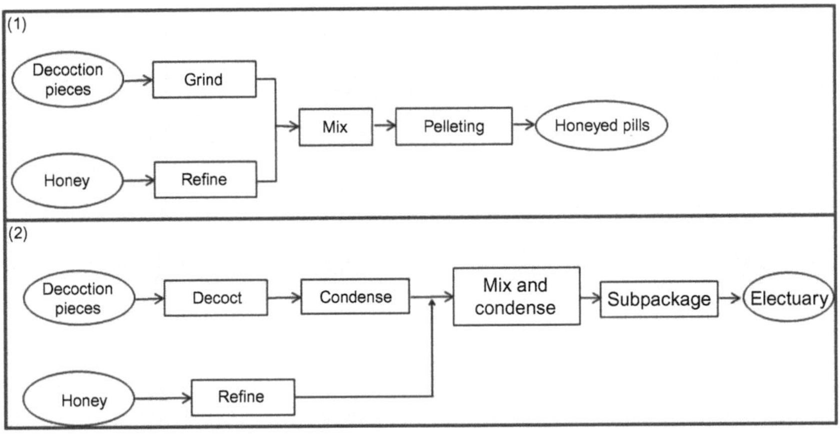

Fig. 6 Typical steps in the preparation of honeyed pills (1) and electuaries (2) (Yang, 2012).

Table 2 Characteristics and applications of different dosage forms with honey as adjuvant/ingredient.

Usage form	Advantage	Disadvantage	Applications
Electuary	High drug concentration, easy to store and convenient to take	Not suitable for heat-sensitive or volatile drugs	Suitable for chronic diseases
Honeyed pills	Slow release in the gastrointestinal tract, long-lasting effect	Large in size	Suitable for antitussive and expectorant drugs, and drugs for tonifying spleen-stomach
Concentrated pill	Small in size, easy to carry and store	Long heating time	Not suitable for heat-unstable drugs
Water-honeyed pill	Save honey, reduce costs, easy to store	Long heating time	Not suitable for heat-unstable drugs
Wax pill	Prolonged efficacy, prevent poisoning or irritation of drug in gastrointestinal tract	Unable to control the rate of drug release	Suitable for toxic or irritative drugs

is an interesting material, not only for formulation but also as drug delivery system. Beeswax is poorly water soluble, neither in vivo, nor in vitro. The application of beeswax significantly prolongs the pharmacological effects of a drug because of the slow release, and prevents irritation of the gastrointestinal tract by the drug and may avoid possible toxic effects (Commission of the Ministry of Health of the People's Republic of China, 2015; Yang, 2012).

Another common dosage form of TCM with honey as additive is an electuary. An electuary is a semi-liquid preparation made by boiling plant materials with water, concentrating the decoction, and adding refined honey or sugar (invert sugar). Electuaries are mostly used for the treatment of some chronic diseases, with the advantages of a high drug concentration, easy production and storage. Due to the boiling process, however, it is recommended that plant material containing heat-sensitive or volatile compounds should not be made into electuaries. The preparation of an electuary consists of several steps: decoction with water; concentration of the extract; mixing with refined honey; paste collecting; and subpackaging (Fig. 6). When the density of the concentrated extract reaches around at 1.21–1.25 $g\,cm^{-3}$ at 80 °C, refined honey is added. Generally, the amount of refined honey or sugar added does not exceed three times of the amount of paste. The paste is continuously stirred and heated until a general relative density of paste reaches about 1.4 $g\,cm^{-3}$. Then the paste should be divided into clean and dry containers.

As pharmaceutical adjuvant, viscous liquid or semi-liquid honey is mostly used to make pills. For this application, honey with higher content of fructose than glucose is required. This kind of honey can not only increase the moisture of honeyed pills, but also take advantage of the sweetness of fructose. Refining process makes honey more viscous and with lower water level, that brings the advantages of strong adhesive and plasticity characteristics as an adjuvant. Also, refined honey can give a special character to the pills, e.g. round shapes, moisture, and softness.

The use of honey in the preparation of traditional medicines, can be explained by the NADES character of honey. The typical characteristics of NADES are non-volatile, liquid in a very broad temperature range, high viscosity, easy to mix with water, antimicrobial activity, good solubility of medium polar poorly water-soluble compounds, and increased stability of phenolic compounds (Dai et al., 2014, 2016, 2020; Dai, Van Spronsen, et al., 2013; Dai, Witkamp, et al., 2013; Dai, Witkamp, Verpoorte, & Choi, 2015). The honeyed pills are slow to disintegrate in the digestive tract, which enables a long-lasting effect. Also, there might be an effect on the absorption of ingredients of TCM. The sustained release effect of honey pills on the bioavailability of some important active and toxic components has been observed in the case of Aconitum roots when comparing honey pills and water pills (Gan et al., 2018).

3. The physico-chemical properties of honey as NADES

Above we discussed the roles of honey in Chinese medicine. The observations made in various studies clearly show that honey has the typical characteristics of a NADES and the studies on stir-fried Astragalis Radix with artificial honey showed similar results as the natural honey. Here we will discuss in some more detail the properties of NADES, and specifically the sugar-based NADES, like honey.

3.1 Hygroscopicity

When the air is humid, honey can absorb water from the air. The absorption capacity of honey, i.e. the hygroscopicity, depends on the percentage of water already present in honey, and the water vapor pressure in the air. When the air is dry, honey evaporates water.

3.2 Viscosity

The main factors that determine the viscosity of honey are water content and temperature. The lower the water content, the higher the viscosity. The viscosity decreases with increasing temperature (Dai et al., 2015; Guo, 2004).

3.3 Refractive index

Determination of the refractive index of honey is a simple and accurate method to identify the water content of honey. The refractive index of honey decreases with the increase of water content and temperature (Guo, 2004).

3.4 Optical rotation

Optical rotation is related to its structure, water content, and other factors. But the optical rotation is mainly due to the sugars in honey, in which glucose and sucrose are dextrorotary, and fructose is levorotary. Normally mature honey is levorotatory. If glucose or sucrose is added, it will cause a change of rotation, and the levorotation will be smaller. Therefore, the rotation can be used to confirm the authenticity of honey (Guo, 2004).

The physical properties of honey are consistent with the reported physical data of artificial honey, a glucose-fructose based NADES (Dai et al., 2015; Dai, Van Spronsen, et al., 2013). It means that the natural deep eutectic solvent's part of honey determines the main physico-chemical properties of honey.

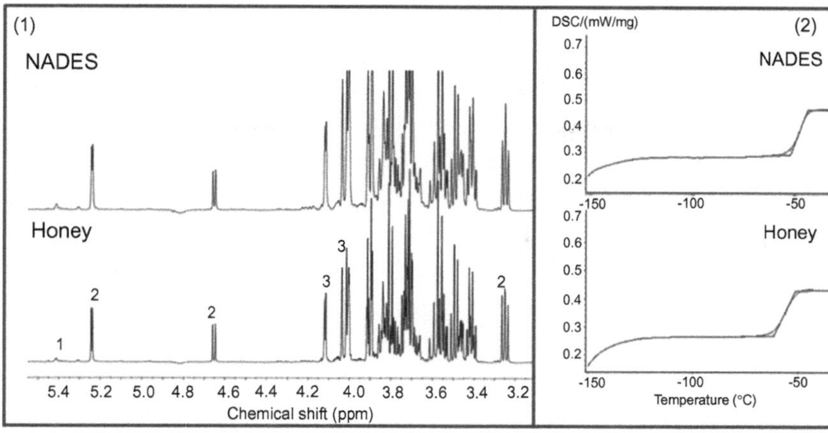

Fig. 7 The ^{1}H NMR spectra (A) and DSC curve (B) of honey and man-made honey, namely, natural deep eutectic solvents (NADES). The honey and NADES are composed of a mixture of glucose, fructose, and sucrose with certain amount of water (1: sucrose; 2: glucose; 3: fructose) (Dai et al., 2020).

To test the hypothesis, that honey might be a NADES, some sugar-based NADES that mimic honey were investigated for their physico-chemical characteristics using several methods. The ^{1}H NMR spectra showed that the NADES and honey have a similar chemical composition and the similar chemical shifts which support the occurrence of strong hydrogen bonding between the constituents, creating a supramolecular structure (Fig. 7). The differential scanning calorimetry (DSC) test showed a glass transition point (from −50 to −60 °C) but no melting point of NADES and honey, which suggests that the NADES and honey have similar supramolecular structures.

Recently the presence of active macromolecules in honey was postulated (Brudzynski et al., 2017). A supramolecular structure of honey was proposed because the UV spectra of honey change upon dilution with water. The changes were thought to be due to light scattering particles present in honey. These particles were supposed to be formed by self-assembly from macromolecules present in honey. By means of scanning electron microscopy it seemed like two phases were present: dense globules in a sugar solution. Dilution with water resulted at a certain point in the microparticles falling apart in nanoparticles. Below this threshold level honey lost its antibacterial activity. Do these observations contradict the NADES hypothesis? Most honeys are rich in phenolic compounds, like flavonoids, anthocyanins and cinnamic acid derivatives which have a low to medium solubility in water, but dissolve well in NADES. These compounds are very sensitive for shifts in their UV spectra, among others, the pH, and metals cause shifts in the UV spectra. In honey itself there is no pH, as by definition the pH is the concentration of protons in water, but dilution of honey with water changes the UV spectra of the phenolics due to the effect of the acids present in honey. Supramolecular structures of NADES were clearly proven by ^{1}H NMR spectroscopy (Choi et al., 2011; Dai et al., 2015), showing that the NADES components and water are strongly interacting through H-bonding. Upon dilution the signals of the components in the ^{1}H NMR start to shift. Below about 50% NADES in water, the ^{1}H NMR spectra do not change anymore and they look like an aqueous solution of the NADES components (Dai et al., 2015). The gradual shift of the proton resonances of the sugars upon dilution can be explained by a fast equilibrium between two phases, one in which the NADES components are organized in small molecular aggregates, and in part are freely moving in the water. The changes in the chemical shifts of the protons in the ^{1}HNMR spectra reflect the time the protons are in one or the other phase. The self-assembly in honey observed by Brudzynski et al. (2017) may involve the sugar molecules that interact with the poorly water-soluble phenolics like flavonoids and anthocyanins (Dai et al., 2014). It is well known that mixtures of anthocyanins, flavones and cinnamic acid derivatives may form complexes, and in a number of plant species various anthocyanic vacuolar inclusions (AVI) have been observed that contain such phenolic compounds. AVI with different structures have been found, often without a membrane, and some types seemed to have a protein matrix (Andersen & Jordheim, 2010). The assumption that active macromolecules are responsible for the formation of colloidal particles in honey is in line with this (Brudzynski et al., 2017). The formation of particles in honey also fits the hypothesis of Choi et al. (2011) that NADES function as a third phase in cells that can dissolve poorly water-soluble phenolics and form some sort of vesicles in which the biosynthesis can occur (e.g. in the ER). Also transport of these compounds to the vacuole could be through these particles that contain the NADES in which the phenolic compounds are dissolved. Further research is needed to integrate the physico-chemical properties of honey with the biological system.

4. Other adjuvants in TCM

Sugars in honey might be the physico-chemical platform for the functions of honey, like to foster or support the roles of minor bioactive metabolites. The NADES may also be formed during the processing from compounds in the herbal medicine and added compounds, like sugars, sugar alcohols, amino acids, organic bases and acids and ethanol. Some common adjuvants applied in processing of TCM are rice wine, vinegar, salt, and oil.

Rice wine is used in preparing pharmaceutical formulations. Rice wine is brewed from sticky rice and contains kojic acid. It is generally a yellow transparent liquid with a mellow fragrance. Stir-frying with rice wine, helps to dissolve the active compounds and thus increases the efficacy of the drug. Drugs stir-fried with rice wine include Scutellaria radix (roots of *Scutellaria baicalensis* Georgi), Rhei radix et rhizoma (rhizomes of *Rheum palmatum* L.), Paeoniae radix alba (roots of *Paeonia lactiflora* Pall.), Angelicae sinensis radix (roots of *Angelica sinensis* (Oliv.) Diels), etc.

Vinegar for pharmaceutical use should be food grade vinegar, and the longer it is stored, the better, yielding the so-called "mature vinegar." Vinegar is produced from oxidation of alcohol in fermented rice, wheat, sorghum or other grains. Vinegar is acidic and can dissolve free alkaloids in water by protonation, and thus improving the therapeutic effect. The traditional medicines that often are stir-fried with vinegar are Phytolaccae radix (roots of *Phytolacca acinosa* Roxb.), Bupleuri Radix (roots of *Bupleurum chinense* DC.), and Cyperi rhizoma (rhizomes of *Cyperus rotundus* L.).

Baking herbal medicines with salt water is a common procedure for Morindae officinalis radix (roots of *Morinda officinalis* F.C. How), Foeniculi fructus (fruits of *Foeniculum vulgare* Mill.), and Plantaginis semen (seeds of *Plantago asiatica* L.).

5. Prospects

Nature is a treasure house for bioactive compounds, but also for adjuvants that can be used in formulating medicines. Honey is a clear example. Although the individual character of a honey is governed by its minor plant-originating metabolites, all types of honey share two common metabolites: glucose and fructose in about equal molar amounts, representing 60–80% of the weight of honey. According to our hypothesis, in Nature the combination of various sugars and other abundant primary metabolites in certain molar ratios will form NADES. This can be on subcellular level but also on the level of a tissue or in plant saps like nectar. This hypothesis generates many ideas for possible applications. In fact our ancestors already discovered the use of honey to improve traditional medicines. With the NADES concept we can now explain the various applications of honey as additive or adjuvant. As shown in numerous studies

NADES are excellent solvents for medium polar bioactive compound. Besides improving the efficacy of herbal medicine, NADES also stabilize the active compounds. Honey has been used since ancient times for this purpose. The same effect can be achieved by a pure NADES, and the principle of improving solubility and stability can be applied to any poorly water-soluble medicine.

References

Al-Waili, N., Al-Ghamdi, A., Ansari, M. J., Al-Attal, Y., & Salom, K. (2012). Synergistic effects of honey and propolis toward drug multi-resistant *Staphylococcus aureus, Escherichia coli* and *Candida albicans* isolates in single and polymicrobial cultures. *International Journal of Medical Sciences, 9*, 793–800.

Andersen, O. M., & Jordheim, M. (2010). Overview and introduction. In L. Mander, & H. W. Liu (Eds.), *Comprehensive natural products II chemistry and biology* (pp. 547–614). Oxford: Elsevier Ltd.

Brudzynski, K., Miotto, D., Kim, L., Sjaarda, C., Maldonado-Alvarez, L., & Fuk, H. (2017). Active macromolecules of honey form colloidal particles essential for honey antibacterial activity and hydrogen peroxide production. *Scientific Reports*, 7, 7637. https://doi.org/10.1038/s41598-017-08072-0.

Chen, C. G., & He, S. Y. (2010). *Bee products*. Beijing: China Agricultural Press.

Chen, L. L., Verpoorte, R., Yen, H. R., Peng, W. H., Cheng, Y. C., Chao, J., et al. (2018). Effects of processing adjuvants on traditional Chinese herbs. *Journal of Food and Drugs Analysis, 26*, S96–S114.

Choi, Y. H., van Spronsen, J., Dai, Y., Verberne, M., Hollmann, F., Arends, I. W., et al. (2011). Are natural deep eutectic solvents the missing link in understanding cellular metabolism and physiology? *Plant Physiology, 156*, 1701–1705.

Codex Alimentarius Commission, 2011. (Adopted in 1981. Revisions 1987 and 2001). CODEX STAN 12-1981 *codex standard for honey*.

Commission of the Ministry of Health of the People's Republic of China. (2015). *Pharmacopoeia of the people's republic of China. Vol. I*. Beijing: Ministry of Public Health of the People's Republic of China.

Cramer, L., & Beuerle, T. (2012). Detection and quantification of pyrrolizidine alkaloids in antibacterial medical honeys. *Planta Medica, 78*, 1976–1982.

Dai, Y., Jin, R., Verpoorte, R., Lam, W., Cheng, Y. C., Xiao, Y., et al. (2020). Natural deep eutectic characteristics of honey improve the bioactivity and safety of traditional medicines. *Journal of Ethnopharmacology, 250*, 112460. https://doi.org/10.1016/j.jep.2019.112460.

Dai, Y., Rozema, E., Verpoorte, R., & Choi, Y. H. (2016). Application of natural deep eutectic solvents to the extraction of anthocyanins from *Catharanthus roseus* with high extractability and stability replacing conventional organic solvents. *Journal of Chromatography A, 1434*, 50–56.

Dai, Y., Van Spronsen, J., Witkamp, G. J., Verpoorte, R., & Choi, Y. H. (2013). Natural deep eutectic solvents as new potential media for green technology. *Analytica Chimica Acta, 766*, 61–68.

Dai, Y., Verpoorte, R., & Choi, Y. H. (2014). Natural deep eutectic solvents providing enhanced stability of natural colorants from safflower (*Carthamus tinctorius*). *Food Chemistry, 159*, 116–121.

Dai, Y., Witkamp, G. J., Verpoorte, R., & Choi, Y. H. (2013). Natural deep eutectic solvents as new extraction media for phenolic metabolites in Safflower. *Analytical Chemistry*, *85*, 6272–6278.

Dai, Y., Witkamp, G. J., Verpoorte, R., & Choi, Y. H. (2015). Tailoring properties of natural deep eutectic solvents with water to facilitate their applications. *Food Chemistry*, *187*, 14–19.

Faggian, M., Sut, S., Perissutti, B., Baldan, V., Grabnar, I., & Dall'Acqua, S. (2016). Natural deep eutectic solvents (NADES) as a tool for bioavailability improvement: Pharmacokinetics of rutin dissolved in proline/glycine after oral administration in rats: Possible application in nutraceuticals. *Molecules*, *21*, 1531.

Gan, J., Wang, C., Song, Z., Liu, Y., Ning, Z., Ma, X., et al. (2018). Effect of refined honey on pharmacokinetics of 9 different alkanoids in Fuzi Lizhongwan. *Chinese Journal of Experimental Traditional Medical Formulae.*, *24*, 90–96.

Guo, F. B. (2004). *Honey and its magical effects*. Beijing: China Agricultural Press.

Hixon, K. R., Klein, R. C., Eberlin, C. T., Linder, H. R., Ona, W. J., & Gonzalez, H. (2019). A critical review and perspective of honey in tissue engineering and clinical wound healing. *Advances in Wound Care*, *8*, 403–415.

Jin, S. Y. (1988). *Science of processing Chinese materia medica*. Jiangsu: Jiangsu Science and Technology Press.

Kurek-Górecka, A., Górecki, M., Rzepecka-Stojko, A., Balwierz, R., & Stojko, J. (2020). Bee products in dermatology and skin care. *Molecules*, *25*, 556.

Kuropatnicki, A. K., Klosek, M., & Kucharzewski, M. (2018). Honey as medicine: Historical perspectives. *Journal of Apicultural Research*, *57*, 113–118.

Kuś, P. M., Jerković, I., Tuberoso, C. I., & Šarolić, M. (2013). The volatile profiles of a rare apple (*Malus domestica* Borkh.) honey: Shikimic acid-pathway derivatives, terpenes, and others. *Chemistry & Biodiversity*, *10*, 1638–1652.

León-Ruiz, V., González-Porto, A. V., Al-Habsi, N., Vera, S., San Andrés, M. P., & Jauregi, P. (2013). Antioxidant, antibacterial and ace-inhibitory activity of four monofloral honeys in relation to their chemical composition. *Food & Function*, *4*, 1617–1624.

Lin, L. Z., He, X. G., Lindenmaier, M., Nolan, G., Yang, J., Cleary, M., et al. (2000). Liquid chromatography-electrospray ionization mass spectrometry study of the flavonoids of the roots of *Astragalus mongholicus* and *A. membranaceus*. *Journal of Chromatography A*, *876*, 87–95.

Liu, Y. Z., Wang, Z. M., & Zhang, J. Z. (2015). *Dietary Chinese herbs*. Springer Vienna.

Ma, C. M., Wei, Y., Wang, Z. G., & Hattori, M. (2009). Triterpenes from *Cynomorium songaricium*—Analysis of HCV protease inhibitory activity, quantification, and content change under the influence of heating. *Journal of Natural Medicine*, *63*, 9–14.

Mao, W. L. (2011). *Study on the components of propolis and honey*. Thesis Henan: Henan University.

Martinotti, S., Bucekova, M., Majtan, J., & Ranzato, E. (2019). Honey: An effective regenerative medicine product in wound management. *Current Medical Chemistry*, *26*, 5230–5240.

Nolan, V. C., Harrison, J., & Cox, J. (2019). Dissecting the antimicrobial composition of honey. *Antibiotics*, *8*, 251.

Tang, D. G., Wang, Y. X., & Ren, X. Y. (2005). *Honey*. Beijing: Beijing Science and Technology Press.

Xia, L. Y., Zhang, X. Y., Wang, T. X., & Ma, Y. S. (2010). Study on the chromatographic fingerprint of volatile constituents from acacia honey. *Agricultural Science & Technology*, *11*, 42–44.

Xie, G. Y. (1987). Honey and traditional Chinese medicine. *Bee Journal*, *02*, 31–32.

Xiong, P. H. (2005). Application of honey in processing and preparation of traditional Chinese medicine. *Li Shizhen Medicine and Materia Medica Research, 16*, 35–36.

Yang, M. (2012). *Traditional Chinese medicine pharmaceutics*. Beijing: China Traditional Chinese Medicine Press.

Yu, Z. H. (2017). *Study on the changes of compositions in honey during maturing*. Thesis Fujian: Fujian Agricultural and Forestry University.

Yuan, Z. L., & Feng, F. (2000). *Bee product processing technology and health care*. Beijing: Scientific and Technical Documents Publishing Press.

Zheng, N., Shi, Y. L., Jie, Y. X., & Lu, W. (2014). Stability and transformation of astragaloside IV, III and I. *Central South Pharmacy, 12*, 1062.

CHAPTER FIFTEEN

Natural deep eutectic systems—A new era of cryopreservation

Ana Rita Jesus, Liane Meneses, Ana Rita Cruz Duarte, and Alexandre Paiva*

LAQV@REQUIMTE, Departamento de Química, Faculdade de Ciências e Tecnologia, Universidade Nova de Lisboa, Caparica, Portugal

*Corresponding author: e-mail address: alexandre.paiva@fct.unl.pt

Contents

Abstract

Human and animal cells are extremely important for research and many medical applications; however, they do not survive long-term storage with simple cooling or freezing. On the other hand, the increase in world's population along with the changes in feeding behaviors has led to the establishment of new plant varieties. Several plant species are endangered and germplasm conservation is far from developed, hence the search for new cryoprotectant agents became crucial. The metabolites produced by animals and plants to survive extreme cold temperatures served as starting point for the development of natural cryoprotectants. Natural Deep Eutectic Systems (NADES) emerged as alternative solvents and several reports show their low toxicity profile and biodegradability. The components present in organisms that survive extreme cold are potential candidates to prepare a wide variety of NADES. This review exploits NADES' potential as well, as their components, as new alternative cryoprotectant agents.

Advances in Botanical Research, Volume 97
ISSN 0065-2296
https://doi.org/10.1016/bs.abr.2020.09.015

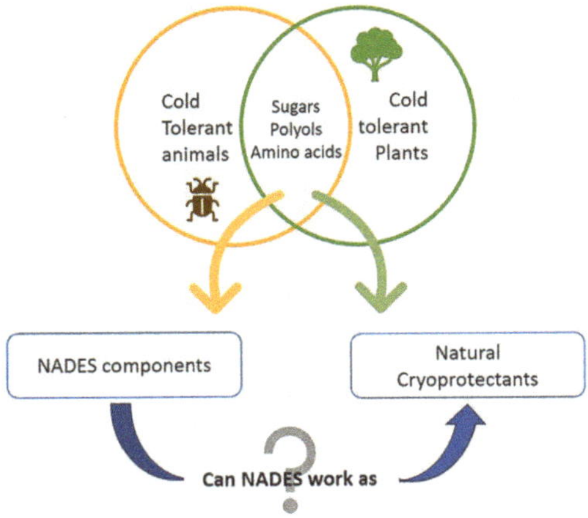

1. Introduction

Cryopreservation is widely used to preserve biological systems and organisms over long periods of time without compromising their structure or function. The continuous maintenance of cell lines in culture is highly impractical because serum and culture media are expensive, there is a high risk of microbial and cross-contamination, and the possibility of genetic drift with the possible consequence of phenotypical changes and the loss of the cell's original characteristics (Stacey & Masters, 2008). Hence, a high post-thawing survival rate is the key factor for an efficacious cell culture.

Cells preservation at extremely low temperatures has allowed for significant advances in science and medicine. However, its effectiveness is strongly dependent on the freezing process, specifically in the formation of ice crystals, since these can cause cell disruption and loss of function. Cells are generally frozen using a slow-cooling process with cooling rates of -1 to $-5\,°C/min$, so that there is a low amount of intracellular water, but they are not completed dehydrated, reducing the amount of ice crystals formation. On the other hand, vitrification can be used, whereby the biological material is directly immersed in liquid nitrogen, resulting in extremely fast cooling rates ($>-400\,°C/min$), avoiding ice formation. This technique is mostly used for specific types of cryopreservation, (e.g., storage of organs, oocytes, follicles) (Ahmadi et al., 2018; Chian, Wang, & Li, 2014). Both methods require a cryoprotective agent (CPA) but vitrification is only achieved using high concentrations of CPA (Fuller, 2004; Taylor, Weegman, Baicu, & Giwa, 2019). Also, to ensure maximum post-thawing viability, cells must be rapidly thawed because intracellular ice crystals can also cause lethal damage during this process.

Cryoprotectants are an essential ingredient for cell freezing, protecting it on its journey from room temperature to −196 °C. The first CPA reported in literature was glycerol, discovered by chance when Polge and colleagues accidentally used a bottle containing a mixture of glycerol during their work on the survival of frozen spermatozoa at cryogenic temperatures in 1949 (Polge, Smith, & Parkes, 1949; Smith & Polge, 1950). This discovery led to the establishment of the modern cryobiology.

1.1 Cryoprotectant agents

To fulfill the requirements as CPA a compound should have high viscosity, which increases for lower temperatures, and the ability to form hydrogen bonds with water, which causes depression of its glass transition temperature (Fuller, 2004; Gertrudes et al., 2017).

Karow established the definition of the two classes of CPAs; penetrating (small molecular weight) and non-penetrating agents (high molecular weight). He has listed 56 chemicals (from both classes) that have promoted a post-thawing survival higher than 40% in different cell lines (Elliott, Wang, & Fuller, 2017; Karow, 1969). Most of those are still used nowadays, including dimethyl sulfoxide (DMSO), propylene and ethylene glycol, methanol, some amino acids, a number of oligosaccharides, and some polymers such as albumin and polyvinylpyrrolidone (PVP). During the 1990s, the Karow's list was reduced to 20 agents by Ashwood-Smith. In Table 1

Table 1 Cryoprotectant agents (Elliott et al., 2017).

Alcohols and derivatives	Sugars and sugar alcohols	Polymers	Sulfoxides and amines	Amines
Methanol	Glucose	Polyethylene glycol	Dimethyl sulfoxide	Proline
Ethanol	Galactose	Polyvinylpyrrolidone	Acetamide	Glutamine
Glycerol	Lactose	Dextrans	Formamide	Betaine
Propylene glycol	Sucrose	Ficoll	Dimethyl acetamide	
Ethylene glycol	Trehalose	Hydroxyethyl starch		
	Raffinose	Serum proteins		
	Mannitol	Milk proteins		
	Sorbitol	Peptones		

are listed the most common CPAs based on both lists. These CPAs have well characterized drawbacks such as high toxicity and low post-thawing survival rate. Since then, no new CPAs classes have been discovered. (Elliott et al., 2017; Fuller, 2004; Hubálek, 2003).

Currently, DMSO is considered the gold-standard CPA due to its effectiveness and ability to penetrate the cell membranes. It was initially used for the preservation of red blood cells and spermatozoa but soon after it was applied for numerous different samples (Hubálek, 2003).

1.2 NADES as cryoprotectants

Deep eutectic systems (DES) emerged in the early 2000s as an alternative to ionic liquids (IL) (Paiva et al., 2014). The term was first described by Abbott, Capper, Davies, Rasheed, and Tambyrajah (2003) as a mixture of compounds with melting point lower than the initial components. The first reported combinations described were mostly based on choline chloride, carboxylic and fatty acids or urea (Paiva et al., 2014). About 10 years later Choi et al. (2011), proved that DES components such as sugars, organic acids, amino acids and polyols are present in large amounts in all types of cells and they hypothesized that these could have a more important role than simple intermediates in metabolic pathways. This led to the introduction of a new class of DES, the so-called, natural deep eutectic systems (NADES). There are countless combinations of molecules known to form NADES, (Espino et al., 2016) meaning that these solvents can be designed in a way that fits the demands of each application (Paiva et al., 2014).

Recently, following Choi's hypothesis on the role of NADES in cellular metabolism, Gertrudes et al. (2017), reviewed the mechanisms responsible for the tolerance to extreme temperature amplitudes that some animal species show, which provided an outlook of the potential of NADES as CPA. Different properties dictate the ability of NADES to act as a CPA, but toxicity is the key factor. Although their toxicity is expected to be low, since they are prepared using naturally occurring molecules, studies have determined that NADES and DES toxicity is mainly dependent on the initial components and on their molar ratio (Paiva et al., 2014; Wen, Chen, Tang, Wang, & Yang, 2015). Recently, Halder and Cordeiro (2019), used an *in-silico* approach to predict the toxicity of DES components, which led to the determination of the likelihood of toxicity of numerous components. As expected, components involved in the preparation of NADES, namely sugars and polyols, were classified with low likelihood for toxicity and sucrose, xylitol, glycerol and sorbitol were regarded as the least toxic.

The component that mainly influences the properties of a NADES is water. It impacts their viscosity, conductivity, toxicity and biocompatibility (Castro et al., 2018; Dai, Witkamp, Verpoorte, & Choi, 2015).

NADES are known as highly viscous liquids, thus, the effect of water content in their viscosity has been described by several authors (Castro et al., 2018; Craveiro et al., 2016; Dai et al., 2013, 2015). For instance, Dai et al. (2013) showed that the addition of 10 wt.% of water to glucose: choline chloride: water (2:5:5) decreased the viscosity of the system by 10-fold.

Considering their thermal behavior, overall, NADES show thermal stability at a wide range of temperatures, even far below 0 °C, strongly supporting the hypothesis on the important role of NADES in cold resistance (Dai et al., 2013; Florindo, Oliveira, Rebelo, Fernandes, & Marrucho, 2014). Craveiro et al. (2016) tested the effect of different concentrations of water in the thermal properties of a selection of NADES. Their work showed that increasing water content lead to a decrease on the glass transition temperature of the mixture, due to a plasticizing effect. This effect was also proved by Castro et al. (2018), for the system trehalose:glycerol (1:30) in which the NADES promoted water vitrification in a large range of temperatures, proving their potential as CPA. In this work, they successfully substituted the gold-standard CPA, DMSO, by this system composed of trehalose and glycerol in the preservation of the L929 model cell line. It was showed that the components alone did not afford the same results of NADES nor DMSO. Furthermore, the viscosity behavior of a mixture of 10% Treh:Gly (1:30) in 90% water was very similar to DMSO in water in the same percentages. On the other hand, the viscosity of 10% of NADES in 90% of fetal bovine serum (FBS) was much higher at lower temperatures than 10% DMSO in 90% of FBS. Only at positive temperatures the viscosity profiles were similar between the two systems, suggesting that besides cytotoxicity, viscosity also plays an important role in cryopreservation. Additionally, thermal behavior was also evaluated and, differential scanning calorimetry (DSC) was performed in pure NADES as well as NADES mixed with different amounts of water. From the results, it was clear that NADES can influence the thermophysical properties of water; the highest wt.% of NADES the lowest melting temperature of the mixture and the highest melting enthalpy. Even a small amount of NADES in water presents a significant decrease in the melting temperature when compared with water alone. It is also suggested that NADES can have some plasticizing effect since the glass transition temperature and the variation on heat capacity of the mixture of water and

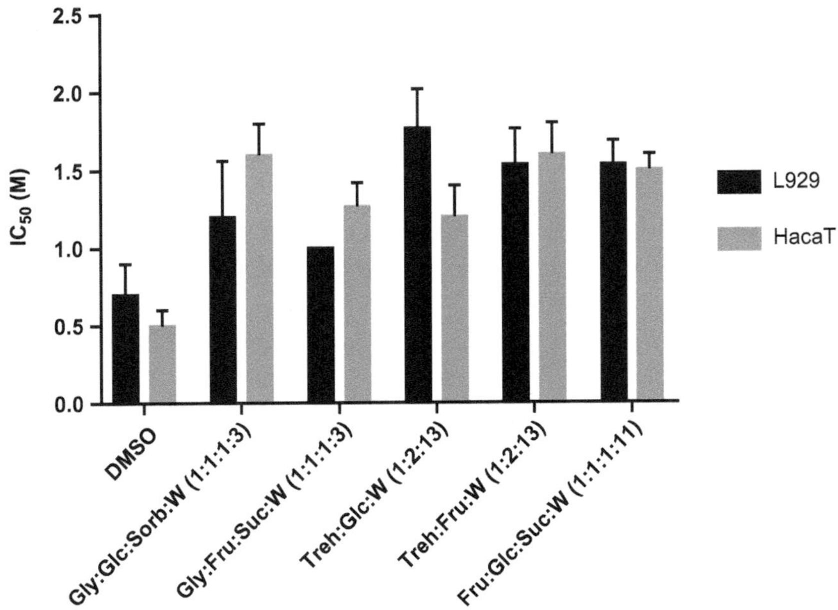

Fig. 1 IC_{50} values of NADES vs DMSO in L929 and HacaT cells.

NADES followed the same trend. These results together may explain the cryoprotective effect of NADES in cryopreservation of mammalian cells.

Preliminary work is being carried out by our research group aiming the discovery of new and more efficient eutectic systems as nontoxic alternatives to DMSO for the cryopreservation of mammalian cells. These studies include the use of NADES mainly composed of sugars, polyols and betaine, in different combinations and proportions. Preliminary results showed that these new NADES are nontoxic for L929 and HacaT cells, with IC_{50} values significantly higher than DMSO (Fig. 1), which supports their potential as CPA.

2. Cryoprotective strategies in nature

2.1 Effect of natural compounds on animal's cold tolerance

Living beings have adapted to life on earth over millions of years. The particular adaptation to extreme low temperatures has been target of several studies since the mechanisms underlying this resistance can be extrapolated to the lab bench in order to improve the methods of cryopreservation and reduce the chill injure caused to cells and tissues during this process.

Animals and insects have two main types of adaptation: they can be freeze-tolerant and survive the freezing process, by freezing only the extracellular fluids, including the interstitial fluid, blood plasma, and other fluids in the body, or non-freeze tolerant and avoid freezing under extremely low temperatures, by entering a super cooled state. Both these survival strategies involve the production of metabolites by the species, such as glycerol, glucose, trehalose and proteins (Brockbank, Campbell, Greene, Brockbank, & Duman, 2011). Examples of animals that have evolved to overcome these extreme situations can be found all around the globe, from north to south pole (Gertrudes et al., 2017). Along this section we will review the key aspects behind the survival of some of these animals.

2.1.1 Tardigrades

Tardigrades are small invertebrate animals, with size in the range on microns. They can survive for a long period in anhydrobiosis, a state in which all metabolic functions are suspended due to extreme dehydration that can be resumed upon hydration. This allows them to survive conditions extremely different from the ones on their natural habitat (Persson et al., 2011; Wełnicz, Grohme, Kaczmarek, Schill, & Frohme, 2011). Tardigrades have been widely investigated for their remarkable adaptation strategies (Boothby et al., 2017; Westh & Ramløv, 1991). Trehalose plays a major role in this process, as many authors have described that tardigrades experience an increase in trehalose accumulation during desiccation and a reduction upon rehydration (Hengherr, Heyer, Köhler, & Schill, 2008; Wełnicz et al., 2011).

Westh and Ramløv (1991) observed that the amount of trehalose in tardigrades exceeded 1.6 wt.% during anhydrobiosis and upon rehydration this value would decrease to c.a. 0.2 wt.%, which corresponded to the concentration found on active animals. The authors also concluded that these animals could increase their trehalose levels up to 2.3 wt.% depending on the conditions under which they were dried. A different study focused on another crucial condition for tardigrades survival, the hydration state in which the authors compared the survival rates of three Tardigrada species from Antarctica subjected to extremely low temperature. They showed a correlation between the hydration state and the ability that tardigrades have to endure low temperatures. Briefly, dehydrated samples remained more stable at the three temperatures tested (−22, −80 and −180 °C) than the hydrated equivalents (Sømme & Meier, 1995). The resistance of tardigrades to freezing temperatures was also observed on a moss sample recovered from Antarctica, in which two specimens were found alive after 30 years of storage and were also able to reproduce (Tsujimoto, Imura, & Kanda, 2016).

The role of trehalose in the tolerance to extremely low temperatures is highly important and more recent studies, showed its production as protection mechanism is not a general trait to all tardigrade's species (Hengherr et al., 2008). In *M. tardigradum*, for instance, the production of trehalose is not detected at any circumstances, and in other species its amount does not differ between hydrated and dehydrated state. In these species, the tolerance to harsh temperature conditions is due to the contribution of proteins, such as the late embryogenesis abundant (LEA) proteins, found in tardigrades during desiccation and hydration periods (Hengherr et al., 2008). There are also the tardigrade-specific intrinsically disordered proteins (TDPs), which are able to form amorphous solids, hence protecting the cellular tissues (Boothby et al., 2017).

2.1.2 Wood frog

Wood frogs (*Rana sylvatica*) are freeze-tolerant amphibians mostly found in the north of Alaska and Canada. They are known to survive being frozen up to 7 months at temperatures of c.a. −20 °C (Larson & Barnes, 2016). As a protective measure to such conditions, the metabolite mainly produced by wood frogs is glucose (Storey & Storey, 1984). However, there are two other components that also contribute for their extreme freeze tolerance, namely antifreeze glycolipids and urea (Costanzo, 2019; Costanzo & Lee, 2013; Larson & Barnes, 2016).

Several papers have focused on the understanding of the mechanisms behind these protective strategies. Storey and Storey (1984) study concluded that glucose is produced from glycogen in the liver only after exposition to sub 0 °C temperatures, reaching the rest of the body later through the blood stream. Lee, Costanzo, Davidson, and Layne (1992), studied the effects of freezing on different organs of wood frogs. They observed that freezing caused dehydration of the animals' organs and the water content was regained during thawing with reduced damage to the tissues. The decrease in water content of the frog's tissues led to an increase of cryoprotective metabolites, such as glucose. The effect of temperature variation is also determinant for cryoprotectant accumulated by wood frogs. Repeated freeze and thaw cycles lead to increased concentrations of glucose from cycle to cycle

(Larson & Barnes, 2016). Under natural conditions, the animals actually go through several freeze and thaw cycles before remaining frozen for the wintertime (Larson et al., 2014).

As for urea, it seems to be involved with metabolism depression, besides the cryoprotective properties. In moderate concentrations, this metabolite has a positive effect on the stabilization of membranes, macromolecules and even reactive oxygen species (ROS) (Costanzo, 2019; Costanzo & Lee, 2013). Antifreeze glycolipids stabilize the cell membranes, and prevent ice from spreading into the cytoplasm and recrystalise (Larson & Barnes, 2016).

2.1.3 Invertebrate

Due to invertebrates reduced size, during winter, their body temperature becomes very close to the environment's temperature. In order to survive they also show survival strategies closely related to the water amount in their bodies and to the presence of cryoprotectant agents (Lee, 2010; Walters, Sformo, Barnes, & Duman, 2009).

The Alaskan stonefly (*Nemoura arctica*), is a freeze-tolerant aquatic insect able to survive over 2 weeks at −15 °C. Walters et al. (2009), observed that the main metabolite produced by the stonefly was glycerol, with a three-fold increase, but they also identified the production of proline. The authors also observed that glycerol concentration on the stonefly kept increasing during the freezing period. Glycerol is known to increase the protection to freezing damage, and proline may be involved in the cold-hardiness process.

Beetles are another example of insects that have developed their own strategies to overcome the winter temperatures (Koštál, Zahradníčková, Šimek, & Zelený, 2007; Sformo et al., 2010). Spruce bark beetle (*Ips typographus*), from Europe is non-freeze-tolerant insect that relies on supercooling to avoid damage at low temperatures. It exhibits an intricate pattern of metabolites production that are adjusted to each season. The system is comprised of several compounds, including sugars and polyols. During fall, when temperatures are above 0 °C, glycerol, trehalose and glucose accumulate. As the temperatures stabilize around 0 °C, trehalose accumulation continues but the concentration of glucose stabilizes and glycerol's decreases. When the winter temperature hits (c.a. −20 °C) trehalose is depleted and sorbitol, mannitol and fructose accumulate in high concentrations. This mechanism allows spruce bark beetles to survive supercooling, since the metabolites increase the viscosity of their body fluids and decrease the available water, avoiding freezing (Koštál et al., 2011, 2007). Red flat bark beetle, from Alaska, survives supercooling at much lower temperatures, around −40 °C, mostly due to glycerol accumulation and dehydration (Sformo et al., 2010). The study of other beetles' species also suggest that, besides sugars and polyols, their survival is also due to free amino acids (Lalouette, Košťál, Colinet, Gagneul, & Renault, 2007).

Furthermore, earthworms show remarkable temperature tolerance capacities, as recently reviewed by Singh, Schädler, Demetrio, Brown, and Eisenhauer (2019). These small invertebrates tolerate freezing through the conversion of glycogen reserves into glucose and small amounts of glycerol, and through the presence of alanine (Holmstrup & Overgaard, 2007; Holmstrup, Slotsbo, Henriksen, & Bayley, 2016).

Some other examples of animals and the cryoprotectant metabolites they produce are listed in Table 2.

2.2 Effect of natural compounds on plant's cold tolerance

Not only animals have the ability to adapt to extreme temperatures. Also, in plants, many molecular, physiological, biochemical and metabolic changes occur when optimal temperature varies as an effort to maximize their growth and developing processes, and to maintain cellular homeostasis during those variations. At the physiological level soluble sugars, amino acids and cold-resistant proteins are accumulated to regulate the osmotic potential

Table 2 Examples of cold tolerance capacities in the animal world and the metabolites produced during exposition to freezing temperatures.

Invertebrate	Metabolites produced	References
Alaska stonefly	Glycerol, proline	Walters et al. (2009)
Beetles	Glucose, trehalose, fructose, glycerol, sorbitol, mannitol, erythritol, amino acids	Košťál et al. (2011), Košťál et al. (2007), Sformo et al. (2010)
Earthworms	Glucose, glycerol, alanine	Holmstrup and Overgaard (2007), Holmstrup et al. (2016)
Rice water weevil	Glucose, trehalose, fructose, sucrose, glycerol, sorbitol, mannitol	Lee, Chang, and Kim (2002)
Codling moth	Trehalose, fructose, glucose, sorbitol, mannitol, alanine	Rozsypal, Košťál, Zahradníčková, and Šimek (2013)

and ice crystal formation, to maintain the stability of cell membranes and to protect the plant against reactive oxygen species (ROS) (Ding, Shi, & Yang, 2019; Sasaki, Ichimura, & Oda, 1996; Tarkowski & Van den Ende, 2015).

Under normal climate conditions, the amount of soluble sugars increases during the onset of winter and decreases in spring when plants are deacclimating (Yuanyuan, Yali, Jiang, & Hongbo, 2009). Sucrose is the main soluble sugar accumulated by plants, but starch and hexoses have also been reported to be accumulated by some species (Rütten & Santarius, 1992; Yuanyuan et al., 2009). Additionally, cereals and woody plants are able to develop high levels of freezing tolerance in winter, and raffinose and stachyose, are commonly found during seasonal cold acclimation.

Sucrose confers cell membrane protection by interacting with the phosphate in their lipid headgroups and decreasing membrane permeability (Strauss & Hauser, 1986). On the other hand, fructans, fructose-based oligo- and polysaccharides, and raffinose family oligosaccharides protect cell membrane by increasing the stability of phospholipidic mono- and bilayers by direct insertion between polar headgroups (Vereyken, Chupin, Demel, Smeekens, & De Kruijff, 2001). The amino acid proline is involved in the cold tolerance response in plants and acts as osmolyte for osmotic adjustment, but it can also contribute for the stabilization of sub-cellular structures such as membranes and proteins, by scavenging ROS and buffering cellular redox potential under stress conditions. It can also act as protein compatible hydrotope alleviating cytoplasmic acidosis and maintaining the appropriate NADP+/NADPH ratios (Botta, 2012; Hare & Cress, 1997; Serraj & Sinclair, 2002). Other metabolites such as polyols, quaternary ammonium compounds (e.g., glycine betaine, alanine betaine, proline betaine and pipecolate betaine) can also be accumulated as response to stress conditions in plants (Hayat et al., 2012).

Interestingly, betaine is associated with the cold tolerance of some plants, such as the winter-type barley, that can accumulate betaine at high levels in its leaves during colder periods to protect itself against freezing injury (Nomura, Muramoto, Yasuda, Takabe, & Kishitani, 1995). These findings have motivated researchers to investigate whether betaine could be beneficial in cryopreservation techniques.

3. Natural cryoprotectants in cryopreservation

3.1 Cryopreservation of mammalian cells

The search for new, practical and more applicable protocols as well as the discovery of new CPA systems could facilitate the conservation of biological materials. The latest developments in terms of CPAs inspired in nature, discussed in the previous section is addressed herein for the cryopreservation of mammalian cells as many studies have been conducted in the past years to overcome the toxicity problems of the standard CPAs, DMSO and ethylene glycol (Kresna, Widjiati, & Damayanti, 2019; Szurek & Eroglu, 2011). Intracellular DMSO cannot be metabolized, so it must be removed immediately after thawing, hence, other cryoprotectants have been suggested as alternatives.

Interestingly, sugars and amines showed in Table 1, are some of the metabolites produced by organisms that survive extreme cold environments, as a protection against freeze injury (as discussed in Section 2). The understanding of these defense mechanisms and the role of such metabolites inspired researchers to studied them as alternative CPAs. Additionally, many of these natural compounds are also known to form NADES (Lee et al., 2018). Hence, to raise the discussion and generate hypothesis on the role of each molecule, the components most discussed in literature will be revised herein.

3.1.1 Trehalose

Trehalose is a nontoxic disaccharide found at high concentrations in a variety of organisms capable of surviving in complete dehydration, including bacteria, yeast, tardigrades and nematodes. There are two main theories behind the role of trehalose in dehydration survival, the water replacement theory and the glassy state theory. In the first theory, trehalose creates hydrogen bonds and replaces the lost water on membranes, preventing dehydration damage (Crowe, 2002). The other hypothesis considers that the disaccharide causes vitrification, immobilizing the cellular structures, avoiding the injuries (Crowe, Carpenter, & Crowe, 1998). Apparently, the protective action of trehalose is related to both osmotic effect and specific interactions with cell membrane phospholipids and labile proteins, which prevents their damage and denaturation due to desiccation and oxidative stress. However, the precise mechanism of such protection remains unknown (Benaroudj, Lee, & Goldberg, 2001).

The addition of trehalose to DMSO freezing solutions was found to be efficient in the improvement of the viability of human embryonic stem cells and primary hepatocytes in suspension which are known to be difficult to cryopreserve (Buchanan et al., 2004; Martinetti et al., 2017; Ntai, La Spada, De Blasio, & Biunno, 2018; Wu et al., 2005). The effect of trehalose and DMSO in the cryopreservation of HepG2 cells in suspension and monolayer formats has been reported by Stokich et al. (2014). According to the authors, cells cryopreserved in monolayers in the presence of trehalose or DMSO did not survive, while cells in suspension presented little viability after thawing. However, the preincubation with 100 mM of trehalose before freezing with DMSO was enough for the cells in monolayer format to present a significantly higher post-thawing viability. Interestingly, the metabolic activity of cells in suspension immediately after thawing did not shown any significant differences compared to unfrozen control cells.

Human hepatocytes for cell transplantation are limited due to the difficulty of their cryopreservation but Katenz et al. (2007) have studied the effect of trehalose on their cryopreservation. Many parameters were studied, and it was found that the use of trehalose as additive in the freezing media almost doubled the post-thawing viability. Moreover, total protein levels in attached cells and secretion of albumin were also higher than in DMSO group, while aspartate aminotransferase (AST) was lower, suggesting that this methodology would be a great advantage for the cryopreservation of hepatocytes.

On the other hand, CPA-free methodologies have also been investigated. The pre-hydration of fibroblasts with trehalose prior the induction of ice formation by a copper wire precooled in liquid nitrogen and plunged into a −196 °C freezer has been successfully reported (Huang et al., 2017). It was observed that ice seeding could minimize the interfacial free energy that drives the devastating ice recrystallization-induced cell injury during warming steps. Apparently, this CPA-free methodology can also be effectively applied to adult stem cells and red blood cells.

Trehalose has also been reported as CPA in the vitrification of mice oocytes (Lestari et al., 2018). The successful embryos development has been achieved using a combination of trehalose or sucrose with DMSO and EG, although the first afforded better results.

3.1.2 *Raffinose*

Raffinose is also a good glass former, and because it is small enough it can directly interact with biomolecules and form hydrogen bonds with the cells' lipid head groups, although weaker than those formed by trehalose (Crowe, Leslie, & Crowe, 1994).

In cryobiology, raffinose is mainly studied for the cryopreservation of oocytes and sperm (Crowe et al., 1994; Eroglu, 2010; Hino, Takabe, Suzuki-Migishima, & Yokoyama, 2007; Tada et al., 1990). It is shown that raffinose affords a remarkable protection against cryopreservation-related stresses when mixed with small amounts of DMSO (Eroglu, 2010). Oocytes microinjected with raffinose, and later cryopreserved in the presence of extracellular raffinose and a small amount of DMSO, presented survival rates higher that 80% and fertilization and embryonic development rates between 70% and 95%, demonstrating a striking cryoprotective effect of intra- and extracellular raffinose with small amounts of DMSO.

In the case of mouse spermatozoa, it has been showed that with only 18wt.% of raffinose a major protection against freeze injury was achieved, and sperm motility and fertilizing ability was significantly higher than those frozen with DMSO or glycerol (Tada et al., 1990).

3.1.3 Betaine

Betaine is abundantly present in many microorganisms, plants and animals (Craig, 2004). It is able to protect cells against osmotically induced inactivation (Courtenay, Capp, Anderson, & Record, 2000) and according to literature, it can form a monolayer of water around proteins, preserving their stability and function (Craig, 2004).

Yang et al. (2016) tested betaine as CPA using an ultra-rapid freezing method and different types of cells in comparison with glucose and DMSO. DSC studies revealed that all compounds showed an endothermic peak due to the melting of solute-water system, but betaine showed the smallest melting peak at comparable concentrations, suggesting its strong ability to inhibit water crystallization. Moreover, when GLC-82 cells were exposed to a hypertonic media containing betaine their viability was significantly improved, suggesting that betaine can in fact regulate the osmotic stress in cells. The post-thawing survival of GLC-82, Hela and MCF-10 cells, frozen in the presence of 4–10% of betaine directly in liquid nitrogen was higher than 75%. The fact that cells recovered well after some days when the freezing media was not removed after thawing and before culture, highlights the major advantage of using betaine as CPA.

3.1.4 Proline

Proline was firstly introduced as osmoprotectant by Withers and King (1979) for the cryopreservation of cultured cells of *Zea mays* L. (maize cells), and was studied for that purpose for many years. Recently, L-proline was used for the cryopreservation of endothelial cells and its effectiveness was compared with mixtures of two other biocompatible solutes (ectoine and hydroxyectoine). Although L-proline was the most efficient CPA, the combination of L-proline and ectoine resulted in a higher recovery and, therefore, this combination could be used as potential CPA to improve the cryopreservation of human endothelial cells in the presence of a very low amount of DMSO (ca. 2.5%) (Sun, Glasmacher, & Hofmann, 2012).

Later, proline was studied to cryopreserve red blood cells (RBC) (Dou, Lu, Sun, & Rao, 2019). Currently, glycerol is the cryoprotectant allowed in clinical therapy for RBCs cryopreservation, but its toxicity and complex deglycerolization process limits its application (Lagerberg, 2015). A simple protocol was developed using the combination of L-proline and trehalose as cryoprotectants for RBCs. Compared with glycerol, this combination presented the lowest post-thawing haemolysis and a normal morphology of RBC cells as well as the activities of Na^{+}/K^{+}-ATPase and Ca^{2+}/Mg^{2+}-ATPase. However, when L-proline and trehalose were used alone as cryoprotectants, the results were not as promising as the mixture, suggesting a synergetic effect between the two compounds. The mechanistic studies suggested that L-proline might decrease the freezing point and inhibit ice crystal growth by permeating into the cells during the freezing process; trehalose, on the other hand, might work as inhibitor of ice crystal growth during the process of freezing and inhibitor of ice recrystallization during the thawing process (Dou et al., 2019).

Recently, L-proline was also reported as CPA for the cryopreservation of mammalian oocytes, as alternative to DMSO or EG. Although DMSO is the most effective CPA there are reports of its negative effect on differentiation of somatic cells (Vajta & Kuwayama, 2006). Since L-proline is a nontoxic membrane-permeating amino acid it has been studied as CPA for the cryopreservation of oocytes by vitrification technique. Different freezing solutions were tested combining different concentrations of L-proline with both DMSO and EG. A mixture of 2M L-proline, 7.5% DMSO and 10% EG afforded very promising results. Osmotic pressure played a crucial role in the viability of post-thaw oocytes, since those treated with high concentrations of L-proline suffered from osmotic stress and did not survived. No changes in global DNA methylation were observed nor adverse effects on embryonic development (Zhang et al., 2016).

3.2 Cryopreservation of plant tissue

In which concerns the plant kingdom, the use of natural cryoprotectants on plant tissue goes back to 1993, when the plant vitrification solution 3 (PVS3) was developed, a solution with 50% (w/v) glycerol and 50% (w/v) sucrose. This was applied to embryogenic cells of asparagus and the obtained survival rate was 80–90% (Nishizawa, Sakai, Amano, & Matsuzawa, 1993). From the three PVS, this is the only one that does not contain DMSO, and still results in high post-thaw survival rates (Wang et al., 2020). For the cryopreservation of *Satureja spicigera* leaves, vitrification with PVS3 lead to the best results of regrowth, compared to DMSO and other solutions (Ghaffarzadeh-Namazi, Keller, Senula, & Babaeian, 2017). Sucrose can also have cryoprotective properties when applied as freezing pre-treatment, as evidenced for potato apical shoot tips, in which this pre-treatment improved the meristem growth and plant recovery upon cryopreservation (Folgado et al., 2015).

Recently, trehalose, sucrose and glycerol, were used to evaluate the effect on dehydration, cryopreservation and reculture conditions of growth recovery of vanilla (*Vanilla planifolia*) shoot tips using a D-cryoplate procedure. The use of these natural compounds was compared using PSV2 and PSV3. The highest recovery was achieved using the D-cryoplate after the preconditioning with trehalose and loading with sucrose/glycerol solution. DSC studies showed that the osmotically active water content in shoot tips was reduced from 77% to 17% after the effective drying duration. Melting endotherms showed crystallization events along with cryopreservation of the tissues. The results reported by Hernández-Ramírez et al. (2020) were, so far, the most promising ones for the *V. planifolia* shoot tips cryopreservation. In the future, further studies using natural cryoprotectants for the conservation of different plant species are expected, contributing to maintain these valuable genetic resources available.

4. Advances and perspectives in cryopreservation

Huge advances have been seen in the field of cryopreservation of animal and human cells, as previously discussed. Despite all efforts from researchers, the natural compounds discussed throughout this chapter are still most of the times combined with small amounts of DMSO, EG or glycerol. Therefore, the problem of cytotoxicity has not been eliminated. There are a few NADES reported in literature using these metabolites that could be potential options for new CPAs and that may be used in the absence of DMSO, EG or glycerol (Table 3). The major advantage of using NADES as CPAs is the fact that they do not need to be removed from the freezing media after thawing as, in general, they present low toxicity and can be easily metabolized.

Table 3 Example of NADES composed of naturally occurring metabolites with function in organisms with cold tolerance capacity.

NADES	References
Glucose:Frutose (1:1) Frutose:Sucrose (1:1) Glucose:Sucrose (1:1) Sucrose:Fructose:Glucose (1:1:1)	Choi et al. (2011)
Betaine:Glucose (5:2) Betaine:Sucrose (1:1, 2:1, 4:1) Betaine:Trehalose (4:1) Betaine:Mannose (5:2) Betaine:Inositol:Raffinose (9:1:1) Betaine:Sucrose:Proline (1:1:1, 5:2:2) Betaine:Glucose:Proline (1:1:1) Proline:Sucrose (1:1, 2:1, 3:1, 4:1) Proline:Sorbitol (1:1) Proline:Glucose (1:1, 5:3)	Dai et al. (2013)
Glycerol:Glucose:Sorbitol:Water (1:1:1:3) Glycerol:Fructose:Sucrose:Water (1:1:1:3) Trehalose:Glucose:Water (1:2:13) Trehalose:Fructose:Water (1:2:13)	This work

Several techniques for the characterization and evaluation of CPAs are being used by the scientific community to answer important questions such as, does the perfect CPA need to be penetrating or non-penetrating? Or should it be a combination of both?

Recently, Pollock et al. (2016) combined the use of Raman spectroscopy and DSC to evaluate the differences in cell recovery and ice crystal formation for two solutions containing multiple CPAs; (1) sucrose, mannitol and creatine and (2) sucrose, glycerol and creatine. Mesenchymal cells (MSC) showed a post-thaw recovery similar or even better than when 10% DMSO was used, with a cooling rate of 1 °C/min. However, when cells were preincubated for a maximum of 2h with these solutions, maximum post-thaw recovery was achieved. Raman images showed that short periods of preincubation afforded larger ice crystals, suggesting the need of longer permeation periods. Interestingly, there was no significant differences in the solidification patterns and glass-forming tendency for solutions with similar osmolarity. The differences in post-thaw recovery imply that biological factors are involved. The results from this study may increase the knowledge of how these CPA solutions can help to enhance post-thaw cell recovery.

Later, the authors developed a differential evolution (DE) algorithm to optimize the composition of these CPA solutions and cooling rates for specific cell lines. Their study included Jurkat cells and MSCs, which were combined with non-DMSO solutions with concentrations and cooling rates dictated by the algorithm. For Jurkat cells the optimal composition of freezing solution was composed of 300 mM trehalose, 10% glycerol and 0.01% ectoine; whereas to cryopreserve MSCs the optimal composition included 300 mM ethylene glycol, 1 mM taurine and 1% ectoine, with a cooling rate of 1 °C/min. Both solutions resulted in significantly higher survival rate than for 10% DMSO. Hence, this method may allow the optimization of CPA composition for other cell types, in a rational and faster way (Pollock, Budenske, McKenna, Dosa, & Hubel, 2017). Further developments could benefit from these findings if it would be possible to use the algorithm developed by Pollock et al. to predict and combine the proper metabolites to prepare NADES for specific cell lines.

From what was pinpointed in this chapter the potential of NADES as CPAs should be pursued for preservation of mammalian cells and plant tissues. Although only one report shows the effect of NADES in cryopreservation, most of the components able to form NADES have been reported for their cryoprotectant properties, therefore the possibilities are tremendous. It is however, of the utmost importance to understand the role of each molecule present in the CPA composition.

There is still a long path ahead and some questions need to be answered before scientists are able to find the perfect CPA to replace DMSO, EG or glycerol as standard CPAs in common cryopreservation techniques. Answering the questions raised in this chapter would surely help scientists to predict and combine the proper compounds to prepare new NADES that would successfully substitute the standard CPAs. In Nature there is seldomly a solution that fits all problems. As seen in this chapter each organism has adapted differently to their environment. One common thing is the presence of natural metabolites that are able to form NADES. There are no coincidences in Nature, so if we follow Nature's clues, we can expect that NADES may have a valuable contribution in cryopreservation. There will not be a NADES that is the perfect cryoprotectant, but rather a case by case solution where specific NADES will be the best solution. Therefore, modeling tools are important to significantly reduce research time. Naturally, that models alone will not provide all the answers and there is still a long road to cover regarding lab research in cryobiology. Our aim is to boost research in this field which could trigger new developments and push forward alternative cryopreservation methodologies that could ensure safety and could secure the performance maintenance of the biological material stored.

Acknowledgments

The authors thank the funding from European Union Horizon 2020 Programme under grant agreement number ERC-2016-CoG 725034 (ERC Consolidator Grant Des.solve). The authors also thank Laboratory for Green Chemistry—LAQV (UIDB/50006/2020), project CryoDES (PTDC/EQU-EQU/29851/2017) and the grants SFRH/BD/148510/2019 and IF/01146/2015 financed by national funds from FCT/MCTES.

References

Abbott, A. P., Capper, G., Davies, D. L., Rasheed, R. K., & Tambyrajah, V. (2003). Novel solvent properties of choline chloride/urea mixtures. *Chemical Communications*, *1*, 70–71. https://doi.org/10.1039/b210714g.

Ahmadi, R., Hemmateenejad, B., Safavi, A., Shojaeifard, Z., Mohabbati, M., & Firuzi, O. (2018). Assessment of cytotoxicity of choline chloride-based natural deep eutectic solvents against human HEK-293 cells: A QSAR analysis. *Chemosphere*, *209*, 831–838. https://doi.org/10.1016/j.chemosphere.2018.06.103.

Benaroudj, N., Lee, D. H., & Goldberg, A. L. (2001). Trehalose accumulation during cellular stress protects cells and cellular proteins from damage by oxygen radicals. *Journal of Biological Chemistry*, *276*(26), 24261–24267. https://doi.org/10.1074/jbc.M101487200.

Boothby, T. C., Tapia, H., Brozena, A. H., Piszkiewicz, S., Smith, A. E., Giovannini, I., et al. (2017). Tardigrades use intrinsically disordered proteins to survive desiccation. *Molecular Cell*, *65*(6), 975–984.e5. https://doi.org/10.1016/j.molcel.2017.02.018.

Botta, A. (2012). Enhancing plant tolerance to temperature stress with amino acids: An approach to their mode of action. *Acta Horticulturae*, *1009*, 29–36. https://doi.org/10.17660/actahortic.2013.1009.1.

Brockbank, K. G. M., Campbell, L. H., Greene, E. D., Brockbank, M. C. G., & Duman, J. G. (2011). Lessons from nature for preservation of mammalian cells, tissues, and organs. *Vitro Cellular and Developmental Biology - Animal*, *47*(3), 210–217. https://doi.org/10.1007/s11626-010-9383-2.

Buchanan, S. S., Gross, S. A., Acker, J. P., Toner, M., Carpenter, J. F., & Pyatt, D. W. (2004). Cryopreservation of stem cells using trehalose: Evaluation of the method using a human hematopoietic cell line. *Stem Cells and Development*, *13*(3), 295–305. https://doi.org/10.1089/154732804323099226.

Castro, V. I. B., Craveiro, R., Silva, J. M., Reis, R. L., Paiva, A., & Duarte, A. R. C. (2018). Natural deep eutectic systems as alternative nontoxic cryoprotective agents. *Cryobiology*, *83*(April), 15–26. https://doi.org/10.1016/j.cryobiol.2018.06.010.

Chian, R. C., Wang, Y., & Li, Y. R. (2014). Oocyte vitrification: Advances, progress and future goals. *Journal of Assisted Reproduction and Genetics*, *31*(4), 411–420. https://doi.org/10.1007/s10815-014-0180-9.

Choi, Y. H., van Spronsen, J., Dai, Y., Verberne, M., Hollmann, F., Arends, I. W. C. E., et al. (2011). Are natural deep eutectic solvents the missing link in understanding cellular metabolism and physiology? *Plant Physiology*, *156*(4), 1701–1705. https://doi.org/10.1104/pp.111.178426.

Costanzo, J. P. (2019). Overwintering adaptations and extreme freeze tolerance in a subarctic population of the wood frog, Rana sylvatica. *Journal of Comparative Physiology B*, *189*(1), 1–15. https://doi.org/10.1007/s00360-018-1189-7.

Costanzo, J. P., & Lee, R. E. (2013). Avoidance and tolerance of freezing in ectothermic vertebrates. *Journal of Experimental Biology, 216*(11), 1961–1967. https://doi.org/10.1242/jeb.070268.

Courtenay, E. S., Capp, M. W., Anderson, C. F., & Record, M. T. (2000). Vapor pressure osmometry studies of osmolyte-protein interactions: Implications for the action of osmoprotectants in vivo and for the interpretation of "osmotic stress" experiments in vitro. *Biochemistry, 39*(15), 4455–4471. https://doi.org/10.1021/bi992887l.

Craig, S. A. (2004). Betaine in human nutrition. *American Journal of Clinical Nutrition, 80*, 539–549. https://doi.org/10.1177/1461444810365020.

Craveiro, R., Aroso, I., Flammia, V., Carvalho, T., Viciosa, M. T., Dionísio, M., et al. (2016). Properties and thermal behavior of natural deep eutectic solvents. *Journal of Molecular Liquids, 215*, 534–540. https://doi.org/10.1016/j.molliq.2016.01.038.

Crowe, L. M. (2002). Lessons from nature: The role of sugars in anhydrobiosis. *Comparative Biochemistry and Physiology, 131*(3), 505–513. https://doi.org/10.1016/S1095-6433(01)00503-7.

Crowe, J. H., Carpenter, J. F., & Crowe, L. M. (1998). The role of vitrification in anhydrobiosis. *Annual Review of Physiology, 60*(1), 73–103. https://doi.org/10.1146/annurev.physiol.60.1.73.

Crowe, J. H., Leslie, S. B., & Crowe, L. M. (1994). Is Vitrification sufficient to preserve liposomes during freeze-drying? *Cryobiology, 31*, 355–366.

Dai, Y., van Spronsen, J., Witkamp, G., Verpoorte, R., Choi, Y. H., Van Spronsen, J., et al. (2013). Natural deep eutectic solvents as new potential media for green technology. *Analytica Chimica Acta, 766*, 61–68. https://doi.org/10.1016/j.aca.2012.12.019.

Dai, Y., Witkamp, G.-J., Verpoorte, R., & Choi, Y. H. (2015). Tailoring properties of natural deep eutectic solvents with water to facilitate their applications. *Food Chemistry, 187*, 14–19. https://doi.org/10.1016/j.foodchem.2015.03.123.

Ding, Y., Shi, Y., & Yang, S. (2019). Advances and challenges in uncovering cold tolerance regulatory mechanisms in plants. *New Phytologist, 222*(4), 1690–1704. https://doi.org/10.1111/nph.15696.

Dou, M., Lu, C., Sun, Z., & Rao, W. (2019). Natural cryoprotectants combinations of L-proline and trehalose for red blood cells cryopreservation. *Cryobiology, 91*, 23–29. https://doi.org/10.1016/j.cryobiol.2019.11.002.

Elliott, G. D., Wang, S., & Fuller, B. J. (2017). Cryoprotectants: A review of the actions and applications of cryoprotective solutes that modulate cell recovery from ultra-low temperatures. *Cryobiology, 76*, 74–91. https://doi.org/10.1016/j.cryobiol.2017.04.004.

Eroglu, A. (2010). Cryopreservation of mammalian oocytes by using sugars: Intra- and extra-cellular raffinose with small amounts of dimethylsulfoxide yields high cryosurvival, fertilization, and development rates. *Cryobiology, 60*(3S), S54–S59. https://doi.org/10.1016/j.cryobiol.2009.07.001.Cryopreservation.

Espino, M., de los Ángeles Fernández, M., Gomez, F. J. V., Silva, M. F., Fernández, M.d.l. Á., Gomez, F. J. V., et al. (2016). Natural designer solvents for greening analytical chemistry. *Trends in Analytical Chemistry, 76*(December), 126–136. https://doi.org/10.1016/j.trac.2015.11.006.

Florindo, C., Oliveira, F. S., Rebelo, L. P. N., Fernandes, A. M., & Marrucho, I. M. (2014). Insights into the synthesis and properties of deep eutectic solvents based on cholinium chloride and carboxylic acids. *ACS Sustainable Chemistry and Engineering, 2*(10), 2416–2425. https://doi.org/10.1021/sc500439w.

Folgado, R., Panis, B., Sergeant, K., Renaut, J., Swennen, R., & Hausman, J. F. (2015). Unravelling the effect of sucrose and cold pretreatment on cryopreservation of potato through sugar analysis and proteomics. *Cryobiology, 71*(3), 432–441. https://doi.org/10.1016/j.cryobiol.2015.09.006.

Fuller, B. J. (2004). Cryoprotectants: The essential antifreezes to protect life in the frozen state. *Cryo-Letters*, *25*(6), 375–388.
Gertrudes, A., Craveiro, R., Eltayari, Z., Reis, R. L., Paiva, A., & Duarte, A. R. C. (2017). How do animals survive extreme temperature amplitudes? The role of natural deep eutectic solvents. *ACS Sustainable Chemistry and Engineering*, *5*(11), 9542–9553. https://doi.org/10.1021/acssuschemeng.7b01707.
Ghaffarzadeh-Namazi, L., Keller, E. R. J., Senula, A., & Babaeian, N. (2017). Investigations on various methods for cryopreservation of callus of the medicinal plant Satureja spicigera. *Journal of Applied Research on Medicinal and Aromatic Plants*, *5*, 10–15. https://doi.org/10.1016/j.jarmap.2016.09.003.
Halder, A. K., & Cordeiro, M. N. D. S. (2019). Probing the environmental toxicity of deep eutectic solvents and their components: An in silico modeling approach. *ACS Sustainable Chemistry and Engineering*, 7(12), 10649–10660. https://doi.org/10.1021/acssuschemeng.9b01306.
Hare, P. D., & Cress, W. A. (1997). Metabolic implications of stress-induced proline accumulation in plants. *Plant Growth Regulation*, *21*(2), 79–102. https://doi.org/10.1023/A:1005703923347.
Hayat, S., Hayat, Q., Alyemeni, M. N., Wani, A. S., Pichtel, J., & Ahmad, A. (2012). Role of proline under changing environments. *Plant Signaling and Behavior*, 7(11), 1456–1466. https://doi.org/10.4161/psb.21949.
Hengherr, S., Heyer, A. G., Köhler, H. R., & Schill, R. O. (2008). Trehalose and anhydrobiosis in tardigrades - Evidence for divergence in responses to dehydration. *FEBS Journal*, *275*(2), 281–288. https://doi.org/10.1111/j.1742-4658.2007.06198.x.
Hernández-Ramírez, F., Dolce, N., Flores-Castaños, O., Rascón, M. P., Ángeles-Álvarez, G., Folgado, R., et al. (2020). Advances in cryopreservation of vanilla (Vanilla planifolia jacks.) shoot-tips: Assessment of new biotechnological and cryogenic factors. *In Vitro Cellular & Developmental Biology - Plant*, *56*, 236–246. https://doi.org/10.1007/s11627-020-10069-w.
Hino, T., Takabe, M., Suzuki-Migishima, R., & Yokoyama, M. (2007). Cryoprotective effects of various saccharides on cryopreserved mouse sperm from various strains. *Reproductive Medicine and Biology*, *6*(4), 229–233. https://doi.org/10.1111/j.1447-0578.2007.00190.x.
Holmstrup, M., & Overgaard, J. (2007). Freeze tolerance in Aporrectodea caliginosa and other earthworms from Finland. *Cryobiology*, *55*(1), 80–86. https://doi.org/10.1016/j.cryobiol.2007.06.001.
Holmstrup, M., Slotsbo, S., Henriksen, P. G., & Bayley, M. (2016). Earthworms accumulate alanine in response to drought. *Comparative Biochemistry and Physiology - Part A*, *199*, 8–13. https://doi.org/10.1016/j.cbpa.2016.04.015.
Huang, H., Zhao, G., Zhang, Y., Xu, J., Toth, T. L., & He, X. (2017). Predehydration and ice seeding in the presence of Trehalose enable cell cryopreservation. *ACS Biomaterials Science and Engineering*, *3*(8), 1758–1768. https://doi.org/10.1021/acsbiomaterials.7b00201.
Hubálek, Z. (2003). Protectants used in the cryopreservation of microorganisms. *Cryobiology*, *46*(3), 205–229. https://doi.org/10.1016/S0011-2240(03)00046-4.
Karow, A. M. (1969). Biological effects of cryoprotectants as related to cardiac cryopreservation. *Cryobiology*, *5*(6), 429–443. https://doi.org/10.1016/S0011-2240(69)80109-4.
Katenz, E., Vondran, F. W. R., Schwartlander, R., Pless, G., Gong, X., Cheng, X., et al. (2007). Cryopreservation of primary human hepatocytes: The benefit of Trehalose as an additional Cryoprotective agent. *Liver Transplantation*, *13*(5), 767–768. https://doi.org/10.1002/lt.
Košťál, V., Doležal, P., Rozsypal, J., Moravcová, M., Zahradníčková, H., & Šimek, P. (2011). Physiological and biochemical analysis of overwintering and cold tolerance

in two central European populations of the spruce bark beetle, Ips typographus. *Journal of Insect Physiology*, *57*(8), 1136–1146. https://doi.org/10.1016/j.jinsphys.2011.03.011.

Košťál, V., Zahradníčková, H., Šimek, P., & Zelený, J. (2007). Multiple component system of sugars and polyols in the overwintering spruce bark beetle, Ips typographus. *Journal of Insect Physiology*, *53*(6), 580–586. https://doi.org/10.1016/j.jinsphys.2007.02.009.

Kresna, A., Widjiati, W., & Damayanti, T. (2019). Cryoprotectant combination ethylene glycol and propanediol on mice blastocyst viability post vitrification. *Journal of Physics: Conference Series*, *1146*(012028), 1–6. https://doi.org/10.1088/1742-6596/1146/1/012028.

Lagerberg, J. W. (2015). Cryopreservation and freeze-drying protocols. In *Cryopreservation and freeze-drying protocols* (Vol. 1257, pp. 477–488). https://doi.org/10.1007/978-1-4939-2193-5.

Lalouette, L., Koštál, V., Colinet, H., Gagneul, D., & Renault, D. (2007). Cold exposure and associated metabolic changes in adult tropical beetles exposed to fluctuating thermal regimes. *FEBS Journal*, *274*(7), 1759–1767. https://doi.org/10.1111/j.1742-4658.2007.05723.x.

Larson, D. J., & Barnes, B. M. (2016). Cryoprotectant production in freeze-tolerant wood frogs is augmented by multiple freeze-thaw cycles. *Physiological and Biochemical Zoology*, *89*(4), 340–346. https://doi.org/10.1086/687305.

Larson, D. J., Middle, L., Vu, H., Zhang, W., Serianni, A. S., Duman, J., et al. (2014). Wood frog adaptations to overwintering in Alaska: New limits to freezing tolerance. *Journal of Experimental Biology*, *217*(12), 2193–2200. https://doi.org/10.1242/jeb.101931.

Lee, R. E. (2010). A primer on insect cold-tolerance. In *Low temperature biology of insects* (pp. 3–34). Cambridge: Cambridge University Press.

Lee, K. Y., Chang, Y. D., & Kim, Y. G. (2002). Trehalose, a major sugar Cryoprotectant of the overwintering Rice water weevil, Lissorhoptrus oryzophilus (Coleoptera: Curculionidae). *Journal of Asia-Pacific Entomology*, *5*(1), 35–41. https://doi.org/10.1016/S1226-8615(08)60130-2.

Lee, R. E., Costanzo, J. P., Davidson, E. C., & Layne, J. R. (1992). Dynamics of body water during freezing and thawing in a freeze-tolerand frog (Rana sylvatica). *Journal of Thermal Biology*, *17*(4/5), 263–266.

Lee, J., Kim, E. J., Kong, H. S., Youm, H. W., Kim, S. K., Lee, J. R., et al. (2018). Establishment of an improved vitrification protocol by combinations of vitrification medium for isolated mouse ovarian follicles. *Theriogenology*, *121*, 97–103. https://doi.org/10.1016/j.theriogenology.2018.07.022.

Lestari, S. W., Ilato, K. F., Pratama, M. I. A., Fitriyah, N. N., Pangestu, M., Pratama, G., et al. (2018). Sucrose "versus" trehalose cryoprotectant modification in oocyte vitrification: A study of embryo development. *Biomedical and Pharmacology Journal*, *11*(1), 97–104. https://doi.org/10.13005/bpj/1351.

Martinetti, D., Colarossi, C., Buccheri, S., Denti, G., Memeo, L., & Vicari, L. (2017). Effect of trehalose on cryopreservation of pure peripheral blood stem cells. *Biomedical Reports*, *6*(3), 314–318. https://doi.org/10.3892/br.2017.859.

Nishizawa, S., Sakai, A., Amano, Y., & Matsuzawa, T. (1993). Cryopreservation of asparagus (Asparagus officinalis L.) embryogenic suspension cells and subsequent plant regeneration by vitrification. *Plant Science*, *91*(1), 67–73. https://doi.org/10.1016/0168-9452(93)90189-7.

Nomura, M., Muramoto, Y., Yasuda, S., Takabe, T., & Kishitani, S. (1995). The accumulation of glycinebetaine during cold acclimation in early and late cultivars of barley. *Euphytica*, *83*(3), 247–250. https://doi.org/10.1007/BF01678137.

Ntai, A., La Spada, A., De Blasio, P., & Biunno, I. (2018). Trehalose to cryopreserve human pluripotent stem cells. *Stem Cell Research, 31*(July), 102–112. https://doi.org/10.1016/j.scr.2018.07.021.

Paiva, A., Craveiro, R., Aroso, I., Martins, M., Reis, R. L., & Duarte, A. R. C. (2014). Natural deep eutectic solvents - Solvents for the 21st century. *ACS Sustainable Chemistry and Engineering, 2*(5), 1063–1071. Retrieved from http://doi.wiley.com/10.1002/chin.201424290.

Persson, D., Halberg, K. A., Jørgensen, A., Ricci, C., Møbjerg, N., & Kristensen, R. M. (2011). Extreme stress tolerance in tardigrades: Surviving space conditions in low earth orbit. *Journal of Zoological Systematics and Evolutionary Research, 49*(Suppl.1), 90–97. https://doi.org/10.1111/j.1439-0469.2010.00605.x.

Polge, C., Smith, A. U., & Parkes, A. (1949). Revival of spermatozoa after vitrification and dehydration. *Nature, 164*, 666.

Pollock, K., Budenske, J. W., McKenna, D. H., Dosa, P. I., & Hubel, A. (2017). Algorithm-driven optimization of cryopreservation protocols for transfusion model cell types including Jurkat cells and mesenchymal stem cells. *Journal of Tissue Engineering and Regenerative Medicine, 11*(10), 2806–2815. https://doi.org/10.1002/term.2175.

Pollock, K., Yu, G., Moller-Trane, R., Koran, M., Dosa, P. I., McKenna, D. H., et al. (2016). Combinations of Osmolytes, including monosaccharides, disaccharides, and sugar alcohols act in concert during cryopreservation to improve mesenchymal stromal cell survival. *Tissue Engineering - Part C: Methods, 22*(11), 999–1008. https://doi.org/10.1089/ten.tec.2016.0284.

Rozsypal, J., Koštál, V., Zahradníčková, H., & Šimek, P. (2013). Overwintering strategy and mechanisms of cold tolerance in the codling moth (Cydia pomonella). *PLoS One, 8*(4), 1–8. https://doi.org/10.1371/journal.pone.0061745.

Rütten, D., & Santarius, K. A. (1992). Relationship between frost tolerance and sugar concentration of various bryophytes in summer and winter. *Oecologia, 91*(2), 260–265. https://doi.org/10.1007/BF00317794.

Sasaki, H., Ichimura, K., & Oda, M. (1996). Changes in sugar content during cold acclimation and deacclimation of cabbage seedlings. *Annals of Botany, 78*(3), 365–369. https://doi.org/10.1006/anbo.1996.0131.

Serraj, R., & Sinclair, T. R. (2002). Osmolyte accumulation: Can it really help increase crop yield under drought conditions? *Plant. Cell and Environment, 25*(2), 333–341. https://doi.org/10.1046/j.1365-3040.2002.00754.x.

Sformo, T., Walters, K., Jeannet, K., Wowk, B., Fahy, G. M., Barnes, B. M., et al. (2010). Deep supercooling, vitrification and limited survival to −100°C in the Alaskan beetle Cucujus clavipes puniceus (Coleoptera: Cucujidae) larvae. *Journal of Experimental Biology, 213*(3), 502–509. https://doi.org/10.1242/jeb.035758.

Singh, J., Schädler, M., Demetrio, W., Brown, G. G., & Eisenhauer, N. (2019). Climate change effects on earthworms-a review. *Soil Organisms, 91*(3), 114–138. https://doi.org/10.25674/so91iss3pp114.

Smith, A. U., & Polge, C. (1950). Survival of spermatozoa at low temperatures. *Nature, 166*(4225), 668–669. https://doi.org/10.1038/166668a0.

Sømme, L., & Meier, T. (1995). Cold tolerance in Tardigrada from Dronning Maud land, Antarctica. *Polar Biology, 15*(3), 221–224. https://doi.org/10.1007/BF00239062.

Stacey, G. N., & Masters, J. R. (2008). Cryopreservation and banking of mammalian cell lines. *Nature Protocols, 3*(12), 1981–1989. https://doi.org/10.1038/nprot.2008.190.

Stokich, B., Osgood, Q., Grimm, D., Moorthy, S., Chakraborty, N., & Menze, M. A. (2014). Cryopreservation of hepatocyte (HepG2) cell monolayers: Impact of trehalose. *Cryobiology, 69*(2), 281–290. https://doi.org/10.1016/j.cryobiol.2014.08.001.

Storey, K. B., & Storey, J. M. (1984). Biochemical adaption for freezing tolerance in the wood frog, Rana sylvatica. *Journal of Comparative Physiology B, 155*(1), 29–36. https://doi.org/10.1007/BF00688788.

Strauss, G., & Hauser, H. (1986). Stabilization of lipid bilayer vesicles by sucrose during freezing. *Proceedings of the National Academy of Sciences of the United States of America, 83*, 2422–2426. https://doi.org/10.1073/pnas.83.8.2422.

Sun, H., Glasmacher, B., & Hofmann, N. (2012). Compatible solutes improve cryopreservation of human endothelial cells. *CryoLetters, 33*(6), 485–493.

Szurek, E. A., & Eroglu, A. (2011). Comparison and avoidance of toxicity of penetrating cryoprotectants. *PLoS One, 6*(11), 1–9. https://doi.org/10.1371/journal.pone.0027604.

Tada, N., Sato, M., Yamanoi, J., Mizorogi, T., Kasai, K., & Ogawa, S. (1990). Cryopreservation of mouse spermatozoa in the presence of raffinose and glycerol. *Journal of Reproduction and Fertility, 89*(2), 511–516. https://doi.org/10.1530/jrf.0.0890511.

Tarkowski, Ł. P., & Van den Ende, W. (2015). Cold tolerance triggered by soluble sugars: A multifaceted countermeasure. *Frontiers in Plant Science, 6*(APR), 1–7. https://doi.org/10.3389/fpls.2015.00203.

Taylor, M. J., Weegman, B. P., Baicu, S. C., & Giwa, S. E. (2019). New approaches to cryopreservation of cells, tissues, and organs. *Transfusion Medicine and Hemotherapy, 46*(3), 197–215. https://doi.org/10.1159/000499453.

Tsujimoto, M., Imura, S., & Kanda, H. (2016). Recovery and reproduction of an Antarctic tardigrade retrieved from a moss sample frozen for over 30 years. *Cryobiology, 72*(1), 78–81. https://doi.org/10.1016/j.cryobiol.2015.12.003.

Vajta, G., & Kuwayama, M. (2006). Improving cryopreservation systems. *Theriogenology, 65*(1), 236–244. https://doi.org/10.1016/j.theriogenology.2005.09.026.

Vereyken, I. J., Chupin, V., Demel, R. A., Smeekens, S. C. M., & De Kruijff, B. (2001). Fructans insert between the headgroups of phospholipids. *Biochimica et Biophysica Acta, 1510*, 307–320. https://doi.org/10.1016/S0005-2736(00)00363-1.

Walters, K. R., Sformo, T., Barnes, B. M., & Duman, J. G. (2009). Freeze tolerance in an arctic Alaska stonefly. *Journal of Experimental Biology, 212*(2), 305–312. https://doi.org/10.1242/jeb.020701.

Wang, M. R., Lambardi, M., Engelmann, F., Pathirana, R., Panis, B., Volk, G. M., et al. (2020). Advances in cryopreservation of in vitro-derived propagules: Technologies and explant sources. *Plant Cell, Tissue and Organ Culture*. https://doi.org/10.1007/s11240-020-01770-0.

Wełnicz, W., Grohme, M. A., Kaczmarek, Ł., Schill, R. O., & Frohme, M. (2011). Anhydrobiosis in tardigrades—The last decade. *Journal of Insect Physiology, 57*, 577–583. https://doi.org/10.1016/j.jinsphys.2011.03.019.

Wen, Q., Chen, J. X., Tang, Y. L., Wang, J., & Yang, Z. (2015). Assessing the toxicity and biodegradability of deep eutectic solvents. *Chemosphere, 132*, 63–69. https://doi.org/10.1016/j.chemosphere.2015.02.061.

Westh, P., & Ramløv, H. (1991). Trehalose accumulation in the tardigrade Adorybiotus coronifer during anhydrobiosis. *Journal of Experimental Zoology, 258*(3), 303–311. https://doi.org/10.1002/jez.1402580305.

Withers, L. A., & King, P. J. (1979). Proline: A novel Cryoprotectant for the freeze preservation of cultured cells of Zea mays L. *Plant Physiology, 64*(5), 675–678. https://doi.org/10.1104/pp.64.5.675.

Wu, C. F., Tsung, H. C., Zhang, W. J., Wang, Y., Lu, J. H., Tang, Z. Y., et al. (2005). Improved cryopreservation of human embryonic stem cells with trehalose. *Reproductive Biomedicine Online, 11*(6), 733–739. https://doi.org/10.1016/S1472-6483(10)61692-6.

Yang, J., Cai, N., Zhai, H., Zhang, J., Zhu, Y., & Zhang, L. (2016). Natural zwitterionic betaine enables cells to survive ultrarapid cryopreservation. *Scientific Reports*, *6*(October), 1–9. https://doi.org/10.1038/srep37458.

Yuanyuan, M., Yali, Z., Jiang, L., & Hongbo, S. (2009). Roles of plant soluble sugars and their responses to plant cold stress. *African Journal of Biotechnology*, *8*(10), 2004–2010. https://doi.org/10.5897/AJB09.177.

Zhang, L., Xue, X., Yan, J., Yan, L. Y., Jin, X. H., Zhu, X. H., et al. (2016). Cryobiological characteristics of L-proline in mammalian oocyte cryopreservation. *Chinese Medical Journal*, *129*(16), 1963–1968. https://doi.org/10.4103/0366-6999.187846.

Kolibri 360 GmbH
Ferdinand-Jühlke-Straße 7
99095 Erfurt, Deutschland
produktsicherheit@kolibri360.de